T0182096

Photodissociation induced by the absorption of single photons permits the detailed study of molecular dymanics such as the breaking of bonds, internal energy transfer and radiationless transitions. The availability, over the last decade, of powerful lasers operating over a wide frequency range has stimulated rapid development of new experimental techniques which make it possible to analyze photodissociation processes in unprecedented detail. At the same time, theorists have developed powerful methods to treat this fundamental process, at least for small molecules, in an essentially exact quantum mechanical way. The confluence of theory and experiment has greatly advanced understanding of molecular motion in excited electronic states.

This text elucidates the achievements in calculating photodissociation cross sections and fragment state distributions from first principles, starting from multi-dimensional potential energy surfaces and the Schrödinger equation of nuclear motion. Following an extended introduction in which the various types of observables are outlined, the next four chapters summarize the basic theoretical tools, namely the time-independent and the time-dependent quantum mechanical approaches as well as the classical picture of photodissociation. The discussions of absorption spectra, diffuse vibrational structures, the vibrational and rotational state distributions of the photofragments form the core of the book. More specific topics such as the dissociation of vibrationally excited molecules, emission during dissociation or nonadiabatic effects are discussed in the last third of the book.

It will be of interest to graduate students as well as senior scientists working in molecular physics, spectroscopy, molecular collisions and molecular kinetics.

Cambridge Monographs on Atomic, Molecular, and Chemical Physics 1

Photodissociation Dynamics

Spectroscopy and Fragmentation of Small Polyatomic Molecules

Cambridge Monographs on Atomic, Molecular, and Chemical Physics

1. R. Schinke: *Photodissociation Dynamics*
2. L. Frommhold: *Collision-induced absorption in gases*
3. T. F. Gallagher: *Rydberg Atoms*
4. M. Auzinsh and R. Ferber: *Optical Polarization of Molecules*

Photodissociation Dynamics

Spectroscopy and Fragmentation of Small Polyatomic Molecules

Reinhard Schinke

Max-Planck-Institut für Strömungsforschung, Göttingen

CAMBRIDGE UNIVERSITY PRESS
Cambridge, New York, Melbourne, Madrid, Cape Town, Singapore, São Paulo

Cambridge University Press
The Edinburgh Building, Cambridge CB2 2RU, UK

Published in the United States of America by Cambridge University Press, New York

www.cambridge.org
Information on this title: www.cambridge.org/9780521383684

First published 1993
First paperback edition 1995

A catalogue record for this publication is available from the British Library

Library of Congress Cataloguing in Publication data
Schinke, Reinhard.
Photodissociation dynamics / Reinhard Schinke.
p. cm. – (Cambridge monographs on atomic, molecular, and
chemical physics ; 1)
Includes bibliographical references (p.) and index.
ISBN 0-521-38368-4 (hardback)
1. Photodissociation. 2. Molecular dynamics. I. Title.
II. Series
QD716.P48S35 1993
541.3′ 5–dc20 91-43849 CIP

ISBN-13 978-0-521-38368-4 hardback
ISBN-10 0-521-38368-4 hardback

ISBN-13 978-0-521-48414-5 paperback
ISBN-10 0-521-48414-6 paperback

Transferred to digital printing 2006

Contents

	Preface	xiii
1	**Introduction**	1
1.1	Types of photodissociation	2
1.2	Why photodissociation dynamics?	6
1.3	Full and half collisions	8
1.4	From the spectrum to state-selected photodissociation	10
	1.4.1 The absorption spectrum	10
	1.4.2 Chemical identities and electronic fragment channels	12
	1.4.3 Vibrational and rotational product state distributions	13
	1.4.4 Angular distributions and vector correlations	15
	1.4.5 Photodissociation of single quantum states	15
	1.4.6 The various levels of photodissociation cross sections	16
1.5	Molecular dynamics and potential energy surfaces	19
	1.5.1 Electronic Schrödinger equation	19
	1.5.2 First example: CH_3ONO	20
	1.5.3 Second example: H_2O	23
2	**Light absorption and photodissociation**	27
2.1	Time-dependent perturbation theory	28
	2.1.1 Coupled equations	28
	2.1.2 Photon-induced transition rate	29
2.2	The absorption cross section	32
2.3	Born-Oppenheimer approximation	33
2.4	Bound-bound transitions in a linear triatomic molecule	37
	2.4.1 Jacobi or scattering coordinates	38
	2.4.2 Variational calculation of bound-state energies and wavefunctions	41
2.5	Photodissociation	42
	2.5.1 Dissociation channels	42

2.5.2 Continuum basis 43
2.5.3 Photodissociation cross sections 48
2.5.4 Total dissociation wavefunction 50

3 Time-independent methods 52
3.1 Close-coupling approach for vibrational excitation 53
3.1.1 Close-coupling equations 53
3.1.2 Bound-free dipole matrix elements 55
3.2 Close-coupling approach for rotational excitation 56
3.3 Approximations 60
3.3.1 Adiabatic approximation, appropriate for vibrational excitation 61
3.3.2 Sudden approximation, appropriate for rotational excitation 67
3.4 Numerical methods 69

4 Time-dependent methods 72
4.1 Time-dependent wavepacket 73
4.1.1 Autocorrelation function and total absorption spectrum 73
4.1.2 Evolution of the wavepacket 76
4.1.3 Energy domain and time domain 78
4.1.4 Partial photodissociation cross sections 81
4.2 Numerical methods 82
4.2.1 Temporal propagation 82
4.2.2 Spatial propagation 83
4.2.3 Time-dependent close-coupling 84
4.3 Approximations 86
4.3.1 Gaussian wavepackets 86
4.3.2 Time-dependent SCF 88
4.3.3 Classical path methods 89
4.4 A critical comparison 90

5 Classical description of photodissociation 93
5.1 Equations of motion, trajectories, and excitation functions 94
5.2 Phase-space distribution function 98
5.3 Classical absorption and photodissociation cross sections 102
5.3.1 Formal definitions 102
5.3.2 Monte Carlo calculations 104
5.4 Examples 105

6 Direct photodissociation: The reflection principle 109
6.1 One-dimensional reflection principle 110
6.1.1 The classical view 110
6.1.2 The time-dependent view 112
6.1.3 The time-independent view 114

6.2 Multi-dimensional reflection principle 115
6.2.1 The time-dependent view 115
6.2.2 The adiabatic view 117
6.2.3 Broad vibrational bands 118
6.3 Rotational reflection principle 120
6.3.1 An ultrasimple classical model 121
6.3.2 Mapping of the potential anisotropy 125
6.3.3 Examples 126
6.4 Vibrational reflection principle 128
6.5 Epilogue 133

7 Indirect photodissociation: Resonances and recurrences 134
7.1 A phenomenological prologue 135
7.2 Decay of excited states 138
7.3 Time-dependent view: Recurrences 143
7.3.1 Resonant absorption and Lorentzian line shapes 143
7.3.2 Example: Internal vibrational excitation 147
7.4 Time-independent view: Resonances 152
7.4.1 Stationary wavefunctions and assignment 152
7.4.2 Adiabatic picture and decay mechanism 155
7.4.3 Relation to resonances in full collisions 159
7.5 Resolution in the energy and in the time domain 160
7.6 Other types of resonant internal excitation 163
7.6.1 Excitation of bending motion 163
7.6.2 Excitation of symmetric and anti-symmetric stretch motion 168
7.7 Epilogue 173

8 Diffuse structures and unstable periodic orbits 177
8.1 Large-amplitude symmetric and anti-symmetric stretch motion 179
8.1.1 A collinear model system 179
8.1.2 Unstable periodic orbits 184
8.1.3 Examples 189
8.2 Large-amplitude bending motion 193
8.2.1 Breakdown of adiabatic separability 193
8.2.2 Dissociation guided by an unstable periodic orbit 196
8.3 Epilogue 200

9 Vibrational excitation 202
9.1 The elastic case: Franck-Condon mapping 203
9.1.1 Franck-Condon distribution 203
9.1.2 Examples 207
9.2 The inelastic case: Dynamical mapping 208
9.2.1 Energy redistribution 208
9.2.2 The photodissociation of CH_3I and CF_3I 210

9.3 Symmetric triatomic molecules 213
9.4 Adiabatic and nonadiabatic decay 217

10 Rotational excitation I 222
Sources of rotational excitation 222
What are the right coordinates? 223
10.1 The elastic case: Franck-Condon mapping 225
10.1.1 Rotational Franck-Condon factors 226
10.1.2 Example: Photodissociation of $H_2O(\tilde{A})$ 230
10.2 The inelastic case: Dynamical mapping 234
10.2.1 Photodissociation of H_2O_2 235
10.2.2 Photodissociation of $H_2O(\tilde{B})$ 238
10.3 Rotational distributions following the decay of long-lived states 241
10.3.1 Mapping of the transition-state wavefunction 241
10.3.2 Statistical limit 250
10.4 The impulsive model 251
10.5 Thermal broadening of rotational state distributions 255
10.5.1 The photodissociation of H_2O_2 256
10.5.2 Other examples 257

11 Rotational excitation II 261
11.1 General theory of rotational excitation for $\mathbf{J} \neq 0$ 262
11.1.1 Hamiltonian, expansion functions, and coupled equations 262
11.1.2 Rotational states of asymmetric top molecules 266
11.1.3 Selection rules and detailed state-to-state cross sections 267
11.2 Population of Λ-doublet and spin-orbit states 270
11.2.1 The eigenstates of $OH(^2\Pi)$ 271
11.2.2 Preferential Λ-doublet population in the photodissociation of $H_2O(\tilde{A}^1B_1)$ 272
11.2.3 Statistical and nonstatistical population of spin-orbit manifolds 275
11.3 Dissociation of single rotational states 277
11.4 Vector correlations 283
11.4.1 $\mathbf{E}_0$–$\boldsymbol{\mu}$–$\mathbf{v}$ correlation 283
11.4.2 $\mathbf{E}_0$–$\boldsymbol{\mu}$–$\mathbf{j}$ correlation 285
11.4.3 $\mathbf{v}$–$\mathbf{j}$ correlation 286
11.5 Correlation between product rotations 287

12 Dissociation of van der Waals molecules 293
12.1 Vibrational predissociation 296
12.2 Rotational predissociation 301
12.2.1 Characterization of rotational eigenstates 302
12.2.2 Decay mechanisms 304
12.3 Product state distributions 307

	12.3.1 Final vibrational state distributions	307
	12.3.2 Final rotational state distributions	308
13	**Photodissociation of vibrationally excited states**	314
13.1	Reflection structures	316
	13.1.1 One-dimensional case	316
	13.1.2 Two-dimensional case	318
13.2	Photodissociation of vibrationally excited H_2O	319
	13.2.1 Calculation and characterization of the bound states of H_2O	319
	13.2.2 Absorption spectra	320
	13.2.3 Final vibrational state distributions	323
13.3	State-selective bond breaking	324
14	**Emission spectroscopy of dissociating molecules**	331
14.1	Theoretical approaches	333
	14.1.1 The time-independent view	334
	14.1.2 The time-dependent view	335
14.2	Emission spectroscopy of dissociating $H_2O(\tilde{A})$	337
14.3	Raman spectra for H_2S	344
15	**Nonadiabatic transitions in dissociating molecules**	347
15.1	The adiabatic representation	349
15.2	The diabatic representation	352
15.3	Examples	356
	15.3.1 The photodissociation of CH_3I	357
	15.3.2 The photodissociation of H_2S	359
16	**Real-time dynamics of photodissociation**	366
16.1	Coherent excitation	368
	16.1.1 Coupled equations	368
	16.1.2 Pulse duration and spectral width	371
16.2	Examples	374
	References	380
	Index	412

Preface

Photodissociation of small polyatomic molecules is an ideal field for investigating molecular dynamics at a high level of precision. The last decade has seen an explosion of many new experimental methods which permit the study of bond fission on the basis of single quantum states. Experiments with three lasers — one to prepare the parent molecule in a particular vibrational-rotational state in the electronic ground state, one to excite the molecule into the continuum, and finally a third laser to probe the products — are quite usual today. *State-specific chemistry* finally has become reality. The understanding of such highly resolved measurements demands theoretical descriptions which go far beyond simple models.

Although the theory of photodissociation has not yet reached the level of sophistication of experiment, major advances have been made in recent years by many research groups. This concerns the calculation of accurate multi-dimensional potential energy surfaces for excited electronic states and the dynamical treatment of the nuclear motion on these surfaces. The exact quantum mechanical modelling of the dissociation of a triatomic molecule is nowadays practicable without severe technical problems. Moreover, simple but nevertheless realistic models have been developed and compared against exact calculations which are very useful for understanding the interrelation between the potential and the nuclear dynamics on one hand and the experimental observables on the other hand.

The aim of this book is to provide an overview of the theoretical methods for treating photodissociation processes in small polyatomic molecules and the achievements in merging *ab initio* calculations and detailed experiments. It is primarily written for graduate students starting research in molecular physics. However, experimentalists working in photochemistry, spectroscopy, unimolecular reactions, or molecular scattering, who are generally not very familiar with quantum mechanical methods for systems with more than one degree of freedom, also might benefit from reading this monograph. The basic equations are confined to a simple level and derived in substantial detail. Rather than outlining the theory in full generality (which leads to vast amounts of indices which usually obscure the simplicity of the underlying molecular process) I concentrate on the discussion of the fundamental dynamical effects and their relation to the multi-dimensional potential energy surface. Elementary knowledge of quantum mechanics should be sufficient to follow the derivations.

Instead of reviewing the many systems which have been investigated experimentally up to now, I will highlight only a few examples in order to elucidate the main dynamical features. This includes H_2O, H_2O_2, H_2S, ClNO, and CH_3ONO. For all these systems more or less complete potential energy surfaces have been calculated allowing rigorous dynamical studies without simplifying assumptions. Although experimental methods, which deserve a book for themselves, will not be explicitly discussed in this monograph, I will present many experimental results in order to illustrate the various aspects of photodissociation and the success of recent theoretical investigations. Despite the fact that this monograph covers mainly small polyatomic molecules, many of the general pictures may also apply to larger molecules. The photodissociation of diatomic molecules will not be considered explicitly except for illustration purposes.

The general theory for the absorption of light and its extension to photodissociation is outlined in Chapter 2. Chapters 3–5 summarize the basic theoretical tools, namely the time-independent and the time-dependent quantum mechanical theories as well as the classical trajectory picture of photodissociation. The two fundamental types of photofragmentation — direct and indirect photodissociation — will be elucidated in Chapters 6 and 7, and in Chapter 8 I will focus attention on some intermediate cases, which are neither truly direct nor indirect. Chapters 9–11 consider in detail the internal quantum state distributions of the fragment molecules which contain a wealth of information on the dissociation dynamics. Some related and more advanced topics such as the dissociation of van der Waals molecules, dissociation of vibrationally excited molecules, emission during dissociation, and nonadiabatic effects are discussed in Chapters 12–15. Finally, we consider briefly in Chapter 16 the most recent class of experiments, i.e., the photodissociation with laser pulses in the femtosecond range, which allows the study of the evolution of the molecular system in real time.

Acknowledgements: I am very grateful to the students who have joined my research group in the last decade for their invaluable contributions to my understanding of photodissociation in general and to this monograph in particular. Without their enthusiasm this book would not have been written in the present form. In chronological order, I am thankful to Dr V. Engel, Dr S. Hennig, Dr K. Weide, Dr A. Untch, K. Kühl, Dr B. Heumann, Dr M. von Dirke, G. Ebel, P. Molnar, and H.-M. Keller. During the last decade I had the privilege of collaborating with several distinguished scientists on several photodissociation systems. The benefit I got from these collaborations cannot be overestimated. In particular, I am very grateful to Prof. P. Andresen, Prof. V. Staemmler, Prof. J.R. Huber, Prof. F.F. Crim, and Prof. H.-J. Werner. The early phase of writing this monograph started in Boulder when I was a Visiting Fellow of the Joint Institute for Laboratory Astrophysics. I appreciate many helpful comments on the manuscript by Prof. A.H. Zewail, Prof. J.P. Simons, Dr N. Shafer, Dr V. Engel, Dr A. Untch, Dr K. Weide, Dr M. von Dirke, and G. Ebel. I am grateful to Prof. R.N. Zare for encouraging me to write a monograph on photodissociation dynamics. I am indebted to H. Teike and G. Barz from the Max-Planck-Institut für Strömungsforschung for drawing almost all the figures. Finally, I am obliged to my wife Ellen for typing the first draft of this monograph, for patiently teaching me the secrets of TEX, and for her continuous help in preparing the final version.

Göttingen, 1992 Reinhard Schinke

Units: Throughout this monograph energies will be measured in eV or in cm^{-1} (1 eV = 8065 cm^{-1} or 96.48 kJ mol^{-1}) and distances will be measured in Å or in a_0 (1 Å = 10^{-10} m ; 1 a_0 = 0.52918 10^{-10} m).

Abbreviations: Frequently used abbreviations are PES for potential energy surface and FC for Franck–Condon.

1
Introduction

The fragmentation of a bound molecule through absorption of one or more photons is called photodissociation. The electromagnetic energy of the light beam is converted into internal energy of the molecule and if the transferred energy exceeds the binding energy of the weakest bond, the molecule will irreversibly break apart. Let us, as an example, consider the dissociation of a parent molecule AB into products A and B, where A and B represent either structureless atoms or molecules with internal degrees of freedom of their own. Formally we write a photodissociation process as

$$\mathrm{AB} + N_{photon}\,\hbar\omega \xrightarrow{(1)} (\mathrm{AB})^* \xrightarrow{(2)} \mathrm{A}(\alpha) + \mathrm{B}(\beta), \qquad (1.1)$$

where $\hbar\omega$ is the energy of one photon with frequency ω and N_{photon} is the number of absorbed photons. $(\mathrm{AB})^*$ represents the excited complex before it breaks apart and the labels α and β specify the particular internal quantum states of the newborn products. The first step indicates absorption of the photons by the parent molecule and the second step represents the fragmentation of the excited complex.

In the center of our discussion of photodissociation dynamics are questions like the following:

1) How does the photon cleave the molecular bond?
2) What is the lifetime of the intermediate complex?
3) What are the primary fragments?
4) How does the absorbed energy partition among the various degrees of freedom of the products?
5) What is the internal quantum state distribution of the fragments?
6) How does the dissociation depend on the initial state or the temperature of the parent molecule?

In this monograph we shall attempt to answer these questions in the light of experimental and theoretical advances achieved in the last decade or so.

1.1 Types of photodissociation

Dissociation energies vary from a few thousandths of an eV for physically bound van der Waals molecules to several eV for chemically bound molecules. Van der Waals molecules are bound by the weak long-range forces and exist only at very low temperatures, either in a supersonic beam or in the interstellar space (Buckingham, Fowler, and Hutson 1988). Typical examples are:†

He $\cdots$ HF ($D_0 = 0.88$ meV; Lovejoy and Nesbitt 1990)

Ar $\cdots$ HCl ($D_0 = 14.2$ meV; Howard and Pine 1985)

HF $\cdots$ HF ($D_0 = 131.7$ meV; Dayton, Jucks, and Miller 1989).

Representative examples of chemically bound molecules which will play vital roles in this monograph are:

ClNO ($D_0 = 1.62$ eV; Bruno, Brühlmann, and Huber 1988)

H_2O_2 ($D_0 = 2.12$ eV; Giguèra 1959)

H_2O ($D_0 = 5.11$ eV; Herzberg 1967).

The extremely wide range of possible dissociation energies necessitates the use of different kinds of light source to break molecular bonds. Van der Waals molecules can be fragmented with single infrared (IR) photons whereas the fission of a chemical bond requires either a single ultraviolet (UV) or many IR photons. The photofragmentation of van der Waals molecules has become a very active field in the last decade and deserves a book in itself (Beswick and Halberstadt 1993). It is a special case of UV photodissociation and can be described by the same theoretical means. In Chapter 12 we will briefly discuss some simple aspects of IR photodissociation in order to elucidate the similarities and the differences to UV photodissociation.

Figure 1.1 illustrates the two basic types of photodissociation of a chemically bound molecule. In Figure 1.1(a) the photon excites the molecule from the ground to a higher electronic state. If the potential of the upper electronic state is repulsive along the intermolecular coordinate R_{AB}, the

† The dissociation energy D_0 is measured from the zero-point level of the parent molecule to the zero-point level(s) of the products.

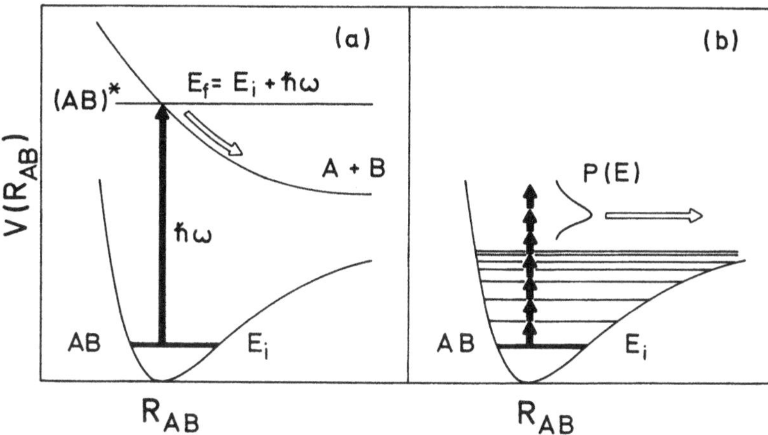

Fig. 1.1. Schematic illustration of photodissociation with a single UV photon (a) and with many IR photons (b). R_{AB} is the intermolecular separation of the two fragments A and B. In (a) the photon creates a single quantum state in the upper electronic state while in (b) the photons produce a comparably broad ensemble of states with energy distribution $P(E)$.

excited complex $(AB)^*$ ultimately dissociates. Part of the photon energy $E_{photon} = \hbar\omega$ is consumed to break the A-B bond and the *excess energy*

$$E_{excess} = E_{photon} - D_0 = E_{trans} + E_{int}$$

partitions between the translational energy E_{trans} and the internal energy E_{int} of the product atoms or molecules (including vibrational, rotational, and electronic energy).

UV photodissociation is usually carried out with long light pulses of low intensity and narrow bandwidth. These conditions guarantee — at least in principle — that the photon creates a single quantum state in the upper electronic manifold with corresponding energy $E_f = E_i + E_{photon}$, where E_i is the energy of the parent molecule. We shall demonstrate below that under such well defined conditions the main observables, i.e., the absorption spectrum and the product state distributions, uniquely "reflect" the molecular wavefunction of the particular quantum state of the parent molecule before the excitation. How the dynamics in the upper electronic state mediates this type of "reflection" is one of the central issues of this monograph.

Multiphoton dissociation takes place in the electronic ground state as illustrated in Figure 1.1(b) (Grunwald, Dever, and Keehn 1978; Schulz et al. 1979; Golden, Rossi, Baldwin, and Barker 1981; Letokhov 1983; Reisler and Wittig 1985; Lupo and Quack 1987). Since the exact number of absorbed photons cannot be controlled, the laser creates an ensemble of quantum states above the dissociation threshold with a distribution of energies $P(E_f)$. As a consequence, multiphoton dissociation is subject to unavoidable averaging over many quantum states. It requires basically different theoretical tools and models than dissociation with a single photon. Therefore, we shall not further discuss it in this monograph.

Figure 1.1(a) illustrates the simplest type of UV photodissociation. Upon excitation the two fragments immediately repel each other and the excited complex (AB)* fragments directly on a very short time scale. We will demonstrate below that this kind of photodissociation can be satisfactorily treated in the framework of classical mechanics. The photodissociation of H_2O in the first absorption band is a prototype of direct photodissociation (see Section 1.5 and the review of Engel et al. 1992). In contrast, Figure 1.2 schematically illustrates two examples of indirect photodissociation. In (a) the photon excites first a binding electronic state which by itself cannot break apart. However, if the molecule undergoes a transition from the binding state to another electronic state whose potential is repulsive, the complex will ultimately decay with a rate that depends on the coupling between the two electronic states. The final fragmentation takes place in a different electronic state than the one originally excited by the photon. This process is called *electronic predissociation* or Herzberg's type I predissociation (Herzberg 1967:ch.IV).[†]

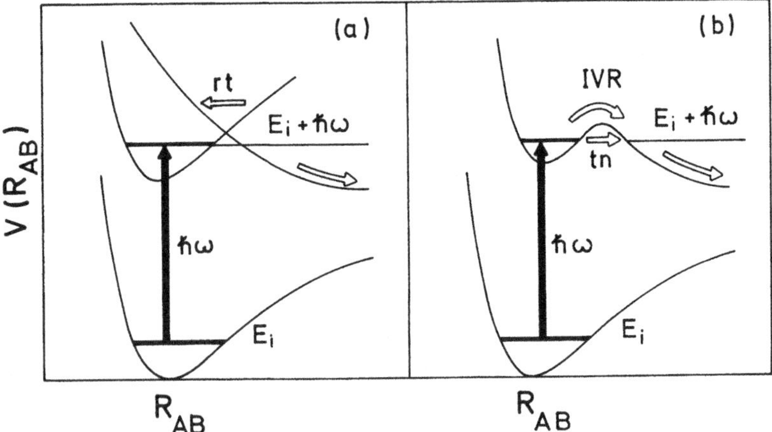

Fig. 1.2. Schematic illustration of electronic (a) and vibrational (b) predissociation. In the first case, the molecule undergoes a radiationless transition (rt) from the binding to the repulsive state and subsequently decays. In the second case, the photon creates a quasi-bound state in the potential well which decays either by tunneling (tn) or by internal energy redistribution (IVR).

[†] The word "predissociation" means that the molecule dissociates before it can decay to the electronic ground state by emission of a photon.

An example is the photodissociation of H_2S which will be discussed in Chapter 15.

Figure 1.2(b) elucidates a different type of predissociation. The potential has a well at close distances and a barrier that blocks the dissociation channel. The barrier might be considered to be the result of an avoided crossing with a higher electronic state. In this case the photon excites quasi-stable (so-called "resonance") states inside the well which are prevented from immediate dissociation by the potential barrier. They can decay either by tunneling through the barrier or by *internal vibrational energy redistribution* between the various nuclear degrees of freedom if more than two atoms are involved. The lifetime of the compound depends on the tunneling rate and/or the efficiency of internal energy transfer. Figure 1.2(b) illustrates *vibrational predissociation* or Herzberg's type II predissociation (Herzberg 1967:ch.IV). The photodissociation of CH_3ONO in the first excited singlet state, which we will discuss in detail in Chapter 7, is an illuminating example.

Figure 1.3 shows two additional types of photodissociation. In (a) the photon excites first a binding electronic state which subsequently decays following a transition to the lower electronic state. This produces a highly excited vibrational-rotational quantum level above the dissociation threshold of the electronic ground state which eventually breaks apart. Alternatively, in (b) a highly excited quantum state above the dissociation threshold is created directly by pumping a large amount of energy into the molecule by single-photon excitation of overtone vibrations. Both processes are known as *unimolecular reactions* (Forst 1973; Hase 1976; Quack and Troe 1977; Crim 1984; Pritchard 1985; Reisler and Wittig 1986; Crim 1987; Pilling and Smith 1987; Gilbert and Smith 1990). Representative examples are the photodissociation of H_2CO (Moore and Weisshaar 1983) and H_2O_2 (Crim 1987), respectively.

The distinction between the various dissociation schemes (with the exception of multiphoton dissociation) is rather artificial from the formal point of view. Common to direct dissociation, predissociation, and unimolecular decay is the possibility of state-specificity, i.e., the dependence of the dissociation on the quantum state of the parent molecule (Manz and Parmenter 1989). The absorption of a single photon uniquely defines the energy in the dissociative state. As we will demonstrate in subsequent chapters, one can treat all three classes of fragmentation with the same basic theoretical tools. However, the underlying molecular dynamics is quite different demanding different interpretation models.

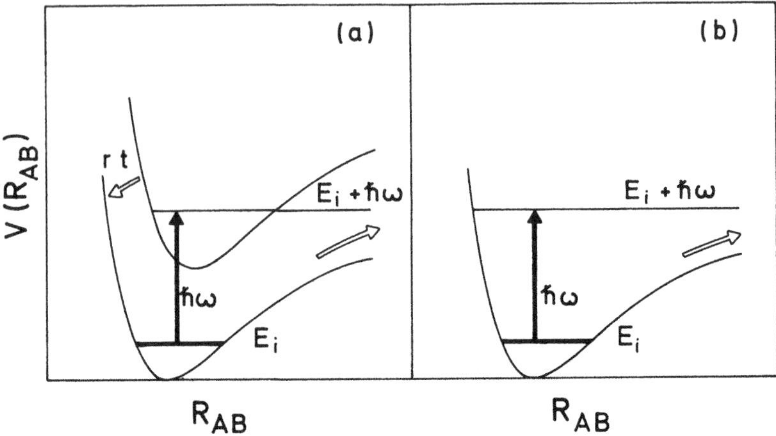

Fig. 1.3. Schematic illustration of unimolecular decay induced by electronic excitation. In (a) the photon creates a bound level in the upper electronic state which subsequently decays as a result of a radiationless transition (rt) to the electronic ground state. In (b) overtone pumping directly creates a quantum state above the threshold of the electronic ground state. In both cases the dissociation occurs in the electronic ground state.

1.2 Why photodissociation dynamics?

Photodissociation is at the heart of *photochemistry* (Turro 1965, 1978; Calvert and Pitts 1966; Ben-Shaul, Haas, Kompa, and Levine 1981; von Bünau and Wolff 1987; Wayne 1988; Klessinger and Michl 1989). In many cases, the photofragments are radicals which rapidly undergo secondary and tertiary reactions. Photodissociation is the motor for many important chain reactions determining the complex chemistry in the atmosphere. The sun supplies sufficient UV radiation to keep the motor going (Okabe 1978). The ozone cycle,

$$\begin{aligned} O_3 + \hbar\omega &\longrightarrow O_2 + O \\ O + O_3 &\longrightarrow 2\,O_2 \\ O + O_2 + M &\longrightarrow O_3 + M, \end{aligned}$$

for example, is vital for all life on earth. It controls the global abundance of O_3 in the upper atmosphere which is essential for the absorption of harmful UV radiation.

Photodissociation is the starting point for many *chemical lasers* such as the iodine laser discovered by Kasper and Pimentel (1964). The reaction

$$CH_3I + \hbar\omega \longrightarrow CH_3 + I$$

yields an inverted population of the two electronic states of iodine, $^2P_{1/2}$ and $^2P_{3/2}$, which allows a laser to be built on the subsequent $^2P_{1/2} \rightarrow {}^2P_{3/2}$ electronic transition. The final state distributions of photofragments are often inverted, which is a necessary prerequisite for the operation of lasers (Ben-Shaul et al. 1981:ch.4). Detailed understanding of the above reaction and similar processes is undoubtedly desirable in order to develop and control new laser sources systematically. Understanding of photodissociation is also very important in astrophysics (Kirby and van Dishoeck 1988).

Besides its practical importance, photodissociation — especially of small polyatomic molecules — provides an ideal opportunity for the study of molecular dynamics on a detailed state-to-state level. We associate with molecular dynamics processes such as energy transfer between the various molecular modes, the breaking of chemical bonds and the creation of new ones, transitions between different electronic states etc. One goal of modern physical chemistry is the microscopical understanding of molecular reactivity beyond purely kinetic descriptions (Levine and Bernstein 1987). Because the "initial conditions" can be well defined (absorption of a single monochromatic photon, preparation of the parent molecule in selected quantum states), photodissociation is ideally suited to address questions which are unprecedented in chemistry. The last decade has witnessed an explosion of new experimental techniques which nowadays makes it possible to tackle questions which before were beyond any practical realization (Ashfold and Baggott 1987).

Photodissociation combines aspects of both molecular spectroscopy and molecular scattering. The spectroscopist is essentially interested in the first step of Equation (1.1), i.e., the absorption spectrum. In the past six decades or so methods of ever increasing sophistication have been developed in order to infer molecular geometries from structures in the absorption or emission spectrum (Herzberg 1967), whereas the fate of the fragments, i.e., the final state distribution is of less relevance in spectroscopy. The decay of the excited complex is considered only inasfar as the widths of the individual absorption lines reflect the finite lifetime in the excited state and therefore the decay rate of the excited molecule.

The second step in (1.1), i.e., the fragmentation of the excited complex belongs to the field of molecular collisions. The actual decay mechanism, the partitioning of the excess energy, and the population of the possible quantum states of the products are the central themes. However, we must emphasize from the outset of this monograph that the distinction between absorption and fragmentation is — in principle at least — artificial. The two steps in (1.1) belong together and must be considered as an entity.

1.3 Full and half collisions

Photodissociation can be viewed as the second half of a full collision. Let us consider an atom A colliding with a diatom BC as illustrated schematically in Figure 1.4. In a full collision, the reactants are prepared separately at infinite distances where the interaction between them is zero. During the collision they form an intermediate complex $(ABC)^*$ which in the end decays to the four possible product channels with the diatomic fragments being created in particular internal quantum states. In photodissociation, on the other hand, the excited complex is generated by electronic excitation of the initially bound parent molecule ABC. The photon promotes ABC to a higher electronic state where it dissociates, either immediately if the potential is repulsive or after a delay if a barrier hinders the fragmentation. In any case, the second step of photodissociation is equivalent to the second step of a full collision which explains the terminology "half collision".

The decay of the complex is described by exactly the same equations of motion independent of its creation, namely Hamilton's equations in classical mechanics and Schrödinger's equations in quantum mechanics. Only the initial conditions for the ultimate fragmentation step are different because they reflect how the intermediate complex was formed. However, these differences lead to several important implications!

Let us denote by $\mathbf{j}$ the rotational angular momentum of the diatomic entity BC and by $\mathbf{l}$ the orbital angular momentum of atom A with respect to BC. The total angular momentum of the entire system is then defined by $\mathbf{J} = \mathbf{j}+\mathbf{l}$ and the corresponding quantum number will be denoted by J. It is, of course, conserved during the collision. The definition of scattering or photodissociation cross sections unavoidably involves an average over total angular momentum states from $J = 0$ to some maximum value J_{max} with each term being weighted by $(2J+1)$ (Child 1974:ch.3; Levine and Bernstein 1987:ch.2; Hirst 1990:ch.7; Child 1991:ch.8; Chapter 11 of

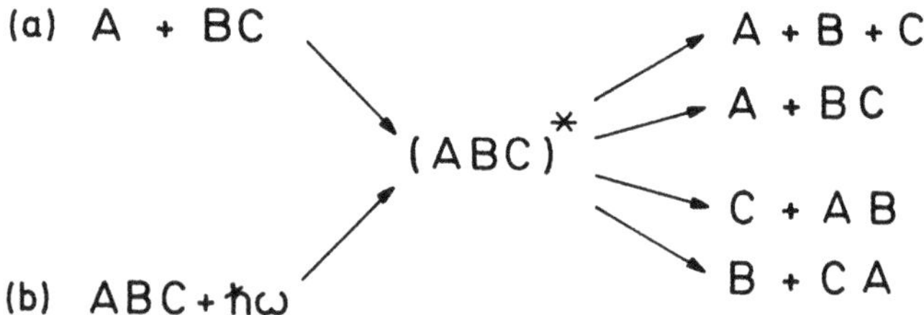

Fig. 1.4 Schematic illustration of a full collision (a) and a half collision (b).

this monograph).[†] This holds equally for full as well as half collisions. However, the consequences are quite different in the two cases.

Under normal conditions for an atom-molecule collision the summation over J extends over many (of the order of one hundred or even more) so-called partial waves which makes the practical calculation rather cumbersome. An even more serious problem is the substantial blurring of distinct dynamical structures such as quantum mechanical interferences or resonances. Since these structures depend parametrically on J, the summation over all possible J values rapidly washes out finer details.

Excitation with a photon, on the other hand, is subject to the dipole selection rule (see Chapter 11)

$$\Delta J = J_i - J_f = 0 \quad \text{or} \quad \pm 1,$$

where J_i and J_f are the total angular momentum quantum numbers before and after the absorption of the photon, respectively. As a consequence of this selection rule the number of angular momentum states, which must be taken into account in the calculation of photodissociation cross sections, is comparatively limited. For a given temperature T the distribution of initial states is determined by the Boltzmann distribution

$$P_T(J_i) \propto (2J_i + 1)\, \mathrm{e}^{-B_{rot} J_i(J_i+1)/k_B T},$$

where B_{rot} and k_B are the rotational constant of the molecule and Boltzmann's constant, respectively. The Boltzmann distribution is under normal experimental conditions confined to a relatively small range of J_i values and because of the selection rule the same holds also for J_f.

Thus, even at room temperature ($T = 300$ K) only very few states, in comparison to full collisions, contribute to the dissociation cross sections which significantly reduces the numerical efforts and suppresses incoherent averaging. The result is a clearer picture of the fragmentation dynamics and the possibility of resolving resonance structures which are very difficult to observe in full collisions. Cooling the gas in a molecular beam down to temperatures of the order of 50 K or below leads to a further reduction of the possible angular momentum states. Therefore, many experiments can be modelled with the restriction that $J = 0$ in both electronic states and almost all theoretical investigations take advantage of this limitation.

Furthermore, the total energy E_f in half collisions can be relatively easily controlled by variation of the photolysis wavelength. This is essential for the detection of resonances and other structures which require high

---

[†] In classical mechanics, the summation over the total angular momentum quantum number, $\sum_J (2J+1)\ldots$, corresponds to an integral over the impact parameter b according to $\int db\, b \ldots$.

energy resolution. On the other hand, controlling the scattering energy in full collisions is undoubtedly more problematic.

1.4 From the spectrum to state-selected photodissociation

In this section we summarize briefly the various cross sections which can be measured in a photodissociation experiment, starting with the least resolved quantity, the absorption spectrum, up to the most detailed ones, final state resolved cross sections following the dissociation of a particular vibrational-rotational state of the parent molecule. We illustrate the hierarchy of possible measurements by an important example, the photodissociation of H_2O sketched in Figure 1.5.† For reviews of modern experimental methods see Leone (1982) and Ashfold and Baggott (1987), for example.

1.4.1 The absorption spectrum

The most averaged quantity is the absorption spectrum or total absorption cross section $\sigma_{tot}(\omega)$ which, loosely speaking, measures the capability of a molecule to absorb radiation with frequency ω (for a rigorous definition see Chapter 2). As the name suggests, the total cross section is defined irrespective of the fate of the excited complex and the population of the possible fragmentation channels.

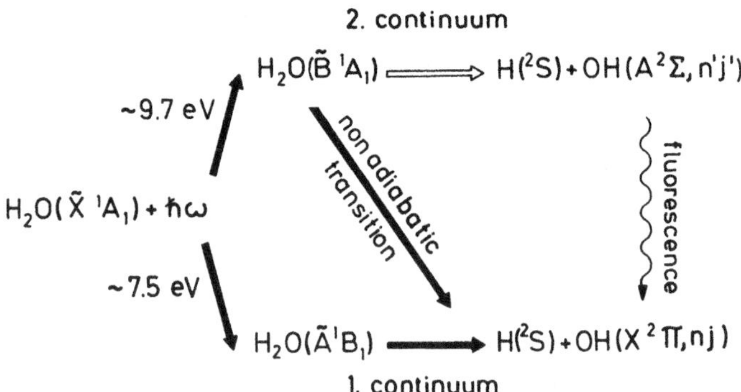

Fig. 1.5. Schematic representation of the photodissociation of H_2O in the first two absorption bands. The corresponding potentials are illustrated in Figures 1.12 and 1.13. Further explanation is given in the text.

† The photodissociation of H_2O in the first two absorption bands will be amply discussed in this monograph. For reviews see Andresen and Schinke (1987), Docker, Hodgson, and Simons (1987), and Engel et al. (1992). H_2O can be considered as a prototype for the photodissociation of triatomic molecules.

Absorption cross sections are relatively easy to measure and have been surveyed for most stable molecules (Herzberg 1967; Rabalais, McDonald, Scherr, and McGlynn 1971; Hudson 1971; Robin 1974, 1975, 1985; Okabe 1978). Figure 1.6 depicts the absorption spectrum of H_2O in the 5–11 eV energy region. The maxima at 7.5 eV ($\lambda \approx 165$ nm; first absorption band) and 9.7 eV ($\lambda \approx 128$ nm; second absorption band) are ascribed to electronic transitions $\tilde{X}^1A_1 \rightarrow \tilde{A}^1B_1$ and $\tilde{X}^1A_1 \rightarrow \tilde{B}^1A_1$, respectively. A third band, $\tilde{X}^1A_1 \rightarrow \tilde{C}^1B_1$, strongly overlaps the second continuum at the blue side of the spectrum.

From the overall shape of the spectrum and possible structures one can draw general conclusions about the dissociation dynamics. The $\tilde{A}$ band of H_2O is (almost) structureless indicating a mainly direct dissociation mechanism. The $\tilde{B}$ band exhibits some weak undulations which can be attributed to a special type of trapped motion with a "lifetime" of the order of one internal vibration (see Section 8.2). However, the broad background indicates that the dissociation via the $\tilde{B}$ state also proceeds primarily in a direct way. Finally, the $\tilde{C}$ band consists of rather pronounced structures which immediately tell us that the excited $H_2O(\tilde{C}^1B_1)$ complex lives on the order of at least several internal vibrations. Although the absorption spectrum is a highly averaged quantity it contains a wealth of dynamical information: more of this in Chapters 6–8.

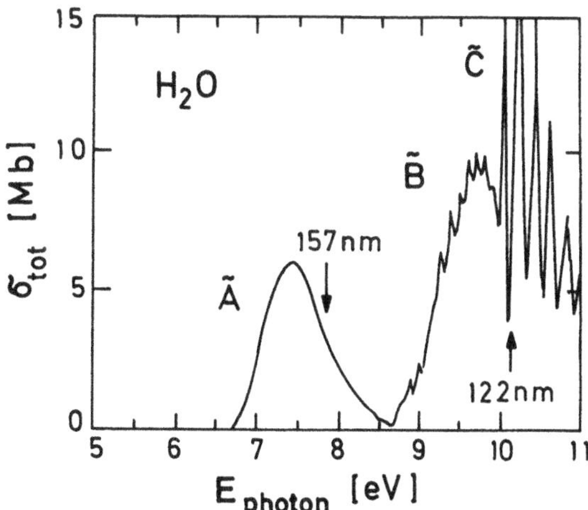

Fig. 1.6. Total absorption cross section of H_2O as a function of photon energy. $\tilde{A}$, $\tilde{B}$, and $\tilde{C}$ indicate the three lowest absorption bands. The two arrows mark the wavelengths at which rotational state distributions of the OH product, depicted in Figure 1.8, have been measured. Adapted from Gürtler, Saile, and Koch (1977).

1.4.2 Chemical identities and electronic fragment channels

Let us consider the photodissociation of HOD in the $\tilde{A}$ band. The excited complex can decay into four distinct chemical channels which we will designate by the index γ: $\mathrm{H + O + D}$ ($\gamma = 1$), $\mathrm{H + OD}$ ($\gamma = 2$), $\mathrm{D + OH}$ ($\gamma = 3$), and $\mathrm{O + HD}$ ($\gamma = 4$) (see also Figure 1.4). If the photon energy is sufficiently high all of them are, in principle, allowed. In reality, however, the probabilities for channels 1 and 4 are negligibly small and therefore they will not be considered hereafter.

Although the H-O and D-O bonds are chemically equivalent (only the atomic masses, m_{H} and m_{D}, respectively, are different) the *chemical branching ratio* for producing OD and OH radicals, $\sigma_{\mathrm{H+OD}}/\sigma_{\mathrm{D+OH}}$, is not unity but varies considerably with energy as shown in Figure 1.7. At all energies the H-O bond breaks more rapidly than the O-D bond. The deviation from unity manifests the asymmetry of the initial wavefunction in the electronic ground state and the relatively large mass ratio $m_{\mathrm{D}}/m_{\mathrm{H}}$ (Engel and Schinke 1988). An ultrasimple kinematic model yields an approximate ratio $\sigma_{\mathrm{H+OD}}/\sigma_{\mathrm{D+OH}} \approx m_{\mathrm{D}}/m_{\mathrm{H}} = 2$. The differences from this limit reflect details of the actual bond breaking mechanism beyond simple kinematical effects.

Figure 1.7 represents a simple example of bond selectivity in photodissociation. More interesting, of course, are cases in which two or more chemically different product channels are possible such as in the dissoci-

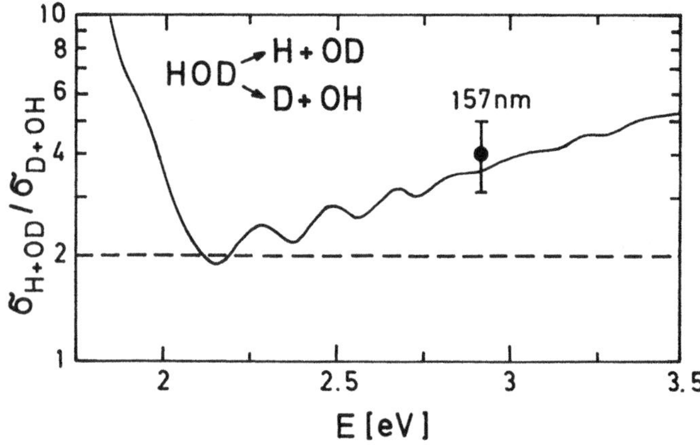

Fig. 1.7. Branching ratio for the production of OD and OH fragments in the photodissociation of HOD in the first absorption band. The energy is measured with respect to $\mathrm{H + OH}(r_e)$, where r_e is the equilibrium bond distance of OH. The dashed curve indicates a simple kinematical limit (see the text) and the data point represents the measured value of Shafer, Satyapal, and Bersohn (1989) for the photolysis at 157 nm.

ation of glyoxal (Burak et al. 1987), for example,

$$\begin{aligned} \mathrm{CHOCHO} + \hbar\omega &\longrightarrow \mathrm{H_2CO} + \mathrm{CO} \\ &\longrightarrow \mathrm{H_2} + 2\mathrm{CO} \\ &\longrightarrow \mathrm{HCOH} + \mathrm{CO}. \end{aligned}$$

Similarly, the fragments can be formed in different electronic channels which we will designate by the index e. The $\tilde{A}^1B_1$ electronic state of H_2O correlates asymptotically with the electronic ground state of OH, $X^2\Pi$, (see Figure 1.5) and therefore this is the only final electronic channel that can be accessed in the dissociation through the $\tilde{A}$ band at moderate energies. Excitation in the second continuum is more interesting because two electronic states of OH can be populated. Although $H_2O(\tilde{B}^1A_1)$ correlates with the excited electronic state of OH, $A^2\Sigma$, most of the products are formed in the $X^2\Pi$ electronic ground state (Vinogradov and Vilesov 1976; Dutuit et al. 1985; Lee and Suto 1986). The *electronic branching ratio* $\mathrm{OH}(A^2\Sigma)/\mathrm{OH}(X^2\Pi)$ is of the order of only 0.1 which implies that during the fragmentation an efficient nonadiabatic transition from $H_2O(\tilde{B}^1A_1)$ to either $H_2O(\tilde{A}^1B_1)$ or $H_2O(\tilde{X}^1A_1)$, which both correlate with $\mathrm{OH}(^2\Pi)$, takes place as indicated schematically in Figure 1.5. The electronic branching ratio reflects the strength of nonadiabatic coupling between different electronic states during the separation of the fragments (see Chapter 15).

1.4.3 Vibrational and rotational product state distributions

The OH radical is produced in a particular vibrational and rotational quantum state specified by the quantum numbers n and j. The corresponding energies are denoted by ϵ_{nj}. The probabilities with which the individual quantum states are populated are determined by the forces between the translational mode (the dissociation coordinate) and the internal degrees of freedom of the product molecule along the reaction path. Final vibrational and rotational state distributions essentially reflect the dynamics in the fragment channel. They are one major source of information about the dissociation process.

Figure 1.8 clearly illustrates this point: the dissociation in the $\tilde{A}$ band yields rotationally cold $\mathrm{OH}(^2\Pi)$ products, whereas $\mathrm{OH}(^2\Sigma)$ produced via excitation of the $\tilde{B}$ state is generated with very high rotational angular momentum, despite the fact that the total available energy is almost three times larger in the first case! No statistical or kinematical model can explain this dramatic difference. It is solely due to the different topologies of the $\tilde{A}$- and $\tilde{B}$-state potential energy surfaces which we will discuss in

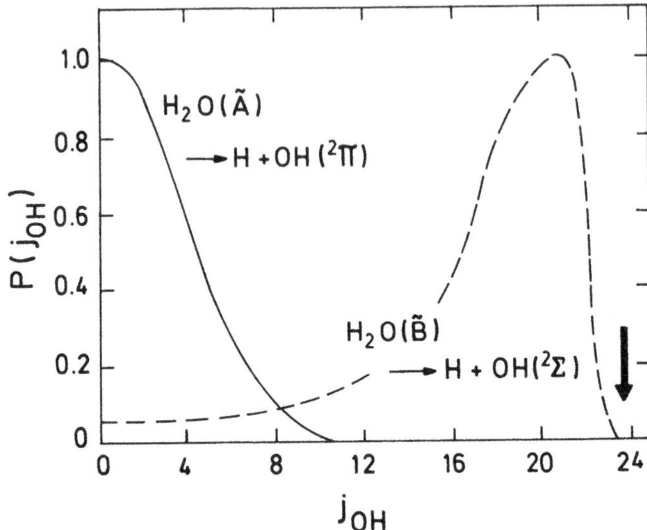

Fig. 1.8. Rotational state distributions of OH($^2\Pi$) and OH($^2\Sigma$) following the photodissociation of H_2O via the $\tilde{A}$ state (λ = 157 nm; Andresen, Ondrey, Titze, and Rothe 1984) and via the $\tilde{B}$ state (λ = 121.6 nm; Carrington 1964), respectively. The heavy arrow marks the highest rotational state of OH($^2\Sigma$) which can be populated at 121.6 nm.

the next section. $H_2O(\tilde{A}^1B_1)$ and $H_2O(\tilde{B}^1A_1)$ are essentially different chemical species and that shows up in the product state distributions.

Using the quantum state distributions one can define *mean energies* by

$$E_{rot}^{(n)} \equiv \sum_j \left(\epsilon_{nj} - \epsilon_{nj=0}\right) P^{(n)}(j) \quad , \quad E_{vib} \equiv \sum_n \epsilon_n P(n), \tag{1.2}$$

where $P^{(n)}(j)$ is the rotational distribution in vibrational state n and $P(n)$ is the distribution of vibrational states. Both are assumed to be normalized to unity. The relative energies

$$\langle f_{rot} \rangle \equiv E_{rot}/E_{excess} \quad , \quad \langle f_{vib} \rangle \equiv E_{vib}/E_{excess} \tag{1.3}$$

specify the fraction of the excess energy which goes into rotation and vibration, respectively.

For example, in the photodissociation of H_2O in the $\tilde{A}$ band the main fraction of E_{excess} is released as translational energy (88%) and only a very small fraction (2%) goes into rotation; the relative vibrational energy of OH($^2\Pi$) is about 10% (Andresen, Ondrey, Titze, and Rothe 1984). The situation is quite different in the dissociation via the $\tilde{B}$ state where the major portion of the excess energy is released as rotational energy. Bersohn (1984) gives a comparison of relative energies for several photodissociation processes. In this monograph we shall exclusively consider final state distributions because they contain obviously more information than mean or relative energies. An extensive analysis of vibrational and rotational state distributions follows in Chapters 6 and 9–11.

1.4.4 Angular distributions and vector correlations

The product state distributions and the populations of the various chemical and electronic channels belong to the category of so-called scalar properties which are defined without reference to a particular coordinate frame. They have a magnitude but no direction. However, since the electromagnetic field vector **E** of the photolysis laser defines a specific direction in the laboratory frame, all vectors inherent to a photodissociation process can be measured relative to **E**. The vectors of interest are:

1) The transition dipole moment function of the parent molecule, $\boldsymbol{\mu}$.
2) The recoil velocity vector of the fragments, **v**.
3) The rotational angular momentum vector of the fragment, **j**.

The oldest known vector correlation is the angular distribution of the photofragments, i.e., the relation between the recoil velocity **v** and the polarization vector **E**,

$$\frac{d\sigma}{d\Omega}(\theta) = \frac{\sigma}{4\pi}\left[1 + \beta P_2(\cos\theta)\right], \tag{1.4}$$

where $-1 \leq \beta \leq 2$ is the *anisotropy parameter*, P_2 is the second-order Legendre polynomial, and θ is the angle between **E** and **v** (Zare and Hershbach 1963; Bersohn and Lin 1969; Zare 1972; Busch and Wilson 1972b). $d\sigma/d\Omega$ is the angle-dependent *differential photodissociation cross section* and σ is the *integral photodissociation cross section* obtained by integration of $d\sigma/d\Omega$ over the full solid angle $d\Omega$. Further vector correlations, which elucidate subtle details of the absorption and the subsequent fragmentation, exist between $\boldsymbol{\mu}$ and **j** and between **v** and **j**, respectively (Chapter 11).

1.4.5 Photodissociation of single quantum states

The absorption spectrum, branching ratios, and final state distributions depend uniquely — at least in principle — on the initial quantum state of the parent molecule before it absorbs the photon. Recent experimental advances have made it possible to prepare with a first laser (wavelength λ_1) the parent molecule in a particular state before a second laser (wavelength λ_2) dissociates it (Häusler, Andresen, and Schinke 1987; Vander Wal, Scott, and Crim 1991). This allows the study of molecular dynamics on a truly state-specific level. Incoherent averaging caused by the simultaneous dissociation of several initial states is thus avoided. The dissociation of selected quantum states in the ground electronic state undisputedly provides a clearer picture than the investigation of the temperature dependence of photodissociation cross sections. In the latter case, one considers averages over many initial states, each weighted by the corresponding temperature-dependent Boltzmann factor.

Figure 1.9 gives an informative example of vibrationally mediated photodissociation. The absorption cross section of H_2O via the first excited electronic state is shown as a function of the photolysis wavelength λ_2. $H_2O(\tilde{X})$ is initially prepared in a highly excited vibrational state containing about four quanta of OH stretching energy. This spectrum must be compared with the first part (labelled $\tilde{\text{A}}$) of Figure 1.6 showing the absorption spectrum for $H_2O(\tilde{X})$ in its vibrational ground state. While the unimodal spectrum in Figure 1.6 manifests the Gaussian-type shape of the zero-point wavefunction of the vibrational ground state of the parent molecule, the oscillations in Figure 1.9 reflect the nodal structures of the highly excited initial wavefunction. Excitation of another state with about the same amount of internal energy, but in which it is differently partitioned between the two OH stretching modes, yields a completely different absorption spectrum (Weide, Hennig, and Schinke 1989). How the absorption spectrum and the final state distributions reflect the nuclear wavefunction of the parent molecule before it absorbs the photon will be elucidated in Chapter 13.

1.4.6 The various levels of photodissociation cross sections

Let us, for illustration purposes, consider the dissociation of a triatomic molecule as sketched in Figure 1.4. In an ideal experiment one would measure differential photodissociation cross sections with full specification of the initial state and full resolution of the final state,

$$\frac{d\sigma^{(i)}}{d\Omega}(\omega, \theta, \gamma, e, n, j, m_j),$$

where ω is the photon frequency and θ is the recoil angle. The indices γ and e specify the chemical and the electronic channels of the products, respectively, and the quantum numbers n and j designate the particular vibrational-rotational state of the diatomic product; m_j is the projection or magnetic quantum number of the angular momentum vector $\mathbf{j}$ defined with respect to a space-fixed axis. The superscript (i) labels the initial state of the parent molecule. Such an experiment has not yet been performed and probably never will be because the desired resolution is too ambitious.

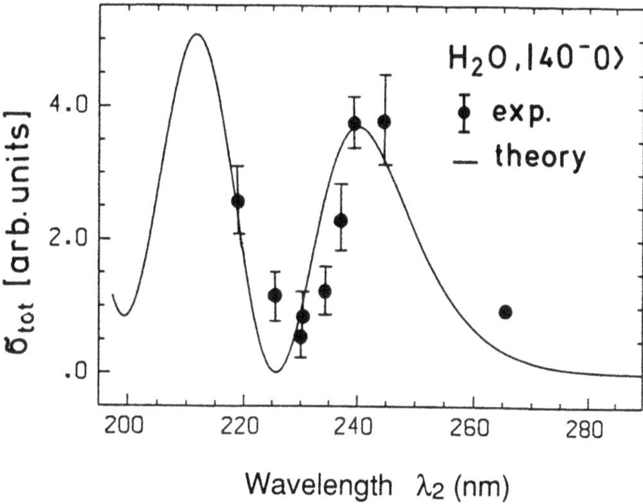

Fig. 1.9. Absorption cross section (arbitrary units) of H_2O initially prepared in the $|04^-0\rangle$ state as a function of the photolysis wavelength λ_2. The first two quantum numbers specify the excitation of the two stretching modes (using a local mode assignment, see Chapter 13), the minus sign indicates the symmetry with respect to the interchange of the two H atoms, and the third quantum number denotes the bending state. Adapted from Vander Wal, Scott, and Crim (1991); calculations by Weide, Hennig, and Schinke (1989).

For most small systems the chemical and electronic channels are rather obvious and therefore the labels γ and e will be omitted hereafter. Normally, one measures integral absorption cross sections,

$$\sigma^{(i)}(\omega, n, j, m_j) = \int d\Omega \, \frac{d\sigma^{(i)}}{d\Omega}(\omega, \theta, n, j, m_j), \tag{1.5}$$

where $d\Omega = \sin\theta \, d\theta \, d\varphi$ with θ and φ being the polar angles of the recoil velocity vector in the laboratory frame. Summation over all magnetic sublevels m_j of the rotor, which are usually not explicitly resolved, is the next step of degradation. Unless special arrangements are made in order to prepare the parent molecule in a single initial state, the measured cross sections represent thermal averages of the form

$$\sigma_T(\omega, n, j) = \sum_i (2J_i + 1) \, \mathrm{e}^{-E_i/k_B T} \, \sigma^{(i)}(\omega, n, j), \tag{1.6}$$

where T is the temperature of the molecular gas, k_B is Boltzmann's constant, and J_i is the total angular momentum quantum number of initial state (i). The summation runs over all initial states with energies E_i. Experiments are performed either in a cell at room temperature ($T \approx$ 300 K) or in a supersonic jet expansion ($T \approx$ 10–50 K). Dissociation in the beam has the advantage that thermal averaging is rather insignificant because only few rotational states J_i of the parent molecule contribute to the cross section.

In the following we will call the $\sigma(\omega, n, j)$ *partial photodissociation cross sections.*[†] They are the cross sections for absorbing a photon with frequency ω *and* producing the diatomic fragment in a particular vibrational-rotational state (n, j). Partial dissociation cross sections for several photolysis frequencies constitute the main body of experimental data and the comparison with theoretical results is based mainly on them. Summation over all product channels (n, j) yields the *total photodissociation cross section* or *absorption cross section*[‡]

$$\sigma_{tot}(\omega) = \sum_{nj} \sigma(\omega, n, j). \tag{1.7}$$

The product state distributions, which we shall amply discuss in this monograph, are defined by

$$P(\omega, n, j) = \sigma(\omega, n, j) \; / \; \sigma_{tot}(\omega). \tag{1.8}$$

Although we are still far away from the ideal experiment, the pace with which experimentalists are tackling this ultimate goal is breathtaking. The experimental data measured over the last ten years or so with ever more sophisticated methods are overwhelming (Gelbart 1977; Simons 1977; Leone 1982; de Vries 1982; Simons 1984; Bersohn 1984; Jackson and Okabe 1986; Dixon et al. 1986; Ashfold and Baggott 1987; Ashfold et al. 1992). Simultaneously, theoretical methods have advanced as well (Freed and Band 1977; Shapiro and Bersohn 1982; Balint-Kurti and Shapiro 1985, Kresin and Lester 1986; Schinke 1988a, 1989b).

The confluence of theory and experiment achieved in recent years has greatly deepened our understanding of molecular photodissociation. At this point, however, it is important to underline that the cornerstone of realistic dynamical investigations is a multi-dimensional potential energy surface (PES). The interrelation between PESs on one hand and the various dissociation cross sections on the other hand is one prominent topic of this book and therefore we think it is useful to elucidate some qualitative aspects of PESs *before* we start with the development of the dynamical concepts.

[†] If necessary, the definition of the partial cross sections must be extended to include the particular chemical and/or electronic channels. Furthermore, the explicit reference to the initial state (i) or the temperature T will be omitted in what follows.

[‡] In the following we will always assume that all molecules excited to the upper electronic state will finally dissociate so that the total dissociation cross section is identical to the absorption cross section.

1.5 Molecular dynamics and potential energy surfaces

The one-dimensional potential curves depicted in Figures 1.1–1.3 represent the dissociation of diatomic molecules for which the potential $V(R_{\mathrm{AB}})$ depends only on the internuclear distance between atoms A and B. However, if one or both constituents are molecules, V is a multi-dimensional object, a so-called potential energy surface which depends on several (at least three) nuclear coordinates denoted by the vector $\mathbf{Q} = (Q_1, Q_2, Q_3, \ldots)$ (Margenau and Kestner 1969; Balint-Kurti 1974; Kuntz 1976; Schaefer III 1979; Kuntz 1979; Truhlar 1981; Salem 1982; Murrell et al. 1984; Hirst 1985; Levine and Bernstein 1987:ch.4; Hirst 1990:ch.3). The *intramolecular* and *intermolecular forces*, defined by

$$F_k(\mathbf{Q}) = -\frac{\partial V(\mathbf{Q})}{\partial Q_k},$$

govern the coupling between the various degrees of freedom, the energy flow from one mode to the others, which bond breaks, and how fast it breaks.

1.5.1 Electronic Schrödinger equation

The potential energy $V(\mathbf{Q})$ for a particular electronic state is defined within the *Born-Oppenheimer approximation* through the *electronic Schrödinger equation* (Daudel, Leroy, Peeters, and Sana 1983:ch.7; Lefebvre-Brion and Field 1986:ch.2)

$$\left[\hat{H}_{el}(\mathbf{q};\mathbf{Q}) - V(\mathbf{Q})\right] \Xi_{el}(\mathbf{q};\mathbf{Q}) = 0, \tag{1.9}$$

where $\mathbf{q}$ collectively represents all electronic coordinates. $V(\mathbf{Q})$ is a particular eigenvalue of the electronic Hamiltonian $\hat{H}_{el}$ for fixed nuclear configuration $\mathbf{Q}$. Since $\mathbf{Q}$ enters (1.9) only as a parameter both the potential and the electronic wavefunction also depend parametrically on the nuclear geometry. The kinetic energy associated with the nuclear degrees of freedom is neglected within the Born-Oppenheimer approximation. For a more detailed discussion see Section 2.3.

In general, (1.9) must be solved numerically by quantum chemical or so-called *ab initio* methods (Lowe 1978; Szabo and Ostlund 1982; Daudel et al. 1983; Dykstra 1988; Hirst 1990:ch.2). The pointwise solution of (1.9) for a set of nuclear geometries and the fitting of all points to an analytical representation yields the PES which is the input to the subsequent dynamics calculations. In principle, one expands $\Xi_{el}(\mathbf{q};\mathbf{Q})$ in a suitable set of electronic basis functions and diagonalizes the corresponding Hamilton matrix, i.e., the representation of $\hat{H}_{el}$ within the chosen basis of electronic wavefunctions. Since the number of electrons is usually large, even for simple molecules like H_2O and ClNO, the solution of

(1.9) is a formidable task and requires sophisticated methods and powerful computers (Lawley 1987).

The solution of the electronic Schrödinger equation has to be repeated for many different nuclear configurations in order to construct a global PES. A triatomic molecule requires three coordinates for a complete description, the three interatomic distances, for example. In general, the number of coordinates necessary for a complete PES is $3N - 6$ with $N(> 2)$ being the number of constituent atoms. As a rule of thumb, we need about 5–10 points per degree of freedom which add up to several hundreds of points to represent a complete PES, even for a triatomic molecule. This large number of nuclear configurations imposes a severe bottleneck for realistic theoretical studies of half as well as full collisions. In UV photodissociation one has the additional problem that the fragmentation takes place in an excited electronic state whose potential energy surface is generally more difficult to determine than that of the electronic ground state (Bruna and Peyerimhoff 1987).

We will not discuss the actual construction of potential energy surfaces. This monograph deals exclusively with the nuclear motion taking place *on* a PES and the relation of the various types of cross sections to particular features of the PES. The investigation of molecular dynamics is — in the context of classical mechanics — equivalent to rolling a billiard ball on a multi-dimensional surface. The way in which the forces $F_k(\mathbf{Q})$ determine the route of the billiard ball is the central topic of this monograph. In the following we discuss briefly two illustrative examples which play key roles in the subsequent chapters.

1.5.2 First example: CH_3ONO

The photodissociation of aliphatic nitrites

$$\mathrm{RONO} + \hbar\omega \longrightarrow (\mathrm{RONO})^* \longrightarrow \mathrm{RO} + \mathrm{NO}$$

with R = H, CH_3 and $(CH_3)_3C$, for example, has received much interest in recent years (Reisler, Noble, and Wittig 1987; Huber 1988). We will discuss methyl nitrite, CH_3ONO, as a prototype. The geometry of *cis*-CH_3ONO in the electronic ground state S_0 (the lowest singlet state) is illustrated in Figure 1.10(a). The calculation of a complete multidimensional PES for this system is, of course, far beyond present day practicability and future developments seem unlikely to bring us significantly closer to this goal. Thus, in order to tackle the photodissociation of CH_3ONO we must necessarily restrict the number of coordinates. The essential modes, which must be retained if one wants to calculate the absorption spectrum and the final state distributions of the NO fragment,

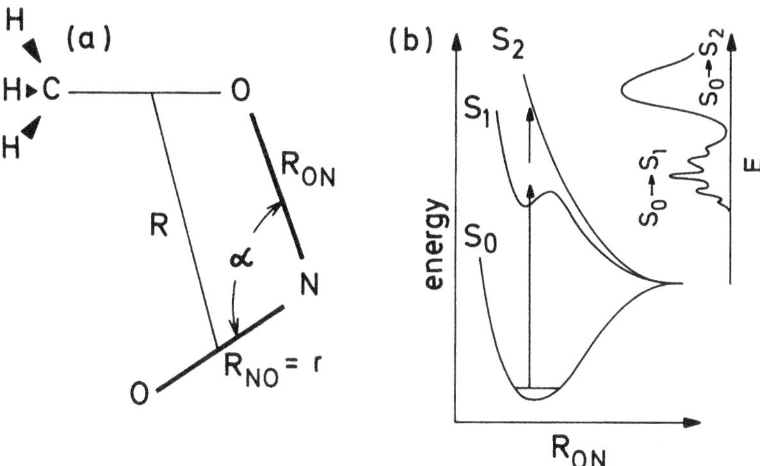

Fig. 1.10. (a) Schematic representation of the geometry of *cis*-CH_3ONO in the electronic ground state S_0. R and r are the Jacobi coordinates used in the dynamics calculations (see Section 2.4). (b) One-dimensional cuts along the minimum energy paths of the potential energy surfaces of the S_0, S_1, and S_2 states. The right-hand side shows schematically the corresponding absorption spectra.

are: the dissociation bond R_{ON}, the NO vibrational bond R_{NO}, and the ONO bending angle α. Although this restriction appears quite severe, the success with which the gross experimental results are reproduced by the calculations, justifies this simplification.

Figure 1.11 depicts the PESs of the three lowest singlet states S_0, S_1, and S_2 as functions of R_{ON} and R_{NO}; the bending angle α and all other coordinates are fixed at their equilibrium values in the ground state S_0. Figure 1.10(b) shows qualitatively cuts along the reaction path when the CH_3O-NO bond is elongated. The lowest PES is binding along both stretching coordinates with a deep potential well at short distances; separating CH_3O and NO requires about 2 eV before the bond breaks.

The S_1 PES, calculated by Nonella and Huber (1986), has a shallow minimum above the ground-state equilibrium, or expressed differently, a small potential barrier hinders the immediate dissociation of the excited S_1 complex. Although the height of the barrier is less than a tenth of an eV, it drastically affects the dissociation dynamics, even at energies which significantly exceed the barrier. The excited complex lives for about 5–10 internal NO vibrational periods before it breaks apart. The photodissociation of CH_3ONO through the S_1 state exemplifies indirect photodissociation or vibrational predissociation (Chapter 7).

In contrast, the PES of the S_2 state calculated by Suter, Brühlmann, and Huber (1990) is repulsive along the CH_3O-NO dissociation coordi-

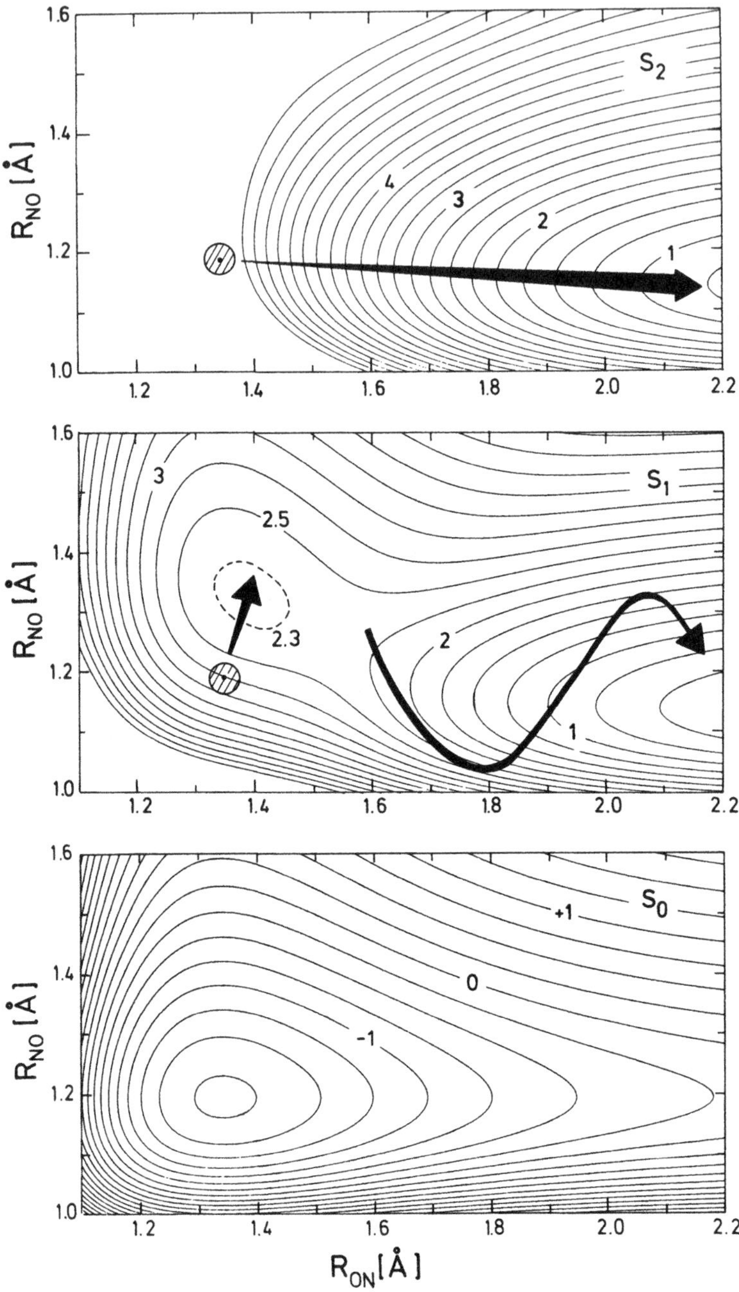

Fig. 1.11. Two-dimensional contour maps of the potential energy surfaces of *cis*-CH_3ONO in the three lowest singlet states S_0, S_1, and S_2. The coordinates are defined in Figure 1.10(a). The energy contours are given in eV and the normalization is such that $CH_3O + NO(r_e)$ corresponds to $E = 0$ in all three cases. (*cont.*)

nate, i.e., there is no barrier hindering the immediate fragmentation once the molecule is excited by the light pulse. In this case, the survival time of the excited CH_3ONO complex is less than an internal NO vibrational period. The dissociation of CH_3ONO via the S_2 state manifests direct photodissociation (Chapter 6).

Fragmentation of CH_3ONO through excitation of the S_1 state drastically differs from fragmentation in the S_2 state. This is reflected by the lifetimes, the vibrational state distributions of the NO fragment, and by the absorption spectra. The latter are schematically shown on the right-hand side of Figure 1.10(b). While the $S_0 \rightarrow S_1$ spectrum consists of a progression of relatively narrow structures, the spectrum for the $S_0 \rightarrow S_2$ transition is broad and structureless.

1.5.3 *Second example:* H_2O

Figure 1.12 depicts the PESs of the three lowest electronic states of H_2O (which below we shall denote by V_X, V_A, and V_B, respectively). One of the O-H bonds is fixed at the equilibrium value in the electronic ground state. Figure 1.12 features the angular dependence of the potential energy surfaces when one hydrogen atom is pulled away. The corresponding absorption spectra for the transitions $\tilde{X} \rightarrow \tilde{A}$ and $\tilde{X} \rightarrow \tilde{B}$ are shown in Figure 1.6.

Profiles of the two-dimensional PESs along the reaction path, indicated by the heavy arrows in Figure 1.12, are illustrated in Figure 1.13(a). The two lowest states correlate asymptotically with the electronic ground state of OH while the third state leads to OH in the first excited state. Figure 1.13(b) shows schematically the dependence on the HOH bending angle α with the two H-O bond distances being fixed at relatively small values. All potentials are symmetric with respect to $\alpha = 180°$ which implies that $\partial V/\partial\alpha = 0$ at linearity. Notice that at short H-O bond distances the $\tilde{A}$ and the $\tilde{B}$ state are degenerate at the linear configuration.

Fig. 1.11. (*cont.*) The S_1 and S_2 potential energy surfaces have been calculated by Nonella and Huber (1986) and Suter, Brühlmann, and Huber (1990), respectively, whereas the PES for the S_0 state is approximated by the sum of two uncoupled Morse oscillators. The shaded circles indicate the equilibrium region of the ground electronic state where the dissociative motion in the excited electronic states starts and the heavy arrows illustrate the subsequent dissociation paths. Detailed discussions of the absorption spectra and the vibrational state distributions of NO follow in Chapters 7 and 9.

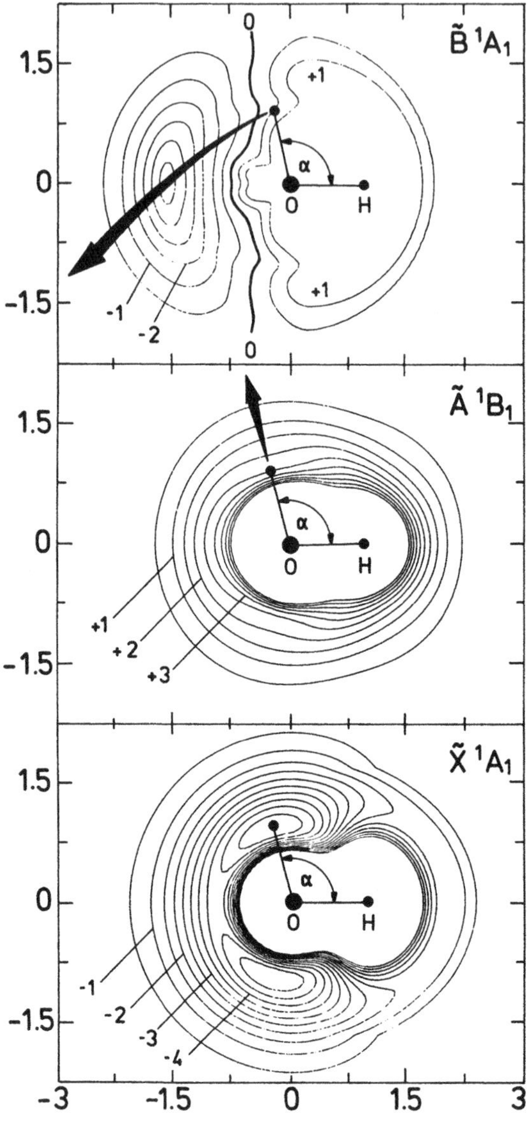

Fig. 1.12. Two-dimensional polar plots of the potential energy surfaces (denoted below by V_X, V_A, and V_B) of the three lowest electronic states of H_2O. One of the O-H bonds is frozen at its equilibrium in the electronic ground state. The contours represent the potential energy as the other H atom swings around the O atom. Energies and distances are given in eV and Å, respectively. The energy is normalized such that $H + OH(^2\Pi, r_e)$ and $H + OH(^2\Sigma, r_e)$, respectively, correspond to $E = 0$. V_X is the empirical fit of Sorbie and Murrell (1975) whereas V_A and V_B have been calculated by Staemmler and Palma (1985) and by Theodorakopoulos, Petsalakis, and Buenker (1985), respectively. The heavy arrows illustrate the main dissociation paths in the excited states.

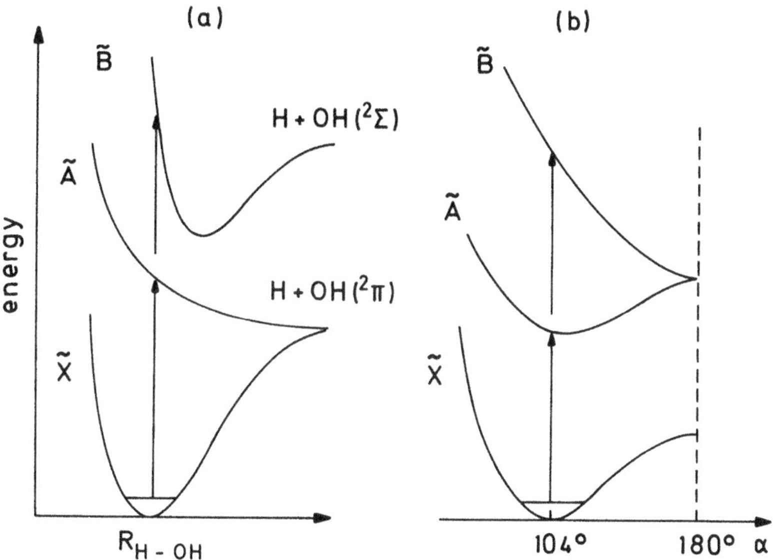

Fig. 1.13. (a) One-dimensional cuts along the minimum energy paths for the three lowest potential energy surfaces of H_2O. (b) Illustration of the dependence of the potential energy surfaces on the HOH bending angle α (schematic).

The PES of the $\tilde{X}^1A_1$ state is strongly binding with an equilibrium angle of $\alpha_e = 104°$. The PES of the $\tilde{A}^1B_1$ state calculated by Staemmler and Palma (1985), on the other hand, is repulsive and therefore the H-OH bond breaks immediately the parent molecule is excited. The dissociation of H_2O in the first absorption band is an example of direct photodissociation. At short H-OH distances V_A has a minimum at roughly the same bending angle as the PES of the $\tilde{X}$ state.† With increasing H-OH bond distances the PES becomes more and more isotropic, i.e., angle-independent, and the *torque* $-\partial V_A/\partial\alpha$ diminishes. The weak anisotropy of V_A around 104° readily explains the experimental observation that $OH(^2\Pi)$ following the excitation of H_2O in the first continuum is predominantly created in low rotational states (Section 10.1).

Photodissociation via the second excited state proceeds in a very different way. The PES of the $\tilde{B}$ state calculated by Theodorakopoulos, Petsalakis, and Buenker (1985) has a deep potential minimum at the linear configuration ($\alpha = 180°$). This well is the result of a conical intersection with the PES of the $\tilde{X}$ state at a H-OH bond distance of about 1.6 Å. The equilibrium configurations of the $\tilde{X}$ and the $\tilde{B}$ states are sig-

† In the case of the $\tilde{A}$ state of H_2O we deliberately avoid using the term "equilibrium angle" because the corresponding PES is purely repulsive. The $\tilde{A}$ state of H_2O does not have an equilibrium!

nificantly displaced so that the excitation point, i.e., the point where the molecule is born in the $\tilde{B}$ state, lies high up at the inner slope of the potential, about 0.5–1.0 eV above the $\mathrm{H} + \mathrm{OH}(^2\Sigma)$ dissociation threshold. The $\tilde{B}$-state PES is highly anisotropic above the ground-state equilibrium of 104° and therefore the bending angle opens immediately under the influence of the strong torque $-\partial V_B/\partial\alpha$ with the result that the OH fragment is generated with an extremely high degree of rotational excitation as indeed observed experimentally (see Figure 1.8). As opposed to the dissociation in the $\tilde{A}$ state, a substantial amount of the available energy is converted into rotational energy.

Figures 1.12 and 1.13 readily explain, without quantitative calculation, some key features of the photodissociation of H_2O through excitation in the $\tilde{A}$ and in the $\tilde{B}$ absorption bands. Multi-dimensional potential energy surfaces are *the* cornerstones for a trustworthy analysis of molecular dynamics. Knowing the general topology of the PES often suffices for a qualitative explanation of the main experimental observations. However, in order to perform realistic calculations we need potential energy surfaces which are as accurate and complete as possible.

2
Light absorption and photodissociation

A beam of light which passes through a gas is attenuated because the gas atoms or molecules absorb part of the electromagnetic energy and convert it into internal (i.e., electronic, translational, vibrational, or rotational) energy. Assuming that the light beam propagates along the z-axis, the intensity decreases according to *Beer's law*

$$I(z) = I_0 \, \mathrm{e}^{-\sigma \varrho z}, \tag{2.1}$$

where I_0 stands for the intensity at $z = 0$ and ϱ is the density. The frequency-dependent absorption cross section $\sigma(\omega)$ is a characteristic quantity of the molecules in the gas; it does not depend on the intensity of the light beam.

The microscopic theory of light-matter interaction relates $\sigma(\omega)$ to a matrix element of the electric dipole operator $\hat{\mathbf{d}}$ of the molecule,

$$\sigma(\omega) \propto \omega \, \delta(\omega_{fi} - \omega) \, |\langle F_f \mid \mathbf{e} \cdot \hat{\mathbf{d}} \mid F_i \rangle|^2, \tag{2.2}$$

where $|F_i\rangle$ and $|F_f\rangle$ denote the initial and final molecular quantum states with energies E_i and E_f, respectively, $\omega_{fi} = (E_f - E_i)/\hbar$ is the corresponding *transition frequency*, and $\mathbf{e}$ is a unit vector in the direction of the polarization of the electric field. Equation (2.2) rests upon two major assumptions: first, the interaction between the light beam and the matter must be weak and second, the duration of the perturbation must be long. These conditions hold for almost all experiments which we will discuss in this monograph. New advances with the excitation by short laser pulses will be briefly mentioned at the very end in Chapter 16.

Equation (2.2) is at the center of spectroscopy and photodissociation. It applies equally to bound-bound, bound-free, and free-free transitions. Although almost every textbook on quantum mechanics contains the derivation of (2.2) (see, for example, Merzbacher 1970:ch.18; Loisell 1973:ch.5; Cohen-Tannoudji, Diu, and Laloë 1977:ch.XIII; Loudon

1983:ch.2), we shall derive it in some detail in order to distinguish clearly the two different time scales inherent to photodissociation: the time dependence of the absorption process and the evolution of the subsequent fragmentation in the excited electronic state.

2.1 Time-dependent perturbation theory

Our derivation of Equations (2.1) and (2.2) follows very closely the presentation of Loudon (1983:ch.2). The basic concept is to describe the molecule quantum mechanically, the photon field classically, and to treat the interaction between them in first-order perturbation theory.

2.1.1 Coupled equations

We start from the time-dependent Schrödinger equation for the wavefunction describing the evolution of the unperturbed molecular system, $\mathcal{F}(\mathbf{Q}, \mathbf{q}; t)$,

$$\left[i\hbar\frac{\partial}{\partial t} - \hat{H}_{mol}(\mathbf{Q}, \mathbf{q})\right] \mathcal{F}(\mathbf{Q}, \mathbf{q}; t) = 0, \tag{2.3}$$

where the *molecular Hamiltonian* $\hat{H}_{mol}(\mathbf{Q}, \mathbf{q})$ includes all nuclear ($\mathbf{Q}$) and all electronic ($\mathbf{q}$) degrees of freedom. In the absence of an external perturbation $\hat{H}_{mol}$ is time-independent and therefore $\mathcal{F}(\mathbf{Q}, \mathbf{q}; t)$ separates into a time-independent part and a time-dependent phase factor,

$$\mathcal{F}_\alpha(\mathbf{Q}, \mathbf{q}; t) = \mathrm{e}^{-iE_\alpha t/\hbar}\, F_\alpha(\mathbf{Q}, \mathbf{q}). \tag{2.4}$$

The stationary wavefunctions $F_\alpha(\mathbf{Q}, \mathbf{q})$ solve the time-independent Schrödinger equation

$$[\hat{H}_{mol}(\mathbf{Q}, \mathbf{q}) - E_\alpha] F_\alpha(\mathbf{Q}, \mathbf{q}) = 0 \tag{2.5}$$

with eigenenergies E_α. We explicitly assume in this section that the spectrum of $\hat{H}_{mol}$ is discrete so that the eigenfunctions F_α are properly normalizable. In the beginning, we consider only bound molecular states and defer the extension to the continuum of $\hat{H}_{mol}$ (which is necessary in order to describe the dissociation of the molecule) to Section 2.5.

The $F_\alpha(\mathbf{Q}, \mathbf{q})$ are orthogonal,

$$\langle F_\alpha \mid F_{\alpha'} \rangle = \int d\mathbf{Q}\, d\mathbf{q}\, F_\alpha^*(\mathbf{Q}, \mathbf{q})\, F_{\alpha'}(\mathbf{Q}, \mathbf{q}) = \delta_{\alpha\alpha'}, \tag{2.6}$$

and they fulfil the closure relation,

$$\sum_\alpha |F_\alpha\rangle\langle F_\alpha| = \sum_\alpha F_\alpha(\mathbf{Q}, \mathbf{q})\, F_\alpha^*(\mathbf{Q}', \mathbf{q}') = \delta(\mathbf{Q} - \mathbf{Q}')\, \delta(\mathbf{q} - \mathbf{q}'). \tag{2.7}$$

The star indicates the complex conjugate. The stationary eigenfunctions $F_\alpha(\mathbf{Q}, \mathbf{q})$ form a basis in the Hilbert space of $\hat{H}_{mol}$, i.e., each function within this space can be uniquely represented in terms of the $F_\alpha(\mathbf{Q}, \mathbf{q})$.

When the light beam is switched on at time $t = 0$ the total Hamiltonian becomes

$$\hat{H}_{tot}(t) = \hat{H}_{mol} + \hat{h}(t), \tag{2.8}$$

where the perturbation $\hat{h}(t)$ represents the time-dependent interaction of the electromagnetic field with the molecule. $\hat{h}(t)$ induces transitions between the eigenstates $|F_\alpha\rangle$ of the molecular system. In order to solve the time-dependent Schrödinger equation, including the perturbation, the molecular wavefunction is expanded in terms of the $F_\alpha(\mathbf{Q}, \mathbf{q})$,

$$\mathcal{F}(\mathbf{Q}, \mathbf{q}; t) = \sum_\alpha a_\alpha(t)\; F_\alpha(\mathbf{Q}, \mathbf{q})\; \mathrm{e}^{-iE_\alpha t/\hbar} \tag{2.9}$$

with unknown time-independent coefficients $a_\alpha(t)$. $|a_\alpha(t)|^2$ with normalization $\sum_\alpha |a_\alpha(t)|^2 = 1$ is the probability of finding at time t the molecule in state $|F_\alpha\rangle$.

In order to determine the coefficients $a_\alpha(t)$ we insert expansion (2.9) into (2.3) with $\hat{H}_{mol}$ replaced by $\hat{H}_{tot}(t)$. Multiplication with $\langle F_\alpha|$ and utilizing (2.6) yields, after some simple manipulations, the following set of coupled equations,

$$i\hbar \frac{d}{dt} a_\alpha(t) = \sum_{\alpha'} h_{\alpha\alpha'}(t)\; a_{\alpha'}(t)\; \mathrm{e}^{i\omega_{\alpha\alpha'} t}, \tag{2.10}$$

where the time-dependent matrix elements of the perturbation operator,

$$h_{\alpha\alpha'}(t) = \langle F_\alpha \mid \hat{h}(t) \mid F_{\alpha'}\rangle = \int d\mathbf{Q}\; d\mathbf{q}\; F_\alpha^*(\mathbf{Q}, \mathbf{q})\; \hat{h}(t)\; F_{\alpha'}(\mathbf{Q}, \mathbf{q}), \tag{2.11}$$

couple state $|F_\alpha\rangle$ with all other states $|F_{\alpha'}\rangle$. The *transition frequencies* in (2.10) are defined by

$$\omega_{\alpha\alpha'} = (E_\alpha - E_{\alpha'})/\hbar. \tag{2.12}$$

Equation (2.10) is a set of first-order differential equations which describe the evolution of the molecular system under the external perturbation $\hat{h}(t)$. It is valid for time-independent as well as time-dependent perturbations and must be solved subject to the initial conditions $a_i(0) = 1$ and $a_{\alpha\neq i}(0) = 0$ if the molecule is initially in state $|F_i\rangle$. The time dependence of the coefficients $a_\alpha(t)$ together with the stationary basis functions $F_\alpha(\mathbf{Q}, \mathbf{q})$ describe completely the state of the molecule at each instant t. When the perturbation is switched off, the coefficients $a_\alpha(t)$ become constant again.

2.1.2 Photon-induced transition rate

Within the *electric dipole approximation* (Loudon 1983:ch.2) the perturbation is given by

$$\hat{h}(t) = \hat{\mathbf{d}} \cdot \mathbf{E}_0 \cos \omega t, \tag{2.13}$$

where $\hat{\mathbf{d}}$ is the electric dipole operator of the molecule and $\mathbf{E}_0$ is the electric field vector of the light beam. Higher-order contributions to the interaction energy are usually much smaller and therefore neglected. Furthermore, we have assumed that the molecular dimensions are much smaller than the wavelength of the photon so that the electric field is approximately constant over the range of the molecule.

Insertion of Equation (2.13) into (2.11) yields

$$h_{\alpha\alpha'}(t) = d_{\alpha\alpha'} \cos \omega t, \tag{2.14}$$

where we defined the time-independent matrix elements as

$$d_{\alpha\alpha'} = \langle F_\alpha \mid \mathbf{E}_0 \cdot \hat{\mathbf{d}} \mid F_{\alpha'} \rangle = E_0 \langle F_\alpha \mid \mathbf{e} \cdot \hat{\mathbf{d}} \mid F_{\alpha'} \rangle \tag{2.15}$$

with $E_0 = |\mathbf{E}_0|$ and $\mathbf{e}$ being a unit vector in the direction of the electric field. Using Equation (2.14) in (2.10), the coupled equations finally become

$$i\hbar \frac{d}{dt} a_\alpha(t) = \sum_{\alpha'} d_{\alpha\alpha'} \; a_{\alpha'}(t) \; \cos \omega t \; e^{i\omega_{\alpha\alpha'} t}. \tag{2.16}$$

Equation (2.16) is exact independent of the strength or the shape of the light pulse.

Under the assumption that the coupling elements $d_{\alpha\alpha'}$ are very small, Equation (2.16) may be solved by first-order perturbation theory: the coefficients $a_{\alpha'}(t)$ on the right-hand side are replaced by their initial values at $t = 0$. The evolution of each final state ($f \neq i$) is then governed by the (uncoupled) equation

$$i\hbar \frac{d}{dt} a_f(t) = d_{fi} \; \cos \omega t \; e^{i\omega_{fi} t}. \tag{2.17}$$

Within this approximation we explicitly assume that the probabilities $|a_\alpha|^2$ do not significantly change while the light beam is switched on, i.e., $|a_i(t)|^2 \approx 1$ and $\sum_{f \neq i} |a_f(t)|^2 \ll 1$. Under these restrictions each final molecular state $|F_f\rangle$ is coupled only to the initial state $|F_i\rangle$.

Equation (2.17) can be readily integrated to yield

$$a_f(t) = \frac{d_{fi}}{2\hbar} \left[\frac{1 - e^{i(\omega_{fi} + \omega)t}}{\omega_{fi} + \omega} + \frac{1 - e^{i(\omega_{fi} - \omega)t}}{\omega_{fi} - \omega} \right]. \tag{2.18}$$

For $\omega_{fi} > 0$ (absorption) the first term is usually much smaller than the second one and therefore it is neglected (*rotating wave approximation*). The reverse holds for $\omega_{fi} < 0$ (stimulated emission). Within this approximation the time-dependent probability for making a transition from initial state $|F_i\rangle$ to final state $|F_f\rangle$ under the influence of the photon beam with frequency ω becomes[†]

$$P_{fi}(t) = |a_f(t)|^2 = \left(\frac{d_{fi}}{\hbar}\right)^2 \frac{\sin^2\left[(\omega_{fi}-\omega)\,t/2\right]}{(\omega_{fi}-\omega)^2}. \tag{2.19}$$

Equation (2.19) is a complicated function of time t, frequency ω, and transition frequency ω_{fi}. Shortly after the perturbation is switched on, all molecular states become populated. However, as the electric field continues to drive the molecule with constant frequency ω only the *resonant transition* with $\omega_{fi} \approx \omega$ prevails while the probabilities for the nonresonant transitions remain negligibly small. The excitation of an atom or molecule by the electric field is formally equivalent to the excitation of an oscillator by an external periodic perturbation with constant frequency. Using the representation

$$\delta(\omega_{fi}-\omega) = \frac{2}{\pi}\lim_{t\to\infty} \frac{\sin^2\left[(\omega_{fi}-\omega)\,t/2\right]}{(\omega_{fi}-\omega)^2\,t}, \tag{2.20}$$

for the Dirac delta-function (Loudon 1983:ch.2) we can rewrite (2.19), in the limit as t goes to infinity, as

$$P_{fi}(t) = \frac{\pi}{2}\left(\frac{d_{fi}}{\hbar}\right)^2 t\,\delta(\omega_{fi}-\omega). \tag{2.21}$$

Equation (2.21) represents the final expression for the probability that by the time t the molecular system has made a transition from state $|F_i\rangle$ to state $|F_f\rangle$.

- The probability rises linearly with time and therefore it leads to a constant *transition rate*, i.e., transition probability per unit time interval,

$$k_{fi} = \frac{d}{dt}P_{fi} = \frac{\pi}{2}\left(\frac{d_{fi}}{\hbar}\right)^2 \delta(\omega_{fi}-\omega). \tag{2.22}$$

It is worthwhile recalling at this stage the basic assumptions made so far. First of all, the perturbation must be sufficiently weak in order to allow first-order treatment. Secondly, the pulse must be so long that the molecule can recognize the electric field as a periodic perturbation. On the other hand the transition probability must remain small compared to unity because otherwise the first-order approximation breaks down. A more detailed and illuminating discussion of the applicability of (2.22) is given by Cohen-Tannoudji, Diu, and Laloë (1977:ch.XIII).

[†] Since the stationary wavefunctions F_α can be considered to be real, the matrix elements of the electric dipole operator are also real.

2.2 The absorption cross section

The microscopic theory derived in the previous section describes the evolution of the molecular system under the influence of the electric field of the light beam. In this section, following Loudon (1983:ch.1), we use (2.22) to deduce an expression for the phenomenological absorption cross section $\sigma(\omega)$ defined in (2.1). See Loudon for a more detailed discussion.

Let us assume that the light beam propagates along the z-axis through a cavity containing the molecular gas. The volume of the cavity is V and the total number of particles within the cavity is N. For simplicity, we assume monochromatic light with frequency $\omega = \omega_{fi}$. Consider a thin slice perpendicular to the z-axis with length dz and area A which is sufficiently thin that the intensity I (energy crossing through a unit area per unit time) as well as the (cycle-averaged) energy density W (energy per volume) are approximately constant within the slice.

The beam energy in the slice is $S = AdzW$. As a result of the molecular transition from state $|F_i\rangle$ to state $|F_f\rangle$ it decreases by

$$
\begin{aligned}
dS &= -N \frac{Adz}{V} \hbar \omega_{fi} k_{fi} \, dt && (2.23a) \\
&= A \, dz \, dW && (2.23b)
\end{aligned}
$$

within a time interval dt. Here, $NAdz/V$ is the number of molecules within the slice, $\hbar\omega_{fi}$ is the absorbed energy per molecule, and k_{fi} is the transition rate. The minus sign indicates the diminution of S and the second line of (2.23) follows simply from the definition of dS. Inserting (2.15) and (2.22) into (2.23) and solving for dW yields

$$
dW = -\omega_{fi} \, \delta(\omega_{fi} - \omega) \, dt \, \frac{\pi N E_0^2}{2\hbar V} \left| \langle F_f \mid \mathbf{e} \cdot \hat{\mathbf{d}} \mid F_i \rangle \right|^2, \qquad (2.24)
$$

where E_0 is the amplitude of the electric field and $\mathbf{e}$ is a unit vector in the direction of $\mathbf{E}_0$. Using the relation (Loudon 1983:ch.1) $E_0^2 = 2W/\epsilon_0$, with ϵ_0 being the electric permitivity, we get

$$
\frac{dW}{dt} = -\omega_{fi} \, \delta(\omega_{fi} - \omega) \, \frac{\pi N W}{\hbar \epsilon_0 V} \left| \langle F_f \mid \mathbf{e} \cdot \hat{\mathbf{d}} \mid F_i \rangle \right|^2. \qquad (2.25)
$$

Equation (2.25) governs the time dependence of the energy density W. However, we are searching for an expression which describes the spatial dependence of the intensity I. Using the relations (Loudon 1983:ch.1) $dW/dt = dI/dz$ and $W = I/c$, where c is the velocity of light, we obtain

$$
\frac{dI}{dz} = -\varrho \, \sigma(\omega) \, I \qquad (2.26)
$$

with $\varrho = N/V$ being the density of the gas in the cavity. The absorption cross section is defined as

$$\sigma(\omega) = \frac{\pi}{\hbar\epsilon_0 c}\,\omega_{fi}\,\delta(\omega_{fi}-\omega)\,|\langle F_f \mid \mathbf{e}\cdot\hat{\mathbf{d}} \mid F_i\rangle|^2. \tag{2.27}$$

Integration of (2.26) readily yields the desired Equation (2.1). We note that $|\hat{\mathbf{d}}|^2/\epsilon_0$ has units of energy times volume and with that it is easy to show that the cross section has units of area.

The absorption spectrum, i.e., the absorption cross section as a function of ω consists of a series of infinitely narrow, discrete lines with amplitudes according to (2.27).

- The gas absorbs the photon only if the frequency of the light beam is in resonance with an atomic or molecular transition frequency, i.e., the resonance condition $\omega \approx \omega_{fi}$ must be fulfilled.

The theory outlined above assumes ideal conditions which are, of course, never met in a real experiment: a monochromatic light beam, sharply defined molecular energies E_i and E_f, and an infinitely long light pulse. However, in reality the frequency is not sharp, the energy levels may be broadened by collision effects, and the pulse has a finite duration. These imperfections together with broadening due to thermal motion of the molecules and losses of molecules in the excited state as a result of spontaneous or stimulated emission lead to a broadening of each individual absorption line. However, all these effects are rather unimportant for the purpose of this book, namely bound-free transitions (photodissociation), because the corresponding absorption spectrum is a continuous function of the photon frequency itself. The broadening caused by dissociation usually by far exceeds the broadening due to the aforementioned effects and therefore we will not elaborate on them.

Let us assume that the upper state is degenerate with substates $|F_f^{(\beta)}\rangle$, all corresponding to the same total energy E_f. The photon excites each of these states simultaneously because the resonance condition $\omega_{fi} \approx \omega$ holds for all of them. The absorption cross section is consequently composed of several *partial absorption cross sections* $\sigma(\omega,\beta)$ each being defined as in (2.27) with $|F_f\rangle$ replaced by $|F_f^{(\beta)}\rangle$. We will come back to this in Section 2.5 when discussing photodissociation.

2.3 Born-Oppenheimer approximation

The stationary states $|F_\alpha\rangle$ in Sections 2.1 and 2.2 represent general molecular states including all electronic ($\mathbf{q}$) and all nuclear ($\mathbf{Q}$) degrees of freedom. In this section we employ the *Born-Oppenheimer approximation* in order to separate the molecular wavefunction into a nuclear part, $\Psi^{nu}(\mathbf{Q})$, and an electronic part, $\Xi^{el}(\mathbf{q};\mathbf{Q})$, with the latter depending *parametrically* on all nuclear coordinates $\mathbf{Q}$. The Born-Oppenheimer approximation is *the* keystone of molecular physics (Weissbluth 1978:ch.25; Daudel et al. 1983:ch.7; Lefebvre-Brion and Field 1986:ch.2). Without the separation of the nuclear motion from the motion of the electrons calculations of scattering and photodissociation cross sections or vibrational-rotational spectra of molecules would be essentially impracticable.

The molecular Hamiltonian may be written as

$$\hat{H}_{mol}(\mathbf{Q}, \mathbf{q}) = \hat{T}_{nu}(\mathbf{Q}) + \hat{H}_{el}(\mathbf{q}; \mathbf{Q}), \tag{2.28}$$

where $\hat{T}_{nu}$ is the nuclear kinetic energy and $\hat{H}_{el}$ is the electronic Hamiltonian. The latter comprises the kinetic energy of the electrons and the electrostatic potential energy, i.e., the electron-electron repulsion, the electron-nuclear attraction, and the nuclear-nuclear repulsion. All differential operators with respect to the nuclear coordinates $\mathbf{Q}$ are contained in $\hat{T}_{nu}$ and therefore $\hat{H}_{el}$ depends only parametrically on $\mathbf{Q}$. In the so-called *adiabatic representation* we expand the time-independent molecular wavefunction according to

$$F(\mathbf{Q}, \mathbf{q}) = \sum_k \Psi_k^{nu}(\mathbf{Q})\ \Xi_k^{el}(\mathbf{q}; \mathbf{Q}), \tag{2.29}$$

where the electronic expansion functions $\Xi_k^{el}(\mathbf{q}; \mathbf{Q})$ are solutions of the electronic Schrödinger equation

$$\left[\hat{H}_{el}(\mathbf{q}; \mathbf{Q}) - V_k(\mathbf{Q})\right] \Xi_k^{el}(\mathbf{q}; \mathbf{Q}) = 0 \tag{2.30}$$

for any fixed nuclear geometry $\mathbf{Q}$. For simplicity, the index α has been omitted. Note, that the electronic wavefunctions Ξ_k^{el} as well as the corresponding electronic energies V_k depend parametrically on all nuclear coordinates. The index k labels the different electronic states, i.e., the different roots of (2.30); $k = 0$ is the electronic ground state, $k = 1$ is the first excited electronic state etc. The eigenenergies $V_k(\mathbf{Q})$ are the potential energy surfaces which we discussed in Section 1.5.

Inserting (2.29) into the time-independent Schrödinger equation (2.5), multiplying with $\langle \Xi_k^{el} |$ from the left, and exploiting the orthogonality of the electronic wavefunctions for each nuclear configuration $\mathbf{Q}$ readily yields a set of coupled equations for the nuclear wavefunctions (Köppel, Domcke, and Cederbaum 1984),

$$\sum_{k'} \langle \Xi_k^{el} \mid \hat{T}_{nu} \mid \Psi_{k'}^{nu}(\mathbf{Q})\ \Xi_{k'}^{el} \rangle + [\ V_k(\mathbf{Q}) - E\ ]\ \Psi_k^{nu}(\mathbf{Q}) = 0, \tag{2.31}$$

where the integration is performed over all electronic coordinates. The different electronic states $k, k' = 0, 1, \ldots$ are coupled through the matrix elements of the nuclear kinetic energy operator.

Equation (2.31) is still exact. In practice, however, it is extremely difficult to solve. Since $\hat{T}_{nu}$ contains first- and second-order derivatives of the form $\partial/\partial Q_i$ and $\partial^2/\partial Q_i^2$ and since the electronic wavefunctions depend parametrically on $\mathbf{Q}$, one needs to evaluate first- and second-order derivatives of the electronic wavefunctions with respect to all nuclear coordinates. The $\Xi_k^{el}(\mathbf{q};\mathbf{Q})$ are very complicated multielectron wavefunctions and therefore it is not difficult to surmise that the determination of the kinetic coupling elements in (2.31) is extremely cumbersome, especially for polyatomic molecules.

However, provided the electronic wavefunctions vary only weakly with the nuclear coordinates $\mathbf{Q}$, we may make the approximation

$$\hat{T}_{nu}\left[\Psi_{k'}^{nu}(\mathbf{Q})\,\Xi_{k'}^{el}(\mathbf{q};\mathbf{Q})\right] \approx \left[\hat{T}_{nu}\,\Psi_{k'}^{nu}(\mathbf{Q})\right]\Xi_{k'}^{el}(\mathbf{q};\mathbf{Q}),$$

i.e., the derivative of the electronic wavefunctions with respect to the nuclear coordinates is assumed to be negligibly small.† With this approximation (2.31) reduces to a set of *uncoupled* equations

$$\left[\hat{T}_{nu}(\mathbf{Q}) + V_k(\mathbf{Q}) - E\right]\Psi_k^{nu}(\mathbf{Q}) = 0, \tag{2.32}$$

where we utilized the orthogonality of the Ξ_k^{el}. Equation (2.32) is the Schrödinger equation for the nuclear motion within the kth electronic state. The coupling to other electronic states is absent and therefore the motion of the nuclei evolves separately in each electronic state, unaffected by the nuclear motion in other electronic states. This is the Born-Oppenheimer approximation.

Within the Born-Oppenheimer approximation the solution of the full problem is thus split into two consecutive problems:

- The solution of the electronic Schrödinger equation (2.30) for fixed nuclear coordinates $\mathbf{Q}$.
- The solution of the nuclear Schrödinger equation (2.32) with the potential $V_k(\mathbf{Q})$ obtained in the first step.

The first problem is the subject of modern quantum chemistry. Many efficient methods and computer programs are nowadays available to solve the electronic Schrödinger equation from first principles without using phenomenological or experimental input data (so-called *ab initio* methods; Lowe 1978; Čársky and Urban 1980; Szabo and Ostlund 1982; Schmidtke 1987; Lawley 1987; Bruna and Peyerimhoff 1987; Werner 1987; Shepard 1987; see also Section 1.5). The potentials $V_k(\mathbf{Q})$ obtained in this way are the input for the solution of the nuclear Schrödinger equation. Solving (2.32) is the subject of spectroscopy (if the motion is bound) and

† In Chapter 15 we will discuss, for a specific example, the precise form of the coupling elements due to the nuclear kinetic energy and which terms are neglected in the Born-Oppenheimer approximation. At this stage a more qualitative discussion is enough.

scattering theory (if the motion is unbound). How to solve (2.32) or its classical analogue is the content of Chapters 3–5.

The physical background of the Born-Oppenheimer approximation is the extreme difference of the electronic and nuclear velocities. Looking from the point of view of the electrons, the heavy atoms move infinitely slowly. The very different time scales allow *adiabatic separation* of the electronic and the nuclear motions. If the electronic wavefunctions vary weakly with $\mathbf{Q}$, the coupling elements due to the nuclear kinetic energy are small and transitions from one electronic state to another are therefore unlikely; the nuclei move on a single potential energy surface $V_k(\mathbf{Q})$. If, on the other hand, two electronic states strongly mix in some region of the nuclear-coordinate space, the characters of the electronic wavefunctions change abruptly with $\mathbf{Q}$, the derivatives with respect to $\mathbf{Q}$ are no longer negligible, and consequently nonadiabatic coupling becomes substantial. This usually happens near the *avoided crossings* of the PESs of different electronic states where the energy separation between them is small. Near the avoided crossing the Born-Oppenheimer approximation breaks down and, as a consequence of the nonadiabatic coupling, transitions between the different electronic states become possible whenever the molecule passes this region (Baer 1983; Köppel, Domcke, and Cederbaum 1984; Lefebvre-Brion and Field 1986; Sidis 1989a,b). Transitions between different electronic states in photodissociation processes will be considered in Chapter 15.

Within the Born-Oppenheimer approximation the time-independent molecular wavefunctions for the various electronic states are written as

$$F_{kl}(\mathbf{Q},\mathbf{q}) = \Psi_{kl}^{nu}(\mathbf{Q})\ \Xi_k^{el}(\mathbf{q};\mathbf{Q}), \tag{2.33}$$

where Ξ_k^{el} is the kth solution of the electronic Schrödinger equation and Ψ_{kl}^{nu} is a distinct solution of the nuclear Schrödinger equation for this particular electronic state with energy E_{kl}. The index l labels the particular solution of the nuclear Schrödinger equation; it comprises a complete set of vibrational and rotational quantum numbers. Inserting (2.33) into (2.27) yields the following expression for the absorption cross section for a transition from initial state $|F_{k_il_i}\rangle$ to final state $|F_{k_fl_f}\rangle$,

$$\sigma(\omega) = \frac{\pi}{\hbar\epsilon_0 c}\ \omega_{k_fl_f,k_il_i}\ \delta(\omega_{k_fl_f,k_il_i} - \omega) \\ \times \left|\langle\Psi_{k_fl_f}^{nu}(\mathbf{Q}) \mid \mathbf{e}\cdot\boldsymbol{\mu}_{k_fk_i}(\mathbf{Q}) \mid \Psi_{k_il_i}^{nu}(\mathbf{Q})\ \rangle\right|^2, \tag{2.34}$$

where the integration is performed over all nuclear coordinates $\mathbf{Q}$. The *transition dipole moment function* is defined as

$$\boldsymbol{\mu}_{k_fk_i}(\mathbf{Q}) = \langle\Xi_{k_f}^{el}(\mathbf{q};\mathbf{Q}) \mid \hat{\mathbf{d}} \mid \Xi_{k_i}^{el}(\mathbf{q};\mathbf{Q})\rangle, \tag{2.35}$$

with the integration extending over all electronic coordinates $\mathbf{q}$. Note that $\boldsymbol{\mu}$ is a vector function.

The transition dipole moment functions are — like the potentials — functions of $\mathbf{Q}$. Their magnitudes determine the overall strength of the electronic transition $k_i \to k_f$. If the symmetry of the electronic wavefunctions demands $\boldsymbol{\mu}_{k_f k_i}$ to be exactly zero, the transition is called electric-dipole forbidden. The calculation of transition dipole functions belongs, like the calculation of the potential energy surfaces, to the field of quantum chemistry. However, in most cases the $\boldsymbol{\mu}_{k_f k_i}$ are unknown, especially their coordinate dependence, which almost always forces us to replace them by arbitrary constants.

Before we discuss a simple example in the next section, let us summarize the ingredients necessary for the calculation of absorption cross sections:

1) The potential energy surfaces $V_{k_i}(\mathbf{Q})$ and $V_{k_f}(\mathbf{Q})$ for the electronic states between which the transition occurs.
2) The transition dipole moment function $\boldsymbol{\mu}_{k_f k_i}(\mathbf{Q})$.
3) The nuclear wavefunctions $\Psi^{nu}_{k_i l_i}(\mathbf{Q})$ and $\Psi^{nu}_{k_f l_f}(\mathbf{Q})$.
4) The overlap of the nuclear wavefunctions and the transition dipole moment function, $\langle \Psi^{nu}_{k_f l_f} \mid \mathbf{e} \cdot \boldsymbol{\mu}_{k_f k_i} \mid \Psi^{nu}_{k_i l_i} \rangle$.

The first two entries belong to the field of quantum chemistry. The calculation of the nuclear wavefunctions and their overlap is the central theme of molecular dynamics; it is the subject of the following two sections.

2.4 Bound-bound transitions in a linear triatomic molecule

First, we describe briefly the calculation of the absorption spectrum for bound-bound transitions. In order to keep the presentation as clear as possible we consider the simplest polyatomic molecule, a linear triatom ABC as illustrated in Figure 2.1. The motion of the three atoms is confined to a straight line; overall rotation and bending vibration are not taken into account. This simple model serves to define the Jacobi coordinates, which we will later use to describe dissociation processes, and to elucidate the differences between bound-bound and bound-free transitions. We consider an electronic transition from the electronic ground state ($k = 0$) to an excited electronic state ($k = 1$) whose potential is also binding (see the lower part of Figure 2.2; the case of a repulsive upper state follows in Section 2.5). The superscripts "*nu*" and "*el*" will be omitted in what follows. Furthermore, the labels k used to distinguish the electronic states are retained only if necessary.

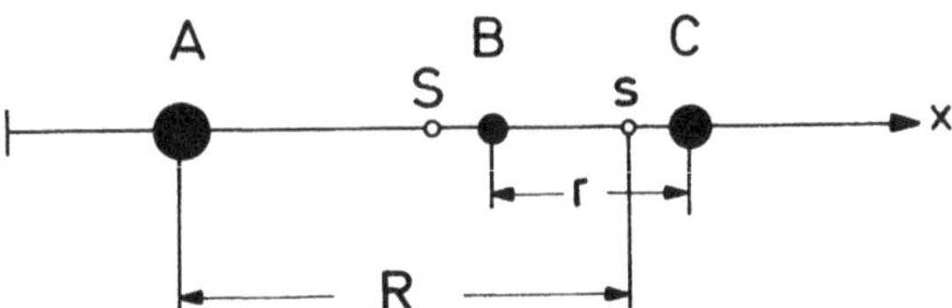

Fig. 2.1. Definition of Jacobi coordinates R and r for the linear triatomic molecule ABC. S denotes the center-of-mass of the total system and s marks the center-of-mass of the diatom BC.

2.4.1 Jacobi or scattering coordinates

For energies below the dissociation threshold we can use various coordinate systems to solve the nuclear Schrödinger equation (2.32). If the displacement from equilibrium is small, *normal coordinates* are most appropriate (Wilson, Decius, and, Cross 1955:ch.2; Weissbluth 1978:ch.27; Daudel et al. 1983:ch.7; Atkins 1983:ch.11). However, if the vibrational amplitudes increase so-called *local coordinates* become more advantageous (Child and Halonen 1984; Child 1985; Halonen 1989). Eventually, the molecular vibration becomes unbound and the molecule dissociates. Under such circumstances, *Jacobi* or so-called *scattering coordinates* are the most suitable coordinates; they facilitate the definition of the boundary conditions of the continuum wavefunctions at infinite distances which we need to determine scattering or dissociation cross sections (Child 1991:ch.10). Normal coordinates become less and less appropriate if the vibrational amplitudes increase; they are completely impractical for the description of unbound motion in the continuum.

In order to simplify the evaluation of overlap integrals between bound and continuum wavefunctions, it is advisable (although not necessary) to describe both wavefunctions by the same set of coordinates. Usually, the calculation of continuum, i.e., scattering, states causes far more problems than the calculation of bound states and therefore it is beneficial to use Jacobi coordinates for both nuclear wavefunctions. If bound and continuum wavefunctions are described by different coordinate sets, the evaluation of multi-dimensional overlap integrals requires complicated coordinate transformations (Freed and Band 1977) which unnecessarily obscure the underlying dynamics.

If $x_{\rm A}$, $x_{\rm B}$, and $x_{\rm C}$ are the coordinates of atoms A, B and C, respectively, the Jacobi coordinates (R, r) are defined by the transformation

$$\begin{aligned} S &= (m_{\rm A} x_{\rm A} + m_{\rm B} x_{\rm B} + m_{\rm C} x_{\rm C})/M \\ R &= (m_{\rm B} x_{\rm B} + m_{\rm C} x_{\rm C})/(m_{\rm B} + m_{\rm C}) - x_{\rm A} \\ r &= x_{\rm C} - x_{\rm B}, \end{aligned} \tag{2.36}$$

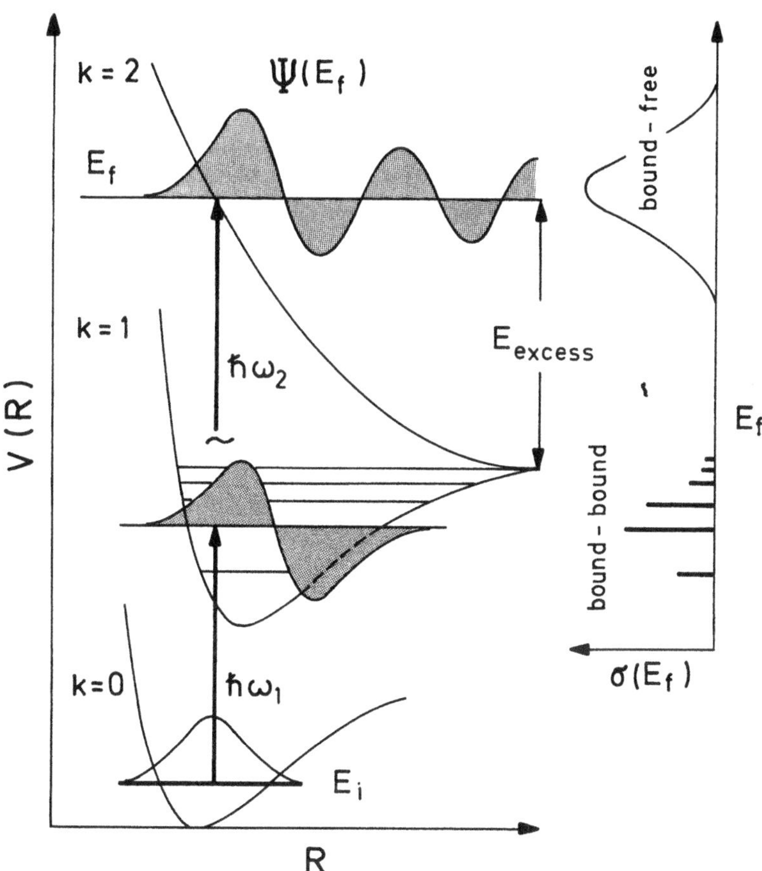

Fig. 2.2. Schematic illustration of a bound-bound ($k = 0 \to k = 1$) and a bound-free ($k = 0 \to k = 2$) transition where k labels the electronic states involved. The right-hand side depicts the corresponding discrete and continuous absorption spectra. $\Psi(E_f)$ is the continuum wavefunction and E_{excess} is the total energy available for distribution among the various degrees of freedom. Note, that the arrow labelled $\hbar\omega_2$ also starts at the ground electronic state.

where $m_{\rm A}$, $m_{\rm B}$, and $m_{\rm C}$ are the atomic masses and $M = m_{\rm A} + m_{\rm B} + m_{\rm C}$. S denotes the coordinate of the total center-of-mass, R describes the motion of A with respect to the center-of-mass of BC, and r describes the internal motion of BC. In photodissociation, R is the bond that is cleaved by the photon, i.e., it becomes the unbound coordinate. The nuclear Hamiltonian $\hat{T}_{nu} + V$ expressed in Jacobi coordinates is given by

$$\hat{H}(S, R, r) = -\frac{\hbar^2}{2M}\frac{\partial^2}{\partial S^2} - \frac{\hbar^2}{2m}\frac{\partial^2}{\partial R^2} - \frac{\hbar^2}{2\mu}\frac{\partial^2}{\partial r^2} + V(R, r), \qquad (2.37)$$

where the reduced masses m and μ are defined by

$$m = \frac{m_{\rm A}(m_{\rm B} + m_{\rm C})}{m_{\rm A} + m_{\rm B} + m_{\rm C}} \quad , \quad \mu = \frac{m_{\rm B} m_{\rm C}}{m_{\rm B} + m_{\rm C}}. \tag{2.38}$$

- The advantage of Jacobi coordinates is that the kinetic energy operator separates into three single terms without mixed derivatives of the form $\partial^2/\partial S \partial R$ etc.
- Kinetic coupling elements are absent if one uses Jacobi coordinates and the coupling between the various degrees of freedom is solely due to the potential V.

Since the potential depends only on the interatomic distances $R_{\rm AB}$ and $R_{\rm BC}$ (or R and r) and not on S, the center-of-mass of the triatomic molecule moves with constant velocity along the x-axis. If we transform to a new system, whose origin moves with the center-of-mass, the first term in (2.37) vanishes and the Hamiltonian reduces to†

$$\hat{H}(R, r) = -\frac{\hbar^2}{2m} \frac{\partial^2}{\partial R^2} + \hat{h}_{vib}(r) + V_I(R, r), \tag{2.39}$$

where

$$\hat{h}_{vib}(r) = -\frac{\hbar^2}{2\mu} \frac{\partial^2}{\partial r^2} + v_{\rm BC}(r) \tag{2.40}$$

describes the internal vibration of BC and the interaction potential defined by

$$V_I(R, r) = V(R, r) - v_{\rm BC}(r) \tag{2.41}$$

represents the coupling between R and r. The Hamiltonian is written in this particular form in order to facilitate the subsequent discussion of half and full collisions. If we choose $v_{\rm BC}(r)$ to be the potential of the free BC oscillator $[\lim_{R\to\infty} V(R, r)]$, it follows that

$$\lim_{R\to\infty} V_I(R, r) = 0, \tag{2.42}$$

which is necessary for the construction of the boundary conditions in scattering and in photodissociation. The first term in (2.39) describes

---

† If we introduce so-called *mass-scaled Jacobi coordinates* $\bar{R} = (m/\mu)^{1/2} R$ and $\bar{r} = r$, the Hamiltonian becomes even simpler,

$$\hat{H}(\bar{R}, \bar{r}) = -\frac{\hbar^2}{2\mu} \left(\frac{\partial^2}{\partial \bar{R}^2} + \frac{\partial^2}{\partial \bar{r}^2} \right) + V(\bar{R}, \bar{r}).$$

In contrast to (2.39) and (2.40), only a single mass occurs in this Hamiltonian. The picture of a billiard ball with mass μ rolling on a two-dimensional PES, which we will frequently adopt in the following chapters, is based on mass-scaled coordinates.

the motion of A with respect to BC, $\hat{h}_{vib}$ specifies the internal vibration of BC, and V_I represents the coupling between them.†

2.4.2 Variational calculation of bound-state energies and wavefunctions

In order to calculate bound-state energies and the corresponding eigenfunctions we expand $\Psi(R,r)$ according to

$$\Psi(R,r) = \sum_{mn} a_{mn}\, \chi_m(R)\, \varphi_n(r), \tag{2.43}$$

where the $\chi_m(R)$ are suitable basis functions to describe the R dependence (harmonic oscillator or Morse-oscillator wavefunctions, for example) and the $\varphi_n(r)$ are the eigenfunctions of $\hat{h}_{vib}$,

$$\left[\hat{h}_{vib}(r) - \epsilon_n\right] \varphi_n(r) = 0, \tag{2.44}$$

with vibrational energies ϵ_n. Both sets are assumed to be orthogonal and normalized to unity, $\langle \chi_m \mid \chi_{m'} \rangle = \delta_{mm'}$ and likewise for the φ_n. Inserting (2.44) into the nuclear Schrödinger equation with Hamiltonian (2.39) and using the orthonormality of the basis yields a set of algebraic equations,

$$(\mathbf{H} - E\,\mathbf{I})\,\mathbf{a} = 0 \tag{2.45}$$

where the elements of the Hamilton matrix are given by

$$\begin{aligned} H_{mn,m'n'} &= \delta_{mm'}\,\delta_{nn'}\,\epsilon_n \\ &- \delta_{nn'} \left\langle \chi_m \left| \frac{\hbar^2}{2m}\frac{\partial^2}{\partial R^2} \right| \chi_{m'} \right\rangle + \langle \chi_m\, \varphi_n \mid V_I \mid \varphi_{n'}\, \chi_{m'} \rangle. \end{aligned} \tag{2.46}$$

I is the unit matrix and **a** is the coefficient vector with elements a_{nm}. If we denote by $\bar{N}$ and $\bar{M}$ the dimensions of the two one-dimensional basis sets, **a** has the dimension $(\bar{N}\bar{M})$ and **H** has the dimension $(\bar{N}\bar{M})\times(\bar{N}\bar{M})$. Diagonalization of **H** yields the energies E_l as well as the corresponding eigenvectors $\mathbf{a}^{(l)}$ and thus the desired eigenfunctions. Problems with this rather general approach may arise if the required basis is impractically large. This happens if too many degrees of freedom are involved, or if a suitable basis is not apparent (floppy molecules), or if one wants to calculate highly excited states near the dissociation threshold.

The calculation of bound-state energies is part of molecular spectroscopy. Many efficient methods and computer codes have been developed

† The Jacobi coordinates R and r are computationally convenient only if the molecule dissociates into a single product channel. If the form of the potential permits dissociation into two channels, A+BC and AB+C, as in the photodissociation of HOD for example, other coordinates are more suitable. We will introduce two additional sets of coordinates appropriate for collinear molecules later in Sections 7.6 and 8.1.

and the interested reader is referred to some recent reviews (Carter and Handy 1986; Tennyson 1986; Bačić and Light 1989). In photodissociation problems one usually considers transitions from the lower vibrational states in the electronic ground state and the simple variational method outlined above does not pose any real problems in calculating the corresponding bound-state wavefunctions. The extension to rotating triatomic molecules follows in Chapter 11.

Using expansion (2.43) for the initial ($\Psi_{k_i l_i}$) and the final ($\Psi_{k_f l_f}$) nuclear states yields with the definition $\mu^{(e)}_{k_f k_i} \equiv \mathbf{e} \cdot \boldsymbol{\mu}_{k_f k_i}$

$$\langle \Psi_{k_f l_f} \mid \mu^{(e)}_{k_f k_i} \mid \Psi_{k_i l_i} \rangle = \sum_{m'n'} \sum_{mn} a^{(f)}_{m'n'} \, a^{(i)}_{mn} \, \langle \chi_{m'} \, \varphi_{n'} \mid \mu^{(e)}_{k_f k_i} \mid \varphi_n \, \chi_m \rangle \tag{2.47}$$

for the two-dimensional overlap integral required in (2.34) where, for simplicity, we use the same vibrational basis sets in both electronic states. The integral over R and r is conveniently performed by two-dimensional numerical integration (Gauss-Legendre quadrature, for example). If the transition dipole function is constant, (2.47) reduces to

$$\langle \Psi_{k_f l_f} \mid \mu^{(e)}_{k_f k_i} \mid \Psi_{k_i l_i} \rangle = \bar{\mu}^{(e)}_{k_f k_i} \sum_{mn} a^{(f)}_{mn} \, a^{(i)}_{mn} \tag{2.48}$$

because of the orthogonality of the vibrational basis functions.

2.5 Photodissociation

In this section we extend the theory to bound-continuum transitions. In order to keep the presentation simple we consider again the triatomic linear molecule. More complicated cases are handled in a similar way.

2.5.1 Dissociation channels

If the photon excites states with energies above the dissociation threshold, i.e., in the continuum of the molecular Hamiltonian, the absorption spectrum becomes a continuous function of the energy $E_f = E_i + \hbar\omega$ as illustrated in the upper part of Figure 2.2. There are two major differences between bound-bound and bound-continuum transitions:

- The nuclear wavefunctions are continuum, i.e., scattering, wavefunctions which asymptotically behave like free waves; rather than decaying to zero like the bound-state wavefunctions, scattering wavefunctions fulfil distinct boundary conditions in the limit $R \rightarrow \infty$.
- For each final energy $E_f = E_i + \hbar\omega$ there are several possible dissociation channels represented by degenerate solutions of the nuclear Schrödinger equation; the corresponding wavefunctions are distinguished by the boundary conditions at large distances.

Continuum states require a completely different treatment than bound states. Their definition and general behavior is the topic of this section.

The possible dissociation channels for the fragmentation of a triatomic molecule were discussed in Section 1.4. The linear ABC molecule can fragment into three chemical channels, A+B+C, A+BC(n), and AB(n')+C with the diatoms being produced in particular vibrational states denoted by quantum numbers n and n', respectively. Furthermore, each of the fragment atoms and molecules can be created in different electronic states. The total energy $E_f = E_i + \hbar\omega$ is the same in all cases and therefore the different channels are simultaneously excited by the monochromatic light pulse. The dissociation channels differ merely in the products and in the way the total energy partitions between translation and vibration.

For simplicity we consider at this point only the case of a single product channel,

$$\mathrm{ABC}(E_i) + \hbar\omega \longrightarrow \mathrm{ABC}(E_f)^* \longrightarrow \mathrm{A} + \mathrm{BC}(n)$$

and both fragments are assumed to be in their electronic ground states. The inclusion of all possible dissociation channels is, at least formally, straightforward and merely requires more indices. From energy conservation it follows that

$$E_{excess} = E_{trans}(n) + E_{vib}(n), \tag{2.49}$$

where E_{trans} is the translational energy associated with the motion of A with respect to BC and E_{vib} is the vibrational energy of BC.

2.5.2 Continuum basis

In Section 2.1 we derived the expression for the transition rate k_{fi} (2.22) by expanding the time-dependent wavefunction $\mathcal{F}(t)$ in terms of orthogonal and complete stationary wavefunctions F_α [see Equation (2.9)]. For bound-free transitions we proceed in the same way with the exception that the expansion functions for the nuclear part of the total wavefunction are continuum rather than bound-state wavefunctions. The definition and construction of the continuum basis belongs to the field of scattering theory (Wu and Ohmura 1962; Taylor 1972). In the following we present a short summary specialized to the linear triatomic molecule.

Equation (2.39) gives the nuclear Hamiltonian in the center-of-mass system. If $v_{\mathrm{BC}}(r)$ represents the vibrational potential of the free BC molecule, the interaction potential V_I vanishes asymptotically [see (2.42)] and the full Hamiltonian becomes

$$\hat{H}_0(R,r) \equiv \lim_{R\to\infty} \hat{H}(R,r) = -\frac{\hbar^2}{2m}\frac{\partial^2}{\partial R^2} + \hat{h}_{vib}(r). \tag{2.50}$$

The translational and the vibrational motions are completely decoupled asymptotically. The eigenfunctions of $\hat{H}_0$ for a total energy E are therefore products of free waves in R and vibrational wavefunctions for the free BC molecule,

$$\Psi_0^{\pm}(R,r;E,n) = \left(\frac{m}{\hbar k_n}\right)^{1/2} \mathrm{e}^{\pm i k_n R}\, \varphi_n(r), \qquad (2.51)$$

where the $\varphi_n(r)$ are the eigenfunctions of $\hat{h}_{vib}(r)$ with energies ϵ_n; the choice of the normalization factor will become apparent below. The *wavenumbers*

$$k_n = \left[\, 2m(E-\epsilon_n)/\hbar^2 \,\right]^{1/2} \qquad (2.52)$$

have units of inverse length. In terms of the wavenumbers the total energy is given by

$$E = \hbar^2 k_n^2\,/2m + \epsilon_n. \qquad (2.53)$$

In the following we denote by n_{max} the highest vibrational state that can be populated for a given total energy E, i.e., for which $k_n^2 > 0$. Product states with $k_n^2 < 0$ cannot be populated because of energy conservation; they are called *closed channels* in contrast to the *open channels*, which are accessible. The total number of open states is $N_{open} = n_{max} + 1$.

The $\Psi_0^{\pm}(R,r;E,n)$ are orthogonal $(s, s' = +/-)$,

$$\langle \Psi_0^s(R,r;E,n) \mid \Psi_0^{s'}(R,r;E',n') \rangle = 2\pi\hbar\; \delta_{ss'}\; \delta_{nn'}\; \delta(E-E') \qquad (2.54)$$

and they fulfil the closure relation

$$\int dE \sum_{n=0}^{n_{max}} |\Psi_0^s(R,r;E,n)\rangle\; \langle \Psi_0^s(R',r';E,n)| = 2\pi\hbar\; \delta(R-R')\; \delta(r-r'). \qquad (2.55)$$

In deriving (2.54) and (2.55) we used the following relations for the Dirac delta-function (Messiah 1972:appendix A)

$$\delta(y) = \frac{1}{2\pi}\int_{-\infty}^{+\infty} dx\; \mathrm{e}^{iyx} \qquad (2.56)$$

and

$$\delta[f(y)] = \sum_\beta \left|\frac{df}{dy}\right|_{y=y_\beta}^{-1} \delta(y-y_\beta), \qquad (2.57)$$

where the sum extends over all solutions y_β of the equation $f(y) = 0$.

In the limit $R \to \infty$ the full Hamiltonian $\hat{H}(R,r)$ reduces to $\hat{H}_0(R,r)$ and therefore the solution of the Schrödinger equation, $\Psi(R,r;E)$, must go over into a linear combination of the free solutions $\Psi_0^{\pm}$,

$$\lim_{R\to\infty} \Psi(R,r;E) = \sum_{n=0}^{n_{max}} \left[A_n^+ \, \Psi_0^+(R,r;E,n) + A_n^- \, \Psi_0^-(R,r;E,n)\right], \tag{2.58}$$

where the $A_n^{\pm}$ are arbitrary complex numbers. Equation (2.58) represents the most general form of boundary conditions for $\Psi(R,r;E)$. Since each term by itself is a solution of the Schrödinger equation for the free Hamiltonian $\hat{H}_0$, the same holds for the linear combination.

The next step is essential: we define N_{open} particular solutions, distinguished by the quantum number n in the list of arguments, by imposing the asymptotic condition

$$\begin{aligned} \lim_{R\to\infty} \Psi(R,r;E,n) &= \Psi_{out}^{(n)} + \sum_{n'=0}^{n_{max}} \Psi_{in}^{(n')} \\ &= \Psi_0^+(R,r;E,n) + \sum_{n'=0}^{n_{max}} S_{nn'}^* \, \Psi_0^-(R,r;E,n'). \end{aligned} \tag{2.59}$$

$S_{nn'}$ is an element of the so-called *scattering matrix* **S** whose meaning will become apparent below. The first term represents a single outgoing free wave in the particular vibrational channel n and the second term represents a sum of incoming free waves in all vibrational channels.

- Each $\Psi(R,r;E,n)$ with $n = 0,1,\ldots,n_{max}$ is a degenerate, yet independent, solution of the full nuclear Schrödinger equation with energy E. They are distinguished by the one and only particular vibrational channel that is associated with an outgoing free wave in the asymptotic region.

For example, $\Psi(E, n = 0)$ has an outgoing free wave in channel $n = 0$, etc. In contrast to (2.58), where the coefficients $A_n^{\pm}$ are arbitrary, the elements of the scattering matrix are uniquely defined through the particular form of the boundary conditions.

In order to further elucidate the choice of boundary conditions let us consider the *probability current* J which is defined for a wavefunction $\psi(R)$ by (Cohen-Tannoudji, Diu, and Laloë 1977:ch.III)

$$J = \frac{\hbar}{m} \, \Re\left[\frac{1}{i}\psi^* \frac{d}{dR} \, \psi\right], \tag{2.60}$$

where $\Re[\cdots]$ denotes the real part of the bracket. With the help of (2.51) and (2.59) we can readily calculate, for large values of R, the outgoing probability current in vibrational channel n [$\psi = \Psi_{out}^{(n)}$] and the incoming probability current in channel n' [$\psi = \Psi_{in}^{(n')}$] and obtain after integration over the vibrational coordinate r

$$J_{out}^{(n)} = 1 \quad , \quad J_{in}^{(n')} = -|S_{nn'}^{*}|^2. \tag{2.61}$$

Since the total wavefunction is stationary, the incoming and the outgoing probability currents must cancel, i.e., $J_{out}^{(n)} + \sum_{n'} J_{in}^{(n')} = 0$, which imposes the unitarity condition

$$\sum_{n'=0}^{n_{max}} |S_{nn'}|^2 = 1 \tag{2.62}$$

for the scattering matrix. Let us summarize:

- The $\Psi(R, r; E, n)$ are solutions of the full Schrödinger equation; in the asymptotic region, where the interaction potential is zero, they represent continuum states with incoming flux in *all* vibrational channels n' but unit outgoing flux in *one and only one* vibrational channel, n.

For fixed total energy E, Equation (2.59) defines one possible set of N_{open} degenerate solutions $\{\Psi(R, r; E, n), n = 0, 1, 2, \ldots, n_{max}\}$ of the full Schrödinger equation. As proven in formal scattering theory they are orthogonal and complete, i.e., they fulfil relations similar to (2.54) and (2.55). Therefore, the $\Psi(R, r; E, n)$ form an orthogonal basis in the continuum part of the Hilbert space of the nuclear Hamiltonian $\hat{H}(R, r)$ and any continuum wavefunction can be expanded in terms of them. Since each wavefunction $\Psi(R, r; E, n)$ describes dissociation into a specific product channel, we call them *partial dissociation wavefunctions.*

In order to get acquainted with the dissociation wavefunctions let us discuss a very simple example. Figure 2.3(a) depicts contours of a two-dimensional PES in R and r; it is purely repulsive along the dissociation coordinate R and in the limit $R \to \infty$ it becomes harmonic in the vibrational coordinate r. Figures 2.3(b)–(d) show partial dissociation wavefunctions $\Psi(R, r; E, n)$ for three selected outgoing vibrational channels, $n = 0, 2$, and 4. They oscillate rapidly in the R-direction because both, the reduced mass m and the energy are large in this particular case. The nodes in the r-direction indicate the extent of vibrational excitation of the BC entity. Since the full wavefunctions consist of an outgoing and an incoming part, the latter containing *all* vibrational states n', the number of nodes in r is not necessarily identical with the quantum number n. For example, the wavefunction for $n = 0$ has one node. The quantum number n merely determines the degree of excitation of the outgoing part $\Psi_{out}^{(n)}$!

The three wavefunctions depicted in Figure 2.3 (as well as those which are not shown in the figure) are solutions of the Schrödinger equation for the same total energy; they differ only by the boundary conditions as

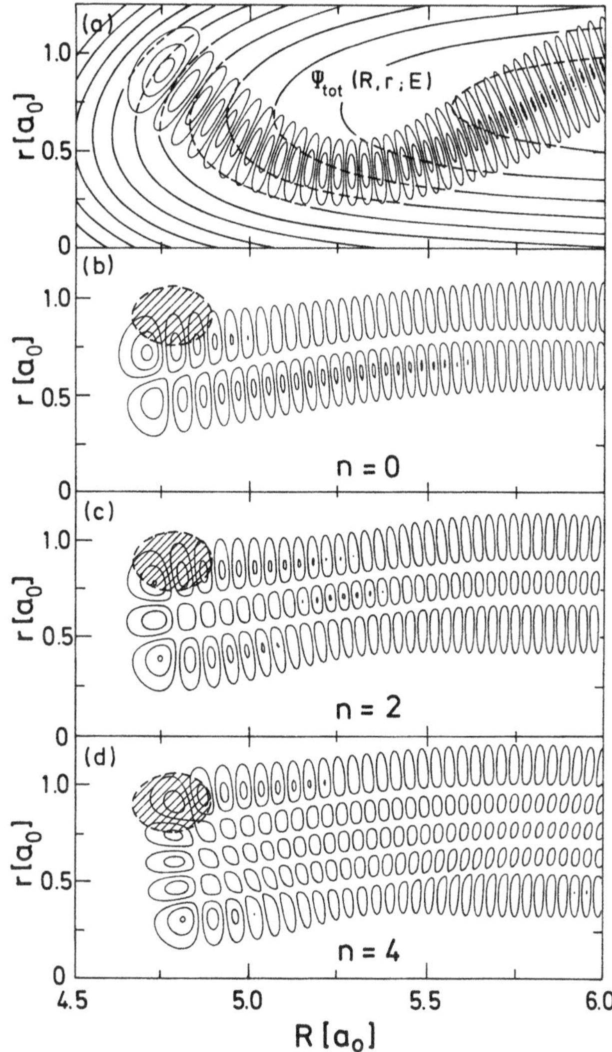

Fig. 2.3. (a) Two-dimensional plot of the potential $V(R,r)$ for an example illustrating direct photodissociation. The potential is purely repulsive in R. Superimposed is the total dissociation wavefunction $\Psi_{tot}(R,r;E)$ defined in (2.70); it is related to the evolving time-dependent wavepacket through a Fourier transformation between the time and the energy domain (see Section 4.1). (b)–(d) Contour plots of three selected partial dissociation wavefunctions $\Psi(R,r;E,n)$ for $n = 0, 2$, and 4. They are solutions of the stationary Schrödinger equation with energy E and fulfil the boundary condition (2.59). The partial photodissociation cross sections for producing the oscillator in vibrational state n are proportional to the squared overlap of these wavefunctions with the ground-state wavefunction indicated by the shaded area.

$R \to \infty$. As we will show below, the partial cross sections for absorbing the photon *and* producing the diatomic fragment in vibrational channel n are proportional to the square modulus of the overlap of these continuum wavefunctions with the nuclear wavefunction in the electronic ground state (indicated by the shaded areas). Since the bound wavefunction of the parent molecule is rather confined, only a very small portion of the continuum wavefunctions is sampled in the overlap integral.

In passing we note that the scattering wavefunctions appropriate for full collisions, $A + BC(n') \to A + BC(n)$, are defined such that the incoming and outgoing probability currents are

$$J_{in}^{(n')} = -1 \quad , \quad J_{out}^{(n)} = |S_{n'n}|^2. \tag{2.63}$$

Scattering wavefunctions represent states with unit incoming flux in *one* particular channel, n', and outgoing flux in *all* vibrational states n. The probability for a transition from initial channel n' to final channel n is given by $|S_{n'n}|^2$, which explains the name scattering matrix.

2.5.3 Photodissociation cross sections

After having defined the partial dissociation wavefunctions $\Psi(R, r; E, n)$ as basis in the continuum, the derivation of the absorption rates and absorption cross sections proceeds in the same way as outlined for bound-bound transitions in Sections 2.1 and 2.2. In analogy to (2.9), the total time-dependent molecular wavefunction $\mathcal{F}(t)$ including electronic ($\mathbf{q}$) and nuclear $[\mathbf{Q} = (R, r)]$ degrees of freedom is expanded within the Born-Oppenheimer approximation as

$$\begin{aligned}\mathcal{F}(t) &= a_i(t)\ \Xi_i(\mathbf{q};\mathbf{Q})\ \Psi_i(\mathbf{Q};E_i)\ \mathrm{e}^{-iE_it/\hbar} \\ &+ \int dE_f \sum_{n=0}^{n_{max}} a_f(t;E_f,n)\ \Xi_f(\mathbf{q};\mathbf{Q})\ \Psi_f(\mathbf{Q};E_f,n)\ \mathrm{e}^{-iE_ft/\hbar}.\end{aligned} \tag{2.64}$$

The first term is the initial state with energy E_i before the light beam is switched on† and the second term represents the total wavefunction in the upper electronic state. The initial conditions for the corresponding time-dependent coefficients are $a_i(0) = 1$ and $a_f(0; E_f, n) = 0$ for all energies E_f and all vibrational channels n. The sum over α in (2.9) is replaced in (2.64) by an integral over E_f and a sum over n. The integral over E_f reflects the fact that the spectrum in the upper electronic state is continuous and the summation over all open vibrational channels n accounts for the degeneracy of the continuum wavefunctions.

† In principle, Equation (2.64) should also include a summation over all nuclear states in the ground electronic state. However, in order to simplify the presentation we display only the initial state of the parent molecule.

The derivation of the time-dependent probability for a transition from nuclear state $|\Psi_i(E_i)\rangle$ in the ground electronic state to state $|\Psi_f(E_f, n)\rangle$ in the excited electronic state, Equation (2.21), and the corresponding transition rate (2.22) proceeds as in Section 2.1. Since a common phase factor $e^{-iE_f t/\hbar}$ governs the time dependence of all degenerate final states, the electromagnetic field excites resonantly *all* states having the same total energy E_f but different outgoing vibrational channels n. This must be taken into account in the energy transfer dS from the electromagnetic field to the gas, Equation (2.23), by summing over all possible n. Therefore, the final expression for the *total photodissociation cross section* is

$$\sigma_{tot}(\omega) = \sum_n \sigma(\omega, n) \tag{2.65}$$

with the *partial photodissociation cross sections* defined by

$$\sigma(\omega, n) = \frac{\rho\pi}{\hbar\epsilon_0 c}\, \omega_{fi}\, \delta(\omega_{fi} - \omega)\, |\langle \Psi_f(E_f, n) \mid \mu_{fi}^{(e)} \mid \Psi_i(E_i)\rangle|^2, \tag{2.66}$$

where all quantities have the same meaning as in (2.34). $\mu_{fi}^{(e)}$ is the component of the transition dipole moment function in the direction of the electric field vector. The additional factor $\rho = (2\pi\hbar)^{-1}$ stems from the normalization of the continuum wavefunctions in (2.54).

Since E_f is a continuous variable, it is — strictly speaking — not meaningful to consider the probability for excitation of a state with sharp energy E_f. Only the probability of exciting a range of continuum states is defined (Cohen-Tannoudji, Diu, and Laloë 1977:ch.VIII). Therefore, we integrate (2.66) over an energy interval dE_f, which is so small that the matrix element can be considered to be constant, and finally we obtain

$$\sigma(\omega, n) = C E_{photon}\, |t(E_f, n)|^2, \tag{2.67}$$

where $C = \rho\pi/\hbar\epsilon_0 c$ is a constant, and

$$t(E_f, n) = \langle \Psi_f(R, r; E_f, n) \mid \mu_{fi}^{(e)}(R, r) \mid \Psi_i(R, r; E_i)\rangle \tag{2.68}$$

is the *partial photodissociation amplitude.* In performing the integration over E_f we used the definition of the delta-function $\delta(\omega_{fi} - \omega)$; the energy in the upper electronic state is always taken as $E_f = E_i + E_{photon}$. Using the normalization (2.51) for the continuum wavefunctions it is straightforward to show that the cross section in (2.67) has units of area. The cross section defined in Equations (2.67) and (2.68) is an example of *Fermi's Golden Rule* (Cohen-Tannoudji, Diu, and Laloë 1977:ch.VIII)

The calculation of absolute cross sections requires knowledge of the transition dipole function which, unfortunately, is rarely known. Therefore, all examples which we will discuss in this monograph are relative cross sections and the constant C will be mostly ignored in what follows. As stressed in Section 1.4, the total photodissociation cross section is the sum of all partial cross sections. In (2.65) we took into account only the final vibrational channels. If necessary the summation must also include all possible chemical and electronic channels as well as all rotational states of the products. Final vibrational state distributions for fixed frequency ω are defined by

$$P(\omega, n) = \sigma(\omega, n) \;/\; \sigma_{tot}(\omega). \tag{2.69}$$

2.5.4 Total dissociation wavefunction

For illustration purposes we find it useful to define the *total dissociation wavefunction*

$$\Psi_{tot}(R, r; E_f) = \sum_{n=0}^{n_{max}} t(E_f, n)\, \Psi_f(R, r; E_f, n), \tag{2.70}$$

where the $t(E_f, n)$ are the partial dissociation amplitudes (2.68).

- Ψ_{tot} represents a particular linear combination of *all* partial dissociation wavefunctions.

With this definition the total absorption cross section becomes

$$\sigma_{tot}(\omega) = C E_{photon}\, \langle \Psi_{tot}(E_f) \mid \mu_{fi}^{(e)} \mid \Psi_i(E_i) \rangle. \tag{2.71}$$

The partial wavefunctions $\Psi(E, n)$ do not depend on the initial state; they are exclusively governed by the dynamics in the excited electronic state. The definition in (2.70), on the the other hand, involves the overlap with the initial wavefunction and therefore $\Psi_{tot}(E_f)$ depends on the particular state of the parent molecule in the electronic ground state. Ψ_{tot} has no practical advantage because its construction requires the dissociation amplitudes $t(E_f, n)$ which already contain all the desired information.

However, the total dissociation wavefunction is useful in order to visualize the overall dissociation path in the upper electronic state as illustrated in Figure 2.3(a) for the two-dimensional model system. The variation of the center of the wavefunction with r intriguingly illustrates the substantial vibrational excitation of the product in this case. As we will demonstrate in Chapter 5, Ψ_{tot} closely resembles a swarm of classical trajectories launched in the vicinity of the ground-state equilibrium. Furthermore, we will prove in Chapter 4 that the total dissociation function is the Fourier transform of the evolving wavepacket in the time-dependent formulation of photodissociation. The evolving wavepacket, the swarm of classical trajectories, and the total dissociation wavefunction all lead to the same general picture of the dissociation process.

Using the completeness of the basis functions $\Psi(R, r; E, n)$ [Equation (2.55)] one readily derives the following sum rule

$$\int dE_f \sum_{n=0}^{n_{max}} \frac{\sigma(\omega, n)}{E_{photon}} = 2\pi\hbar\, C\, \langle \Psi_i(E_i) \mid [\mu_{fi}^{(e)}]^2 \mid \Psi_i(E_i) \rangle. \qquad (2.72)$$

Equation (2.72) expresses the surprising fact that:

- The integrated total absorption cross section, divided by E_{photon}, is independent of the nuclear wavefunction in the excited electronic state and therefore it is independent of the nuclear dynamics in the upper state.

Let us summarize this section. The calculation of photodissociation cross sections requires the overlap of the bound-state wavefunction in the ground electronic state, which describes the initial state of the parent molecule, and the continuum wavefunctions in the excited electronic state, which describe the breakup of the molecule. Dissociation into the different final states of the products is represented by different degenerate wavefunctions. Each wavefunction is a solution of the full Schrödinger equation; they are distinguished by the asymptotic boundary conditions in the limit $R \rightarrow \infty$. Determination of the continuum wavefunctions is the crucial step. In Chapter 3 we will present practicable methods for the exact and approximate calculation of the $\Psi(R, r; E_f, n)$ based on the time-independent Schrödinger equation. By changing the energy E_f in the excited state one calculates point by point the entire absorption spectrum, as in a spectrometer where one tunes continuously the frequency of the light beam. An alternative method, which starts from the time-dependent Schrödinger equation for the nuclear motion, will be outlined in Chapter 4.

3
Time-independent methods

In this chapter we outline the evaluation of multi-dimensional bound-free matrix elements of the type (2.68),

$$t(E_f, \beta) = \langle \Psi_f(\mathbf{Q}; E_f, \beta) \mid \mu_{fi}^{(e)}(\mathbf{Q}) \mid \Psi_i(\mathbf{Q}; E_i) \rangle, \tag{3.1}$$

where $\Psi_i(\mathbf{Q}; E_i)$ is the nuclear wavefunction in the lower electronic state with energy E_i, $\Psi_f(\mathbf{Q}; E_f, \beta)$ represents a continuum wavefunction in the upper electronic state with energy $E_f = E_i + E_{photon}$, and $\mu_{fi}^{(e)}(\mathbf{Q})$ is the component of the transition dipole function (2.35) in the direction of the electric field vector. The vector $\mathbf{Q}$ represents collectively the nuclear coordinates and the index β comprises all degenerate dissociation channels which can be populated. $\Psi_f(E_f, \beta)$ and $\Psi_i(E_i)$ are both full solutions of the time-independent Schrödinger equation

$$[\hat{H}(\mathbf{Q}) - E] \, \Psi(\mathbf{Q}) = 0. \tag{3.2}$$

$\hat{H}(\mathbf{Q})$ is the nuclear Hamiltonian in the corresponding electronic state; at short distances it describes the motion of the complex and at large intermolecular separations it describes the free fragments. The matrix elements (3.1) are needed for the calculation of photodissociation cross sections. In this chapter we discuss numerically exact and approximate methods that are directly based on the solution of (3.2). The complementary time-dependent view follows in the next chapter.

In order to keep the formulation as simple as possible we confine the discussion to systems with only two degrees of freedom. The extension to more complex problems is — formally at least — straightforward. We will treat triatomic molecules ABC dissociating into products A+BC. First, we again consider in Section 3.1 the linear model, outlined in Sections 2.4 and 2.5, in which the diatomic fragment vibrates while its rotational degree of freedom is frozen. Subsequently, we treat in Section 3.2 the opposite limit, i.e., the rigid rotor model in which the fragment is allowed to rotate while its vibrational coordinate is kept fixed. These two cases represent simple but nevertheless generic examples of vibrational and rotational excitation in photodissociation. Important approximations and some numerical techniques are discussed in Sections 3.3 and 3.4, respectively.

3.1 Close-coupling approach for vibrational excitation

We consider the photodissociation of a linear triatomic molecule, $\mathrm{ABC} \rightarrow \mathrm{A} + \mathrm{BC}(n)$, where n specifies the final vibrational state. The appropriate Jacobi coordinates R and r are defined in Figure 2.1 and the nuclear Hamiltonian is given in (2.39). For simplicity, we assume that only one chemical dissociation channel exists. Figures 1.11 and 2.3 depict typical potential energy surfaces appropriate for this section.

3.1.1 Close-coupling equations

The time-independent Schrödinger equation for the bound as well as the continuum wavefunction is

$$\left[-\frac{\hbar^2}{2m}\frac{\partial^2}{\partial R^2} + \hat{h}_{vib}(r) + V_I(R,r) - E\right]\Psi(R,r;E) = 0, \qquad (3.3)$$

where the internal Hamiltonian $\hat{h}_{vib}$ describes the vibrational motion of BC and V_I is the interaction potential (see Section 2.4). In order to evaluate the matrix elements in Equation (3.1) we need to find the wavefunctions $\Psi_i(E_i)$ and $\Psi_f(E_f,n)$. The bound-state wavefunction $\Psi_i(E_i)$ can easily be calculated by a variational procedure employing a two-dimensional basis of square-integrable functions $\chi_m(R)$ and $\varphi_n(r)$ (as described in Section 2.4). Because of the nontrivial boundary conditions in the limit $R \rightarrow \infty$, the computation of the continuum wavefunctions $\Psi_f(E_f,n)$ requires a completely different numerical approach.

The standard technique for solving (3.3) for energies in the continuum is the so-called close-coupling method (Lester 1976; Gianturco 1979:ch.3; Light 1979). It rests on transforming the partial differential equation into a set of coupled ordinary differential equations. This is accomplished by expanding the wavefunction in all "internal" coordinates — r in the present example — in a basis with expansion functions $\chi(R)$ depending only on the "external" coordinate R,

$$\Psi_f(R,r;E_f,n) = \sum_{n'} \chi_{n'}(R;E_f,n)\,\varphi_{n'}(r), \qquad (3.4)$$

where the $\varphi_n(r)$ are the eigenfunctions of $\hat{h}_{vib}(r)$ with eigenvalues ϵ_n as defined in (2.44). The argument n specifies the particular solution that has incoming flux in all channels but outgoing flux only in the vibrational channel designated by n. $\Psi_f(E_f, n)$ is the wavefunction that describes dissociation into A+BC(n).

Differential equations for the $\chi_{n'}$ are obtained in the usual way: inserting (3.4) into (3.3), multiplying with $\varphi_{n''}(r)$ from the left, and integration over r yields the set of coupled equations,†

$$\left(\frac{d^2}{dR^2} + k_{n'}^2\right) \chi_{n'}(R; E_f, n) = \frac{2m}{\hbar^2} \sum_{n''} V_{n'n''}(R)\, \chi_{n''}(R; E_f, n) \qquad (3.5)$$

for each $n' = 0, 1, \ldots$. The wavenumbers k_n are defined in (2.52) and the *potential matrix* $\mathbf{V}$ is given by‡

$$V_{nn'}(R) = \langle \varphi_n \mid V_I(R, r) \mid \varphi_{n'} \rangle = \int_0^\infty dr\, \varphi_n(r)\, V_I(R, r)\, \varphi_{n'}(r). \qquad (3.6)$$

Equation (3.5) is a set of ordinary differential equations of second order for each radial wavefunction $\chi_{n'}(R)$; the different expansion functions $\chi_{n'}(R; E_f, n)$ are coupled to all other functions by the real and symmetric potential matrix $\mathbf{V}$.

In principle, Equation (3.5) represents an infinite set of coupled equations. In practice, however, we must truncate the expansion (3.4) at a maximal channel $\bar{n}$ which turns (3.5) into a finite set that can be numerically solved by several, specially developed algorithms (Thomas et al. 1981). The required basis size depends solely on the particular system. The convergence of the close-coupling approach must be tested for each system and for each total energy by variation of $\bar{n}$ until the desired cross sections do not change when additional channels are included. Expansion (3.4) should, in principle, include all open channels ($k_n^2 > 0$) as well as some of the closed vibrational channels ($k_n^2 < 0$). Note, however, that because of energy conservation the latter cannot be populated asymptotically.

In order to get acquainted with the potential matrix let us assume an interaction potential of the simple form

$$V_I(R, r) = A\, \mathrm{e}^{-\alpha\left[(R-\bar{R})+\epsilon(r-\bar{r})\right]}, \qquad (3.7)$$

where the parameter ϵ controls the coupling between R and r. For small values of ϵ, the interaction potential can be approximated by

$$V_I(R, r) \approx A\, \mathrm{e}^{-\alpha(R-\bar{R})} \left[1 - \alpha\epsilon(r - \bar{r})\right]. \qquad (3.8)$$

† Equation (3.5) is the set of coupled equations in the so-called diabatic representation in which the vibrational expansion functions $\varphi_n(r)$ are independent of the scattering coordinate R. The corresponding adiabatic representation will be presented in Section 3.3.

‡ The vibrational expansion functions $\varphi_n(r)$ can be considered as real functions.

Insertion of (3.8) into (3.6) yields $\mathbf{V} \approx \mathbf{V}^{(0)} + \mathbf{V}^{(1)}$ where we have defined

$$V_{nn'}^{(0)}(R) = A\, \mathrm{e}^{-\alpha(R-\bar{R})}\, \delta_{nn'} \tag{3.9a}$$

$$V_{nn'}^{(1)}(R) = -\alpha\epsilon A\, \mathrm{e}^{-\alpha(R-\bar{R})} \int_0^\infty dr\, \varphi_n(r)\, (r-\bar{r})\, \varphi_{n'}(r). \tag{3.9b}$$

Several important aspects follow immediately from (3.9):

1) The zeroth-order potential matrix $\mathbf{V}^{(0)}$ is diagonal and therefore it does not couple different vibrational channels.
2) The coupling is proportional to the strength parameter ϵ; it increases at short distances, just like the potential.
3) If the expansion functions $\varphi_n(r)$ are harmonic, the coupling elements $V_{nn'}^{(1)}$ fulfil the selection rule $\Delta n = n' - n = \pm 1$, i.e., only adjacent channels couple directly; (this does not, however, rule out coupling between states with $\Delta n > 1$ caused by higher-order effects.)

By inserting (3.4) into (2.59) and employing the orthogonality of the $\varphi_n(r)$ one easily derives the appropriate boundary conditions for the functions $\chi_{n'}(R; E_f, n)$,

$$\lim_{R\to\infty} \chi_{n'}(R; E_f, n) = \left(\frac{m}{\hbar k_{n'}}\right)^{1/2} \left(\delta_{nn'}\, \mathrm{e}^{ik_{n'}R} + S_{nn'}^{*}\, \mathrm{e}^{-ik_{n'}R}\right). \tag{3.10}$$

Closed channels, which have imaginary wavenumbers $k_n = i|k_n|$, must decay exponentially for large R. Asymptotically, only the diagonal terms $(n = n')$ consist of an incoming as well as an outgoing free wave whereas the off-diagonal terms $(n \neq n')$ are purely incoming free waves. Furthermore, all wavefunctions $\chi_{n'}(E_f, n)$ must decay to zero in the limit $R \to 0$. Numerical methods for solving the coupled equations (3.5) subject to the boundary conditions (3.10) will be outlined in Section 3.4.

3.1.2 Bound-free dipole matrix elements

In analogy to (3.4) the bound nuclear wavefunction in the electronic ground state, multiplied by the transition dipole function, may be expanded according to

$$\mu_{fi}^{(e)}(R, r)\, \Psi_i(R, r; E_i) = \sum_{n''} \chi_{n''}(R; E_i)\, \varphi_{n''}(r), \tag{3.11}$$

where the $\varphi_{n''}(r)$ are appropriate vibrational wavefunctions which are not necessarily identical with the basis functions in the upper electronic state. Using the orthogonality of the vibrational basis the radial wavefunctions are calculated from

$$\chi_{n''}(R; E_i) = \int_0^\infty dr\, \varphi_{n''}(r)\, \mu_{fi}^{(e)}(R, r)\, \Psi_i(R, r; E_i). \tag{3.12}$$

Inserting Equations (3.4) and (3.11) into (3.1) yields for the bound-free matrix elements

$$t(E_f, n) = \sum_{n',n''} \langle \chi_{n'}(R; E_f, n) \mid \chi_{n''}(R; E_i) \rangle \, \langle \varphi_{n'}^{(f)}(r) \mid \varphi_{n''}^{(i)}(r) \rangle, \quad (3.13)$$

where $\varphi_{n''}^{(i)}$ and $\varphi_{n'}^{(f)}$ are the vibrational wavefunctions in the ground and in the excited electronic state, respectively. Choosing the vibrational functions to be identical reduces (3.13) to

$$t(E_f, n) = \sum_{n'} \langle \chi_{n'}(R; E_f, n) \mid \chi_{n'}(R; E_i) \rangle. \quad (3.14)$$

The radial functions $\chi_{n'}(R; E_f, n)$ have the qualitative behavior illustrated in the upper part of Figure 2.2. Each $\chi_{n'}(R)$ rises smoothly out of the classically forbidden region at small distances, becomes maximal near the classical turning point, and oscillates with constant amplitude and wavelength in the asymptotic region. The radial functions for the closed channels must decay to zero as R goes to infinity. The actual R-interval over which the integration in (3.14) extends is rather narrow because the ground-state wavefunction is well localized around the equilibrium bond distance. Nevertheless, one must integrate the coupled equations out into the interaction-free region in order to construct the proper solutions with the correct boundary conditions.

The close-coupling method provides a rather universal technique for the calculation of cross sections. It has been amply exploited in full collisions (Secrest 1979a,b; Dickinson 1979; Clary 1987; Gianturco 1989) as well as in half collisions (Shapiro 1977; Atabek and Lefebvre 1977; Shapiro and Bersohn 1980; Clary 1986a; Pernot, Atabek, Beswick, and Millié 1987; Untch, Hennig, and Schinke 1988). The close-coupling approach can be straightforwardly extended to more degrees of freedom; only the number of channels limits its applicability. In the next section we outline how it is implemented for rotational excitation.

3.2 Close-coupling approach for rotational excitation

We now discuss the dissociation of a triatomic molecule ABC into an atom A and a rotating diatom BC(j) where j denotes the final rotational quantum number. The vibrational degree of freedom of BC is frozen at its equilibrium distance r_e (rigid rotor model). The general theory for the dissociation of a triatomic molecule, which we defer to Section 11.1, is rather complicated because of the addition of the angular momentum of the diatomic fragment BC, $\mathbf{j}$, and the orbital angular momentum of the A–BC complex, $\mathbf{l}$, yielding the total angular momentum $\mathbf{J}$. In this section we restrict the discussion to a simple model which nevertheless contains the essential features of rotational excitation. We explicitly assume that the total angular momentum in the ground state, $\mathbf{J}_i$, and in the excited electronic state, $\mathbf{J}_f$, are both zero; this assumption significantly simplifies the theory. The resulting close-coupling equations and cross section expressions closely resemble the corresponding equations for the dissociation of the linear, rotationless molecule.

For zero total angular momentum the three atoms form a plane which remains fixed in space throughout the entire dissociation (Figure 3.1). The diatom BC rotates around its center-of-mass within this plane and likewise the triatomic complex A–BC rotates in the same plane around the center-of-mass S of the total system. Both $\mathbf{j}$ and $\mathbf{l}$ are perpendicular to the plane and angular momentum conservation dictates that $\mathbf{l} = -\mathbf{j}$ for $\mathbf{J}$=0, i.e., BC and A–BC rotate in opposite directions at all instants. The total angular momentum $\mathbf{J}$ is conserved during the dissociation. When A recoils from BC it imparts a torque which, in general, leads to rotational excitation of the molecular fragment.

If we fix the intramolecular distance of the diatom, the system is described by the two Jacobi coordinates R and γ, where R is the distance from A to the center-of-mass of BC and γ is the orientation angle of the diatom with respect to the scattering vector $\mathbf{R}$. The appropriate Hamiltonian is given by (see Section 11.1 for a more detailed discussion)

$$\hat{H}(R,\gamma) = -\frac{\hbar^2}{2m}\frac{1}{R}\frac{\partial^2}{\partial R^2}R + \frac{\hat{\mathbf{l}}^2}{2mR^2} + \frac{\hat{\mathbf{j}}^2}{2\mu r_e^2} + V(R,\gamma). \tag{3.15}$$

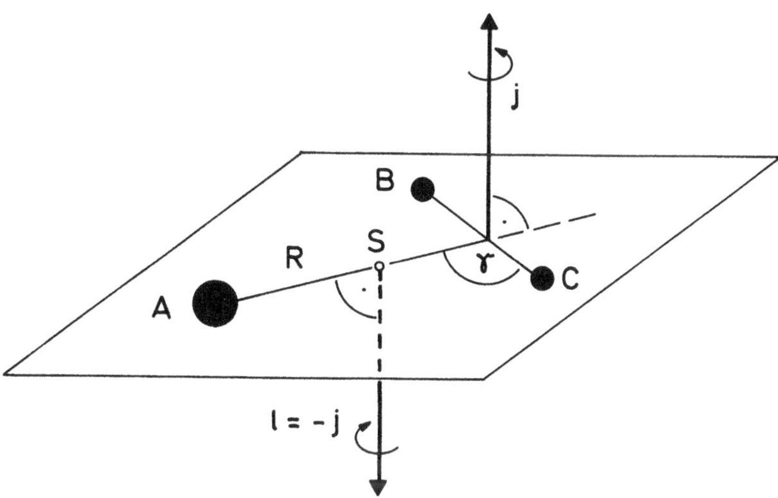

Fig. 3.1. Schematic illustration of the Jacobi coordinates R and γ for the atom-rigid rotor model with the restriction $\mathbf{J} = 0$.

The first term is the translational energy of A with respect to BC and the second and the third terms represent the rotational energies of A–BC and BC, respectively, with $\hat{\mathbf{l}}$ and $\hat{\mathbf{j}}$ being the corresponding rotational angular momentum operators. $V(R,\gamma)$ is the potential energy surface (PES) depending only on R and the orientation angle γ. The reduced masses are defined in (2.38).

In analogy to (2.39) we rewrite the Hamiltonian as

$$\hat{H}(R,\gamma) = -\frac{\hbar^2}{2m}\frac{1}{R}\frac{\partial^2}{\partial R^2}R + \hat{h}_{rot}(\gamma;R) + V(R,\gamma), \tag{3.16}$$

where we defined, using $\hat{\mathbf{l}}^2 = \hat{\mathbf{j}}^2$,

$$\hat{h}_{rot}(\gamma;R) = \left(\frac{1}{2\mu r_e^2} + \frac{1}{2mR^2}\right)\hat{\mathbf{j}}^2. \tag{3.17}$$

In the limit $R \to \infty$, $\hat{h}_{rot}$ goes over into the rotational energy of the free fragment. The eigenfunctions of $\hat{h}_{rot}$ are the rotor wavefunctions, i.e., the spherical harmonics $Y_{j\Omega}(\gamma,\psi)$,

$$\hat{h}_{rot}(\gamma;R)\, Y_{j\Omega}(\gamma,\psi) = \left(B_{rot} + \frac{\hbar^2}{2mR^2}\right) j(j+1)\, Y_{j\Omega}(\gamma,\psi) \tag{3.18}$$

with $B_{rot} = \hbar^2/2\mu r_e^2$ being the rotational constant of the free BC molecule (Cohen-Tannoudji, Diu, and Laloë 1977:ch.VI; Zare 1988:ch.1). The azimuthal angle ψ describes rotation out of the fixed scattering plane. If we choose the quantization axis of the angular momentum vector $\mathbf{j}$ to lie in the collision plane, the projection quantum number is restricted to $\Omega = 0$ and the azimuthal angle ψ can be set to zero.

The partial dissociation wavefunctions $\Psi(R,\gamma;E,j)$, which describe dissociation into A+BC(j), are now expanded in terms of the $Y_{j\Omega}(\gamma,0)$ according to†

$$\Psi_f(R,\gamma;E_f,j) = \frac{1}{R}\sum_{j'} \chi_{j'}(R;E_f,j)\, Y_{j'0}(\gamma,0). \tag{3.19}$$

Inserting (3.19) into the Schrödinger equation and proceeding as described in Section 3.1 yields the set of coupled equations

$$\left[\frac{d^2}{dR^2} + k_{j'}^2 - \frac{j'(j'+1)}{R^2}\right]\chi_{j'}(R;E_f,j) = \frac{2m}{\hbar^2}\sum_{j''} V_{j'j''}(R)\,\chi_{j''}(R;E_f,j) \tag{3.20}$$

† The factor $1/R$ is introduced for convenience; without this factor the close-coupling equations would have a more complicated form. The wavefunctions $\Psi_f(E_f,j)$ are defined in analogy to the $\Psi_f(E_f,n)$ in Section 2.5.

with the definitions

$$k_j = \{2m\,[E_f - B_{rot}j(j+1)]\,/\hbar^2\}^{1/2} \tag{3.21}$$

for the wavenumbers and

$$\begin{aligned} V_{jj'}(R) &= \langle Y_{j0} \mid V(R,\gamma) \mid Y_{j'0}\rangle \\ &= \int_0^{2\pi} d\psi \int_0^{\pi} d\gamma\ \sin\gamma\ Y_{j0}^*(\gamma,0)\ V(R,\gamma)\ Y_{j'0}(\gamma,0) \end{aligned} \tag{3.22}$$

for the potential matrix, respectively. Because the potential does not depend on ψ, the integration over ψ is redundant and merely yields a factor 2π. In the calculations the potential is usually expanded in terms of Legendre polynomials $P_\lambda(\cos\gamma)$,

$$V(R,\gamma) = \sum_\lambda V_\lambda(R)\ P_\lambda(\cos\gamma), \tag{3.23}$$

which have the correct symmetry at $\gamma = 0$ and π. With (3.23) inserted in (3.22) and noting that

$$P_\lambda(\cos\gamma) = [4\pi/(2\lambda+1)]^{1/2}\ Y_{\lambda 0}(\gamma,0) \tag{3.24}$$

the potential matrix elements become essentially an integral over three spherical harmonics, which can be evaluated analytically (Edmonds 1974: ch.4). It yields

$$V_{jj'}(R,\gamma) = [(2j+1)(2j'+1)]^{1/2} \sum_\lambda V_\lambda(R) \begin{pmatrix} j & \lambda & j' \\ 0 & 0 & 0 \end{pmatrix}^2 , \tag{3.25}$$

where $\begin{pmatrix} \cdot & \cdot & \cdot \\ \cdot & \cdot & \cdot \end{pmatrix}$ is a Wigner $3j$-symbol. If the potential is isotropic only the $\lambda = 0$ term is different from zero and the potential matrix is diagonal in j and j', i.e., $V_{jj'}(R) = V(R)\,\delta_{jj'}$.

The product of the ground-state wavefunction and the transition dipole function is also expanded in terms of spherical harmonics,

$$\mu_{fi}^{(e)}(R,\gamma)\ \Psi_i(R,\gamma;E_i) = \frac{1}{R}\sum_{j''} \chi_{j''}(R;E_i)\ Y_{j''0}(\gamma,0). \tag{3.26}$$

Using the orthogonality of the spherical harmonics the radial expansion functions are given by

$$\frac{1}{R}\chi_{j''}(R;E_i) = \int_0^{2\pi} d\psi \int_0^{\pi} d\gamma\ \sin\gamma\ Y_{j''0}^*(\gamma,\psi)\ \mu_{fi}^{(e)}(R,\gamma)\ \Psi_i(R,\gamma;E_i). \tag{3.27}$$

Inserting (3.19) and (3.26) into the expression for the bound-free matrix element, Equation (3.1), we obtain[†]

$$t(E_f, j) = \sum_{j'} \langle \chi_{j'}(R; E_f, j) \mid \chi_{j'}(R; E_i) \rangle \tag{3.28}$$

for the partial photodissociation amplitudes.

In order to get acquainted with rotational excitation in photodissociation we discuss briefly the dissociation of ClCN into Cl and CN(j). Figure 3.2 depicts the corresponding potential energy surface $V(R, \gamma)$ in the excited electronic state. It is steeply repulsive in the Cl-CN dissociation coordinate R and has a maximum at the linear configuration $\gamma = 0$. Since the PES is symmetric with respect to 0° it follows that $\partial V/\partial\gamma = 0$ at the linear configuration. A billiard ball put on this excited-state PES, slightly displaced from the linear configuration, will rapidly roll down the hill following the heavy arrow. The photolysis of ClCN serves as an example of direct dissociation.

ClCN is linear in the electronic ground state and bent in the upper state (linear-bent transition) with the consequence that the potential is rather anisotropic above the excitation point. The relatively strong torque $-\partial V/\partial\gamma$, imparted to CN after the parent molecule is excited, immediately pushes ClCN out of the linear configuration with the result that CN starts to revolve rapidly. The total photodissociation wavefunction $\Psi_{tot}(R, \gamma; E_f)$ elucidates clearly the motion in the upper state. The substantial variation of the expectation value of the angle as R increases is synonymous with strong rotational excitation of the product CN along the dissociation path and the final rotational state distribution, depicted in Figure 3.3, indeed reflects this. It peaks markedly at very high angular momentum states. The characteristic shape of $P(j)$ and its relation to the PES will be analyzed in Chapters 6 and 10.

3.3 Approximations

In this section we introduce two approximations which are very advantageous for the interpretation as well as the computation of photodissociation cross sections. Both approximations require that the external (i.e., the translational) and the internal (i.e., the vibrational or rotational) motion evolve on substantially different time scales.

[†] The factor $1/R^2$, arising from the multiplication of (3.19) and (3.26), is canceled if one bears in mind that $\mathbf{R}$ is, in principle, a three-dimensional vector and that the corresponding volume element $d\mathbf{R}$ in polar coordinates is proportional to R^2.

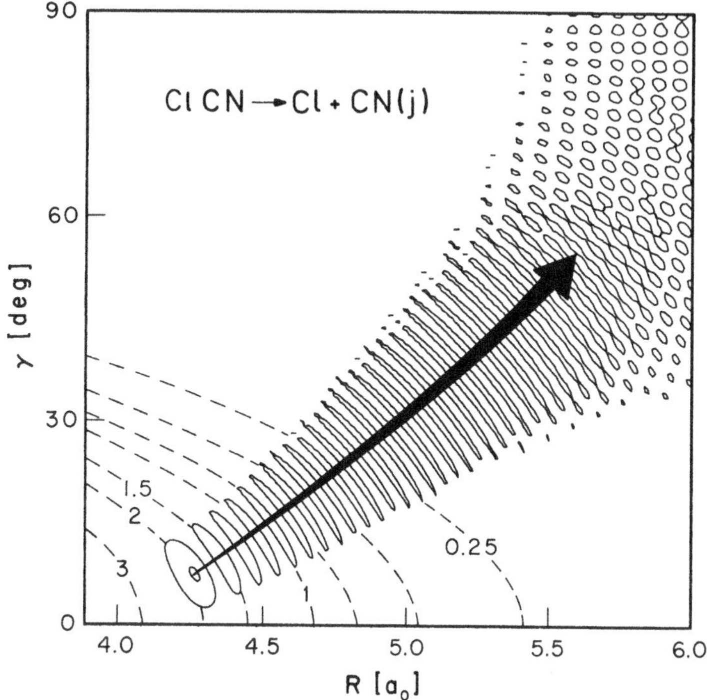

Fig. 3.2. Two-dimensional potential energy surface $V(R,\gamma)$ (dashed contours) for the photodissociation of ClCN, calculated by Waite and Dunlap (1986); the energies are given in eV. The closed contours represent the total dissociation wavefunction $\Psi_{tot}(R,\gamma;E)$ defined in analogy to (2.70) in Section 2.5 for the vibrational problem. The energy in the excited state is $E_f = 2.133$ eV. The heavy arrow illustrates a classical trajectory starting at the maximum of the wavefunction and having the same total energy as in the quantum mechanical calculation. The remarkable coincidence of the trajectory with the center of the wavefunction elucidates Ehrenfest's theorem (Cohen-Tannoudji, Diu, and Laloë 1977:ch.III). Reprinted from Schinke (1990).

3.3.1 Adiabatic approximation, appropriate for vibrational excitation

Within the *adiabatic approximation*, we consider the translational motion to be slow and the internal motion to be fast. The most familiar example is the Born-Oppenheimer approximation employed in Section 2.3 to decouple the fast electronic motion from the slow motion of the heavy nuclei. In the same spirit, the adiabatic approximation may be utilized to decouple two nuclear degrees of freedom within the same electronic state. To be specific, we discuss the photodissociation of the linear triatom ABC as defined in Figure 2.1. The translational motion associated with R is

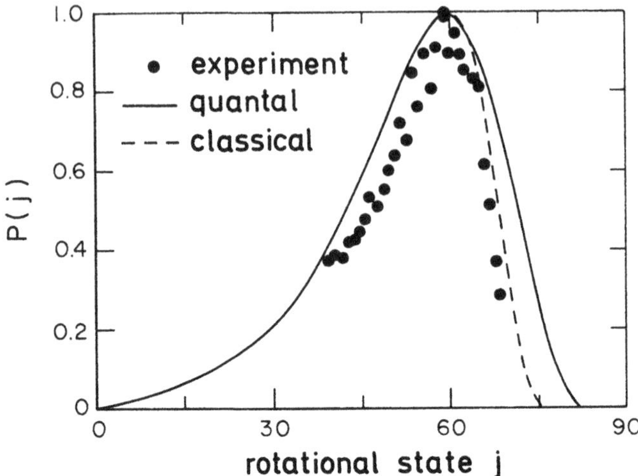

Fig. 3.3. Measured (Barts and Halpern 1989) and calculated rotational state distributions following the photolysis of ClCN. The energy in the excited state is 2.133 eV. The quantal results are calculated with the close-coupling method described in Section 3.2 and the classical distribution has been obtained by Barts and Halpern using the ultrasimple model which we shall present in Section 6.3. Reprinted from Schinke (1990).

considered to be slow while the vibrational motion of BC associated with r is assumed to be fast.

In Section 3.1 we expanded the continuum wavefunctions in terms of a so-called *diabatic basis*, i.e., the vibrational wavefunctions of the free oscillator. The diabatic basis is, of course, independent of R. Alternatively, we may represent them in terms of adiabatic basis functions which continuously vary with the dissociation coordinate R. Such a basis can better account for changes in the r dependence of the PES along the reaction path. The S_1 state of CH_3ONO is an example for which the adiabatic representation is obviously more useful than the diabatic basis. Since the dependence of the PES on the NO vibrational coordinate r changes strongly with the CH_3O-NO separation (Figure 1.11), a basis which is suitable at large separations is inappropriate in the well region and *vice versa.* As a consequence, a diabatic basis would require too many basis functions before convergence is achieved.

To define the adiabatic basis, (2.39) is rewritten as

$$\hat{H}(R,r) = -\frac{\hbar^2}{2m}\frac{\partial^2}{\partial R^2} + \hat{h}_{intern}(r;R), \tag{3.29}$$

where the internal Hamiltonian

$$\hat{h}_{intern}(r;R) = -\frac{\hbar^2}{2\mu}\frac{\partial^2}{\partial r^2} + V(R,r) \tag{3.30}$$

depends parametrically on the translational coordinate R. It serves to define the adiabatic basis through the one-dimensional Schrödinger equation

$$\left[\hat{h}_{intern}(r;R) - \epsilon_n(R)\right] \varphi_n(r;R) = 0, \tag{3.31}$$

where both the eigenvalues ϵ_n and the eigenfunctions φ_n also depend parametrically on R. The adiabatic basis turns smoothly into the diabatic basis in the limit $R \to \infty$ and the energies $\epsilon_n(R)$ eventually become the vibrational energies of the free oscillator. Equation (3.31) corresponds formally to the electronic Schrödinger equation (2.30) in Section 2.3. The energies $\epsilon_n(R)$ are the vibrationally adiabatic potential energy curves or potential energy surfaces if more than two degrees of freedom are involved. They correspond to the electronically adiabatic potential curves or surfaces in the Born-Oppenheimer approximation.

Expanding the continuum wavefunctions as

$$\Psi_f(R,r;E_f,n) = \sum_{n'} \chi_{n'}(R;E_f,n)\, \varphi_{n'}(r;R), \tag{3.32}$$

inserting (3.32) into the Schrödinger equation, and utilizing the orthogonality of the adiabatic wavefunctions yields the following set of coupled equations (Child 1974:ch.6)

$$\left[\frac{d^2}{dR^2} + k_{n'}^2(R) - U_{n'n'}(R)\right] \chi_{n'}(R;E_f,n)$$
$$= \sum_{n''\neq n'} \left[U_{n'n''}(R) + Q_{n'n''}(R)\frac{d}{dR}\right] \chi_{n''}(R;E_f,n) \tag{3.33}$$

with R-dependent wavenumbers

$$k_n = \left\{2m\left[E - \epsilon_n(R)\right]/\hbar^2\right\}^{1/2}. \tag{3.34}$$

The coupling matrix elements due to the kinetic energy operator,

$$U_{n'n''}(R) = -\left\langle \varphi_{n'}(r;R) \left| \frac{\partial^2}{\partial R^2} \right| \varphi_{n''}(r;R) \right\rangle \tag{3.35}$$

$$Q_{n'n''}(R) = -2\left\langle \varphi_{n'}(r;R) \left| \frac{\partial}{\partial R} \right| \varphi_{n''}(r;R) \right\rangle, \tag{3.36}$$

reflect how strongly the internal potential and therefore the vibrational functions $\varphi_n(r;R)$ change with the dissociation coordinate R. The diagonal elements of the coupling matrix $\mathbf{Q}$ are exactly zero (Child 1974:ch.6).

Equations (3.5) and (3.33) are completely equivalent. In the diabatic representation, it is the potential matrix (3.6) that couples the different vibrational channels whereas in the adiabatic representation the matrices (3.35) and (3.36) impart the coupling between translation and vibration. Computationally more advantageous is the diabatic representation because the calculation of the matrices **U** and **Q** is in practice laborious.

In the adiabatic approximation one neglects the coupling elements on the right-hand side of (3.33) so that each wavefunction χ_n is determined by a one-dimensional Schrödinger equation†

$$\left[\frac{d^2}{dR^2} + k_n^2(R) - U_{nn}(R)\right] \chi_n(R; E_f, n) = 0. \tag{3.37}$$

Coupling to other vibrational channels is zero and transitions from one vibrational state to another are therefore prohibited. Within the adiabatic approximation the partial photodissociation wavefunctions separate into a translational and an internal part, the latter depending parametrically on R, i.e.,

$$\Psi(R, r; E_f, n) = \chi_n(R; E_f, n)\, \varphi_n(r; R). \tag{3.38}$$

Inserting (3.38) into the expression for the partial photodissociation amplitudes yields

$$t(E_f, n) = \langle \chi_n(R; E_f, n) \mid \chi_n(R; E_i) \rangle, \tag{3.39}$$

where we have assumed that the product $\mu_{fi}^{(e)} \Psi_i(E_i)$ is expanded in the same adiabatic basis as the excited electronic state. Equation (3.39) is the adiabatic analogue of (3.14). Since coupling between the adiabatic channels is neglected, the sum in (3.14) has collapsed to a single term. The adiabatic approximation is expected to be accurate if the eigenfunctions of $\hat{h}_{intern}$ change smoothly (i.e., adiabatically) with the external coordinate R such that the coupling elements neglected in Equation (3.33) are indeed small.

The main virtue of the adiabatic approximation is its interpretative power. The separation of the various degrees of freedom leads to a factorization of the photodissociation cross sections into a one-dimensional bound-bound and a one-dimensional bound-free overlap integral. Let us assume, for simplicity, a separable ground-state wavefunction of the form

$$\Psi_i(R, r) = \varphi_R(R)\, \varphi_r(r). \tag{3.40}$$

† Very often it is advisable to retain the diagonal elements of **U** on the left-hand side of (3.33). They do not couple different states but merely act as additional energy terms. This can substantially improve, for example, the success of the adiabatic approximation in predicting resonance energies (Römelt 1983).

With the excited-state wavefunction given by (3.38) the partial photodissociation cross sections approximately reduce to

$$\sigma(E_f, n) \approx g(E_f, n)\, f(n) \tag{3.41}$$

where we defined

$$g(E_f, n) = \left|\langle \chi_n(R; E_f, n) \mid \varphi_R(R)\rangle\right|^2 \tag{3.42a}$$

and

$$f(n) = \left|\langle \varphi_n(r; R) \mid \varphi_r(r)\rangle\right|^2_{R=R_e}. \tag{3.42b}$$

For simplicity we have assumed a constant transition dipole function and neglected multiplication with E_{photon}. The $f(n)$ are conventional one-dimensional Franck-Condon (FC) factors for the (bound) vibrational wavefunctions of the BC entity (Graybeal 1988:ch.15.6). Note, that the vibrational wavefunction in the upper state is evaluated at the equilibrium R_e of the electronic ground state.

Since the FC factors $f(n)$ are important for the discussions in the following chapters we illustrate their general behavior in Figure 3.4 for two cases. If the equilibrium separations r_e are roughly equal in the two states, the overlap integral is largest for the vibrational ground state whose wavefunction $\varphi_0(r)$ has the best overlap with the ground-state wavefunction. With increasing n the oscillatory part of $\varphi_n(r)$ overlaps $\varphi_r(r)$ with the consequence that the FC factor diminishes monotonically. By the same token, the FC distribution is broad with a maximum at relatively high quantum numbers n if the two potentials are significantly displaced. Note the similarity of the latter case with the one-dimensional reflection principle which we shall discuss in Section 6.1.

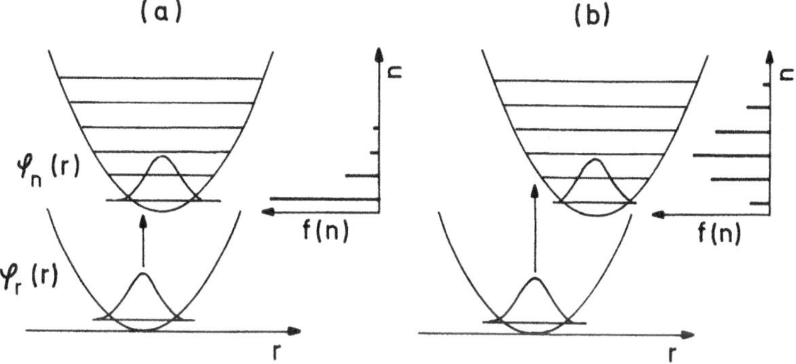

Fig. 3.4. Schematic illustration of the n dependence of the FC factors $f(n)$ defined in Equation (3.42b).

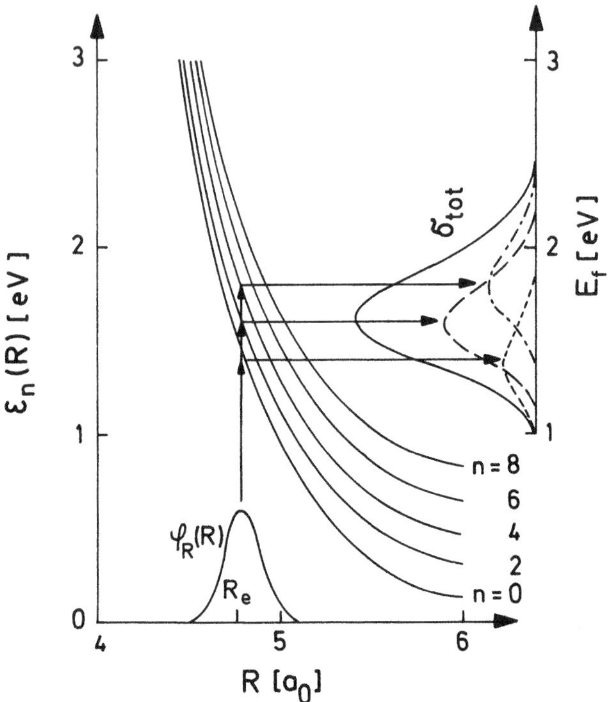

Fig. 3.5. Adiabatic potential curves $\epsilon_n(R)$, defined in (3.31), for the model system illustrated in Figure 2.3. The right-hand side depicts three selected partial photodissociation cross sections $\sigma(E_f, n)$ for the vibrational states $n = 0$ (short dashes), $n = 2$ (long dashes), and $n = 4$ (long and short dashes). The vertical and the horizontal arrows illustrate the reflection principle (see Chapter 6). Also shown is the total cross section $\sigma_{tot}(E_f)$.

The adiabatic potential curves $\epsilon_n(R)$ often provide a simple, yet accurate, explanation of the energy dependence of the partial photodissociation cross sections. Figure 3.5 illustrates this point for the model system of Section 2.5. The left-hand side depicts the potential curves $\epsilon_n(R)$ and the right-hand side shows some of the partial photodissociation cross sections $\sigma(E_f, n)$ calculated by the close-coupling method.† These are the cross sections for absorbing the photon and producing the fragment molecule in vibrational state n. They vary smoothly with energy and each has the general shape of a Gaussian function. The partial cross sections gradually shift to higher energies with increasing quantum number

† Very often we will consider the photodissociation cross section as a function of the energy in the excited state, E_f, rather than the photon energy $E_{photon} = \hbar\omega$. This has the advantage that the overlap with the ground-state wavefunction is independent of the energy E_i of the initial state of the parent molecule.

n. Their peak positions correlate well with the vertical energies $\epsilon_n(R_e)$ above the ground-state equilibrium R_e and the peak intensities vary according to the FC factors $f(n)$. In the present case, the equilibrium distances in the ground and in the excited states are rather displaced so that the FC distribution $f(n)$ is broad and inverted. The result is that $\sigma(E_f, n = 2)$ is the largest cross section. Each $\sigma(E_f, n)$ can be interpreted as a reflection of the ground-state radial wavefunction $\varphi_R(R)$ mediated by the respective potential energy curve $\epsilon_n(R)$ as schematically indicated in Figure 3.5. A more detailed discussion of the reflection principle and adiabatic decay in direct photodissociation follows in Chapters 6 and 9. We will frequently exploit the adiabatic limit in subsequent chapters in order to assign structures in absorption spectra.

There are two reasons why the adiabatic approximation often works so well for vibrational excitation, at least qualitatively: first, the relatively large spacing between the energy levels and second, the weak coupling between the translational and the vibrational modes. Generally, the opposite conditions apply for rotational excitation, relatively small energy level spacings and strong translational-rotational coupling. Rotational excitation in full as well as half collisions is often successfully described by the so-called sudden approximation which is just the counterpart of the adiabatic approximation.

3.3.2 Sudden approximation, appropriate for rotational excitation

As opposed to the adiabatic limit, we assume in the sudden approximation that the internal motion is slow compared to the external (i.e., translational) motion. Most familiar is the rotational sudden approximation which is frequently exploited in energy transfer studies in full collisions (Pack 1974; Secrest 1975; Parker and Pack 1978; Kouri 1979; Gianturco 1979:ch.4). Its application to photodissociation is straightforward and will be outlined below for the model discussed in Section 3.2.

Within the rotational sudden approximation we assume that the interaction time is much smaller than the rotational period of the fragment molecule so that the diatom BC does not appreciably rotate from its original position while the two fragments separate. In terms of energies, this requires the rotational energy, E_{rot}, to be much smaller than the total available energy. If that is true, the operator for the rotational motion of BC, $\hat{h}_{rot}$, can be neglected in (3.16). The partial differential equation thus becomes an ordinary differential equation,

$$\left[-\frac{\hbar^2}{2m}\frac{d^2}{dR^2} + V(R,\gamma) - E_f \right] \chi(R;\gamma;E_f) = 0, \qquad (3.43)$$

where we have defined $\Psi = \chi/R$. However, equation (3.43) and therefore the radial function $\chi(R;\gamma;E_f)$ depend, as a consequence of the anisotropy of the potential energy surface $V(R, \gamma)$, parametrically on the orientation angle γ.

One readily proves, using the completeness of the spherical harmonics, that the radial functions $\chi_{j'}(R; E_f, j)$ defined by

$$\chi_{j'}(R; E_f, j) = \langle Y_{j'0}(\gamma, \psi) \mid \chi(R; \gamma; E_f) \mid Y_{j0}(\gamma, \psi) \rangle \tag{3.44}$$

solve the close-coupling equations (3.20) provided the diagonal terms $k_{j'}^2 - j'(j'+1)/R^2$ are replaced by an average $k_{\bar{j}}^2 - \bar{j}(\bar{j}+1)/R^2$ which is the same for all rotational states. $\bar{j} = 0$ is the simplest choice but other values may lead to more accurate results. Inserting (3.44) into (3.28) one readily obtains approximate expressions for the partial photodissociation amplitudes $t(E_f, j)$.

If we assume, in analogy to (3.40), a separable ground-state wavefunction of the form

$$\Psi_i(R, \gamma) = \frac{1}{R} \, \varphi_R(R) \, \varphi_\gamma(\gamma), \tag{3.45}$$

the expressions become simpler and more illuminating. The partial photodissociation amplitudes are given by

$$t(E_f, j) = 2\pi \int_0^\pi d\gamma \sin\gamma \; Y_{j0}^*(\gamma, 0) \; \varphi_\gamma(\gamma) \; t(E_f; \gamma) \tag{3.46}$$

with the definition

$$t(E_f; \gamma) = \langle \chi(R; \gamma; E_f) \mid \varphi_R(R) \rangle . \tag{3.47}$$

Employing the completeness of the spherical harmonics, one derives the following expression for the total photodissociation cross section

$$\sigma_{tot}(E_f) \propto 2\pi \int_0^\pi d\gamma \; \sin\gamma \; \varphi_\gamma^2(\gamma) \; \sigma(E_f; \gamma), \tag{3.48}$$

with the γ-dependent cross section defined by

$$\sigma(E_f; \gamma) \propto E_{photon} \, |t(E_f; \gamma)|^2 . \tag{3.49}$$

Expression (3.48) is very useful for interpretation purposes. Since the bending wavefunction of the electronic ground state, $\varphi_\gamma(\gamma)$, is usually quite narrow for chemically bound molecules ($\Delta\gamma \approx 10°$–$20°$), it explains why often only a relatively narrow strip of the PES is sampled in the fragmentation process. The bending wavefunction acts like a weighting function for the angle coordinate.

The sudden approximation is easy to implement. One solves the one-dimensional Schrödinger equation (3.43) for several fixed orientation angles γ, evaluates the γ-dependent amplitudes (3.47), and determines the partial photodissociation amplitudes (3.46) by integration over γ. Because of the spherical harmonic $Y_{j0}(\gamma, 0)$ on the right-hand side of (3.46), the integrand oscillates rapidly as a function of γ if the rotational quantum number is large and therefore a semiclassical, so-called stationary phase approximation may be employed (Schinke and Bowman 1983; Schinke 1986a). This approximation is very helpful for explaining the shape of final rotational state distributions in terms of classical trajectories and possible "interferences" between them (Miller 1975, 1985).

The necessary condition for the sudden approximation to be valid is a small ratio E_{rot}/E_{excess}. While the sudden approximation violates energy conservation, the coupling between the rotational states caused by the anisotropy of the PES is exactly taken into account. Because of its convenience, the rotational sudden approximation is widely used in photodissociation (Segev and Shapiro 1983; Atabek, Beswick, and Delgado-Barrio 1985; Schinke, Engel, and Staemmler 1985; Kulander and Light 1986; Delgado-Barrio et al. 1986; Grinberg, Freed, and Williams 1987; Engel, Schinke, and Staemmler 1988).

3.4 Numerical methods

Continuum wavefunctions required for the calculation of bound-free matrix elements of the type (3.1) can be determined by several methods. In principle, all techniques which have been developed in the field of atomic and molecular collisions [see Thomas et al. (1981) for a comprehensive overview] can be employed with only slight modifications and extensions. Since many readers are probably not familiar with the calculation of multi-dimensional continuum wavefunctions, we shall briefly describe in the following one particular method, which is rather universal.

Within the close-coupling approach each partial photodissociation wavefunction $\Psi(R, r; E_f, n)$ is represented by the expansion functions $\chi_{n'}(R; E_f, n)$ and the vibrational basis functions $\varphi_n(r)$ with n and $n' = 0, 1, 2, \ldots, \bar{n}$. Here, $\bar{n}$ denotes the highest state considered in expansion (3.4). It is not necessarily identical with n_{max}, the highest state that can be populated for a given total energy. In order to simplify the subsequent notation we consider the total of the radial functions as the elements $\chi_{n'n}(R; E_f)$ of a $(\bar{n}+1) \times (\bar{n}+1)$ matrix

$$\boldsymbol{\chi} = \begin{pmatrix} \chi_{00} & \chi_{01} & \cdots & \chi_{0\bar{n}} \\ \chi_{10} & \chi_{11} & \cdots & \chi_{1\bar{n}} \\ \chi_{20} & \chi_{21} & \cdots & \chi_{2\bar{n}} \\ \vdots & \vdots & & \vdots \\ \chi_{\bar{n}0} & \chi_{\bar{n}1} & \cdots & \chi_{\bar{n}\bar{n}} \end{pmatrix},$$

where the first index, n', labels the channels of the expansion and the second index, n, indicates the particular solution with unit outgoing flux in channel n. The dimension of the close-coupling system is $\bar{N} = \bar{n} + 1$. Each column of $\boldsymbol{\chi}$ represents a particular solution of the coupled

Equations (3.5) satisfying the boundary conditions (3.10). In matrix notation, the boundary conditions read

$$\lim_{R\to\infty} \boldsymbol{\chi} = \mathbf{A} + \mathbf{B}\mathbf{S}^*, \tag{3.50}$$

where **A** and **B** are diagonal matrices with elements

$$A_{n'n} = B^*_{n'n} = \left(\frac{m}{\hbar k_{n'}}\right)^{1/2} \mathrm{e}^{+ik_{n'}R}\,\delta_{nn'} \tag{3.51}$$

for open channels ($k_{n'}^2 > 0$) and

$$A_{n'n} = B_{n'n} = \left(\frac{m}{\hbar |k_{n'}|}\right)^{1/2} \mathrm{e}^{-|k_{n'}|R}\,\delta_{nn'} \tag{3.52}$$

for closed channels ($k_{n'}^2 < 0$); **S** is the unitary scattering matrix.

The set of coupled equations together with the conditions for $R \to 0$ and $R \to \infty$ constitute a boundary value problem. It does not suffice merely to find a solution of the coupled equations; the solution must also have a particular behavior as R goes to zero and to infinity. One possible way of solving the boundary value problem is the transformation into a computationally more convenient initial value problem (Lester 1976). From the theory of linear differential equations we know that Equation (3.5) has $\bar{N}$ regular solutions, i.e., solutions which diminish as $R \to 0$, and which are linearly independent. Each solution has a distinct behavior in the limit $R \to \infty$ which, however, is not necessarily the behavior imposed by (3.50). However, if we have found one particular set of solutions, we can easily construct new solutions by taking linear combinations so that the new wavefunctions fulfil the required boundary conditions.

The numerical recipe is the following: first, we calculate, for a specific energy E_f, $\bar{N}$ linearly independent solutions (denoted by $\tilde{\boldsymbol{\chi}}$) by pointwise numerical propagation from R_{start} to R_{end}, where R_{start} must be sufficiently deep inside the nonclassical region and R_{end} must be large enough to assure that the interaction potential is zero. The propagation starts with $\tilde{\boldsymbol{\chi}}(R_{start}) = 0$ and any nonsingular matrix of derivatives $\tilde{\boldsymbol{\chi}}'(R_{start}) \neq 0$, where the prime indicates derivation with respect to R. This particular solution matrix $\tilde{\boldsymbol{\chi}}$ will certainly not satisfy the required boundary conditions (3.50) but it can be used to construct a new solution matrix $\boldsymbol{\chi} \equiv \tilde{\boldsymbol{\chi}}\mathbf{T}$ by demanding that

$$\boldsymbol{\chi}(R_{end}) \equiv \tilde{\boldsymbol{\chi}}(R_{end})\mathbf{T} = \mathbf{A}(R_{end}) + \mathbf{B}(R_{end})\mathbf{S}^* \tag{3.53a}$$

$$\boldsymbol{\chi}'(R_{end}) \equiv \tilde{\boldsymbol{\chi}}'(R_{end})\mathbf{T} = \mathbf{A}'(R_{end}) + \mathbf{B}'(R_{end})\mathbf{S}^*, \tag{3.53b}$$

where **T** is the transformation matrix which is independent of R. Equations (3.53) are nothing other than a 2×2 set of linear matrix equations for the two unknown matrices, **T** and **S**. Solving (3.53) yields the scattering matrix **S** and at the same time it determines the required transformation matrix $\mathbf{T}$. Since $\mathbf{T}$ is independent of R, the prescription $\chi(R_i) = \tilde{\chi}(R_i)\mathbf{T}$ yields the correct radial functions at each propagation point from R_{start} through R_{end}. Finally, the overlap with the radial functions of the ground state, $\langle \chi_{n'}(R; E_f, n) \mid \chi_{n''}(R; E_i) \rangle$, can be determined by simple one-dimensional quadrature schemes.

Once we have determined the radial expansion functions $\chi_{n'}(R; E_f, n)$ it is rather straightforward to construct the full partial photodissociation wavefunctions $\Psi(R, r; E_f, n)$ or the total photodissociation wavefunction $\Psi_{tot}(R, r; E_f)$. Examples are shown in Figures 2.3 and 3.2. Plots of the stationary wavefunctions for a given energy E_f are useful for illustrating the overall dissociation path, assessing the region of the multi-dimensional PES sampled in the fragmentation, and in assigning resonance structures in the absorption spectrum. More examples will be shown in subsequent chapters.

Continuum wavefunctions can also be generated by solving the partial differential equation (3.3) directly without first transforming it into a set of ordinary differential equations. One possible scheme is the *finite elements method* (Askar and Rabitz 1984; Jaquet 1987). Another method, which has been applied for the calculation of multi-dimensional scattering wavefunctions, is the S-matrix version of the *Kohn variational principle* (Zhang and Miller 1990).

In addition to the direct methods, in which one calculates first the continuum wavefunctions and subsequently the overlap integrals with the bound-state wavefunction, there are also indirect methods, which encompass the separate computation of the continuum wavefunctions: the *artificial channel method* (Shapiro 1972; Shapiro and Bersohn 1982; Balint-Kurti and Shapiro 1985) and the *driven equations method* (Band, Freed, and Kouri 1981; Heather and Light 1983a,b). Kulander and Light (1980) applied another method, in which the overlap of the bound-state wavefunction with the continuum wavefunction is directly propagated. The desired photodissociation amplitudes are finally obtained by applying the correct boundary conditions for $R \to \infty$.

4
Time-dependent methods

In this chapter we shall discuss exact and approximate methods for the calculation of photodissociation cross sections that are based on the solution of the time-dependent nuclear Schrödinger equation,

$$\left[i\hbar\frac{\partial}{\partial t} - \hat{H}(\mathbf{Q})\right] \Phi(\mathbf{Q};t) = 0, \tag{4.1}$$

where Φ is a time-dependent wavepacket evolving on the potential energy surface (PES) of the excited electronic state. The time-dependent method is an alternative to the time-independent theory outlined in Chapter 3. The two approaches are completely equivalent and merely provide different views of the dissociation dynamics and alternative numerical means.

Quantum mechanical studies of energy transfer in molecular collisions are traditionally performed in the time-independent formalism (Gianturco 1979). The last decade, however, has witnessed an explosion of time-dependent applications in molecular dynamics studies which became possible with the availability of new generations of high-performance computers (Gerber, Kosloff, and Berman 1986; Kosloff 1988; Mohan and Sathyamurthy 1988; Kulander 1991). The wavepacket corresponds, at least for short times, to a swarm of classical trajectories and therefore it leads to a very intuitive picture of the molecular motion. Classical trajectories essentially form the basis for the general understanding of molecular collisions. A wavepacket evolving in time and space more closely resembles our vision of a collision process than the solution of the time-independent Schrödinger equation in conjunction with particular boundary conditions.

The time-dependent approach is particularly profitable for the study of absorption and dissociation processes (Heller 1981a,b). At the center is the relation between the total absorption cross section $\sigma_{tot}(\omega)$ and the autocorrelation function $S(t)$,

$$\sigma_{tot}(\omega) \propto \int_{-\infty}^{+\infty} dt\, S(t)\, e^{i\omega t}. \tag{4.2}$$

Expression (4.2) relates the energy dependence of the spectrum to the evolution of the molecular system in the excited electronic state as it is reflected by the autocorrelation function (Lax 1952; Gordon 1968; Cederbaum and Domcke 1977; Kulander and Heller 1978; Heller 1981a,b; Köppel, Domcke, and Cederbaum 1984; Zare 1988:ch.3; Weissbluth 1989: ch.V). Structures in the spectrum can thus be explained in terms of features of $S(t)$ which in turn can be traced back to the evolution of the molecular system.

In Section 4.1 we will use the time-independent continuum basis $\Psi_f(\mathbf{Q}; E, \beta)$, defined in Section 2.5, to construct the wavepacket in the excited state and to derive (4.2). Numerical methods are discussed in Section 4.2 and quantum mechanical and semiclassical approximations based on the time-dependent theory are the topic of Section 4.3. Finally, a critical comparison of the time-dependent and the time-independent approaches concludes this chapter.

4.1 Time-dependent wavepacket

For simplicity and clarity of presentation we restrict the following derivation to the photodissociation of the linear triatomic molecule ABC as outlined in Section 2.4. The Jacobi coordinates R and r are defined in Figure 2.1 and (2.39) gives the corresponding Hamiltonian $\hat{H}$ for the motion of the nuclei.

4.1.1 Autocorrelation function and total absorption spectrum

In Section 2.5 we have constructed the degenerate continuum wavefunctions $\Psi_f(R, r; E_f, n)$, which describe the dissociation of the ABC complex into A+BC(n). They solve the time-independent Schrödinger equation for fixed energy E_f subject to the boundary conditions (2.59). Furthermore, the $\Psi_f(R, r; E_f, n)$ are orthogonal and complete and thus they form a basis in the corresponding Hilbert space, i.e., any function can be represented as a linear combination of them.

A wavepacket is nothing other than a *coherent superposition of stationary states*, each being multiplied by the time-evolution factor $e^{-iEt/\hbar}$.† In the present case, a most general time-dependent wavepacket is con-

† Cohen-Tannoudji et al. (1977:ch.III) provide an illuminating discussion of the motion of a wavepacket in a potential well; see also Child (1991:ch.11).

structed according to

$$\Phi_f(R,r;t) = \int dE_f \sum_{n=0}^{n_{max}} c(E_f,n)\ \Psi_f(R,r;E_f,n)\ \mathrm{e}^{-iE_ft/\hbar}, \tag{4.3}$$

where n_{max} denotes the highest vibrational channel that can be populated at energy E_f. The coefficients $c(E_f,n)$ are independent of the coordinates and the time. Equation (4.3) is the formal analogue of Equation (2.9). $\Phi_f(R,r;t)$ is obviously a solution of (4.1) because each stationary wavefunction $\Psi_f(R,r;E_f,n)$ is an eigenfunction of $\hat{H}$ with energy E_f.

The next step is the essential one: we fix the coefficients $c(E_f,n)$ by imposing the initial condition

$$\Phi_f(R,r;t=0) = \mu_{fi}^{(e)}\ \Psi_i(R,r;E_i) \tag{4.4}$$

at $t=0$. Thus, we demand that:

- The wavepacket at its start in the upper electronic state equals the wavefunction of the parent molecule, $\Psi_i(R,r;E_i)$, multiplied by the transition dipole function $\mu_{fi}^{(e)}(R,r)$.

Inserting the right-hand side of (4.3) for $t=0$ into (4.4) and using the orthogonality of the basis functions $\Psi_f(E_f,n)$ yields the relation

$$c(E_f,n) = \rho\ \langle \Psi_f(E_f,n) \mid \mu_{fi}^{(e)} \mid \Psi_i(E_i)\rangle = \rho\ t(E_f,n), \tag{4.5}$$

where the amplitudes $t(E_f,n)$ are defined in (2.68) and $\rho = (2\pi\hbar)^{-1}$.

Multiplying (4.3) from the left by $\Phi_f(0)$ and integrating over all nuclear coordinates gives

$$\begin{aligned} S(t) &\equiv \langle \Phi_f(0) \mid \Phi_f(t)\rangle \\ &= \rho \int dE_f \sum_{n=0}^{n_{max}} t(E_f,n)\ \langle \Phi_f(0) \mid \Psi_f(E_f,n)\rangle\ \mathrm{e}^{-iE_ft/\hbar} \\ &= \rho \int dE_f \sum_{n=0}^{n_{max}} |t(E_f,n)|^2\ \mathrm{e}^{-iE_ft/\hbar}. \end{aligned} \tag{4.6}$$

In deriving this relation we used (4.4) and (4.5). $S(t)$ on the left-hand side of (4.6) is the *autocorrelation function.* It is the central quantity in the time-dependent formulation of spectroscopy and photodissociation. Multiplying (4.6) by $\mathrm{e}^{iE'_ft/\hbar}$, integrating over t, and using Equations (2.56), (2.57), and (2.67) yields the final expression for the total absorption cross section,

$$\sigma_{tot}(E_f) = C\ E_{photon} \int_{-\infty}^{+\infty} dt\ \mathrm{e}^{iE_ft/\hbar}\ S(t), \tag{4.7}$$

where $C = \rho\pi/\hbar\epsilon_0 c$ and E_{photon} is the photon energy. One can easily verify that σ_{tot} has units of area.

Let us summarize the essential points:

- The total absorption cross section as a function of energy is proportional to the Fourier transformation of the time-dependent autocorrelation function $S(t)$.
- $S(t)$ reflects the evolution of the wavepacket in the excited electronic state and therefore it reflects the fragmentation dynamics and the forces $-\partial V/\partial Q_i$ in the upper state.
- Equation (4.7) establishes a unique relation between the time-resolved molecular motion in the excited state on one hand and the frequency-resolved absorption spectrum on the other hand.

The formal solution of the time-dependent Schrödinger equation is given by

$$\Phi_f(t) = \mathrm{e}^{-i\hat{H}t/\hbar}\, \Phi_f(0), \tag{4.8}$$

where $\Phi_f(0)$ is the wavepacket at $t = 0$ and $\mathrm{e}^{-i\hat{H}t/\hbar}$ is the time-evolution operator. With (4.8) inserted into (4.6) the autocorrelation function becomes

$$S(t) = \langle \Phi_f(0) \mid \mathrm{e}^{-i\hat{H}t/\hbar} \mid \Phi_f(0)\rangle. \tag{4.9}$$

In practice one does not proceed as we did in the above derivation. Instead of calculating first all stationary wavefunctions and then constructing the wavepacket according to (4.3), one solves the time-dependent Schrödinger equation (4.1) with the initial condition (4.4) directly. Numerical propagation schemes will be discussed in the next section. Since $\Phi_f(0)$ is real[†] the autocorrelation function fulfills the symmetry relation

$$S(-t) = S(t)^*, \tag{4.10}$$

which guarantees that $\sigma_{tot}(E_f)$ is real as well. If the wavepacket is propagated in time until $S(t)$ is essentially zero and stays zero for all later times, σ_{tot} is positive as required for a cross section. Truncating the integration in (4.7) too early causes unphysical structures and even negative cross sections.[‡] Because of (4.10) the integration can be restricted to the interval $[0, +\infty)$.

[†] Both the initial wavefunction Ψ_i and the transition dipole function $\mu_{fi}^{(e)}$ are real functions.

[‡] The integral representation of the delta-function in (2.56), which was used to derive (4.7), requires integration from $-\infty$ to $+\infty$.

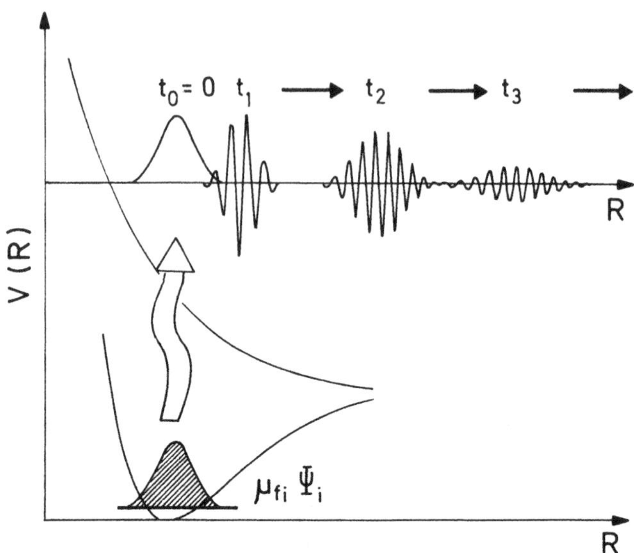

Fig. 4.1. Schematic illustration of the evolution of a one-dimensional time-dependent wavepacket in the upper electronic state. The wavepacket is complex for $t > 0$; only its real part is shown here. Note that *the upper horizontal axis does not correspond to a particular energy*! The wavepacket is a superposition of stationary states corresponding to a broad range of energies, which are all simultaneously excited by the infinitely short light pulse indicated by the vertical arrow.

4.1.2 Evolution of the wavepacket

Figure 4.1 illustrates schematically the evolution of a one-dimensional wavepacket in the upper electronic state. At $t = 0$ the light pulse *instantaneously* promotes the molecule from the ground to the excited electronic state. Since $\Phi_f(0) = \mu_{fi}^{(e)} \Psi_i$ is not an eigenfunction of the upper-state Hamiltonian, the wavepacket immediately starts to move away from its origin. When it has reached the asymptotic region where the potential is zero, the center of the wavepacket travels with constant velocity to infinity. The oscillations in R-space reflect the momentum gained during the breakup.

Figure 4.2 depicts "snapshots" of the evolving wavepacket for the two-dimensional model system that we discussed in Section 2.5 to illustrate vibrational excitation (see Figure 2.3). Because the PES is quite steep in the proximity of the ground-state equilibrium, the molecule dissociates at once. Immediately after the wavepacket is released it slides down the potential slope and disappears into the asymptotic channel for good. Since $\Phi_f(0)$ starts significantly displaced from the minimum energy path of the PES, the evolving wavepacket oscillates along the r-direction, which in-

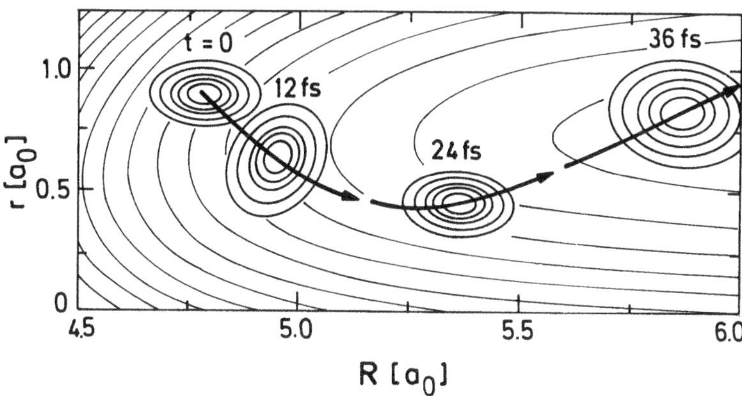

Fig. 4.2. Evolution of a two-dimensional wavepacket $\Phi(R, r; t)$. $|\Phi(t)|$ is shown at several instants. The model system is the same as the one in Figure 2.3. According to (4.11), the wavepacket is related to the time-independent total dissociation wavefunction $\Psi_{tot}(E_f)$, depicted in Figure 2.3(a), by a Fourier transformation between the time and the energy domains. The arrows represent the classical trajectory in the upper state that begins at the equilibrium of the electronic ground state with both momenta being initially zero.

dicates that BC is created with substantial vibrational excitation. The overlap with the initial wavepacket becomes rapidly unfavorable, thus $|S(t)|$ decreases smoothly to zero within a very short time (Figure 4.3). There are no recurrences of the wavepacket to its origin and the corresponding absorption spectrum, which was shown in Figure 3.5, varies smoothly with energy. $S(t)$ is a complex function with a real and an imaginary part; as we shall show below, the phase of the autocorrelation function is important and must not be ignored. A more detailed analysis of the relation between $S(t)$ and $\sigma_{tot}(E)$ follows in Chapters 6–8.

The wavepacket in Figure 4.2 follows essentially the same route as the time-independent total dissociation wavefunction $\Psi_{tot}(E_f)$, defined in Equation (2.70), which is shown for a particular energy in Figure 2.3(a). This coincidence does not come as a surprise, however. If we multiply $\Phi_f(t)$ with $e^{iE_f t/\hbar}$ and integrate over t we obtain

$$\begin{aligned}
&\int dt\ e^{iE_f t/\hbar}\ \Phi_f(t) \\
&\quad = \rho \int dE'_f \sum_{n=0}^{n_{max}} t(E'_f, n)\ \Psi_f(E'_f, n) \int dt\ e^{i(E_f - E'_f)t/\hbar} \\
&\quad = \sum_{n=0}^{n_{max}} t(E_f, n)\ \Psi_f(E_f, n) \\
&\quad = \Psi_{tot}(E_f).
\end{aligned} \tag{4.11}$$

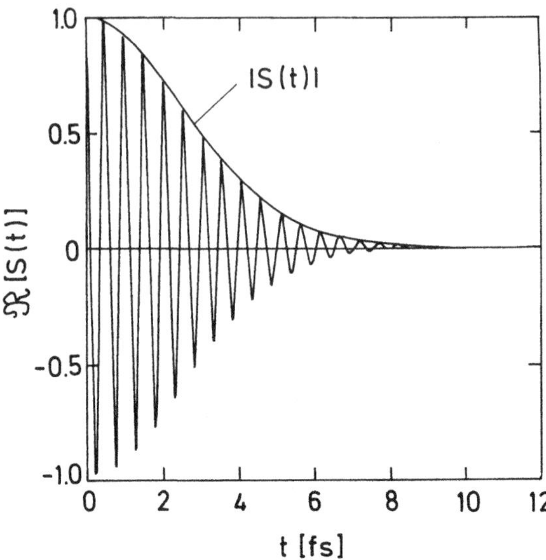

Fig. 4.3. Time-dependence of the autocorrelation function $S(t)$ (real part) of the wavepacket shown in Figure 4.2

In deriving (4.11) we used (4.3) and (4.5) to represent Φ_f, (2.56) and (2.57) to evaluate the integral over t, and (2.70) for the total dissociation wavefunction. For the subsequent discussion we note the following important points:

- The time-independent total dissociation wavefunction $\Psi_{tot}(E_f)$ is the Fourier transform of the time-dependent wavepacket $\Phi_f(t)$.
- The wavefunction $\Psi_{tot}(E_f)$ is a solution of the time-independent Schrödinger equation and therefore corresponds to a particular energy E_f. According to the uncertainty relation, it contains all times, i.e., the entire history of the dissociation process (see also Section 16.1).
- The wavepacket $\Phi_f(t)$ on the other hand is a function of time and therefore contains all energies, weighted by the matrix elements $t(E_f, n)$ defined in Equation (2.68). It is a solution of the time-dependent but not the time-independent Schrödinger equation.

4.1.3 Energy domain and time domain

Structures in the absorption spectrum are related to structures in the autocorrelation function. There are many cases in which an unambiguous assignment of structures in the absorption spectrum is essentially impossible in the energy domain. Figure 4.4(a) depicts an intriguing example. The corresponding autocorrelation function, however, obtained by Fourier transformation of the measured spectrum has a much simpler appearance. It exhibits a series of pronounced spikes, so-called recurrences, which show that the wavepacket or at least some part of it recurs to the vicinity of its origin so that the overlap with $\Phi(0)$ increases rapidly. The recurrences, in turn, can be related to so-called unstable periodic orbits which reveal a wealth of information about the internal motion of the atomic or molecular system. Recurrences and periodic orbits are the subjects of Chapters 7 and 8. Let us bear in mind for the discussion in the following chapters:

- The autocorrelation function $S(t)$ provides the link between the spectrum $\sigma_{tot}(\omega)$ on one hand and the (classical or quantum) dynamics in the excited electronic state on the other hand.

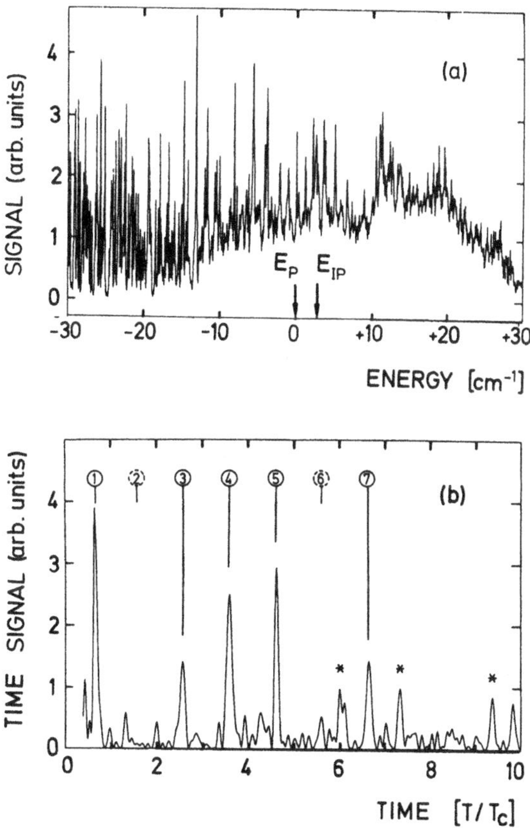

Fig. 4.4. (a) Excitation-ionization spectrum of the H atom Balmer series around the ionization limit in a static homogenous magnetic field. (b) Fourier-transformed time domain spectrum of the spectrum shown in (a). The square of the absolute value is plotted. The time scale is given in units of the cyclotron period $T_c = 2\pi/\omega_c$. Reprinted from Main, Holle, Wiebusch, and Welge (1987).

There is one additional, very important point which we must stress at this stage:

- The wavepacket $\Phi_f(t)$ that we constructed by means of Equations (4.3)–(4.5) is not the real state prepared by the *long* and *monochromatic* laser pulse in the excited electronic state.

The real molecular state generated by the external field in the excited state is given by Equations (2.9) and (2.64) with the expansion coefficients $a_\alpha(t)$ respectively $a_f(t; E, n)$ solving the time-dependent coupled equations (2.16). The light pulse excites — in principle — *all* stationary states $\Psi_f(E, n)$. However, in the limit that the pulse is monochromatic and infinitely long only the resonant state with energy $E_f = E_i + \hbar\omega$ becomes significantly excited whereas the probabilities for all other states, which do not fulfil the resonance condition, remain negligibly small for all times. Thus, after the external field has excited the molecule for a sufficiently long time, the nuclear wavefunction in the upper electronic state is essentially equal to the stationary wavefunction corresponding to the resonant energy E_f, multiplied by the time-evolution factor $e^{-iE_f t/\hbar}$. According to (2.21), the norm of this state rises linearly with time. Recall that the expressions for the photodissociation cross sections defined in Section 2.5, Equations (2.65)–(2.68), are explicitly based on an infinitely long light pulse.

The wavepacket $\Phi_f(t)$, on the other hand, is constructed in a completely different way. In view of (4.4), the initial state multiplied by the transition dipole function is *instantaneously* promoted to the excited electronic state. It can be regarded as the state created by an *infinitely short* light pulse. This picture is essentially classical (Franck principle): the electronic excitation induced by the external field does not change the coordinate and the momentum distributions of the parent molecule. As a consequence of the instantaneous excitation process, the wavepacket $\Phi_f(t)$ contains the stationary wavefunctions for *all* energies E_f, weighted by the amplitudes $t(E_f, n)$ [see Equations (4.3) and (4.5)]. When the wavepacket attains the excited state, it immediately begins to move under the influence of the intramolecular forces. The time dependence of the excitation of the molecule due to the external perturbation and the evolution of the nuclear wavepacket $\Phi_f(t)$ on the excited-state PES must not be confused (Rama Krishna and Coalson 1988; Williams and Imre 1988a,b)

The calculation of cross sections, which are defined in the limit of an infinitely long laser pulse, by means of a wavepacket, which is created by an infinitely short pulse, seems contradictory on the first glance. On the other hand, note that the wavepacket created by the infinitely short light pulse is merely a coherent superposition of all stationary wavefunctions with expansion coefficients proportional to the amplitudes $t(E_f, n)$.

Therefore, we can use this somewhat artificial wavepacket, when it has been determined by direct integration of the time-dependent Schrödinger equation, to extract the transition amplitudes and hence the photodissociation cross sections for *all* energies.

4.1.4 Partial photodissociation cross sections

The time-dependent formalism would be rather limited if it yielded only the total cross section. However, that is not the case; all partial photodissociation cross sections $\sigma(E_f, n)$ can be also extracted from the time-dependent wavepacket. We assume that for large times the wavepacket has completely left the interaction zone and moves entirely in the asymptotic region where the interaction potential $V_I(R, r)$ is zero. Then, the asymptotic conditions (2.59) for the stationary continuum wavefunctions can be inserted into (4.3) yielding

$$\lim_{t\to\infty} \Phi_f(t) = \rho \int dE_f \sum_{n=0}^{n_{max}} t(E_f, n)\, \mathrm{e}^{-iE_f t/\hbar} \times \left[\Psi_0^+(E_f, n) + \sum_{n'=0}^{n_{max}} S_{nn'}^* \Psi_0^-(E_f, n') \right]. \tag{4.12}$$

The $\Psi_0^\pm$ are the eigenfunctions of the asymptotic Hamiltonian $\hat{H}_0$ defined in (2.51).

In order to extract the amplitudes $t(E_f, n)$ one proceeds in the usual way: multiplying (4.12) with $\langle \Psi_0^+(E_f', n')|$ and utilizing the orthogonality condition (2.54), we readily obtain

$$t(E_f, n) = \lim_{t\to\infty} \mathrm{e}^{iE_f t/\hbar} \langle \Psi_0^+(E_f, n) \mid \Phi_f(t) \rangle. \tag{4.13}$$

The partial cross sections (2.67) are finally given by

$$\sigma(E_f, n) = C\, E_{photon} \frac{m}{\hbar k_n} \lim_{t\to\infty} |\langle \mathrm{e}^{ik_n R} \varphi_n(r) \mid \Phi_f(R, r; t) \rangle|^2, \tag{4.14}$$

where we have explicitly used the definition of the asymptotic wavefunctions in Equation (2.51). The calculation of the partial cross sections is slightly more involved than the calculation of the total cross section because, in general, the wavepacket must be propagated over a longer time interval and therefore a larger spatial region. The autocorrelation function usually decays to zero faster than the wavepacket leaves the interaction zone. For an alternative procedure for the calculation of partial cross sections see Balint-Kurti, Dixon, and Marston (1990). The total cross section obtained by Fourier transforming the autocorrelation function and the cross section obtained by summing the partial cross sections over all fragment channels must be identical. This can be used to check the accuracy of the numerical calculation.

Let us summarize the essential steps of the time-dependent calculation:

1) Propagation of the wavepacket $\Phi_f(t)$ in the upper electronic state with initial condition $\Phi_f(0) = \mu_{fi}^{(e)} \Psi_i(E_i)$.
2) Calculation and Fourier transformation of the autocorrelation function $S(t)$ in order to yield the total dissociation cross section.
3) Propagation of the wavepacket until it has completely left the interaction zone and calculation of the partial cross sections by means of (4.14).

4.2 Numerical methods

One of the virtues of the time-dependent approach is the ease with which it can be implemented. In this section we shall briefly outline several methods that are nowadays routinely employed for the propagation of wavepackets. Rather than being exhaustive we attempt to give a more general understanding of such calculations.

4.2.1 Temporal propagation

Let us again consider the photodissociation of the linear triatomic molecule with coordinates R and r (Figure 2.1). We want to solve the time-dependent Schrödinger equation (4.1) with the Hamiltonian given in (2.39) and the initial condition (4.4).

Since the nuclear Hamiltonian is time-independent the wavepacket at time $(t + dt)$ follows (formally) from the wavepacket at time t according to

$$\Phi(t + dt) = \mathrm{e}^{-i\hat{H}dt/\hbar} \, \Phi(t). \tag{4.15}$$

Several accurate approximations of the time-evolution operator $\mathrm{e}^{-i\hat{H}dt/\hbar}$ and efficient algorithms for the propagation of the wavepacket have been developed in recent years. For a comprehensive overview and an extensive list of references see Gerber, Kosloff, and Berman (1986), Kosloff (1988), Leforestier et al. (1991), and Kulander (1991), for example. Here, we review only one of these methods.

Following Tal-Ezer and Kosloff (1984) the time evolution-operator is expanded in terms of Chebychev polynomials ψ_k according to

$$\mathrm{e}^{\hat{X}} \approx \sum_{k=0}^{K} a_k \psi_k(\hat{X}) \tag{4.16}$$

with the abbreviation $\hat{X} = -i\hat{H}dt/(\hbar s)$, where s is an appropriate scaling factor and the expansion coefficients a_k are special complex numbers. The Chebychev polynomials satisfy the recursion relation

$$\psi_k(\hat{X}) = 2\hat{X}\psi_{k-1}(\hat{X}) + \psi_{k-2}(\hat{X}), \tag{4.17}$$

starting with $\psi_0(\hat{X}) = 1$ and $\psi_1(\hat{X}) = \hat{X}$. Inserting (4.16) into (4.15) yields for the wavepacket at $(t + dt)$

$$\Phi(t + dt) \approx \sum_{k=0}^{K} a_k \, \Phi^{(k)}(t + dt) \tag{4.18}$$

where we have defined

$$\Phi^{(k)}(t + dt) = \psi_k(\hat{X}) \, \Phi(t). \tag{4.19}$$

The $\Phi^{(k)}$ fulfil the same recursion relation as the Chebychev polynomials, namely

$$\Phi^{(k)}(t + dt) = 2\hat{X}\Phi^{(k-1)}(t + dt) + \Phi^{(k-2)}(t + dt) \tag{4.20}$$

with $\Phi^{(0)}(t + dt) = \Phi(t)$ and $\Phi^{(1)}(t + dt) = \hat{X}\Phi(t)$.

The order of the expansion, K, depends on the time interval dt and on the scaling factor s. The method of Tal-Ezer and Kosloff does not require dt to be small; however, the larger we choose dt, the larger must be the order K. Equations (4.18)–(4.20) provide a very convenient iteration scheme for the wavepacket at each coordinate point (R, r). The Chebychev method is, however, only applicable for time-independent Hamiltonians. If the Hamiltonian is time-dependent, one can, for example, use the split operator method of Feit, Fleck, and Steiger (1982) and Feit and Fleck (1983, 1984). In passing we note, that under favorable conditions one can use very large time intervals if one propagates the wavepacket in the so-called interaction picture (Das and Tannor 1990; Zhang 1990; Williams, Qian, and Tannor 1991).

4.2.2 Spatial propagation

The essential step in (4.20) is the repeated application of the Hamiltonian. While the application of the potential is simply a multiplication, application of the kinetic energy operator, which for the linear triatomic molecule has the form

$$\hat{T} = \hat{T}_R + \hat{T}_r = -\frac{\hbar^2}{2m}\frac{\partial^2}{\partial R^2} - \frac{\hbar^2}{2\mu}\frac{\partial^2}{\partial r^2}, \tag{4.21}$$

needs special attention; it is *the* time-determining step. The second-order derivatives can be conveniently evaluated if we expand the wavepacket at each instant in a Fourier series according to (Feit and Fleck 1983, 1984; Kosloff and Kosloff 1983)

$$\Phi(R, r; t) = \sum_{kl} a_{kl}(t) \, e^{ikR/L_R} \, e^{ilr/L_r}, \tag{4.22}$$

where L_R and L_r are constants having units of length. Then, operation of $\hat{T}$ gives

$$\hat{T}\ \Phi(R,r) = \sum_{kl} a_{kl}\left[\frac{\hbar^2}{2m}\left(\frac{k}{L_R}\right)^2 + \frac{\hbar^2}{2\mu}\left(\frac{l}{L_r}\right)^2\right] \mathrm{e}^{ikR/L_R}\ \mathrm{e}^{ilr/L_r}. \quad (4.23)$$

In practice, the wavepacket is represented by its values $\Phi_{mn}(t) \equiv \Phi(R_m, r_n; t)$ at the points (R_m, r_n) of a two-dimensional grid with interval sizes ΔR and Δr. The expansion coefficients a_{kl} are calculated by discrete Fourier transformation employing the Fast Fourier Transformation (FFT) algorithm. Application of the kinetic energy operator then amounts to multiplying the old coefficients by the term in brackets on the right-hand side of (4.23). The new values of the wavepacket at the grid points are subsequently calculated by the inverse Fourier transformation. This completes one step of the recursion in (4.20). The heart of the computer program is a FFT routine.

If the excited complex lives for a long time, the wavepacket will spread over a very wide range in coordinate-space. While one part of the wavefunction has already left the interaction zone and travels freely in the asymptotic region, the remaining part is still trapped at short intermolecular distances. This might cause technical problems because the number of grid points (R_m, r_n) becomes too large. If the autocorrelation function and the total spectrum are the only desired quantities, the outgoing flux may be "absorbed" by an imaginary potential wall placed in the asymptotic region in order to prevent spurious reflection of the wavepacket at the boundary of the grid. If, however, the complete wavepacket is required in order to extract the partial cross sections, $\Phi(t)$ may be split into an inner part $\Phi_<(R < \bar{R})$ and an outer part $\Phi_>(R > \bar{R})$ as suggested by Heather and Metiu (1987). $\bar{R}$ must be large enough to ensure that the interaction potential is zero. $\Phi_>$ can be propagated analytically while $\Phi_<$ must be propagated numerically as described above. Eventually, $\Phi_<$ becomes zero when the wavepacket has completely left the interaction zone (Engel and Metiu 1990; Untch, Weide, and Schinke 1991b).

The Fourier method is best suited to cartesian coordinates because the expansion functions $\mathrm{e}^{ikR/L_R}\ \mathrm{e}^{ilr/L_r}$ are just the eigenfunctions of the kinetic energy operator. For problems including the rotational degree of freedom other propagation methods have been developed (Mowrey, Sun, and Kouri 1989; Le Quéré and Leforestier 1990; Dateo, Engel, Almeida, and Metiu 1991; Dateo and Metiu 1991).

4.2.3 Time-dependent close-coupling

Expansion of the wavepacket in terms of time-independent basis functions, in the same way as described in Section 3.1, provides another approach to solve the time-dependent Schrödinger equation (Mowrey and Kouri 1985; Kouri and Mowrey 1987; Sun, Mowrey, and Kouri 1987; Weide, Kühl, and Schinke 1989; Gray and Wozny 1989, 1991; Untch, Weide, and Schinke 1991a,b; Manthe and Köppel 1991; Manthe, Köppel, and Cederbaum 1991). Let us consider, as an example, the two-dimensional model for rotational excitation (Section 3.2). The wavepacket is expanded as

$$\Phi(R,\gamma;t) = \frac{1}{R}\sum_j \chi_j(R;t)\, Y_{j0}(\gamma,0), \tag{4.24}$$

where, in contrast to (3.19), the expansion functions $\chi_j(R;t)$ depend on R and on t. Inserting (4.24) into (4.1) with Hamiltonian (3.16) and exploiting the orthogonality of the Y_{j0} yields the following set of close-coupling equations,

$$i\hbar\frac{\partial}{\partial t}\chi_j(R;t) = \left[-\frac{\hbar^2}{2m}\frac{\partial^2}{\partial R^2} + \left(\frac{1}{2\mu r_e^2} + \frac{1}{2mR^2}\right) j(j+1)\right]\chi_j(R;t) + \sum_{j'} V_{jj'}(R)\,\chi_{j'}(R;t). \tag{4.25}$$

Each rotational state is coupled to all other states through the potential matrix $\mathbf{V}$ defined in (3.22). Initial conditions $\chi_j(R;0)$ are obtained by expanding — in analogy to (3.26) — the ground-state wavefunction multiplied by the transition dipole function in terms of the Y_{j0}. The total of all one-dimensional wavepackets $\chi_j(R;t)$ forms an R- and t-dependent vector $\boldsymbol{\chi}$ whose propagation in space and time follows as described before for the two-dimensional wavepacket, with the exception that multiplication by the potential is replaced by a matrix multiplication $\mathbf{V}\boldsymbol{\chi}$. The close-coupling equations become computationally more convenient if one makes an additional transformation to the so-called *discrete variable representation* (Bačić and Light 1986). The autocorrelation function is simply calculated from

$$S(t) = \sum_j \langle \chi_j(0) \mid \chi_j(t)\rangle. \tag{4.26}$$

The description of a triatomic molecule requires three coordinates: the two linear coordinates R and r and the orientation angle γ. Treating the motion in R and r on a two-dimensional grid as in (4.22) and representing the angular motion by expansion into rotor eigenfunctions as described in (4.24) is one possible way of propagating the wavepacket for a triatomic molecule in three dimensions. This hybrid method yields a set of coupled two-dimensional partial differential equations which can be solved without serious problems (Untch, Weide, and Schinke 1991a,b; Guo 1991).

4.3 Approximations

The time-dependent approach is the starting point for several useful approximations, which are particularly important for systems with more than three degrees of freedom as well as for gaining physical insight into the fragmentation process. The following discussion is rather limited; the review article of Kosloff (1988), for example, contains an extensive list of appropriate references.

One of the main assets of the time-dependent theory is the possibility of treating some degrees of freedom quantum mechanically and others classically. Such composite methods necessarily lead to time-dependent Hamiltonians which obviously exclude time-independent approaches. We briefly outline three approximations that are frequently used in molecular dynamics studies. To be consistent with the previous sections we consider the collinear triatomic molecule ABC with Jacobi coordinates R and r.

4.3.1 Gaussian wavepackets

Gaussian wavepackets remain Gaussians if they are propagated in a potential containing at most quadratic terms (Feynman and Hibbs 1965: ch.3). This remarkable feature of a Gaussian wavepacket is the starting point for an approximate propagation scheme introduced by Heller (1975, 1976). In most photodissociation studies the parent molecule is in its lowest vibrational state before it is promoted to the excited electronic state. Therefore, it is reasonable to presume that the initial wavepacket at $t = 0$ is a multi-dimensional Gaussian. If the (exact) wavepacket remains localized during its journey from the excitation region out into the asymptotic region (see Figure 4.2 for a simple example), it might be a good approximation to expand the true potential locally in a Taylor series up to second order around the center of the wavepacket. This ensures that the wavepacket remains a Gaussian for all times. The center of the wavepacket, its width, and its phase are then functions of time.

The general form of a Gaussian wavepacket is given by (Heller 1975; Child 1991:ch.11)

$$\begin{aligned}\Phi(R,r;t) = \exp\Big\{\frac{i}{\hbar}\,[\alpha_{RR}\,(R-\bar{R})^2 + \alpha_{rr}\,(r-\bar{r})^2 \\ + \alpha_{Rr}\,(R-\bar{R})\,(r-\bar{r}) + \bar{P}\,(R-\bar{R}) + \bar{p}\,(r-\bar{r}) + \alpha_0]\Big\}, \quad (4.27)\end{aligned}$$

where $(\bar{R},\bar{r})$ is its center in coordinate-space, and $(\bar{P},\bar{p})$ specifies the center in the momentum-space. Note that all parameters in (4.27) are functions of time. The potential energy surface $V(R,r)$ is *locally* approximated by

$$V(R,r) \approx V_0 + V_R\,(R-\bar{R}) + V_r\,(r-\bar{r}) + \frac{1}{2}V_{RR}\,(R-\bar{R})^2 + \frac{1}{2}V_{rr}\,(r-\bar{r})^2 + V_{Rr}\,(R-\bar{R})\,(r-\bar{r}), \quad (4.28)$$

where $V_0 = V(\bar{R},\bar{r})$, $V_R = \partial V/\partial R$ evaluated at $(\bar{R},\bar{r})$, etc. Inserting (4.27) into the time-dependent Schrödinger equation with $V(R,r)$ approximated by (4.28) and comparing coefficients of like powers in $(R-\bar{R})$ and $(r-\bar{r})$, one finds coupled differential equations for the time-dependent parameters. Here, we list only those for $\bar{R},\bar{r},\bar{P}$, and $\bar{p}$,

$$\begin{aligned} \frac{d}{dt}\bar{R} &= \frac{\bar{P}}{m} \quad , \quad \frac{d}{dt}\bar{r} = \frac{\bar{p}}{\mu} \\ \frac{d}{dt}\bar{P} &= -V_R \quad , \quad \frac{d}{dt}\bar{p} = -V_r, \end{aligned} \quad (4.29)$$

which resemble Hamilton's equations in classical mechanics (see Chapter 5).

The propagation of the wavepacket is thereby reduced to the solution of coupled first-order differential equations for the parameters representing the Gaussian wavepacket, with the true potential being expanded about the instantaneous center of the wavepacket $[\bar{R}(t),\bar{r}(t)]$. This propagation scheme is very appealing and efficient provided the basic assumptions are fulfilled. The essential prerequisite is that the locally quadratic approximation of the PES is valid over the spread of the wavepacket. This rules out bifurcation of the wavepacket, resonance effects, or strong anharmonicities.

Nevertheless, this simple propagation method provides an intriguing picture of the evolution of the quantum mechanical wavepacket, at least for short times. It readily demonstrates that for short times the center of the wavepacket follows essentially a classical trajectory (Ehrenfest's theorem, Cohen-Tannoudji, Diu, and Laloë 1977:ch.III). Figure 4.2 depicts an example; the evolution of the two-dimensional wavepacket follows very closely the classical trajectory that starts initially with zero momenta at the Franck-Condon point.

The Gaussian wavepacket approach works best for direct and fast dissociation (Lee and Heller 1982); several extensions and improvements of the original idea have been proposed (Coalson and Karplus 1982; Skodje 1984; Sawada, Heather, Jackson, and Metiu 1985; Heather and Metiu 1986; Huber and Heller 1988; Henriksen and Heller 1988, 1989; Kučar and Meyer 1989; Huber, Ling, Imre, and Heller 1989).

4.3.2 Time-dependent SCF

Another, purely quantum mechanical approximation is the so-called time-dependent self-consistent field (TDSCF) method. For general reviews see Kerman and Koonin (1976), Goeke and Reinhard (1982), and Negele (1982). For applications to molecular systems see, for example, Gerber and Ratner (1988a,b). In the TDSCF method the wavepacket is separated according to

$$\Phi(R, r; t) = \xi_R(R; t)\, \xi_r(r; t) \tag{4.30}$$

with the normalization $\langle \xi_R \mid \xi_R \rangle = \langle \xi_r \mid \xi_r \rangle = 1$ for all times. The Hamiltonian (2.39) for the two-dimensional problem is written as

$$\hat{H}(R, r) = \hat{T}_R(R) + \hat{T}_r(r) + V(R, r), \tag{4.31}$$

with $\hat{T}_R$ and $\hat{T}_r$ being defined as

$$\hat{T}_R = -\frac{\hbar^2}{2m} \frac{\partial^2}{\partial R^2} \quad , \quad \hat{T}_r = -\frac{\hbar^2}{2\mu} \frac{\partial^2}{\partial r^2}. \tag{4.32}$$

Inserting (4.30) into the time-dependent Schrödinger equation and multiplication with $\langle \xi_r |$ yields

$$i\hbar \frac{\partial \xi_R(R; t)}{\partial t} = \hat{H}_R^{SCF}(R; t)\, \xi_R(R; t), \tag{4.33}$$

where we have used that $\partial \langle \xi_r \,|\, \xi_r \rangle / \partial t = 0$. The effective time-dependent or mean field Hamiltonian is defined by

$$\hat{H}_R^{SCF}(R; t) = \hat{T}_R(R) + \langle \xi_r \mid \hat{T}_r(r) + V(R, r) \mid \xi_r \rangle. \tag{4.34}$$

Similar equations hold for $\xi_r(r; t)$ and $\hat{H}_r^{SCF}(r; t)$. In this way the single partial differential equation in R and r is split into two ordinary differential equations, one for each degree of freedom.

The one-dimensional Schrödinger equations for $\xi_R(R; t)$ and $\xi_r(r; t)$ are coupled to each other and therefore they must be integrated simultaneously in time. The evolution of ξ_R, for example, depends on ξ_r and *vice versa*. Because of the coupling, energy can flow between the two degrees of freedom. However, although the two wavepackets are coupled, much of the correlation between the R and the r motions is discarded by the separable ansatz in Equation (4.30). In contrast to the TDSCF approximation, the close-coupling expansion (4.24) retains the full correlation between the two degrees of freedom for all times. The TDSCF approach is therefore best suited for systems with weak coupling between the modes as in the dissociation of weakly bound van der Waals molecules, for example. The TDSCF method can easily be extended to systems with several (more than three) degrees of freedom which cannot be treated by exact approaches. While in an exact method the number of grid points or the number of expansion functions rapidly grows with each additional degree of freedom, the TDSCF equations are augmented by only one extra one-dimensional Schrödinger equation.

Several improvements of the TDSCF approach have been proposed in the recent literature (Kučar, Meyer, and Cederbaum 1987; Makri and Miller 1987; Meyer, Kučar, and Cederbaum 1988; Kotler, Nitzan, and Kosloff 1988; Meyer, Manthe, and Cederbaum 1990; Campos-Martínez and Coalson 1990; Waldeck, Campos-Martínez, and Coalson 1991).

4.3.3 Classical path methods

One of the main assets of the time-dependent theory is the possibility of combining quantum and classical mechanics. In many problems some degrees of freedom behave classically while others require a fully quantum mechanical treatment. If, for example, the de Broglie wavelength associated with the dissociation coordinate, $2\pi\hbar/(2mE)^{1/2}$, is small the translational motion may be well described by an average classical trajectory. The vibrational or rotational degree of freedom of the fragment molecule, on the other hand, is quantized and therefore may need a quantum mechanical description, especially if only few quantum states are populated (Child 1976; Tully 1976; Billing 1984; Swaminathan, Stodden, and Micha 1989).

The TDSCF approximation is a good starting point for a mixed quantum mechanical/classical treatment. Let us assume that R is the classical and r the quantum mechanical mode. Then, the wavefunction $\xi_r(r;t)$ describing the vibration of the fragment molecule is a solution of the time-dependent Schrödinger equation

$$i\hbar\frac{\partial\xi_r(r;t)}{\partial t} = \{\hat{h}_{vib}(r) + \langle\xi_R \mid \hat{T}_R(R) \mid \xi_R\rangle + \langle\xi_R \mid V_I(R,r) \mid \xi_R\rangle\}\, \xi_r(r;t), \tag{4.35}$$

where $\hat{h}_{vib}(r)$ is the vibrational Hamiltonian (2.40) and $V_I(R,r)$ is the interaction potential (2.41). Treating the motion in R classically amounts to replacing $\langle\xi_R \mid \hat{T}_R(R) \mid \xi_R\rangle$ by the kinetic energy $P^2/2m$, where P is the classical momentum corresponding to R, and replacing $\langle\xi_R \mid V_I \mid \xi_R\rangle$ by $V_I(\bar{R},r)$ where $\bar{R}(t)$ is the position of the "classical particle". Equation (4.35) therefore reduces to

$$i\hbar\frac{\partial\xi_r(r;t)}{\partial t} = \{h_{vib}(r) + V_I[\bar{R}(t),r]\}\, \xi_r(r;t), \tag{4.36}$$

where we have neglected the kinetic energy $P^2/2m$ because it yields only an unimportant phase factor. The trajectory $\bar{R}(t)$ is determined by the classical path Hamiltonian (CP)

$$H_R^{CP}(R;t) = \frac{\bar{P}_R^2(t)}{2m} + \langle\xi_r(t) \mid \hat{h}_{vib}(r) + V_I[\bar{R}(t),r] \mid \xi_r(t)\rangle. \tag{4.37}$$

Equation (4.37) is identical to (4.34) with the exception that the kinetic energy operator $\hat{T}_R$ is replaced by its classical analogue.

Expanding $\xi_r(r;t)$ in terms of vibrational states $\varphi_n(r)$ of the free oscillator with energies ϵ_n,

$$\xi_r(r;t) = \sum_n a_n(t)\, \varphi_n(r)\, \mathrm{e}^{-i\epsilon_n t/\hbar}, \tag{4.38}$$

yields a set of coupled first-order differential equations for the expansion coefficients $a_n(t)$,

$$i\hbar \frac{d}{dt} a_n(t) = \sum_{n'} a_{n'}(t)\, \langle \varphi_n \mid V_I[\bar{R}(t), r] \mid \varphi_{n'} \rangle\, \mathrm{e}^{i(\epsilon_n - \epsilon_{n'})t/\hbar} \tag{4.39}$$

which is a special example of (2.10). Equation (4.39) describes the evolution of the fragment under the influence of the external perturbation $V_I[\bar{R}(t), r]$ which depends on time through the average trajectory (the classical path) $\bar{R}(t)$.

The combination of quantum mechanical and classical modes is particularly important for larger systems which prohibit a complete quantal treatment. It is most suitable for direct processes with short interaction times and it is less applicable for long-lived intermediate complexes. The ultimate step in the hierarchy of time-dependent approximations is a complete classical treatment of *all* degrees of freedom. This is the topic of the next chapter.

4.4 A critical comparison

The time-independent and the time-dependent approaches are completely equivalent. Equation (4.11) documents this correspondence in the clearest way: the time-dependent wavepacket $\Phi_f(t)$, which contains the stationary states for *all* energies, and the time-independent wavefunction $\Psi_{tot}(E_f)$, which embraces the entire "history" of the fragmentation process, are related to each other by a Fourier transformation between the time and the energy domains,

$$\int_{-\infty}^{+\infty} dt\, \mathrm{e}^{iE_f t/\hbar}\, \Phi_f(t) = \Psi_{tot}(E_f).$$

Both approaches rest upon the Golden Rule expression for the photodissociation cross section and comprise the same basic assumption, namely the weak interaction between the light pulse on one hand and the molecule on the other hand (Henriksen 1988).

The time-independent and time-dependent approaches merely provide different views of the dissociation process and different numerical tools for the calculation of photodissociation cross sections. The time-independent approach is a *boundary value problem*, i.e., the stationary wavefunction must satisfy special conditions in the limits $R \to 0$ and $R \to \infty$. From our own experience we know that it is difficult to visualize the behavior of multi-dimensional wavefunctions at large intermolecular separations and that is certainly one major drawback of the time-independent theory. The time-dependent approach, on the other hand, is an *initial value problem* and as such it agrees better with our vision of full or half collisions. One starts with a well defined initial wavepacket at $t = 0$ and follows its evolution in time and space. The evolving wavepacket can be qualitatively interpreted as a swarm of classical trajectories.

Which approach is more advantageous for numerical applications? The time-dependent wavepacket propagation is easier to implement, especially if one uses the Fourier expansion of the wavepacket. Systems which cannot be described by a global expansion basis are difficult to treat by the close-coupling technique outlined in Sections 3.1 and 3.2 because the basis functions must be frequently changed along the dissociation path. On the other hand, such systems can be straightforwardly treated by two- or even three-dimensional grid propagation methods, because the wavepacket is represented by its discrete values on a grid, rather than an expansion in terms of basis functions. This is particularly important for the dissociation of symmetric molecules like H_2O or CO_2 with more than one product channel. In fairness we must emphasize, however, that there are also methods for the calculation of time-independent wavefunctions which do not employ an expansion in terms of basis functions (see, for example, Jaquet 1987 and Zhang and Miller 1990). But even such methods are somewhat more complicated than the propagation of a wavepacket, because one has to construct wavefunctions which fulfil the proper boundary conditions.

Solving the time-independent Schrödinger equation for a single energy is certainly faster than solving the time-dependent Schrödinger equation which includes the time as an additional variable. On the other hand, the time-dependent wavepacket embraces *all* energies and thus the entire absorption spectrum as well as all partial cross sections for any energy can be extracted from *one* solution of the time-dependent Schrödinger equation, while the time-independent Schrödinger equation must be solved for each energy separately. The total computer time in the time-independent approach scales linearly with the number of energies necessary to survey the spectrum. In the time-dependent approach, on the other hand, the computer time is proportional to the elapsed time until the wavepacket has completely reached the asymptotic region.

If the fragmentation is direct and fast the absorption spectrum is structureless like the example shown in Figure 3.5. Under such circumstances calculations for a few energies suffice to specify the spectrum fully. By the same token, the wavepacket rapidly leaves its origin, the autocorrelation function decays to zero, and the dissociation is over in a very short time (Figures 4.2 and 4.3). Therefore, the time-independent as well as the time-dependent calculations are straightforward and usually do not cause severe technical problems.

However, if the fragmentation is indirect, the absorption spectrum exhibits narrow resonance structures and we have to calculate cross sections for many energies in order to resolve the finer details and the widths of the individual absorption lines. Although the computer time for solving the time-independent Schrödinger equation is basically the same for each energy, irrespective of whether the dissociation is direct or indirect, the total computer time drastically increases in comparison to direct processes. In the time-dependent approach, on the other hand, the wavepacket is trapped for a long time in the interaction region with only a small fraction leaking out per unit time interval. The energy resolution ΔE and the propagation time ΔT are related by the time-energy uncertainty relation $\Delta E \, \Delta T \geq h$ (Cohen-Tannoudji, Diu, and Laloë 1977:ch.III) which has the consequence that a better and better energy resolution requires an ever increasing propagation time. This may cause technical problems because the accuracy gradually diminishes with time and in the end numerical errors may become unacceptably large. Thus, in the case of indirect dissociation both the time-independent and the time-dependent approaches become very time consuming.

In the time-independent approach one has to calculate *all* partial cross sections before the total cross section can be evaluated. The partial photodissociation cross sections contain all the desired information and the total cross section can be considered as a less interesting by-product. In the time-dependent approach, on the other hand, one usually first calculates the absorption spectrum by means of the Fourier transformation of the autocorrelation function. The final state distributions for any energy are, in principle, contained in the wavepacket and can be extracted if desired. The time-independent theory favors the state-resolved partial cross sections whereas the time-dependent theory emphasizes the spectrum, i.e., the total absorption cross section. If the spectrum is the main observable, the time-dependent technique is certainly the method of choice.

Which approach offers the better interpretation of the photodissociation dynamics in general and of the dissociation cross sections in particular? The answer to this question depends very much on the particular system and cannot be decided in a unique way. There are cases which are better described by the time-independent picture while other systems may be better understood within the time-dependent view. In the course of this monograph we will always attempt to combine both views in order to reveal all facets of a particular system.

5
Classical description of photodissociation

Our general understanding of molecular collisions and energy transfer rests mainly upon classical mechanics. Except for particular quantum mechanical processes, such as transitions between different electronic states, and quantum mechanical features like resonances or interferences classical mechanics is a very useful tool for the study of molecular encounters (Porter and Raff 1976; Pattengill 1979; Truhlar and Muckerman 1979; Schatz 1983; Raff and Thompson 1985; Levine and Bernstein 1987:ch.4). This holds true for photodissociation as well (Goursaud, Sizun, and Fiquet-Fayard 1976; Heller 1978a; Brown and Heller 1981; Schinke 1986c; Goldfield, Houston, and Ezra 1986; Schinke 1988b; Guo and Murrell 1988a,b).

Classical mechanics is the limit of quantum mechanics as the de Broglie wavelength $\lambda_B = 2\pi\hbar/(2mE)^{1/2}$ becomes small. The total energy released as translational and internal energy in UV photodissociation often exceeds 1 eV and therefore λ_B is of the order of 0.1 Å or shorter. On the other hand, the range of the potential is typically much larger so that the quantum mechanical wavefunction performs many oscillations over the entire interaction region (see Figures 2.3 and 3.2, for example). Furthermore, in many cases the fragments are produced with high internal excitation (Figure 3.3) which additionally favors a classical description.

The classical picture of photodissociation closely resembles the time-dependent view. The electronic transition from the ground to the excited electronic state is assumed to take place *instantaneously* so that the internal coordinates and corresponding momenta of the parent molecule remain unchanged during the excitation step (vertical transition). After the molecule is promoted to the potential energy surface (PES) of the upper electronic state it starts to move subject to the classical equations of motion

$$\frac{d}{dt}Q_i = \frac{\partial H}{\partial P_i} \quad , \quad \frac{d}{dt}P_i = -\frac{\partial H}{\partial Q_i}, \tag{5.1}$$

where $H(\mathbf{Q}, \mathbf{P})$ is the *Hamilton function*, i.e., the classical analogue of the Hamilton operator $\hat{H}$ in the excited state. The vector $\mathbf{Q}(t)$ comprises all nuclear coordinates and the vector $\mathbf{P}(t)$ represents the corresponding momenta (Goldstein 1951:ch.7; Daudel et al. 1983:ch.7; Cohen-Tannoudji, Diu, and Laloë 1977:appendix III).

The classical approach consists of three parts: the solution of the equations of motion (5.1) in the excited electronic state with initial values $\mathbf{Q}_0$ and $\mathbf{P}_0$; the weighting of each set of initial values according to the distributions of coordinates and momenta in the electronic ground state *before* the photon is absorbed; and finally the calculation of photodissociation cross sections by averaging over the phase-space. These three points will be discussed in Sections 5.1–5.3. Examples which demonstrate the usefulness and reliability of classical mechanics are presented in Section 5.4.

5.1 Equations of motion, trajectories, and excitation functions

In order to keep the expressions transparent we, once again, restrict the discussion to the dissociation of the triatomic molecule ABC into products A and BC. Furthermore, the total angular momentum is limited to $\mathbf{J} = 0$. In this chapter we consider the vibration and the rotation of the fragment molecule simultaneously. The corresponding Hamilton function, i.e., the total energy as a function of all coordinates and momenta, using action-angle variables (McCurdy and Miller 1977; Smith 1986), reads

$$H = \frac{P^2}{2m} + \frac{p^2}{2\mu} + \frac{\mathbf{j}^2}{2\mu r^2} + \frac{\mathbf{l}^2}{2mR^2} + V(R, r, \gamma). \tag{5.2}$$

Equation (5.2) is a combination of the two two-dimensional Hamiltonians (2.39) and (3.15) which describe the vibrational and rotational excitations of BC separately. The Jacobi coordinates R, r, and γ are defined in Figures 2.1 and 3.1 and P and p denote the linear momenta corresponding to R and r, respectively. $\mathbf{j}$ is the classical angular momentum vector of BC and $\mathbf{l}$ stands for the classical orbital angular momentum vector describing the rotation of A with respect to BC. For zero total angular momentum $\mathbf{J} = \mathbf{j} + \mathbf{l} = 0$ we have $\mathbf{l} = -\mathbf{j}$ and the Hamilton function reduces to

$$H = \frac{P^2}{2m} + \frac{p^2}{2\mu} + \left(\frac{1}{2\mu r^2} + \frac{1}{2mR^2}\right) j^2 + V(R, r, \gamma), \tag{5.3}$$

where $j = |\mathbf{j}|$.[†] The classical angular momentum is a continuous function of time having units of action.

At each instant t the classical system is uniquely specified by the vector $\boldsymbol{\tau}(\mathrm{t}) = \{R(t), r(t), \gamma(t), P(t), p(t), j(t)\}$ in the six-dimensional phase-space. The total of all coordinates and momenta as a function of time is called a *classical trajectory*. The evolution of the trajectory $\boldsymbol{\tau}(t)$ is determined by the *Hamilton equations* (Goldstein 1951:ch.7; Arnold 1978)

$$\frac{dR}{dt} = \frac{\partial H}{\partial P} = \frac{P}{m} \tag{5.4a}$$

$$\frac{dr}{dt} = \frac{\partial H}{\partial p} = \frac{p}{\mu} \tag{5.4b}$$

$$\frac{d\gamma}{dt} = \frac{\partial H}{\partial j} = 2\left(\frac{1}{2mR^2} + \frac{1}{2\mu r^2}\right) j \tag{5.4c}$$

$$\frac{dP}{dt} = -\frac{\partial H}{\partial R} = \frac{j^2}{mR^3} - \frac{\partial V_I}{\partial R} \tag{5.4d}$$

$$\frac{dp}{dt} = -\frac{\partial H}{\partial r} = \frac{j^2}{\mu r^3} - \frac{\partial v_{BC}}{\partial r} - \frac{\partial V_I}{\partial r} \tag{5.4e}$$

$$\frac{dj}{dt} = -\frac{\partial H}{\partial \gamma} = -\frac{\partial V_I}{\partial \gamma}, \tag{5.4f}$$

where $V_I(R, r, \gamma) = V(R, r, \gamma) - v_{\mathrm{BC}}(r)$ is the interaction potential defined in (2.41).[‡] Each trajectory is uniquely determined by its initial values $\tau_0 = \{R_0, r_0, \gamma_0, P_0, p_0, j_0\}$ at $t = 0$. Once τ_0 is specified, the fate of the system is fully defined for all subsequent times.

Hamilton's equations form a set of coupled first-order differential equations which under normal conditions can be numerically integrated without any problems. The forces $-\partial V_I/\partial R$ and $-\partial V_I/\partial r$ and the torque $-\partial V_I/\partial \gamma$, which reflect the coordinate dependence of the interaction potential, control the coupling between the translational (R, P), the vibrational (r, p), and the rotational (γ, j) degrees of freedom . Due to this coupling energy can flow between the various modes. The translational mode becomes decoupled from the internal motion of the diatomic fragment (i.e., $dP/dt = 0$ and dR/dt =constant) when the interaction potential diminishes in the limit $R \to \infty$. As a consequence, the translational energy

$$E_{trans} = \frac{P^2}{2m} \tag{5.5}$$

[†] Since at all instants the angular momentum vector $\mathbf{j}$ is perpendicular to the collision plane defined by the three atoms, it suffices to consider only its magnitude. Positive and negative values indicate rotation in opposite directions.

[‡] Note, that the equations of motion (5.4) are valid for $\mathbf{J} = 0$ only; for the general case $\mathbf{J} \neq 0$ see Smith (1986), for example.

as well as the internal energy of the molecular fragment,

$$E_{intern} = \frac{p^2}{2\mu} + \frac{j^2}{2\mu r^2} + v_{BC}(r), \tag{5.6}$$

become constant. The total energy given by $E = E_{trans} + E_{intern}$ is a constant of motion, because the Hamilton function is independent of time.

A typical example of a classical trajectory is depicted in Figure 5.1 which shows $R(t)$, $\gamma(t)$, and $j(t)$ for the system ClCN in the rigid-rotor limit. The corresponding PES is shown in Figure 3.2. The dissociation of ClCN is an example of direct dissociation, i.e., Cl and CN strongly repel each other with the consequence that the Cl-CN bond distance $R(t)$ immediately increases. At the same time, the recoiling Cl atom imparts a strong torque to CN so that the diatom begins to rotate rapidly. Therefore, the orientation angle $\gamma(t)$ as well as the rotational angular momentum $j(t)$ of CN increase with time. Note that as a consequence

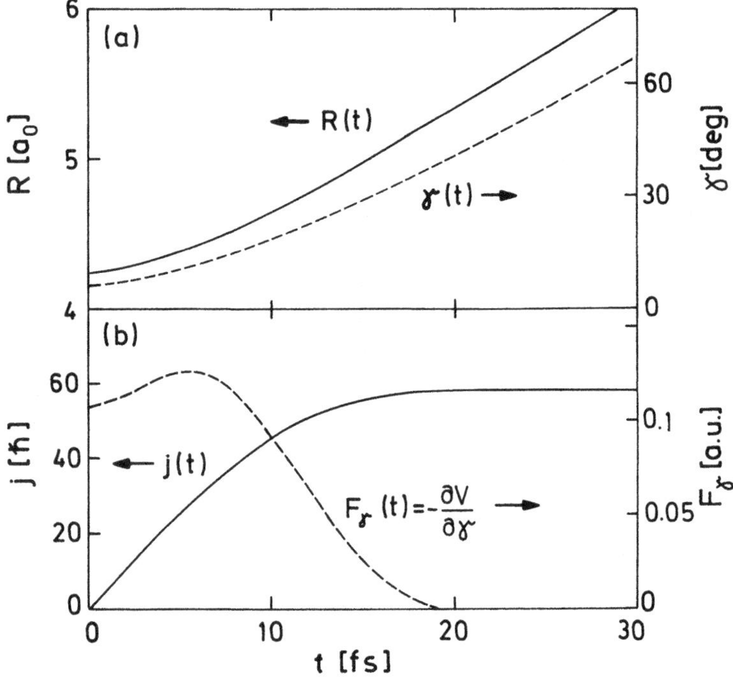

Fig. 5.1. Time dependence of a typical classical trajectory in the photodissociation of ClCN → Cl + CN(j). (a) $R(t)$ (solid line) is the Cl-CN separation and $\gamma(t)$ (dashed line) is the orientation angle of CN with respect to the scattering vector **R** (for the definition of coordinates see Figure 3.1). (b) The angular momentum $j(t)$ of CN (solid line) and the torque $F_\gamma = -\partial V/\partial\gamma$ (dashed line).

of Equation (5.4c) $\gamma(t)$ and $j(t)$ are interrelated according to $d\gamma(t)/dt \propto j(t)$. After about 20 fs the two fragments have separated so far that the interaction between them is practically zero with the consequence that $j(t)$ becomes constant and the angle $\gamma(t)$ increases linearly in time.

Figure 3.2 shows the same trajectory as in Figure 5.1 in a different representation together with the time-independent total dissociation wavefunction $\Psi_{tot}(R,\gamma)$. The energy for the trajectory is the same as for the wavefunction. It comes as no surprise that, on the average, the wavefunction closely follows the classical trajectory. Another example of this remarkable coincidence between quantum mechanical wavefunctions on one hand and classical trajectories on the other hand is shown in Figure 4.2. There, the time-dependent wavepacket $\Phi(t)$ follows essentially the trajectory which starts at the equilibrium geometry of the ground electronic state. This is not unexpected in view of Equations (4.29) which state that the parameters of the center of the wavepacket obey the same equations of motion as the classical trajectory provided anharmonic effects are small. Figures 3.2 and 4.2 elucidate the correspondence between classical and quantum mechanics, at least for short times and fast evolving systems:

- Classical trajectories are the "backbones" for the quantum mechanical wavefunctions (Ehrenfest's theorem). If the dissociation is direct, a single trajectory, which starts near the equilibrium of the parent molecule, illustrates in a clear way the overall fragmentation mechanism.

The force $-\partial V_I/\partial r$ and the torque $-\partial V_I/\partial\gamma$ control the internal excitation of the fragments. According to Equation (5.4f), the rate of change of the angular momentum $j(t)$ is directly proportional to the torque $F_\gamma = -\partial V_I/\partial\gamma$ which in turn reflects the anisotropy of the excited-state PES. Figure 5.1(b) illustrates the close relation between rotational excitation on one hand and the torque $F\gamma(t)$ on the other hand. Initially, the diatom is at rest, i.e., $j(0) = 0$. However, immediately after the trajectory is launched on the upper PES the torque strongly excites the rotor. F_γ is largest at short intermolecular separations and decreases rapidly as the fragments separate because the interaction potential becomes gradually more isotropic with increasing Cl-CN distance. The final rotational angular momentum of the product molecule is relatively high in this example; it is a measure of the net torque along the dissociation path imparted by the recoiling Cl atom.

In order to calculate final rotational state distributions it is useful to define the so-called rotational excitation function (McCurdy and Miller 1977; Schinke and Bowman 1983; Schinke 1986a,c, 1988a,b)

$$J(\boldsymbol{\tau}_0) \equiv \lim_{t\to\infty} j(t;\boldsymbol{\tau}_0)/\hbar. \tag{5.7}$$

$J(\tau_0)$ is the classical counterpart of the rotational quantum number j of the fragment molecule. (Note that $J(\tau_0)$ is defined as a dimensionless quantity.) It represents the final angular momentum of the fragment molecule as a function of all initial variables τ_0. In the same way, we define the vibrational excitation function (Miller 1974, 1975, 1985)

$$N(\tau_0) \equiv \lim_{t\to\infty} n(t;\tau_0), \tag{5.8}$$

where $n(t;\tau_0)$ is the classical counterpart of the vibrational quantum number. Since $n(t)$ is not an independent variable in the equations of motion (5.4), it must be extracted by inverting the quantum mechanical relation between the vibrational energy E_{vib}, which can be easily calculated at the end of each trajectory, and the vibrational quantum number n,

$$E_{vib} = \hbar\omega_{HO}\left(n+\frac{1}{2}\right) \tag{5.9a}$$

for a harmonic oscillator with frequency ω_{HO} and

$$E_{vib} = \hbar\left(\frac{2D\beta^2}{\mu}\right)^{1/2}\left(n+\frac{1}{2}\right) - \hbar^2\frac{\beta^2}{2\mu}\left(n+\frac{1}{2}\right)^2 \tag{5.9b}$$

for a Morse oscillator with dissociation energy D, exponent β, and reduced mass μ. Examples of excitation functions will be shown in Chapter 6. $J(\tau_0)$ and $N(\tau_0)$ are *the* keystones for the interpretation of final rotational and vibrational state distributions and their correlation with the torque $-\partial V_I/\partial\gamma$ and the force $-\partial V_I/\partial r$.

5.2 Phase-space distribution function

Absorption and photodissociation cross sections are calculated within the classical approach by running swarms of individual trajectories on the excited-state PES. Each trajectory contributes to the cross section with a particular weight $P^{(i)}(\tau_0)$ which represents the distribution of all coordinates and all momenta *before* the vertical transition from the ground to the excited electronic state. $P^{(i)}(\tau_0)$ should be a state-specific, quantum mechanical distribution function which reflects, as closely as possible, the initial quantum state (indicated by the superscript i) of the parent molecule before the electronic excitation. The theory pursued in this chapter is actually a hybrid of quantum and classical mechanics: the parent molecule in the electronic ground state is treated quantum mechanically while the dynamics in the dissociative state is described by classical mechanics.

In quantum mechanics, the coordinate distribution is given by the square of the modulus of the wavefunction in coordinate-space, $|\Psi_i(E_i)|^2$.

Alternatively, we can work in momentum-space with the momentum distribution given by the square of the modulus of the momentum wavefunction. However, because of Heisenberg's uncertainty relation it is impossible to specify uniquely the coordinates *and* the momenta simultaneously. Either the coordinates or the momenta can be defined without uncertainty. In classical mechanics, on the other hand, the coordinates as well as the momenta are simultaneously measurable at each instant. In particular, both the coordinates and the momenta must be specified at $t = 0$ in order to start the trajectory. Thus, we have the problem of defining a distribution function in the classical phase-space which simultaneously weights coordinates and momenta and which, at the same time, should mimic the quantum mechanical distributions as closely as possible.

This dilemma is common to all theories that combine quantum and classical mechanics; however, it is particularly severe in photodissociation. Because the motion in the excited state begins at short distances, where the interaction between the various modes is usually strongest, a proper definition of the distribution function $P^{(i)}(\boldsymbol{\tau_0})$ is very crucial. A slight variation of the initial distribution function might lead to significant changes in the final state distributions, for example. This is different from full collisions where the reactants are completely separated at the beginning of the trajectory so that slight variations in the initial conditions are probably "forgotten" by the time they collide.

The quantum mechanical definition of a distribution function in the classical phase-space is an old theme in theoretical physics. Most frequently used is the so-called *Wigner distribution function* (Wigner 1932; Hillery, O'Connell, Scully, and Wigner 1984). Let us consider a one-dimensional system with coordinate R and corresponding classical momentum P. The Wigner distribution function is defined as

$$P_W(R,P) = (\pi\hbar)^{-1} \int d\eta \; \psi^*(R+\eta) \; \psi(R-\eta) \; e^{2i\eta P/\hbar}, \tag{5.10}$$

where $\psi(R)$ is the wavefunction in coordinate-space. It satisfies the conditions

$$\int dP \; P_W(R,P) = |\psi(R)|^2 \tag{5.11}$$

$$\int dR \int dP \; P_W(R,P) = 1, \tag{5.12}$$

which one intuitively expects for an acceptable phase-space distribution function. However, $P_W(R,P)$ is not necessarily positive definite, especially for excited vibrational states and therefore it cannot be interpreted as a probability distribution function.

For a harmonic oscillator in the nth vibrational state with mass m and frequency ω_{HO} the integral in (5.10) can be evaluated analytically (e.g.,

Dahl 1983),

$$P_W^{(n)}(R,P) = (-1)^n(\pi\hbar n!)^{-1}\, L_n(\rho^2)\, \mathrm{e}^{-\rho^2/2}, \tag{5.13}$$

where L_n is a Laguerre polynomial of order n and ρ^2 is defined by

$$\rho^2 = \frac{4}{\hbar\omega_{HO}}\left[\frac{P^2}{2m} + \frac{1}{2}m\,\omega_{HO}^2(R-R_e)^2\right]. \tag{5.14}$$

For the vibrational ground state ($n = 0$) Equation (5.13) reduces to

$$P_W^{(0)}(R,P) = (\pi\hbar)^{-1}\, \mathrm{e}^{-2\alpha(R-R_e)^2/\hbar}\, \mathrm{e}^{-P^2/(2\alpha\hbar)} \tag{5.15}$$

with the definition

$$\alpha = \frac{m\,\omega_{HO}}{2}. \tag{5.16}$$

For the subsequent discussion we emphasize that:

- The Wigner distribution function for the vibrational ground state of the harmonic oscillator is the product of two Gaussians, one Gaussian in R-space centered at the equilibrium distance R_e and one Gaussian in P-space localized at $P = 0$.

The widths of the coordinate and the momentum distribution are inversely related to each other as required by the uncertainty relation. The wider the coordinate distribution the narrower is the momentum distribution and *vice versa.* Figure 5.2 depicts two examples.

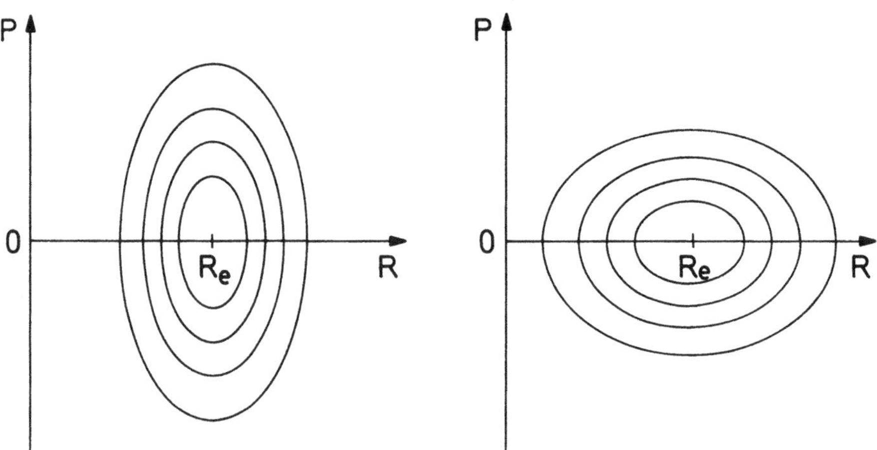

Fig. 5.2. Contour plots of two representative Wigner distribution functions $P_W(R,P)$ for two harmonic oscillators in their ground vibrational states, Equation (5.15), in the two-dimensional phase-space (R,P). The widths in the R- and in the P-directions are inversely related.

It is interesting to note that Equation (5.15) can also be written as

$$P_W^{(0)}(R,P) = |\psi_{n=0}(R)|^2 \, |\xi_{n=0}(P)|^2, \tag{5.17}$$

where $\xi_n(P)$ is the wavefunction in momentum-space corresponding to $\psi_n(R)$. Equation (5.17) is only valid for the lowest vibrational state! However, guided by the general form of Equation (5.17) we can define an alternative to the Wigner distribution function, which is applicable to excited vibrational states as well as unharmonic oscillators, i.e., $|\psi_n(R)|^2 \, |\xi_n(P)|^2$. This distribution function obviously fulfils conditions (5.11) and (5.12) and it is positive definite. However, the coordinate and the momentum are not correlated.

If the parent molecule is described by normal modes with coordinates q_k and momenta p_k, the multi-dimensional wavefunction is simply a product of uncoupled one-dimensional harmonic wavefunctions $\psi_k^{(n_k)}(q_k)$ (Wilson, Decius, and Cross 1955:ch.2; Weissbluth 1978:ch.27) and the corresponding Wigner distribution function reads

$$P_W(q_1, p_1, \ldots, q_N, p_N) = \prod_{k=1}^{N} P_W^{(n_k)}(q_k, p_k) \tag{5.18}$$

with the one-dimensional distribution functions defined in (5.13). Since the equations of motion for the fragmentation in the excited state are most conveniently solved in terms of Jacobi coordinates rather than normal coordinates, a complicated transformation becomes compulsory.

In numerous applications, however, we have found it sufficiently accurate, at least for the vibrational ground state of the parent molecule, to assume a separable wavefunction directly in terms of Jacobi coordinates,

$$\Psi(R, r, \gamma) = \varphi_R(R) \, \varphi_r(r) \, \varphi_\gamma(\gamma). \tag{5.19}$$

The six-dimensional Wigner distribution function for the ABC molecule in its lowest state is then a product of six Gaussians,

$$\begin{aligned} P_W^{(000)}(R, r, \gamma, P, p, j) = (\pi\hbar)^{-3} \, & \mathrm{e}^{-2\alpha_R(R-R_e)^2/\hbar} \, \mathrm{e}^{-P^2/(2\alpha_R\hbar)} \\ & \times \mathrm{e}^{-2\alpha_r(r-r_e)^2/\hbar} \, \mathrm{e}^{-p^2/(2\alpha_r\hbar)} \\ & \times \mathrm{e}^{-2\alpha_\gamma(\gamma-\gamma_e)^2/\hbar} \, \mathrm{e}^{-j^2/(2\alpha_\gamma\hbar)}, \end{aligned} \tag{5.20}$$

where the exponents α_R, α_r, and α_γ are approximately related to the three fundamental frequencies in analogy to (5.16).[†] The trajectory that

[†] In Equations (5.19) and (5.20) we treated the orientation angle γ as a cartesian coordinate which is strictly speaking not correct. This inconsistency, however, becomes less and less significant with decreasing width of the bending wavefunction $\varphi_\gamma(\gamma)$. The exponent α_γ is given by $\alpha_\gamma = \tilde{m}\omega/2$ with $\tilde{m}$ being defined by $1/\tilde{m} = 1/mR_e^2 + 1/\mu r_e^2$ and where ω is the bending frequency [see also Equation (10.8)].

starts at the equilibrium geometry (R_e, r_e, γ_e) of the electronic ground state with zero momenta $(P = p = j = 0)$ always has the largest weight. If the harmonic ansatz for the bound-state wavefunction is too crude, one might either perform the integration in (5.10) numerically or expand the anharmonic wavefunction in terms of harmonic wavefunctions and evaluate the integrals analytically (Dahl 1983; Henriksen, Engel, and Schinke 1987).

Finally, we must stress that a classical distribution function is inadequate. The classical coordinate distribution is proportional to $1/P$; therefore it diverges at the inner and the outer turning points and has a minimum at the equilibrium. Outside the classically allowed region it is exactly zero [Figure 5.3(a)]. The quantum mechanical distribution function [Figure 5.3(b)] behaves entirely differently: it peaks at the equilibrium, decays exponentially toward the turning points, and tunnels into the nonclassical region. The absorption cross sections for direct dissociation "reflect" the wavefunction of the parent molecule (see Chapter 6) and therefore employing the classical distribution function would produce spurious structures (Heller 1978a).

5.3 Classical absorption and photodissociation cross sections

First, we provide the formal definitions of classical absorption and photodissociation cross sections and subsequently we describe a practical way in order to calculate them.

5.3.1 Formal definitions

We define the total classical absorption cross section, save for an unimportant normalization constant, as

$$\sigma_{tot}^{(i)}(\omega) \propto E_{photon} \int d\tau_0 \; P_W^{(i)}(\tau_0) \; [\mu_{fi}^{(e)}]^2 \; \delta[H_f(\tau_0) - E_f], \qquad (5.21)$$

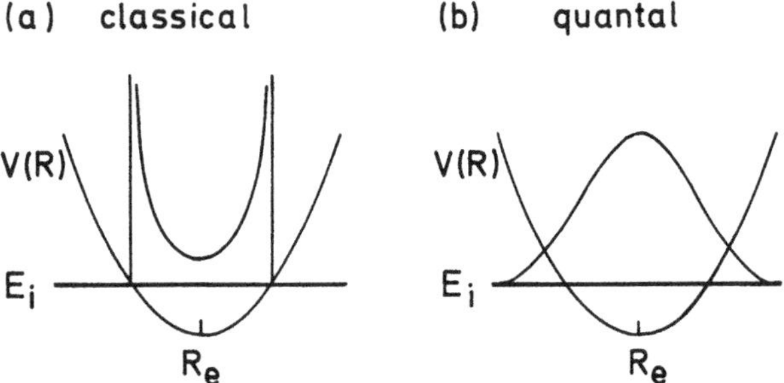

Fig. 5.3. Schematic illustration of the classical and the quantum mechanical coordinate distribution functions.

where $P_W^{(i)}(\boldsymbol{\tau}_0)$ is the distribution function for the electronic ground state (index i), H_f is the Hamilton function in the upper electronic state (index f), $\mu_{fi}^{(e)}$ is the component of the transition dipole function in the direction of the electric field vector, and $E_f = E_i + E_{photon}$ is the resonance energy in the upper state. The integration extends over the entire phase-space with $d\boldsymbol{\tau}_0$ being the appropriate volume element. For the triatomic ABC molecule (with zero total angular momentum) $d\boldsymbol{\tau}_0$ is given by

$$d\boldsymbol{\tau}_0 = \sin\gamma_0 \, dR_0 \, dr_0 \, d\gamma_0 \, dP_0 \, dp_0 \, dj_0. \tag{5.22}$$

The $\sin\gamma_0$ term reflects the fact that the intramolecular vector $\mathbf{r}$ is expressed in polar coordinates; it is especially important for the dissociation of linear molecules and must not be forgotten! The delta-function $\delta(H_f - E_f)$ selects only those points (i.e., trajectories) in the multi-dimensional phase-space that have the correct energy E_f. It ensures that the quantum mechanical resonance condition $E_f = E_i + E_{photon}$ is fulfilled.

The formal definition comprises the following assumptions:

1) Absorption of the photon takes place only if the energy in the excited state, $H_f(\boldsymbol{\tau}_0)$, exactly equals $E_f = E_i + E_{photon}$.
2) The probability for absorbing the photon at a particular point in phase-space is proportional to $[\mu_{fi}^{(e)}]^2$.
3) Each phase-space point has a weight $P_W^{(i)}(\boldsymbol{\tau}_0)$, which reflects the particular quantum mechanical state of the parent molecule in the electronic ground state.

Since the total absorption cross section is independent of the fate of the excited complex, the equations of motion actually need not be integrated.

The partial photodissociation cross sections for absorbing a photon with energy E_{photon} and producing the fragment in a particular vibrational-rotational state (n, j) are similarly defined as

$$\sigma^{(i)}(\omega, n, j) \propto E_{photon} \int d\boldsymbol{\tau}_0 \, P_W^{(i)}(\boldsymbol{\tau}_0) \, [\mu_{fi}^{(e)}]^2 \, \delta[H_f(\boldsymbol{\tau}_0) - E_f] \\ \times \delta[N_f(\boldsymbol{\tau}_0) - n] \, \delta[J_f(\boldsymbol{\tau}_0) - j], \tag{5.23}$$

where $N_f(\boldsymbol{\tau}_0)$ and $J_f(\boldsymbol{\tau}_0)$ are the vibrational and rotational excitation functions, respectively. The latter are solely governed by the dynamics in the upper electronic state as emphasized by the index f. The two additional delta-functions select only those trajectories which lead to the specific vibrational-rotational fragment state (n, j). In contrast to the total cross section, the calculation of the partial cross sections requires integration of the equations of motion until the internal energy is constant and the classical quantum numbers $n(\boldsymbol{\tau}_0)$ and $j(\boldsymbol{\tau}_0)$ can be determined.

If the excitation functions are reasonably smooth functions of all initial coordinates (R_0, r_0, γ_0) and initial momenta (P_0, p_0, j_0), the partial cross sections $\sigma^{(i)}(\omega, n, j)$ can be calculated by means of (5.23). Because of the delta-functions the multi-dimensional integral reduces to a finite sum over all those trajectories $\tau_{0,\alpha}$ defined through the set of equations (Miller 1974, 1975)

$$H_f(\tau_{0,\alpha}) = E_f \quad , \quad N_f(\tau_{0,\alpha}) = n \quad , \quad J_f(\tau_{0,\alpha}) = j \tag{5.24}$$

with $n = 0, 1, 2, \ldots$ and $j = 0, 1, 2, \ldots$. These trajectories, distinguished by the index α, have the correct total energy and at the same time they lead to the specified final quantum numbers n and j. In general, however, this approach is not useful because *all* roots of (5.24) must be found which, in practice, is rather cumbersome in a multi-dimensional variable space. For direct processes, the integral over the entire phase-space often can be reduced to one or at most two important initial variables and then Equations (5.23) and (5.24) are excellent starting points for the calculation and interpretation of final state distributions (Chapter 6).

5.3.2 Monte Carlo calculations

The workhorse for the calculation of cross sections in full collisions is the so-called *Monte Carlo* technique (Schreider 1966; Porter and Raff 1976; Pattengill 1979). The application to photodissociation proceeds in an identical fashion. Within the Monte Carlo method an integral over a function $f(x)$ is approximated by the average of the function over N values x_k randomly selected from a uniform distribution,

$$\int_0^1 f(x)dx \approx \frac{1}{N} \sum_{k=1}^{N} f(x_k). \tag{5.25}$$

The extension to multi-dimensional integrals is straightforward. Then the three delta-functions $\delta(H_f - E_f)$, $\delta(N_f - n)$, and $\delta(J_f - j)$ are approximated by square boxes with widths ΔE_f, $\Delta n = 1$, and $\Delta j = 1$ centered around E_f, n, and j (boxing method). The energy tolerance is typically of the order of 0.01 eV .

The partial photodissociation cross sections are calculated by a sum over N trajectories with initial conditions $\tau_{0,k}$ which are randomly selected from a uniform distribution in the multi-dimensional phase-space,

$$\sigma^{(i)}(\omega, n, j) \propto E_{photon} \sum_{k=1}^{N} \sin\gamma_{0,k}\; P_W^{(i)}(\tau_{0,k})\; [\mu_{fi}^{(e)}]^2\; \Theta(\tau_{0,k}), \tag{5.26}$$

where $\Theta(\tau_{0,k}) = 1$ if

$$E_f - \Delta E_f/2 \leq H_f(\tau_{0,k}) \leq E_f + \Delta E_f/2 \tag{5.27a}$$

$$n - 1/2 \leq N_f(\tau_{0,k}) \leq n + 1/2 \qquad (5.27b)$$

$$j - 1/2 \leq J_f(\tau_{0,k}) \leq j + 1/2 \qquad (5.27c)$$

and $\Theta(\tau_{0,k}) = 0$ otherwise. Equation (5.27) is the "boxing" version of (5.24).

In practice the Monte Carlo calculation proceeds in the following way:

1) Select a set of initial values τ_0 for a trajectory in the upper state and calculate the energy $H_f(\tau_0)$; if the energy does not satisfy (5.27a) choose another set of initial values.
2) If (5.27a) is fulfilled, propagate the trajectory in time until the interaction is zero and the internal energy of the fragment is constant.
3) Determine $N_f(\tau_0)$ and $J_f(\tau_0)$ and find the appropriate "box" according to (5.27b) and (5.27c); the cross section $\sigma^{(i)}(\omega, n, j)$ is increased by $\sin\gamma_0 \; P_W^{(i)}(\tau_0) \; [\mu_{fi}^{(e)}]^2$.
4) Select the next trajectory τ_0 and repeat the cycle until the desired quantities converge.

If one wants to calculate cross sections for a specific energy, it is useful to select all but one initial variables randomly and fix the last variable, R_0 for example, by the condition $H_f(\tau_0) = E_f$. In the light of Equation (2.57) the weighting in (5.26) must be augmented by an additional factor $|\partial H_f / \partial R_0|^{-1}$ which arises from the delta-function in energy. This procedure ensures that all trajectories have the same energy.

Convergence of the Monte Carlo calculation can be greatly improved by choosing the initial conditions from a weighted rather than a uniform distribution (importance sampling). If, for example, the initial distribution function is a product of Gaussians as in (5.20), it is advisable to select the starting values from a *normal distribution*. In this way more trajectories start near the ground-state equilibrium, where the weighting is largest, and fewer trajectories begin with large displacements from equilibrium, where the weighting diminishes exponentially. The total number of trajectories depends crucially on the number of quantum states which are populated. Typically one needs 10^3-10^4 trajectories to get reasonably smooth final state distributions for the dissociation of a triatomic molecule.

5.4 Examples

The classical theory of molecular photodissociation is simple to implement and it can be extended to systems with more than two or three degrees of freedom without severe problems. Moreover, it is invaluable for interpretation purposes. We can rigorously test the validity of the classical approach only by comparison with exact quantum mechanical calculations using the same PES. In the following we demonstrate the usefulness of the classical theory with three realistic examples.

Figure 5.4 depicts the classical and the quantum mechanical absorption spectra of H_2O in the second absorption band which is attributed to the $\tilde{X}^1A_1 \rightarrow \tilde{B}^1A_1$ electronic transition (see Figures 1.5 and 1.6). The corresponding (two-dimensional) PES depicted in Figure 1.12 is repulsive in the transition region leading to a broad continuous spectrum which, on the average, is well reproduced by the classical Monte Carlo calculation. Recall that the calculation of $\sigma_{tot}(\omega)$ does not require the integration of any trajectory and therefore the calculation of the absorption spectrum does not require much computer time. The diffuse structures superimposed on the background have a quantum mechanical origin and a purely classical theory inevitably fails to reproduce them. In Chapter 8, however, we will demonstrate that, despite the inability of classical mechanics to reproduce these diffuse structures, classical trajectories provide an intriguing explanation of their origin.

Figure 5.5 compares quantum mechanical and classical vibrational state distributions of the OH and OD products following the photodissociation of HOD in the first absorption band which is attributed to the $\tilde{X}^1A_1 \rightarrow \tilde{A}^1B_1$ electronic transition (see Figures 1.5 and 1.6). The accuracy of the classical approximation is astonishing, especially if we bear

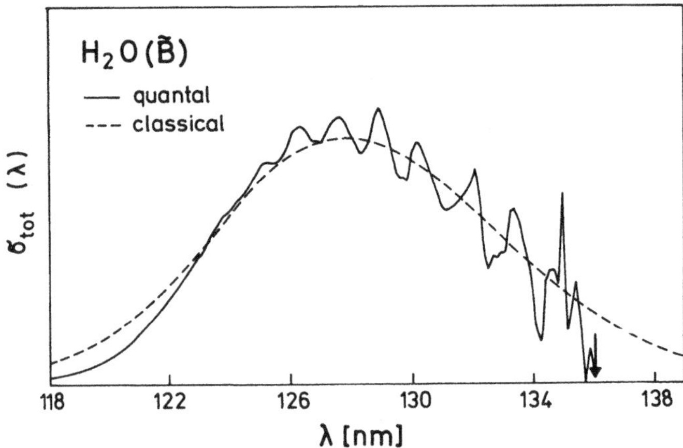

Fig. 5.4. Comparison of the quantum mechanical and the classical absorption spectra for H_2O in the second continuum. The quantal result is calculated by means of the time-independent close-coupling method and the classical curve is obtained in a Monte Carlo simulation. Both cross sections are normalized to the same area. The arrow indicates the threshold for $H + OH(^2\Sigma)$. Reproduced from Weide and Schinke (1989).

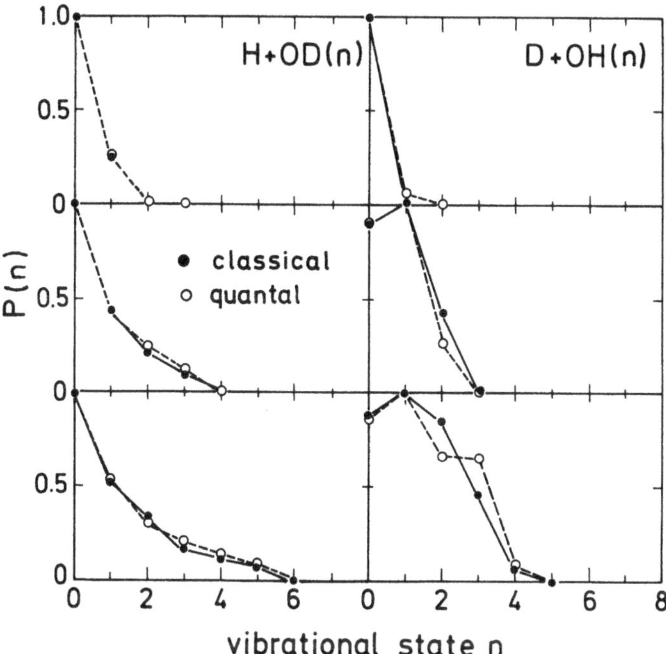

Fig. 5.5. Comparison of quantum mechanical and classical vibrational state distributions of the OD and OH products following the dissociation of HOD in the first continuum. The energies are $E = -2.5$ eV (top), -2.2 eV (middle), and -1.9 eV (bottom), respectively. Energy normalization is such that $E = 0$ corresponds to H + O + D. Reproduced from Engel and Schinke (1988).

in mind that only a few vibrational states are involved (see also Guo and Murrell 1988a,b). Usually, classical mechanics is expected to work if many quantum states are populated. Without showing the results here we note that the classical calculation also reproduces the $\sigma_{H+OD}/\sigma_{D+OH}$ branching ratio (Figure 1.7) very well and at the same time it provides a simple explanation for the preference of the OD channel (Engel and Schinke 1988).

The last example considers rotational excitation. Figure 5.6 shows the final rotational state distribution of NO following the photodissociation of ClNO via the S_1 state (Schinke, Nonella, Suter, and Huber 1990). The strong repulsion between chlorine and nitrogen generates a substantial torque which excites NO to very high rotational levels. The classical Monte Carlo calculation reproduces the exact quantum mechanical results with only minor deviations in the tails of the distribution. More important, however, is the elegant explanation of the characteristic shape which is observed for many similar systems (see also Figure 3.3

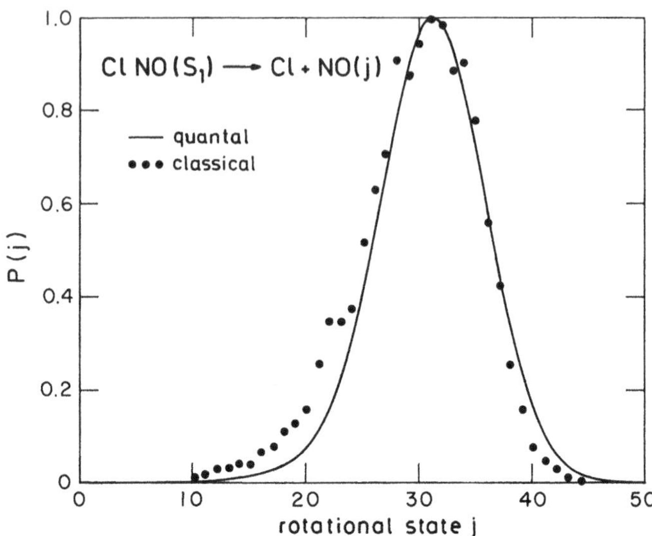

Fig. 5.6. Comparison of the quantum mechanical and the classical rotational state distribution of NO following the photodissociation of ClNO in the S_1 absorption band. Reproduced from Schinke et al. (1990).

for the photodissociation of ClCN). An ultrasimple model based on the rotational excitation function follows in Chapter 6.

The classical theory is a valuable complement of the quantum mechanical approaches. It is best suited for fast and direct photodissociation. Quantum mechanical effects, however, such as resonances or interferences inherently cannot be described by classical mechanics. The obvious extension is a semiclassical theory (Miller 1974, 1975) which incorporates the quantum mechanical superposition principle without the complexity of full quantum mechanical calculations. All ingredients are derived solely from classical trajectories. For an application in photodissociation see Gray and Child (1984).

6
Direct photodissociation: The reflection principle

Photodissociation can be roughly classified as either direct or indirect dissociation. In a direct process the parent molecule dissociates immediately after the photon has promoted it to the upper electronic state. No barrier or other dynamical constraint hinders the fragmentation and the "lifetime" of the excited complex is very short, less than a vibrational period within the complex. For comparison, the period of an internal vibration typically ranges from 30 to 50 fs. The photodissociation of CH_3ONO via the S_2 state is a typical example; the corresponding potential energy surface (PES) is depicted in the upper part of Figure 1.11. A trajectory or a quantum mechanical wavepacket launched on the S_2-state PES immediately leads to dissociation into products CH_3O and NO.

In indirect photofragmentation, on the other hand, a potential barrier or some other dynamical force hinders direct fragmentation of the excited complex and the lifetime amounts to at least several internal vibrational periods. The photodissociation of CH_3ONO via the S_1 state is a representative example. The middle part of Figure 1.11 shows the corresponding PES. Before $CH_3ONO(S_1)$ breaks apart it first performs several vibrations within the shallow well before a sufficient amount of energy is transferred from the N-O vibrational bond to the O-N dissociation mode, which is necessary to surpass the small barrier.

Direct dissociation is the topic of this chapter while indirect photofragmentation will be discussed in the following chapter. Both categories are investigated with the same computational tools, namely the exact solution of the time-independent or the time-dependent Schrödinger equation. The underlying physics, however, differs drastically and requires different interpretation models. Direct dissociation is basically a classical process while indirect dissociation needs a fully quantum mechanical description.

Because of the very short "lifetime" in a direct process the energy dependence of the absorption spectrum as well as the final state distributions of the products is directly related to the initial coordinate distribution of the parent molecule in the electronic ground state. This is called the *reflection principle.* Its general form is the central topic of this chapter. We start in Section 6.1 with the one-dimensional reflection principle which is known from the early days of spectroscopy and extend it in Section 6.2 to more dimensions. Rather new is the explanation of final rotational and vibrational state distributions in terms of a dynamical reflection principle which we will outline in Sections 6.3 and 6.4.

6.1 One-dimensional reflection principle

Let us consider the dissociation of a diatomic molecule with internuclear distance R, linear momentum P, and reduced mass m, as illustrated in Figure 6.1. The corresponding classical Hamilton function in the dissociative state is

$$H(R,P) = \frac{P^2}{2m} + V(R). \tag{6.1}$$

We assume the ground-state potential to be harmonic and the parent molecule to be in the lowest vibrational state. Despite its simplicity, we shall describe the one-dimensional reflection principle in detail because the subsequent extension to more than one dimension follows along the same lines.

6.1.1 The classical view

According to (5.15) and (5.21) the classical approximation of the absorption cross section, as function of the energy in the excited state, is given by

$$\sigma(E) \propto \int dP \int dR \, \mathrm{e}^{-2\alpha_R (R-R_e)^2/\hbar} \, \mathrm{e}^{-P^2/(2\alpha_R \hbar)} \, \delta(H-E), \tag{6.2}$$

where we have assumed a coordinate-independent transition dipole function and all prefactors, including the photon energy, are omitted. The exponential parameter is related to the frequency ω_{HO} of the harmonic oscillator in the ground electronic state by $\alpha_R = m\omega_{HO}/2$ and R_e is the equilibrium separation of the parent molecule. Using property (2.57) for the Dirac delta-function and exploiting the fact that it is mainly $P = 0$ which contributes to the integral, (6.2) reduces to

$$\sigma(E) \approx \mathrm{e}^{-2\alpha_R (R_t - R_e)^2/\hbar} \left| \frac{dV}{dR} \right|^{-1}_{R=R_t(E)}, \tag{6.3}$$

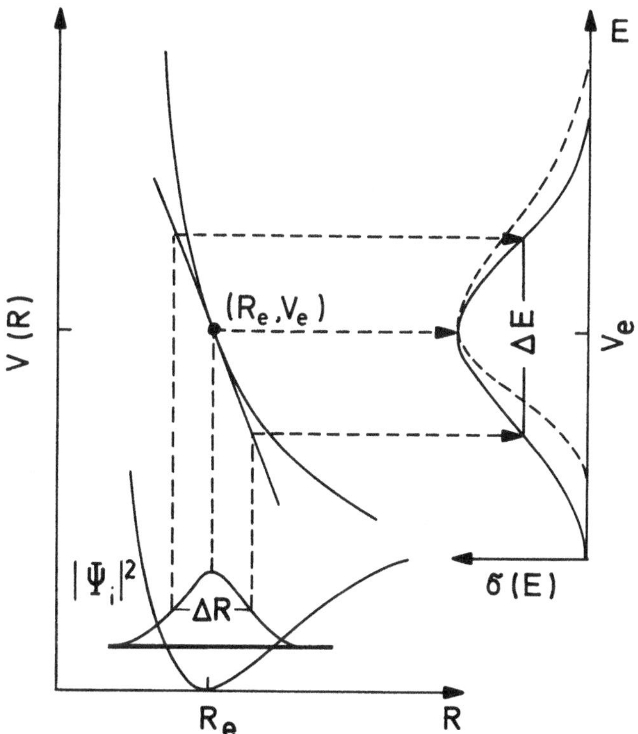

Fig. 6.1. Schematic illustration of the one-dimensional reflection principle. The solid curve on the right-hand side shows the spectrum for a linear potential whereas the dashed curve represents a more realistic case. V_e is the vertical energy defined as $V(R_e)$.

where $R_t(E, P)$ stands for the *classical turning point* defined by[†]

$$H(R_t, P) = E. \tag{6.4}$$

Approximating the potential by

$$V(R) \approx V_e - V_R\,(R - R_e), \tag{6.5}$$

with $V_e = V(R_e)$ and $V_R = -dV/dR|_{R=R_e}$, the cross section becomes

$$\sigma(E) \approx \frac{\mathrm{e}^{-2\beta(E-V_e)^2/\hbar}}{V_R}, \tag{6.6}$$

where $\beta = (V_R^2\,/\alpha_R)^{-1}$. The classical absorption spectrum in the limit of a linear potential is a Gaussian function of energy, centered at $V_e = V(R_e)$

[†] In deriving (6.3) we used that a monotonic potential has only one turning point for each set (E, P); the summation in (2.57) reduces therefore to a single term.

with full width at half maximum of

$$\Delta E = V_R \, \Delta R, \tag{6.7}$$

where ΔR is the width of the coordinate distribution $|\Psi_i(R; E_i)|^2$ in the electronic ground state. This is the one-dimensional reflection principle illustrated in Figure 6.1:

- The absorption spectrum reflects the coordinate distribution of the parent molecule in the ground electronic state with the reflection being mediated by the upper-state potential.
- Equation (6.4) establishes the unique relation between energy E on one hand and coordinate R on the other hand; for a monotonic potential, one point in the coordinate domain exactly corresponds to one point in the energy domain.
- The width of the spectrum is proportional to the width of the initial coordinate distribution, ΔR, and to the steepness $V_R = -dV/dR$ of the excited-state potential above the ground-state equilibrium.

The spectrum becomes broader with increasing steepness and at the same time its magnitude decreases. The integrated spectrum, however, is independent of V_R as readily verified using (6.6) and as predicted by the sum rule in (2.72).

A linear approximation of the potential is certainly too sweeping a simplification. In reality, V_R varies with the internuclear separation and usually rises considerably at short distances. This disturbs the perfect (mirror) reflection in such a way that the "blue side" of the spectrum ($E > V_e$) is amplified at the expense of the "red side" ($E < V_e$).† For a general, nonlinear potential one should use Equations (6.3) and (6.4) instead of (6.6) for an accurate calculation of the spectrum. The reflection principle is well known in spectroscopy (Herzberg 1950:ch.VII; Tellinghuisen 1987); the review article of Tellinghuisen (1985) provides a comprehensive list of references. For a semiclassical analysis of bound-free transition matrix elements see Child (1980, 1991:ch.5), for example.

6.1.2 The time-dependent view

It is informative to study the same problem in the time-dependent, quantum mechanical approach (Heller 1978a, 1981a,b). We start with the (unnormalized) wavepacket at $t = 0$,

$$\Phi(R; t = 0) \propto \mathrm{e}^{-\alpha_R (R - R_e)^2/\hbar}, \tag{6.8}$$

† This follows from the fact that the turning point R_t varies more slowly with energy when the potential is steeper.

whose evolution in the upper electronic state is governed by the time-evolution operator $e^{-i\hat{H}t/\hbar}$ according to (4.8). For very short times we neglect the kinetic energy and replace the Hamiltonian by

$$\hat{H}(R) \approx V(R) \approx V_e - V_R\,(R - R_e). \tag{6.9}$$

Inserting (6.9) into (4.8) yields for the short-time behavior of the wavepacket

$$\Phi(R;t) \approx e^{-iV_e t/\hbar}\; e^{iV_R(R-R_e)t/\hbar}\; e^{-\alpha_R(R-R_e)^2/\hbar}, \tag{6.10}$$

which consists of the original Gaussian multiplied by two complex phase factors. Within the short time interval for which this approximation is valid, the center of the wavepacket has not yet moved and its overall width is still the initial one!

The resulting autocorrelation function $\langle\Phi(0)|\Phi(t)\rangle$,

$$S(t) \approx \left(\frac{\pi\hbar}{2\alpha_R}\right)^{1/2} e^{-iV_e t/\hbar}\; e^{-V_R^2 t^2/(8\hbar\alpha_R)}, \tag{6.11}$$

is a Gaussian in time centered at $t = 0$ and multiplied by a complex phase factor (an example is depicted in Figure 4.3). According to Equation (4.7), the spectrum is obtained by a Fourier transformation of the autocorrelation function. Since the Fourier transformation of a Gaussian yields another Gaussian (Cohen-Tannoudji, Diu, and Laloë 1977:appendix I), it is not difficult to derive that the absorption cross section in the approximate time-dependent treatment becomes identical to the classical result (6.6). The phase of $S(t)$ determines the peak position of the spectrum on the energy axis.

This derivation contains several interesting aspects which are rather general and which are valuable for the overall understanding of dissociation dynamics:

- The rapid decay of the autocorrelation function at very short times is mainly due to a "dephasing" of the wavepacket rather than a displacement in coordinate space.

Within the short-time approximation, the center of the wavepacket remains at R_e while its center in momentum space, $V_R\, t$, moves outward with constant "velocity" $V_R = -dV/dR$.

- The width Δt of the autocorrelation function is inversely proportional to the steepness of the potential.

The steeper the potential the faster the autocorrelation function diminishes and the broader is the spectrum. The width of $S(t)$ is related to the breadth of the spectrum by

$$\Delta t\, \Delta E = 8\,\hbar \ln 2 \approx h, \tag{6.12}$$

which is known as the fourth Heisenberg uncertainty relation (Cohen-Tannoudji, Diu, and Laloë 1977:ch.III).

6.1.3 The time-independent view

In the time-independent formulation, the absorption cross section is proportional to $|\langle\Psi_f(R;E) \mid \Psi_i(R;E_i)\rangle|^2$. Approximate expressions may be derived in several ways. One possibility is to employ the semiclassical WKB approximation of the continuum wavefunction (Child 1980; Tellinghuisen 1985; Child 1991:ch.5). Alternatively, one may linearly approximate the excited-state potential around the turning point and solve the Schrödinger equation for the continuum wavefunction in terms of Airy functions (Freed and Band 1977). Both approaches yield rather accurate but quite involved expressions for bound-free transition matrix elements. Therefore, we confine the subsequent discussion to a merely qualitative illustration as depicted in Figure 6.2.

The absorption cross section is essentially determined by the overlap of the bound wavefunction $\Psi_i(E_i)$ with the first maximum of the continuum wavefunction $\Psi_f(E)$ in the classically allowed region ($R > R_t$). The turning point $R_t(E)$ for a monotonic potential moves gradually to shorter distances as the energy increases. For low energies, the two wavefunctions do not yet overlap and $\sigma(E)$ is negligibly small. With increasing energy, the continuum wavefunction shifts inward and the overlap increases. It

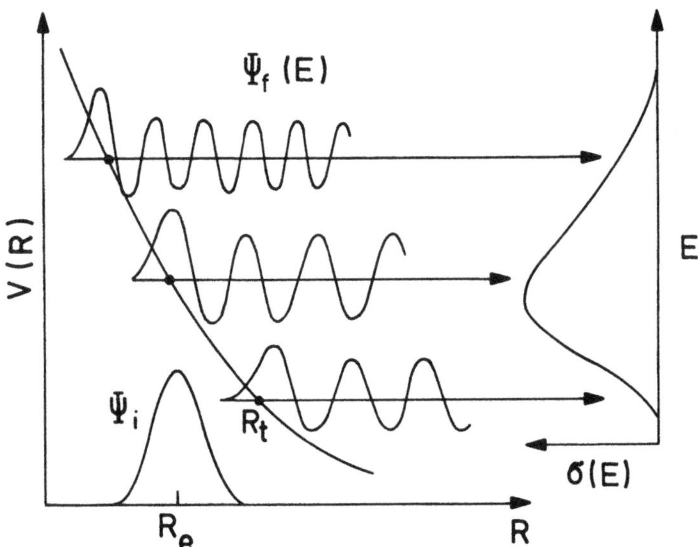

Fig. 6.2. Schematic illustration of the overlap of a bound and a continuum wavefunction as a function of energy. R_t is the classical turning point.

becomes largest when $\Psi_i(R)$ and the main maximum of $\Psi_f(R;E)$ fully overlap. At still higher energies the bound wavefunction overlaps the oscillatory part of the scattering wavefunction and the cross section again diminishes because positive and negative contributions mutually cancel.†

If the potential is linearly expanded about the turning point, $V(R) \approx E - V_R\,(R - R_t)$, the stationary wavefunction in the vicinity of R_t is well approximated by (Freed and Band 1977)

$$\Psi_f(R;E) \approx Ai\left[-\left(2mV_R/\hbar^2\right)^{1/3}(R-R_t)\right], \tag{6.13}$$

where $Ai(x)$ is the Airy function (Abramowitz and Stegun 1970:ch.10). It is maximal for $x_{max} = -1.0188$ and therefore the first maximum of the wavefunction occurs at

$$R_{max} = R_t + 1.0188\left(\frac{\hbar^2}{2mV_R}\right)^{1/3}, \tag{6.14}$$

i.e., at a distance slightly outside the nonclassical region. The shift between R_t and R_{max} implies that the spectrum does not peak exactly at the vertical energy $V_e = V(R_e)$ but at some higher energy. According to (6.14), this shift is largest for light systems and flat potentials.

6.2 Multi-dimensional reflection principle

The extension to more than one dimension is rather straightforward within the time-dependent approach (Heller 1978a, 1981a,b). For simplicity we restrict the discussion to two degrees of freedom and consider the dissociation of the linear triatomic molecule ABC into A and BC(n) as outlined in Section 2.5 where n is the vibrational quantum number of the free oscillator. The Jacobi coordinates R and r are defined in Figure 2.1, Equation (2.39) gives the Hamiltonian, and the transition dipole function is assumed to be constant. The parent molecule in the ground electronic state is represented by two uncoupled harmonic oscillators with frequencies ω_R and ω_r, respectively.

6.2.1 The time-dependent view

In analogy to (6.8) we write the (unnormalized) wavepacket at $t = 0$ as

$$\Phi(R,r;t=0) \propto \mathrm{e}^{-\alpha_R(R-R_e)^2/\hbar}\,\mathrm{e}^{-\alpha_r(r-r_e)^2/\hbar} \tag{6.15}$$

† In the limit of very short de Broglie wavelengths, the continuum wavefunction can be approximated by a delta-function in R localized at the classical turning point, i.e., $\delta(R - R_t)$. This allows us readily to perform the integration over R yielding $\sigma(E) \approx |\Psi_i(R_t;E_i)|^2$, which is another way of expressing the reflection principle.

with $\alpha_R = m\omega_R/2$ and $\alpha_r = \mu\omega_r/2$; m and μ are the corresponding reduced masses. The derivation of the absorption cross section proceeds exactly as described in the one-dimensional case. Expanding the potential about the equilibrium (R_e, r_e) as

$$V(R,r) \approx V_e - V_R\,(R - R_e) - V_r\,(r - r_e) \tag{6.16}$$

with $V_R = -\partial V/\partial R$ and $V_r = -\partial V/\partial r$ and replacing the Hamiltonian for short times by the potential yields the approximate autocorrelation function

$$S(t) \approx \left(\frac{\pi\hbar}{2\alpha_R}\right)^{1/2} \left(\frac{\pi\hbar}{2\alpha_r}\right)^{1/2} \times \mathrm{e}^{-iV_e t/\hbar}\, \mathrm{e}^{-V_R^2 t^2/(8\hbar\alpha_R)}\, \mathrm{e}^{-V_r^2 t^2/(8\hbar\alpha_r)}. \tag{6.17}$$

Finally, the cross section becomes (save for some unimportant constants)

$$\sigma(E) \approx \left(\frac{1}{\alpha_R\alpha_r}\right)^{1/2} \frac{\mathrm{e}^{-2\beta(E-V_e)^2/\hbar}}{(V_R^2/\alpha_R + V_r^2/\alpha_r)^{1/2}}, \tag{6.18}$$

where β is defined as

$$\beta = (V_R^2/\alpha_R + V_r^2/\alpha_r)^{-1}. \tag{6.19}$$

Equation (6.18) is the two-dimensional analogue of (6.6).

As in the one-dimensional case, the general form of the spectrum is of Gaussian-type and can be regarded as a reflection of the two-dimensional bound-state wavefunction at the upper-state PES. It is noteworthy that:

- The width of the absorption spectrum is determined not only by the steepness of the upper-state potential along the dissociation coordinate R but also by its steepness along the internal vibrational coordinate r.

Acceleration of the wavepacket in both directions leads to a faster dephasing and hence to a faster decay of the autocorrelation function, which, in turn, implies a broader absorption spectrum. The extension to even more degrees of freedom is obvious. Figure 6.3 illustrates how the absorption spectrum depends on the steepness of the excited-state PES at the transition point. In case (a), the photon promotes the system to a region where $V_r = -\partial V/\partial r$ is small so that the spectrum is relatively narrow and intense. In case (b), on the other hand, the fragmentation starts significantly displaced from the equilibrium in r, in a region where V_r is large. This implies a relatively broad but less intense spectrum. By construction of the potential the steepness $V_R = -\partial V/\partial R$ is the same in both cases. According to the sum rule (2.72), the integrated intensities are identical. The two arrows illustrate the corresponding dissociation

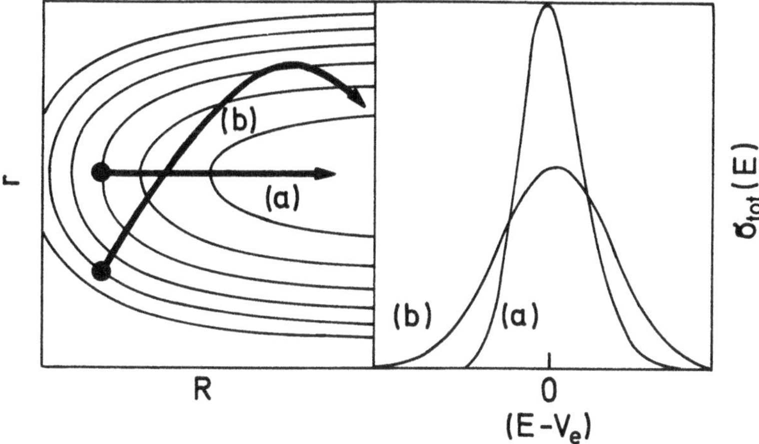

Fig. 6.3. Schematic illustration of the two-dimensional reflection principle and the dependence of the width of the spectrum on the gradients $|\partial V/\partial R|$ and $|\partial V/\partial r|$. The dots indicate the two different excitation points. The vertical energies $V_e = V(R_e, r_e)$ are different in the two cases and therefore the spectra are plotted as functions of $E - V_e$ rather than E.

paths, i.e., the routes of the wavepackets, or simply the classical trajectories starting at the respective transition points. It is easy to surmise that case (a) leads to very weak vibrational excitation of the fragment while case (b) leads to substantial excitation of BC.

6.2.2 The adiabatic view

The separability within the adiabatic limit of the time-independent approach provides an interesting alternative picture. The left-hand side of the upper part of Figure 6.4 shows the vibrationally adiabatic potential curves $\epsilon_n(R)$, which were defined in Equation (3.31), for a two-dimensional PES of the same generic type as the PES in Figure 6.3. The individual potential curves are vertically shifted by the vibrational energy of the free diatomic fragment BC, but otherwise they have roughly the same shape. According to Equation (3.41), the partial photodissociation cross sections approximately factorize as

$$\sigma(E, n) \approx g(E, n)\ f(n),$$

where the $g(E, n)$ are one-dimensional absorption cross sections, as discussed in Section 6.1, and the $f(n)$ are one-dimensional Franck-Condon (FC) factors [for definitions see (3.42)].

The energy dependence of each one-dimensional cross section $g(E, n)$ can be easily explained by the one-dimensional reflection principle as illustrated in Figure 6.4. The vertical energy $\epsilon_n(R_e)$ determines the peak energy of each $g(E, n)$ and the steepness of the corresponding potential curve at the transition point controls its width. Except for the vertical shift, which is the result of the internal excitation of the BC oscillator, the adiabatic potential curves do not significantly vary with the quantum number n and therefore the individual partial cross sections have a similar energy dependence. The relative magnitudes are governed by the energy-independent FC factors $f(n)$.

The total absorption spectrum is the sum of all partial spectra. Therefore, its width is determined by the widths of the individual $\sigma(E; n)$ *and* by the number of partial cross sections which have an appreciable magnitude, i.e., the width of the FC distribution $f(n)$. For case (a) of Figure 6.3, the equilibrium separation r_e in the excited state is roughly the same as in the ground state. The corresponding FC distribution is therefore comparatively limited; it peaks at $n = 0$ and rapidly diminishes with increasing n. Consequently, only $\sigma(E, n = 0)$ has an appreciable magnitude and the total spectrum is relatively narrow. The FC distribution for case (b) of Figure 6.3, on the other hand, is broader because r_e in the upper state evidently differs from the equilibrium in the ground state. Ergo, several partial cross sections contribute to the total spectrum and the total spectrum is broader than for case (a).

Let us summarize:

- In the time-dependent picture, the overall width of the total spectrum is determined by the two slopes $|\partial V/\partial R|$ and $|\partial V/\partial r|$ at the transition point (R_e, r_e).
- In the time-independent adiabatic picture, the slope $|d\epsilon_n/dR|$, which is roughly equal to $|\partial V/\partial R|$, determines the width of each partial cross section and the width of the FC distribution $f(n)$ determines the breadth of $\sigma_{tot}(E)$.

Note, the width of the one-dimensional FC distribution $f(n)$ depends directly on the steepness $|\partial V/\partial r|$ in the FC region and therefore both pictures are equivalent; they merely favor different points of view.

6.2.3 Broad vibrational bands

If the partial spectra $\sigma(E, n)$ are broader than the spacing between them, the total spectrum $\sigma_{tot}(E)$ is structureless; the upper part of Figure 6.4 provides a typical example. However, if the widths of the $\sigma(E, n)$ are smaller than the separations, $\sigma_{tot}(E)$ exhibits broad, so-called *diffuse vibrational structures*. The necessary condition for this to happen is that, according to (6.7), the upper-state PES in the FC region is relatively flat along the dissociation coordinate R. The lower part of Figure 6.4 illustrates this case.

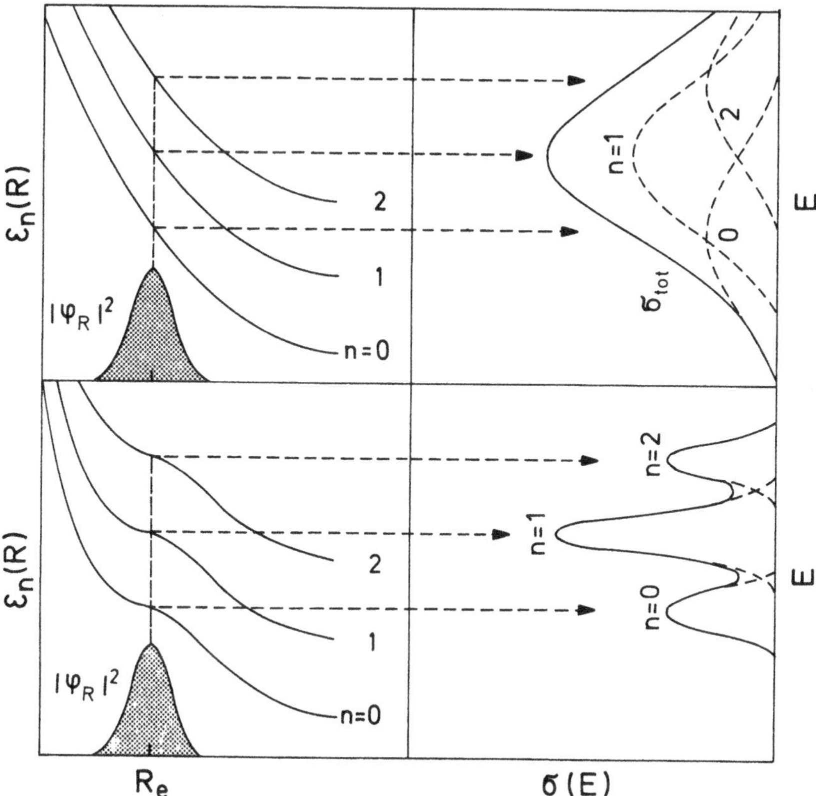

Fig. 6.4. Schematic illustration of the multi-dimensional reflection principle in the adiabatic limit. The left-hand side shows the vibrationally adiabatic potential curves $\epsilon_n(R)$. The R-dependent part of the bound-state wavefunction in the ground electronic state is denoted by $\varphi_R(R)$. The right-hand side depicts the corresponding partial photodissociation cross sections $\sigma(E;n)$ (dashed curves) and the total cross section $\sigma_{tot}(E)$ (solid curve) with the arrows illustrating the one-dimensional reflection principle. Upper part: In this case, the steepness of the PES leads to comparatively broad partial photodissociation cross sections with the result that the total spectrum is structureless. Lower part: In this case, the potential is rather flat near R_e so that the partial cross sections are relatively narrow, and as a result the total cross section shows broad vibrational structures.

Examples of diffuse vibrational structures are the photodissociation of CH_3SNO via the S_1 state (Schinke et al. 1989) and ClNO via the S_1 state (Schinke, Nonella, Suter, and Huber 1990; Bai et al. 1989; Ogai, Qian, and Reisler 1990; Untch, Weide, and Schinke 1991b) . The experimental as well as the calculated spectra for $CH_3SNO(S_1)$ depicted in Figure 6.5 clearly manifest this type of vibrational structure. The calculated (two-dimensional) PES for $CH_3SNO(S_1)$ is indeed rather flat in the FC region; the corresponding adiabatic potential curves $\epsilon_n(R)$ are shown in the lower part of Figure 7.10 in comparison with those for $CH_3ONO(S_1)$. Decomposition of the total spectrum into the partial photodissociation cross sections clearly shows that the first peak corresponds to $NO(n = 0)$, the second peak corresponds to $NO(n = 1)$, etc. A slightly steeper slope of the potential along the reaction path would lead to broader partial cross sections and therefore blot out the vibrational structures. Thus, the energy dependence of the absorption spectrum is a sensitive probe of the R dependence of the PES. The vibrational structures in Figure 6.5 can be considered as very broad or short-lived resonances. See Chapters 7 and 8 for more detailed discussions of resonances and diffuse structures.

6.3 Rotational reflection principle

The reflection principle, outlined in Sections 6.1 and 6.2, explains the energy dependence of absorption spectrum as a mapping of the initial coordinate distribution in the electronic ground state onto the energy axis. Rotational state distributions of diatomic photofragments in direct dissociation can be explained in an analogous manner.

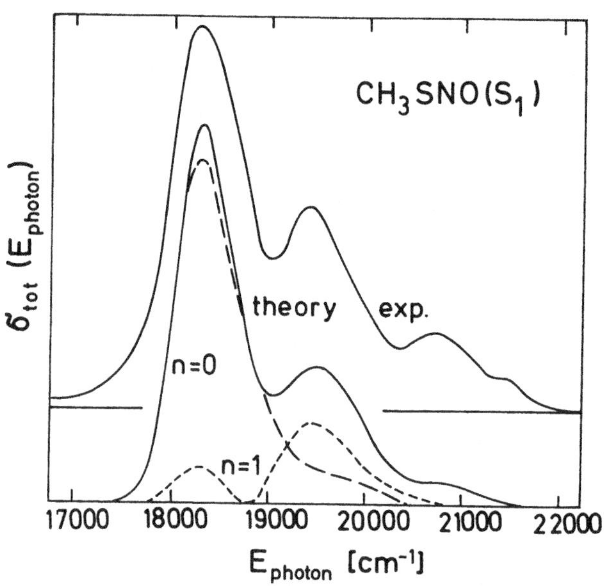

Fig. 6.5. Comparison of the measured and the calculated absorption spectra for $CH_3SNO(S_1)$. The theoretical absorption cross section is decomposed into the various partial cross sections $\sigma(E, n)$ (dashed curves) for producing NO in final vibrational states $n = 0$ and 1. The experimental and the theoretical curves are vertically shifted for clarity. Adapted from Schinke et al. (1989).

6.3.1 An ultrasimple classical model

We consider the photofragmentation of a triatomic molecule, ABC $\longrightarrow$ A + BC(j), within the model outlined in Section 3.2. The vibrational coordinate of BC is fixed and the total angular momentum is zero. According to (5.23), the classical approximation of the partial photodissociation cross section for producing BC in rotational state j is given by

$$\sigma(E,j) \propto \int d\boldsymbol{\tau}_0 \; P_W(\boldsymbol{\tau}_0) \; \delta[H(\boldsymbol{\tau}_0) - E] \; \delta[J(\boldsymbol{\tau}_0 - j], \tag{6.20}$$

where $\boldsymbol{\tau}_0 = (R_0, \gamma_0, P_0, j_0)$, $d\boldsymbol{\tau}_0$ is the volume element according to (5.22), P_W is the four-dimensional Wigner distribution function (5.20) of the parent molecule, $H(\boldsymbol{\tau}_0)$ represents the Hamilton function in the upper electronic state, and $J(\boldsymbol{\tau}_0)$ denotes the rotational excitation function defined in Equation (5.7). The transition dipole moment function is assumed to be constant and all unimportant prefactors are omitted because we shall consider only product state distributions $P(j)$.

As in Section 6.1 we set the initial momenta, P_0 and j_0, to zero and with the help of (2.57) we eliminate R_0 as initial variable. The (unnormalized) final rotational state distribution for a given energy is thus transformed into a one-dimensional integral,

$$P(j) = \int_0^{\pi} d\gamma_0 \; \sin\gamma_0 \; P_W(R_t, \gamma_0) \left| \frac{\partial V}{\partial R} \right|_{R=R_t}^{-1} \delta[J(\gamma_0) - j], \tag{6.21}$$

where the classical turning point $R_t(E, \gamma_0)$ depends on the energy as well as the initial orientation angle γ_0. For fixed energy, γ_0 is the only independent variable in this model; the other variables are subordinate for the determination of rotational state distributions. Using again (2.57) yields

$$P(j) = \sum_{\alpha} W(\gamma_{0,\alpha}) \left| \frac{dJ}{d\gamma_0} \right|_{\gamma_0=\gamma_{0,\alpha}(j)}^{-1}, \tag{6.22}$$

where we defined the weighting function by

$$\begin{aligned} W(\gamma_0) &= \sin\gamma_0 \; P_W[R_t(\gamma_0), \gamma_0] \left| \frac{\partial V}{\partial R} \right|_{R=R_t}^{-1} \\ &= \sin\gamma_0 \; \mathrm{e}^{-2\alpha_R (R_t - R_e)^2/\hbar} \; \mathrm{e}^{-2\alpha_\gamma (\gamma_0 - \gamma_e)^2/\hbar} \left| \frac{\partial V}{\partial R} \right|_{R=R_t}^{-1}. \end{aligned} \tag{6.23}$$

R_e and γ_e specify the equilibrium of the electronic ground state and the exponents α_R and α_γ are related to the frequencies according to (5.16). Throughout this section we will assume that the parent molecule is in its lowest bending state before the excitation takes place. The summation in (6.22) extends over all trajectories which start with initial angle $\gamma_0(j)$ and end with a specified final angular momentum j, i.e., which solve the equation[†]

$$J(\gamma_0) = j. \tag{6.24}$$

In principle, j can be any real number; in order to make contact with quantum mechanics and experiment, however, j is confined to the integers 0,1,2,.... Equations (6.22) and (6.24) are the analogues of (6.3) and (6.4), respectively.

Figure 6.6 illustrates the relation between the anisotropy of the upper-state PES and the final rotational state distribution for a homonuclear fragment molecule. The rotational excitation function is the link between them. A typical PES in the form of an energy contour map is displayed in Figure 6.6(b) and Figure 6.6(c) shows the resulting excitation function $J(\gamma_0)$. Trajectories which start from the linear geometries ($\gamma_0 = 0°$ and $180°$) do not experience any torque throughout the entire breakup and therefore $J(0°) = J(180°) = 0$. Because of symmetry the same holds true for the perpendicular geometry ($\gamma_0 = 90°$) provided the rotor consists of like atoms. Rotational excitation is most efficient if the trajectory begins near $\gamma_0 \approx 45°$ where the imparted torque is largest. The maximum of $J(\gamma_0)$ reflects the point of inflection (with respect to γ) of the dissociative PES. The excitation function consists of two branches of opposite sign.[‡] Because the two branches are anti-symmetric for a homonuclear molecule, it suffices to consider only the range $0 \leq \gamma_0 \leq 90°$ in the following discussion.

For each rotational state below the maximum value J_{max} there are two independent trajectories, one starting at a small angle and the other starting at a larger angle, which both lead to the same final rotational state, i.e., Equation (6.24) has two *independent* solutions. However, the weighting function $W(\gamma_0)$, which intrinsically reflects the distribution of the bending angle in the electronic ground state, is confined to a relatively small angular region, typically of the order of $\Delta\gamma \approx 10$–$20°$ so that only one trajectory contributes to the cross section while the other one can be ignored. As a consequence, the sum in (6.22) collapses to a single term,

$$P(j) = W[\gamma_0(j)] \left| \frac{dJ}{d\gamma_0} \right|^{-1}_{\gamma_0=\gamma_0(j)}. \tag{6.25}$$

For simplicity, let us assume that the weighting function is identical to the square of the bending wavefunction $\varphi_\gamma(\gamma)$ and secondly, that $J(\gamma_0)$

[†] Recall that, according to Equation (5.7), $J(\gamma_0)$ is defined as dimensionless quantity.

[‡] The sign of $j(t)$ specifies the sense of rotation within the collision plane. It is irrelevant for the rotational state distribution and therefore only the magnitude of $j(t)$ is important.

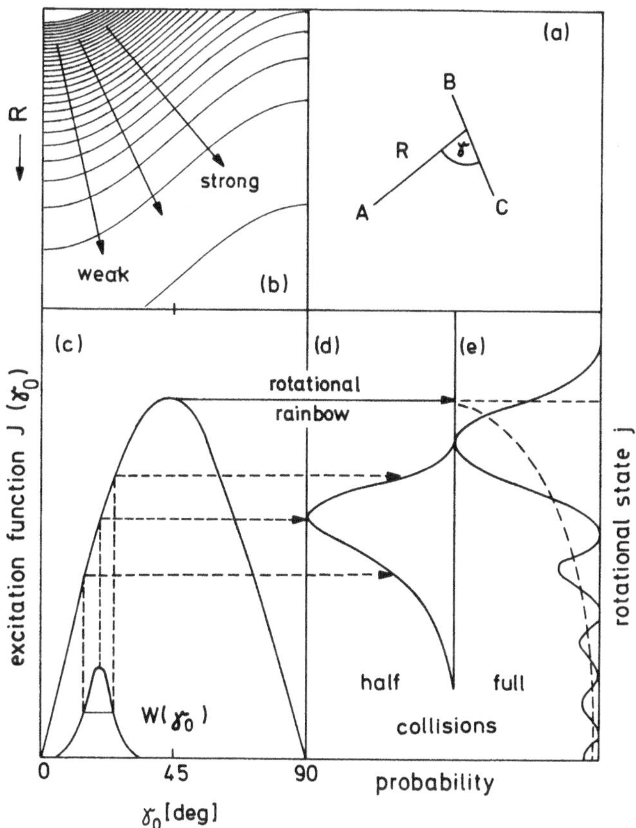

Fig. 6.6. (a) Jacobi coordinates R and γ for the dissociation of a triatomic molecule into products A and BC. (b) A typical potential energy surface as function of R and γ for a direct process. The arrows illustrate schematically three classical trajectories leading to low, moderate, and high rotational excitation of the BC product. (c) The rotational excitation function $J(\gamma_0)$ as a function of the initial orientation angle γ_0. $W(\gamma_0)$ is the weighting function defined in (6.23); it is mainly determined by the bending wavefunction in the ground electronic state. (d) The final rotational state distribution following from photodissociation (half collision). The broken arrows illustrate the rotational reflection principle. (e) The final rotational state distribution as the result of a full collision. The dashed curve shows the classical distribution with a rotational rainbow singularity at the cut-off at $j = J_{max}$. The corresponding quantum mechanical distribution shows pronounced supernumerary rotational rainbows caused by the interference of the two trajectories which lead to the same final state.

is linear around the equilibrium angle γ_e,

$$J(\gamma_0) \approx J_e + J_\gamma \, (\gamma_0 - \gamma_e), \tag{6.26}$$

where $J_e = J(\gamma_e)$ and $J_\gamma = dJ/d\gamma_0|_{\gamma_0=\gamma_e}$. In complete analogy to Section 6.1.1 the (unnormalized) final rotational state distribution thus becomes

$$P(j) \approx \frac{e^{-2\beta(j-J_e)^2/\hbar}}{|J_\gamma|}, \qquad (6.27)$$

where $\beta = (J_\gamma^2/\alpha_\gamma)^{-1}$; Equation (6.27) is the analogue of (6.6). Under these simplifying assumptions the final state distribution is a Gaussian function with width

$$\Delta j = J_\gamma \, \Delta\gamma, \qquad (6.28)$$

where $\Delta\gamma$ is the breadth of the angular distribution function of the parent molecule, i.e., $|\varphi_\gamma(\gamma_0)|^2$.

Equation (6.27) manifests the rotational reflection principle (Schinke 1986c; Schinke and Engel 1986) as illustrated in Figure 6.6:

- The final rotational state distribution essentially reflects the square of the bending wavefunction of the parent molecule; the mapping is mediated by the rotational excitation function $J(\gamma_0)$.
- Equation (6.24) defines the unique relation between the initial orientation angle γ_0 on one hand and the final angular momentum j on the other hand.
- The value of the excitation function at the ground-state equilibrium angle γ_e determines approximately the maximum of the rotational distribution; the breadth of the initial coordinate distribution together with the gradient $|dJ/\gamma_0|$ control the width of $P(j)$.

The assumptions leading to (6.27) are in most cases too simplistic for a quantitative evaluation. In several applications we have found that the R-dependent part of $W(\gamma_0)$ is also important, especially in order to predict the correct energy dependence of $P(j)$. Furthermore, the curvature of $J(\gamma_0)$ can significantly affect the reflection and therefore we recommend the use of (6.22)–(6.24) for quantitative applications.

The ultrasimple classical theory not only provides an intriguing interpretation of final state distributions, it is also very easy to implement and remarkably accurate. The excitation function is the central quantity. For direct processes, $J(\gamma_0)$ behaves smoothly as a function of γ_0 and therefore we can construct it with very few (less than ten or so) trajectories. By fitting the excitation function to a simple analytical expression we can readily calculate $P(j)$ by means of (6.22)–(6.24). Figure 3.3 shows a comparison with exact quantum mechanical calculations for the dissociation of ClCN and the agreement is indeed astonishing. In contrast to the exact calculations, the required computer time is insignificant which makes this simple classical model ideal for testing the influence of the PES and for adjusting potential parameters. Inclusion of nonzero initial momenta in a full Monte Carlo study merely broadens the distributions without changing their overall shape (Schinke 1988b).

There are two additional points that we must stress at this juncture. First, the highly inverted rotational state distributions are not so-called *rotational rainbows.* Rotational rainbows stem from the maximum of the excitation function. The normalization factor $|dJ/d\gamma_0|^{-1}$ in (6.25) diverges for $j \approx J_{max}$ causing a singularity in the classical cross section. Quantum mechanics turns the classical singularity into a pronounced maximum as illustrated in Figure 6.6(e). Rotational rainbows are common features in full collisions where *all* initial angles, including the region around the maximum of the excitation function, are equally weighted (Schinke and Bowman 1983; Korsch and Wolf 1984; Buck 1986; Kleyn and Horn 1991). In photodissociation, however, rotational rainbows exist only if the weighting function $W(\gamma_0)$ maps out the maximum of $J(\gamma_0)$ (Schinke 1985, 1986b).

Second, the calculated (as well as the measured) distributions are remarkably smooth although often more than fifty or so rotational states are populated. If so many quantum states take part in a collision, one intuitively expects pronounced interference oscillations. The reason for the absence of interferences is the uniqueness between γ_0 and j: one and only one trajectory contributes to the cross section for a specific final rotational state. If two trajectories that lead to the same j had comparable weights, the constructive and destructive interference, within a semiclassical picture, would lead to pronounced oscillations (Miller 1974, 1975; Korsch and Schinke 1980; Schinke and Bowman 1983). These so-called *supernumerary rotational rainbows* are well established in full collisions (Gottwald, Bergmann, and Schinke 1987). If the weighting function $W(\gamma_0)$ is sufficiently wide that both trajectories contribute to the dissociation cross section, similar oscillations may also exist in photodissociation (see, for example, Philippoz, Monot, and van den Bergh 1990 and Miller, Kable, Houston, and Burak 1992).

6.3.2 Mapping of the potential anisotropy

The excitation function evinces the angular dependence, i.e., the anisotropy, of the excited-state PES along the route of the classical trajectories or the quantum mechanical wavepacket. Particularly important, of course, is the region around the ground-state equilibrium where the fragmentation begins. By changing γ_e — if that were possible in real life — we could change the starting point and therefore modify the torque exerted during the breakup. In this way one could continuously shift the maximum of the rotational state distribution between the limits $j = 0$ and $j \approx J_{max}$.

Let us make the relation between the potential anisotropy and the final state distribution more quantitative. In first-order perturbation theory Equation (5.4f) can be directly integrated yielding the (approximate) expression

$$J(\gamma_0) \approx -\int_0^\infty \frac{dt}{\hbar} \frac{\partial V}{\partial \gamma} = -\int_{R_t}^\infty \frac{dR}{\hbar v_R} \frac{\partial V}{\partial \gamma}, \tag{6.29}$$

where v_R is the radial velocity dR/dt and R_t is the classical turning point. The energy transfer to the rotational degree of freedom must be small compared to the total energy for (6.29) to be valid (sudden limit). The radial velocity is zero at the turning point and quickly increases as the fragments separate. Rotational excitation therefore takes place mainly at short internuclear distances where, in addition, the anisotropy is usually largest (see Figure 5.1). Equation (6.29) readily explains the overall behavior of the excitation function as illustrated in Figure 6.6(c). For symmetry reasons $\partial V/\partial \gamma = 0$ for $0°$ and $180°$ and likewise for $90°$ if the diatom is symmetric.

In order to relate the rotational excitation function directly to the potential parameters we consider a potential of the form

$$V(R, \gamma) = A\, \mathrm{e}^{-\alpha[R-a(\gamma)]}, \tag{6.30}$$

where $a(\gamma)$ represents the anisotropy. Inserting (6.30) into (6.29) yields, after some straightforward algebra,

$$J(\gamma_0) \approx \frac{(2mE)^{1/2}}{\hbar} \frac{da}{d\gamma_0}. \tag{6.31}$$

However, we must underline that this simple relation is only valid in the sudden limit, $E_{rot} \ll E$. Equation (6.31) emphasizes in a simple way how the excitation function and therefore the final state distribution depends on the energy E, the reduced mass m, and last but not least the anisotropy parameter $a(\gamma)$.[†] More of the interrelation between the anisotropy of the PES and the final rotational state distribution follows in Chapter 10.

6.3.3 Examples

Highly excited rotational distributions of the kind illustrated in Figure 6.6(d) have been measured for many systems:

[†] According to (6.23) the weighting function $W(\gamma_0)$ depends not only on the bending wavefunction; through the turning point R_t it depends also on the energy as well as the anisotropy of the upper-state PES. This makes the correct prediction of the energy dependence of rotational state distributions rather cumbersome, because an increase of $J(\gamma_0)$ as a consequence of increasing the energy may be completely compensated by a shift of $W(\gamma_0)$ along the γ_0-axis.

1) ClCN (Barts and Halpern 1989); BrCN (Russel, McLaren, Jackson, and Halpern 1987); ICN (Fisher et al. 1984; Marinelli, Sivakumar, and Houston 1984; Nadler, Mahgerefteh, Reisler, and Wittig 1985).
2) ClNO (Ticktin, Bruno, Brühlmann, and Huber 1988, Bai et al. 1989).
3) O_3 (Moore, Bomse, and Valentini 1983; Levene, Nieh, and Valentini 1987).
4) OCS (Sivakumar, et al. 1985, 1988).
5) H_2O_2 (Docker, Hodgson, and Simons 1986a,b; Jacobs, Wahl, Weller, and Wolfrum 1987; Grunewald, Gericke, and Comes 1987; Klee, Gericke, and Comes 1988; Gericke, Gölzenleuchter, and Comes 1988).
6) HNCO (Spiglanin, Perry, and Chandler 1987; Spiglanin and Chandler 1987).
7) HONO (Dixon and Rieley 1989).
8) H_2CO (Bamford et al. 1985).
9) HN_3 (Chu, Marcus, and Dagdigian 1990).
10) CH_3ONO (Benoist d'Azy, Lahmani, Lardeux, and Solgadi 1985; Brühlmann and Huber 1988).
11) $Co(CO)_3NO$ (Georgiou and Wight 1990).

This list, which is by no means complete, clearly demonstrates that the generic type of final state distribution is not only observed for atom-diatom systems but also if the recoiling partner is a large polyatomic molecule. In contrast to the many experimental examples, there are only a few systems for which rotational excitation has been analyzed by means of *ab initio* potential energy surfaces and exact quantum mechanical or classical calculations. In the following we discuss two of them.

The photodissociation of ClCN $\longrightarrow$ Cl + CN at wavelengths around 200 nm (Barts and Halpern 1989) beautifully illustrates the rotational reflection principle. The two-dimensional PES, calculated by Waite and Dunlap (1986) and depicted in Figure 3.2, is steeply repulsive along the Cl-CN coordinate which implies very fast bond cleavage. ClCN is linear in the ground state as opposed to the excited state where the minimum with respect to γ occurs at a bent geometry. The photodissociation of ClCN therefore represents a linear-bent transition.[†] Immediately after the classical trajectories or the quantum mechanical wavepacket are launched on the excited-state PES near the ground-state equilibrium they slide down the potential slope and quickly accumulate strong rotational excitation (see Figure 5.1). Figure 3.3 depicts the measured distribution

[†] The terminology "bent excited state" should be used with caution in this case because, strictly speaking, there is no equilibrium in the excited state of ClCN.

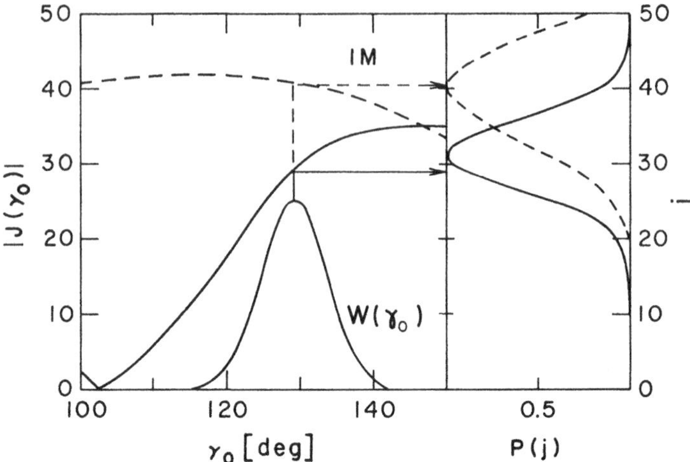

Fig. 6.7. Left-hand side: Rotational excitation function $J(\gamma_0)$ and weighting function $W(\gamma_0)$ for the dissociation of ClNO(S_1). Right-hand side: Calculated final rotational state distribution of NO for an excess energy of 1 eV. The dashed curves represent the same quantities calculated, however, within the so-called impulsive model (IM) which we will discuss in Section 10.4. Reproduced from Schinke et al. (1990).

of CN in comparison to exact close-coupling calculations as well as the classical distribution obtained by the ultrasimple model discussed above.

The dissociation of ClNO $\longrightarrow$ Cl + NO via the S_1 electronic state is an example of a bent-bent transition. The left-hand side of Figure 6.7 depicts the excitation function calculated with a three-dimensional *ab initio* PES (Schinke et al. 1990) and the right-hand side shows the corresponding rotational distribution. $P(j)$ reflects the weighting function $W(\gamma_0)$ which, in turn, is intrinsically determined by the square of the bending wavefunction of ClNO(S_0). It is nearly a perfect Gaussian. Figure 6.8 shows the corresponding experimental distribution decomposed into the four electronic sublevels of NO($X^2\Pi$). Like the theoretical distribution all curves are rather well described by Gaussians. A third example, the photodissociation of H_2O_2, will be discussed in Chapter 10.

6.4 Vibrational reflection principle

Vibrational state distributions in direct dissociation can be described in the same way as rotational state distributions. For simplicity we consider the dissociation of the collinear triatomic molecule, ABC $\longrightarrow$ A + BC, as outlined in Section 2.4.

Proceeding as in Section 6.3 yields the classical probability for produc-

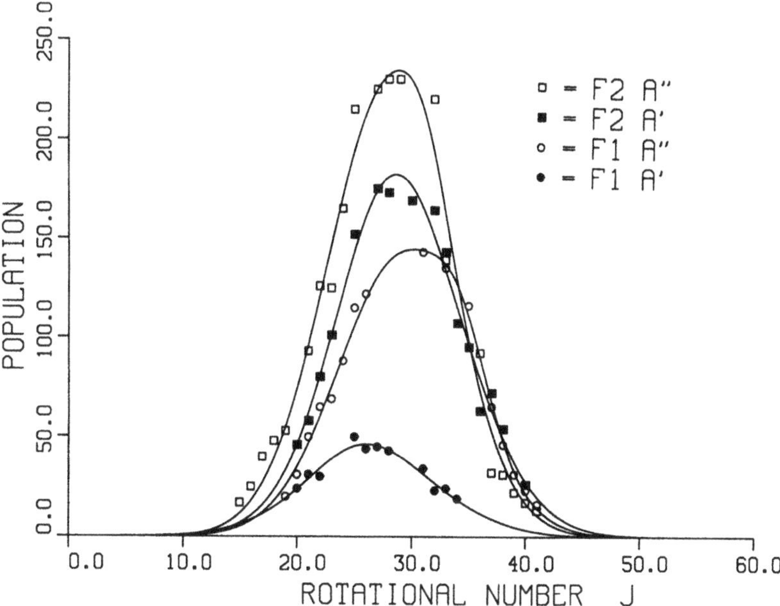

Fig. 6.8. Measured rotational state distributions of $NO(X^2\Pi)$ produced in the photodissociation of $ClNO(S_1)$. The various curves represent the distributions for the four different electronic sublevels of NO. The fitted lines are drawn as guidelines. Reproduced from Ticktin et al. (1988).

ing BC in vibrational state n,

$$P(n) = \sum_{\alpha} W(r_{0,\alpha}) \left| \frac{dN}{dr_0} \right|^{-1}_{r_0 = r_{0,\alpha}(n)}, \tag{6.32}$$

where $N(r_0)$ is the vibrational excitation function defined in Equation (5.8),

$$\begin{aligned} W(r_0) &= P_W[R_t(r_0), r_0] \left| \frac{\partial V}{\partial R} \right|^{-1}_{R=R_t} \\ &= e^{-2\alpha_R (R_t - R_e)^2/\hbar} \, e^{-2\alpha_r (r_0 - r_e)^2/\hbar} \left| \frac{\partial V}{\partial R} \right|^{-1}_{R=R_t} \end{aligned} \tag{6.33}$$

is the weighting function, and $R_t(E, r_0)$ is the classical turning point. The summation runs over all trajectories which lead to the specified final vibrational quantum number n; they are determined through the equation

$$N(r_0) = n. \tag{6.34}$$

The left hand-side of Figure 6.9 depicts three examples of $N(r_0)$ for a potential of the form

$$V(R, r) = A \, e^{-\alpha[R - \epsilon(r - \bar{r})]} + \frac{k}{2}(r - \bar{r})^2. \tag{6.35}$$

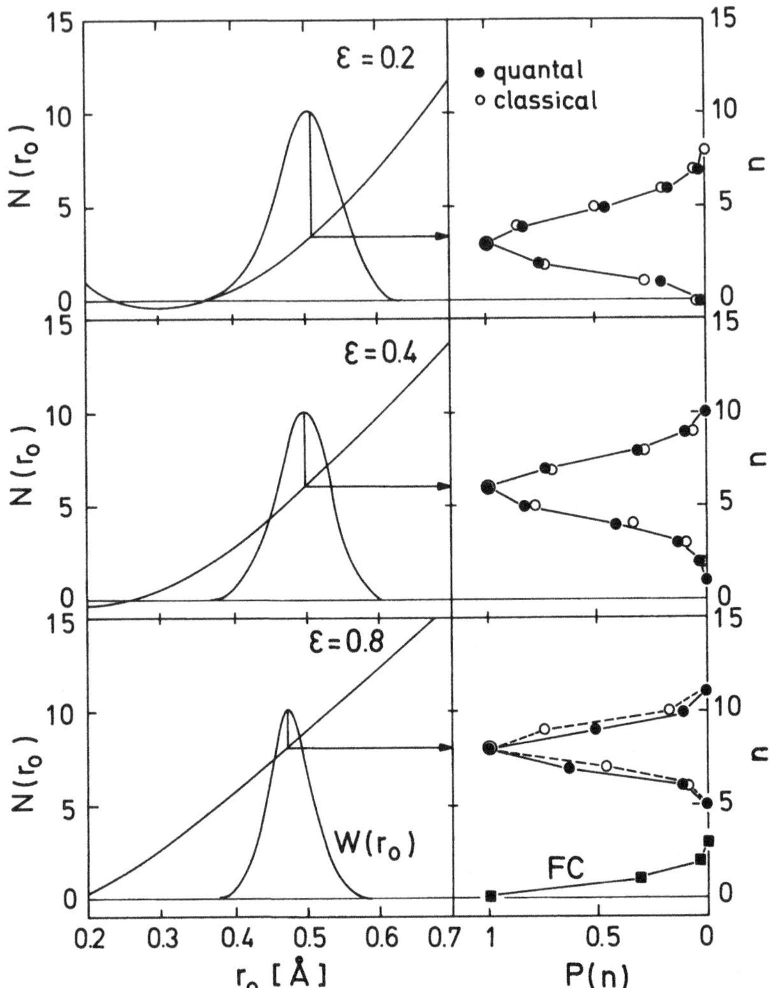

Fig. 6.9. Left-hand side: Vibrational excitation function $N(r_0)$ and weighting function $W(r_0)$ versus the initial oscillator coordinate r_0 for three values of the coupling parameter ϵ. The equilibrium separation of the free BC molecule is $\bar{r} = 0.403$ Å and the equilibrium value within the parent molecule is $r_e = 0.481$ Å. Right-hand side: Final vibrational state distributions $P(n)$ for fixed energy E; the quantum mechanical and the classical distributions are normalized to the same height at the maxima. The classical distributions are obtained with the help of (6.32). The lowest part of the figure contains also the pure Franck-Condon (FC) distribution $\langle \varphi_n(r) \,|\, \varphi_r(r) \rangle$, where φ_n is the nth vibrational wavefunction of the free BC molecule and φ_r is the r-dependent part of the initial wavefunction in the electronic ground state. The parameters correspond roughly to the dissociation of CF_3I. Reproduced from Untch, Hennig, and Schinke (1988).

The free BC oscillator is assumed to be harmonic with force constant k and equilibrium separation $\bar{r}$; the parameter ϵ controls the coupling between the dissociation coordinate R and the vibrational coordinate r. For $\epsilon = 0$ (elastic limit) the equations of motion for (R, P) and (r, p) decouple and energy *cannot* flow from one degree of freedom to the other. As a consequence, the vibrational energy of the oscillator remains constant throughout the dissociation and the corresponding vibrational excitation function, which for zero initial momentum p_0 is given by

$$N(r_0) = (\mu k)^{1/2} \frac{(r_0 - \bar{r})^2}{2\hbar} - \frac{1}{2}, \tag{6.36}$$

simply reflects the harmonic potential of BC. With increasing coupling parameter, energy exchange between the two modes becomes more and more efficient and the excitation function gradually shifts to smaller values of r_0 while its general form remains the same.

As in Section 6.3, the weighting function $W(r_0)$ is confined to a narrow region around the ground-state equilibrium r_e such that the sum in (6.32) reduces to a single term. Approximating $N(r_0)$ around r_e according to

$$N(r_0) \approx N_e + N_r \, (r_0 - r_e) \tag{6.37}$$

with $N_e = N(r_e)$ and $N_r = dN/dr_0|_{r_0=r_e}$ yields the (unnormalized) final vibrational state distribution

$$P(n) = \frac{e^{-2\beta(n-N_e)^2/\hbar}}{|N_r|} \tag{6.38}$$

with $\beta = (N_r^2/\alpha_r)^{-1}$. This is the *vibrational reflection principle* (Untch, Hennig, and Schinke 1988):

- The final vibrational state distribution essentially reflects the initial distribution of the vibrational coordinate mediated by the excitation function $N(r_0)$.

The distributions on the right-hand side of Figure 6.9 clearly illustrate the vibrational reflection principle. As the coupling between translation and vibration increases the degree of excitation increases as well and the distribution gradually shifts to higher vibrational quantum numbers. The width is approximately given by

$$\Delta N = N_r \, \Delta r, \tag{6.39}$$

where Δr denotes the width of the vibrational coordinate distribution of the parent molecule, $|\varphi_r(r_0)|^2$, and N_r is the steepness of the excitation function. In the same way as the rotational state distribution reflects the upper-state anisotropy, $P(n)$ reflects the r-dependence of the PES in the excited electronic state, especially near the equilibrium of the parent molecule. Equation (6.34) establishes, via the excitation function, the unique relation between the initial coordinate r_0 and the final vibrational quantum number n.

The right-hand side of Figure 6.9 compares exact quantum mechanical distributions with the results of purely classical calculations obtained from (6.32)–(6.34). The perfect agreement demonstrates the validity of the simple classical theory. Since the excitation function has a simple shape, only very few trajectories need to be calculated in order to construct the excitation function. In contrast to other models, the potential is not approximated and the equations of motion are solved exactly.

The photodissociation of CF_3I into CF_3 and I (Felder 1990) illustrates the vibrational reflection principle. Figure 6.10 shows the population of the C-F_3 umbrella mode following the photolysis with a 248-nm laser. These data are extracted from time-of-flight measurements in which the distribution of the arrival time at the detector is transformed into the internal state distribution. Since the energy of the fragment can, in principle, be distributed among several degrees of freedom, an analysis which takes into account only the umbrella mode is not unambiguous. Nevertheless, the general shape of the measured distribution is very similar to those shown in Figure 6.9. Employing a two-dimensional, pseudo-collinear model suggested by van Veen, Baller, de Vries, and Shapiro (1985) together with potential (6.35) and $\epsilon = 0.25$ yields a rather good agreement with the measurement. However, whether such a limited model is realistic cannot be decided on the basis of only one final state distribution.

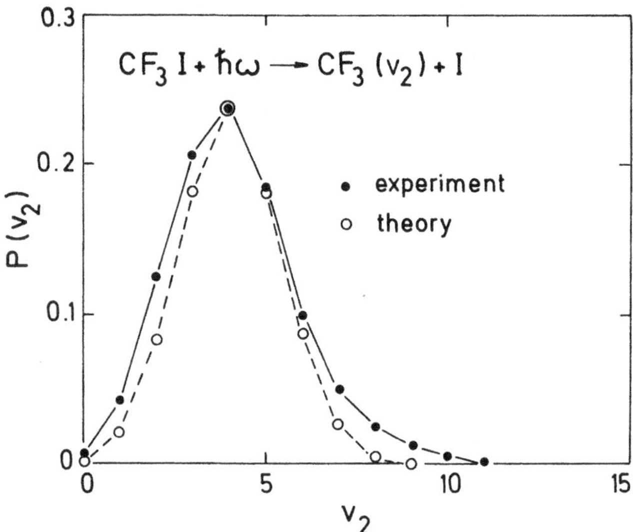

Fig. 6.10. Measured (Felder 1990) and calculated vibrational state distributions for the C-F_3 umbrella mode (ν_2) in the dissociation of CF_3I at 248 nm.

6.5 Epilogue

In contrast to indirect dissociation, which is the topic of Chapter 7, direct photodissociation is relatively simple to understand. The reflection principle describes *qualitatively* the fully state-resolved photofragmentation cross sections $\sigma(E, n, j)$ as a multi-dimensional mapping of the initial coordinate distribution in the electronic ground state:

$$|\Psi_i(R, r, \gamma; E_i)|^2 \approx |\varphi_R(R)|^2 \; |\varphi_r(r)|^2 \; |\varphi_\gamma(\gamma)|^2 \Longrightarrow \sigma(E, n, j). \qquad (6.40)$$

This mapping is unique, i.e., each partial cross section is related to exactly one point in the three-dimensional coordinate space of the parent molecule. Equations (6.4), (6.24), and (6.34) establish the relation between the coordinates (R, r, γ) on one hand and (E, n, j) on the other:

$$\begin{aligned} R_0 &\longrightarrow E &:& \quad V(R_0) = E \\ \gamma_0 &\longrightarrow j &:& \quad J(\gamma_0) = j \\ r_0 &\longrightarrow n &:& \quad N(r_0) = n. \end{aligned}$$

The mapping between (R, r, γ) and (E, n, j) is direct because long-lived "snarled" trajectories, which destroy the simple relation between initial coordinates and final "momenta", do not occur.

The general shape of the cross sections as functions of E, n, and j resembles the initial distribution function in the electronic ground state. If the parent molecule is initially in its lowest vibrational state, the partial cross section behaves qualitatively like a three-dimensional bell-shape function of E, n, and j. If the photodissociation starts from an excited vibrational state, $\sigma(E, n, j)$ will exhibit undulations which reflect the nodal structures of the corresponding bound-state wavefunction [more of this in Chapter 13; see also Shapiro (1981) and Child and Shapiro (1983)].

Relation (6.40) represents a dynamical mapping which is mediated by Hamilton's equations of motion in the upper state and ultimately by the forces $-\partial V/\partial R$ and $-\partial V/\partial r$ and the torque $-\partial V/\partial \gamma$. The energy dependence is mainly determined by the slope $\partial V/\partial R$ of the potential in the direction of the dissociation path while $\partial V/\partial r$ and $\partial V/\partial \gamma$ control the vibrational and rotational state distribution of the fragment.

Because of the lack of quantum mechanical interference effects classical mechanics is well suited for the treatment of direct dissociation. Very few trajectories actually suffice to construct the rotational and the vibrational excitation functions which establish the unique relation between (r_0, γ_0) and (n, j). $J(\gamma_0)$ and $N(r_0)$ are the links between the multi-dimensional PES on one hand and the final state distributions on the other.

7

Indirect photodissociation: Resonances and recurrences

In indirect photodissociation a potential barrier or some other dynamical constraint hinders immediate dissociation of the complex that the light pulse has created in the excited electronic state. Figure 7.1 shows a typical one-dimensional example. The barrier may be due to an avoided crossing with another electronic state. The potential energy surface (PES) of the S_1 state of CH_3ONO (Figure 1.11) is a typical two-dimensional example. Depending on the efficiency of internal energy redistribution between the various degrees of freedom the lifetime of the complex may range from a few to several thousand internal vibrational periods, in contrast to direct processes where the fragmentation finishes in less than one internal period. Because of the long lifetime, the final state distributions of the photofragments no longer reflect the initial coordinate distribution of the parent molecule in the ground electronic state like they do in direct dissociation. The coupling between the various modes in the complex gradually erases the memory of the initial state before the molecule finally breaks apart.

The main characteristics of indirect dissociation are *resonances* in the time-independent picture and *recurrences* in the time-dependent approach. Resonances and recurrences are the two sides of one coin; they reveal the same dynamical information but provide different explanations and points of view. To begin this chapter we discuss in Section 7.1, on a qualitative level, indirect photodissociation of a one-dimensional system. A more quantitative analysis follows in Section 7.2. The time-dependent and the time-independent views of indirect photodissociation are outlined and illustrated in Sections 7.3 and 7.4, respectively, with emphasis on vibrational excitation of the NO moiety in the photodissociation of $CH_3ONO(S_1)$. Section 7.5 accentuates the relation between

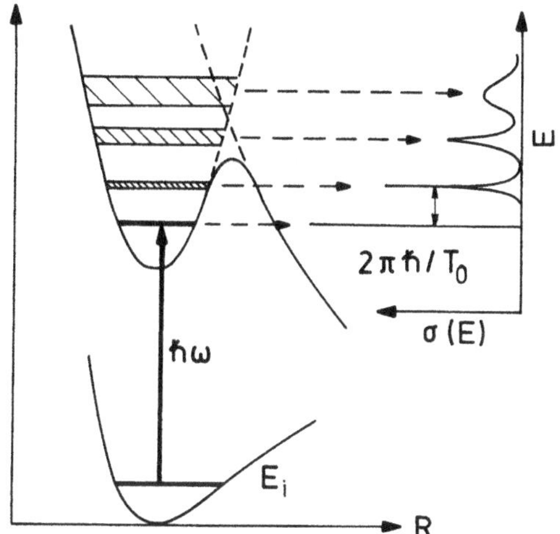

Fig. 7.1. Schematic illustration of indirect photodissociation for a one-dimensional system. The two dashed potential curves represent so-called *diabatic potentials* which are allowed to cross. The solid line represents the lower member of a pair of *adiabatic potential* curves which on the contrary are prohibited to cross. The other adiabatic potential, which would be purely binding, is not shown here. More will be said about the diabatic and the adiabatic representations of electronic states in Chapter 15. The right-hand side shows the corresponding absorption spectrum with the shaded bars indicating the resonance states embedded in the continuum. The lighter the shading the broader the resonance and the shorter its lifetime.

resolution in the time and in the energy domains. Finally, we discuss in Section 7.6 two additional kinds of internal excitation.

7.1 A phenomenological prologue

Let us consider the photodissociation of a diatomic molecule as illustrated in Figure 7.1. A classical particle with an energy below the barrier energy is trapped for ever in the potential well. However, if the energy exceeds the barrier the molecule dissociates at once without recurring to its origin. The situation is more complex if the molecule is described quantum mechanically. A quantum mechanical wavepacket, promoted by a delta-function pulse to the upper potential, embraces an appreciable range of energies and therefore comprises the two classical extremes. Figure 7.2 illustrates its motion in the upper electronic state. The wavepacket first moves toward the barrier without changing its overall shape. Near the barrier, however, it breaks up into two pieces. One piece continues to move outward and when it reaches the steep exit channel it passes quickly out. The other part turns round, returns to its origin where it is reflected at the inner wall of the potential, and a new cycle begins. Each

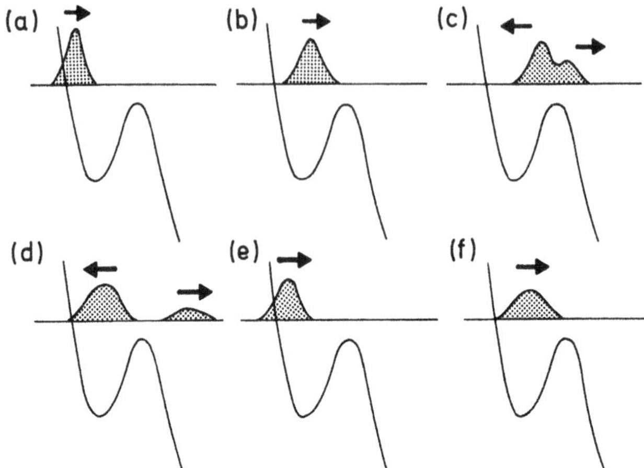

Fig. 7.2. Cartoon of the evolution of a one-dimensional wavepacket, $|\Phi(t)|$, in a potential with barrier. Remember that the horizontal line does not represent a particular energy! The wavepacket is a superposition of stationary wavefunctions for a whole range of energies.

time the wavepacket reaches the barrier some fraction escapes from the inner region while the remaining part is reflected back into the well.

The periodic motion of the wavepacket in the potential well naturally shows up in the autocorrelation function $S(t)$ as depicted schematically in Figure 7.3. Each return of $\Phi(t)$ to its origin leads to a maximum in the autocorrelation function, a so-called *recurrence.* Since the part that is temporarily trapped in the inner region gradually diminishes the overall amplitude of $S(t)$ decays in time. Eventually, the entire wavepacket leaks

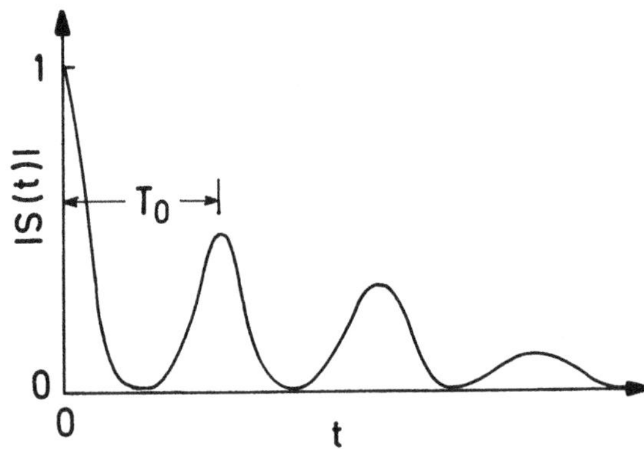

Fig. 7.3. Autocorrelation function for the temporarily trapped wavepacket illustrated in Figure 7.2.

from the intermediate potential well and travels freely in the asymptotic region. According to the basic properties of the Fourier transformation the periodic structures of the autocorrelation function imply structures in the energy dependence of the absorption spectrum $\sigma(E)$ which is shown on the right-hand side of Figure 7.1. The spectrum is composed of several narrow lines approximately separated by $2\pi\hbar/T_0$ where T_0 is the recurrence time, i.e., the average period of the oscillation in the potential well. A more quantitative analysis of the autocorrelation function and its relation to the spectrum follows in Section 7.3.

Within the time-independent approach, one considers the energy dependence of the stationary wavefunction $\Psi(E)$ in order to explain distinct structures in the absorption spectrum. It is well known from scattering theory that $\Psi(E)$ can abruptly change its overall shape in certain regions of energy (Wu and Ohmura 1962:ch.5; Taylor 1972:ch.13; Child 1974:ch.4; Fano and Rao 1986:ch.8; Satchler 1990:ch.4.8; Child 1991:ch.10). This behavior is called a resonance and Figure 7.4 schematically illustrates this fundamental phenomenon. For energies off resonance, $\Psi(E)$ behaves like an ordinary continuum wavefunction that does not significantly penetrate through the barrier into the well region; it is reflected at the outer wall of the potential. On resonance, however, $\Psi(E)$ appears qualitatively as a bound-state wavefunction that accumulates mainly in the potential well. Because of the normalization of the continuum wavefunction, Equation (2.51), the amplitude in the asymptotic region scales as $E^{-1/4}$ irrespective of whether the energy is on or off resonance. It is the amplitude in the inner region that changes abruptly in the vicinity of a resonance. Since the overlap of the bound wavefunction of the electronic ground state with the inner part of the continuum wavefunction determines the absorption cross section, the dramatic variation of $\Psi(E)$ as a function of

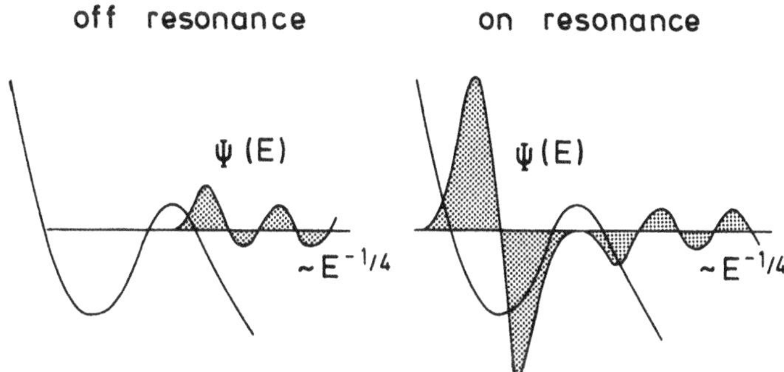

Fig. 7.4. Schematic illustration of the stationary wavefunction $\Psi(E)$ for energies off and on resonance. In contrast to Figure 7.2, here the horizontal line marks a particular energy.

energy obviously shows up in the energy dependence of the spectrum as well.

Resonances occur in the neighborhood of the discrete energy levels of the "binding" part of the potential, i.e., the binding diabatic potential in Figure 7.1. This is schematically indicated by the horizontal lines in Figure 7.1. In the vicinity of these "zeroth-order" energies the molecule — loosely speaking — experiences the underlying bound states and the continuum wavefunction turns qualitatively into a bound-state-like wavefunction:

- Resonances in the absorption spectrum are the fingerprints of the eigenenergies of the "binding" part of the zeroth-order Hamiltonian (i.e., the Hamiltonian in the absence of coupling between the two diabatic states).
- The coupling to the "dissociative" part of the Hamiltonian (represented by the repulsive diabatic potential curve in Figure 7.1) broadens the discrete energy levels with the widths depending on the coupling strength between the two manifolds.
- Resonances are quasi-bound or metastable states embedded in the continuum.

Resonances exist below as well as above the barrier. Their width usually increases with energy, i.e., with increasing excitation of the dissociation mode.†

Resonances are common and unique features of elastic and inelastic collisions, photodissociation, unimolecular decay, autoionization problems, and related topics. Their general behavior and formal description are rather universal and identical for nuclear, electronic, atomic, or molecular scattering. Truhlar (1984) contains many examples of resonances in various fields of atomic and molecular physics. Resonances are particularly interesting if more than one degree of freedom is involved; they "reflect" the quasi-bound states of the Hamiltonian and reveal a great deal of information about the multi-dimensional PES, the internal energy transfer, and the decay mechanism. A quantitative analysis based on time-dependent perturbation theory follows in the next section.

7.2 Decay of excited states

Let us consider a typical situation as sketched in Figure 7.5. V_1 and V_2 represent the (diabatic) potential energy curves (or surfaces) of two

† A very elementary but yet informative discussion of resonance phenomena for a square-well potential, which features both the time-independent and the time-dependent picture, can be found in the textbook by Messiah (1972:ch.III.)

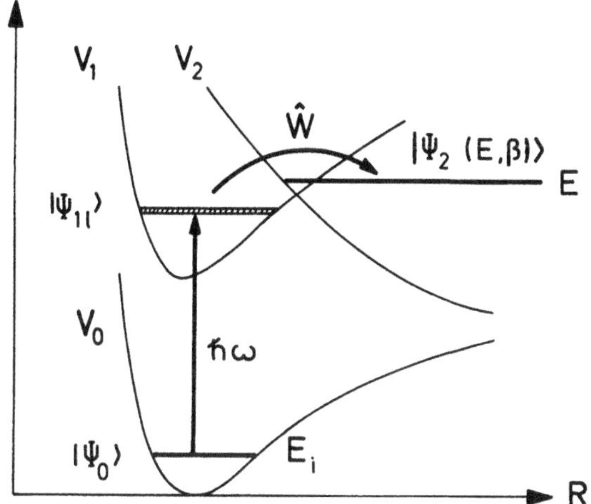

Fig. 7.5. Schematic illustration of the decay of an excited state $|\Psi_{1l}\rangle$, prepared by the absorption of a UV photon, due to the coupling to a continuum of states $|\Psi_2(E,\beta)\rangle$ belonging to the repulsive electronic state.

excited electronic states. V_1 is binding with bound states $|\Psi_{1l}\rangle$ and energies E_l while V_2 is repulsive with outgoing continuum states $|\Psi_2(E,\beta)\rangle$ which we defined in Section 2.5. The index l comprises a set of rotational-vibrational quantum numbers and the index β distinguishes the possible degenerate final states of the fragments. For the dissociation of a triatomic molecule, ABC → A + BC, β represents a particular vibrational-rotational state of the molecular fragment BC or a particular electronic state of the atomic fragment, for example. In the absence of any coupling between the bound and the dissociative manifolds the $|\Psi_{1l}\rangle$ and the $|\Psi_2(E,\beta)\rangle$ are the true eigenstates of the respective zeroth-order Hamiltonians, i.e.

$$(\hat{T}_{nu} + V_1 - E_l)\Psi_{1l}(E_l) = 0$$
$$(\hat{T}_{nu} + V_2 - E)\Psi_2(E,\beta) = 0,$$

where $\hat{T}_{nu}$ is the nuclear Hamiltonian. They fulfil the orthogonality relations[†]

$$\langle \Psi_{1l} \mid \Psi_{1l'} \rangle = \delta_{ll'} \tag{7.1a}$$
$$\langle \Psi_{1l} \mid \Psi_2(E,\beta) \rangle = 0 \tag{7.1b}$$
$$\langle \Psi_2(E,\beta) \mid \Psi_2(E',\beta') \rangle = 2\pi\hbar\ \delta_{\beta\beta'}\ \delta(E-E') \tag{7.1c}$$

[†] The wavefunctions include, in principle, all nuclear and electronic degrees of freedom which guarantees the orthogonality of the two sets [Equation (7.1b)]. For simplicity of the presentation, however, we display here only the nuclear part.

and form complete sets in the respective Hilbert spaces.

In the presence of coupling between the two excited manifolds, represented by the operator $\hat{W}$, the bound states $|\Psi_{1l}\rangle$ generated, for example, either by absorption of a photon (as illustrated in Figure 7.5), by electron impact, or in an atom-molecule collision will decay because they undergo transitions to the continuum states. $\hat{W}$ is assumed to be time-independent and for the discussion in this section its origin and particular form is not pertinent. It may represent nonadiabatic coupling between two electronic or two vibrational states, for example. We explicitly assume that $\hat{W}$ couples only the bound and the continuum states and that there is no coupling between the bound or between the continuum states, i.e.,

$$\langle \Psi_2(E,\beta) \mid \hat{W} \mid \Psi_2(E',\beta') \rangle = \langle \Psi_{1l} \mid \hat{W} \mid \Psi_{1l'} \rangle = 0. \tag{7.2}$$

If the coupling is zero, the bound states will live forever. However, immediately after we have switched on the coupling they start to decay as a consequence of transitions to the continuum states until they are completely depopulated. Our goal is to derive explicit expressions for the depletion of the bound states $|\Psi_{1l}\rangle$ and the filling of the continuum states $|\Psi_2(E,\beta)\rangle$. The method we use is time-dependent perturbation theory in the same spirit as outlined in Section 2.1, with one important extension. In Section 2.1 we explicitly assumed that the perturbation is sufficiently weak and also sufficiently short to ensure that the population of the initial state remains practically unity for all times (first-order perturbation theory). In this section we want to describe the decay process until the initial state is completely depleted and therefore we must necessarily go beyond the first-order treatment. The subsequent derivation closely follows the detailed presentation of Cohen-Tannoudji, Diu, and Laloë (1977:ch.XIII).

As in Section 2.5 we expand the total time-dependent wavefunction as

$$\Phi(t) = a_{1l}(t)\ \Psi_{1l}\ \mathrm{e}^{-iE_l t/\hbar} + \sum_\beta \int dE\ a_2(t;E,\beta)\ \Psi_2(E,\beta)\ \mathrm{e}^{-iEt/\hbar}. \tag{7.3}$$

To simplify the notation we have assumed that the light pulse has prepared the system in a single bound state. The probabilities for finding the system in states $|\Psi_{1l}\rangle$ and $|\Psi_2(E,\beta)\rangle$ at time t are $|a_{1l}(t)|^2$ and $|a_2(t;E,\beta)|^2$, respectively. Inserting (7.3) into the time-dependent Schrödinger equation with the full Hamiltonian which also includes the coupling $\hat{W}$ and utilizing (7.1) and (7.2) yields the coupled equations

$$i\hbar\frac{d}{dt}a_{1l} = \sum_\beta \int dE\ \langle \Psi_{1l} \mid \hat{W} \mid \Psi_2(E,\beta)\rangle\ \mathrm{e}^{i(E_l-E)t/\hbar}\ a_2(E,\beta) \tag{7.4}$$

$$i\hbar \frac{d}{dt} a_2(E,\beta) = \rho \, \langle \Psi_2(E,\beta) \mid \hat{W} \mid \Psi_{1l} \rangle \, e^{i(E-E_l)t/\hbar} \, a_{1l}, \tag{7.5}$$

where $\rho = (2\pi\hbar)^{-1}$. Equations (7.4) and (7.5) are completely analogous to (2.10). They must be solved subject to the initial conditions $a_{1l}(t = 0) = 1$ and $a_2(t = 0; E, \beta) = 0$.

Integration of (7.5) yields the formal solution

$$i\hbar a_2(t; E,\beta) = \rho \, \langle \Psi_2(E,\beta) \mid \hat{W} \mid \Psi_{1l} \rangle \int_0^t dt' \, e^{i(E-E_l)t'/\hbar} \, a_{1l}(t') \tag{7.6}$$

and after inserting (7.6) into (7.4) we obtain an integro-differential equation for the coefficient $a_{1l}(t)$,

$$\frac{d}{dt} a_{1l}(t) = -\frac{\rho}{\hbar^2} \sum_\beta \int dE \, |\langle \Psi_{1l} \mid \hat{W} \mid \Psi_2(E,\beta) \rangle|^2 \times \int_0^t dt' \, e^{i(E_l-E)(t-t')/\hbar} \, a_{1l}(t'). \tag{7.7}$$

Thus, $da_{1l}(t)/dt$ depends on the entire "history" of the system from the beginning at $t = 0$ until the actual time t. Equation (7.7) is still exact within the limiting assumptions discussed above.

As a consequence of the oscillatory phase factor in (7.7), contributions to the integral over t' stem mainly from the region around $t \approx t'$ so that $a_{1l}(t')$ on the right-hand side may be approximated by its value at time t. With this substitution Equation (7.7) becomes an ordinary differential equation,†

$$\frac{d}{dt} a_{1l}(t) \approx -\frac{1}{\hbar^2} \, a_{1l}(t) \sum_\beta I(t; E,\beta), \tag{7.8}$$

where we defined

$$I(t; E,\beta) = \rho \int dE \, |\langle \Psi_{1l} \mid \hat{W} \mid \Psi_2(E,\beta) \rangle|^2 \int_0^t dt' \, e^{i(E_l-E)(t-t')/\hbar}. \tag{7.9}$$

In the limit $t \to \infty$ the integral over t' gives

$$\lim_{t\to\infty} \int_0^t dt' \, e^{i(E_l-E)(t-t')/\hbar} = \hbar \left[\pi \, \delta(E_l - E) + i \, \mathcal{P} \left(\frac{1}{E_l - E} \right) \right], \tag{7.10}$$

where $\mathcal{P}(f)$ is the principal value of the function f. Finally, the rate equation for $a_{1l}(t)$ becomes, in the limit of large t,

$$\frac{d}{dt} a_{1l}(t) \approx -a_{1l}(t) \sum_\beta \left[\frac{\Gamma_\beta}{2} + \frac{i}{\hbar} \delta E_\beta \right]. \tag{7.11}$$

† For a more detailed affirmation of this argument consult the textbook by Cohen-Tannoudji, Diu, and Laloë (1977:ch.XIII).

The *partial decay rates* Γ_β are defined by

$$\begin{aligned}\Gamma_\beta &= \frac{2\pi}{\hbar}\rho \int dE \, |\langle \Psi_{1l} \mid \hat{W} \mid \Psi_2(E,\beta)\rangle|^2 \, \delta(E_l - E) \\ &= \frac{2\pi}{\hbar}\rho \, |\langle \Psi_{1l} \mid \hat{W} \mid \Psi_2(E_l,\beta)\rangle|^2\end{aligned} \tag{7.12}$$

and the *partial shifts* δE_β are given by

$$\delta E_\beta = \rho \, \mathcal{P} \left[\int dE \, \frac{|\langle \Psi_{1l} \mid \hat{W} \mid \Psi_2(E,\beta)\rangle|^2}{E_l - E} \right]. \tag{7.13}$$

With the definition of the continuum wavefunctions in (2.51) it is straightforward to show that the partial rates have units of frequency and the partial shifts have units of energy.

Integration of (7.11) readily yields

$$a_{1l}(t) \approx \mathrm{e}^{-\Gamma t/2} \, \mathrm{e}^{-i\delta E \, t/\hbar}, \tag{7.14}$$

with the *total decay rate* defined as

$$\Gamma = \sum_\beta \Gamma_\beta \tag{7.15}$$

and the *total shift* given by

$$\delta E = \sum_\beta \delta E_\beta. \tag{7.16}$$

The meaning of the name "shift" will become apparent in Equation (7.22). From (7.14) it follows immediately that

$$|a_{1l}(t)|^2 \approx \mathrm{e}^{-\Gamma t}. \tag{7.17}$$

Equations (7.15)–(7.17) sum up the main result of this section:

- The population of the initially excited bound state decays exponentially in time with the rate given by $\Gamma = \sum_\beta \Gamma_\beta$.
- The *lifetime* τ defined by $|a_{1l}(\tau)|^2 = 1/\mathrm{e}$ is, according to (7.17), the inverse of the total decay rate, i.e., $\tau = 1/\Gamma$.
- Both the decay rates and the shifts are proportional to the square of the strength of the coupling $\hat{W}$.

In the above derivation we did not explicitly specify the coupling operator $\hat{W}$ nor the zeroth-order states $|\Psi_{1l}\rangle$ and $|\Psi_2(E,\beta)\rangle$. In this monograph we will mainly use the results of this section to *qualitatively* interpret exact quantum mechanical calculations or experimental observations. A reasonable zeroth-order picture is well defined only if the coupling is weak and only under such conditions does Equation (7.12) provide an accurate prescription for calculating the partial rates Γ_β. The photodissociation of physically bound van der Waals molecules falls into this category (Chapter 12). For many molecular systems, however, a zeroth-order model is not obvious and the expression for the partial rates is of no particular use from the computational point of view. Nevertheless, even then the general results of the perturbation treatment will help us to explain the gross features of indirect photodissociation.

7.3 Time-dependent view: Recurrences

In this section we consider indirect photodissociation of systems with more than one degree of freedom in the time-dependent approach. We will use the results of Section 7.2 to derive approximate expressions for the wavepacket evolving in the upper electronic state, the corresponding autocorrelation function, and the various photodissociation cross sections.

7.3.1 Resonant absorption and Lorentzian line shapes

In order to simplify the general picture we assume that the photodissociation separates into two consecutive steps as illustrated in Figure 7.5. The light pulse promotes the molecule from the initial nuclear state $|\Psi_0\rangle$ in the ground electronic state to the vibrational-rotational states $|\Psi_{1l}\rangle$ with energies E_l in the (binding) excited electronic state. In the second step, the $|\Psi_{1l}\rangle$ couple to the manifold of continuum states and eventually they dissociate. Only the binding state is assumed to be dipole-allowed $[\mu_{10}^{(e)} \neq 0]$ whereas the dissociative state is "dark" $[\mu_{20}^{(e)} = 0]$.

In analogy to Equation (4.3) we expand the time-dependent wavepacket created by the delta-pulse in the excited electronic state in terms of the bound-state wavefunctions Ψ_{1l}. Using (4.5) we obtain†

$$\Phi_1(t) = \sum_l \langle \Psi_{1l} \mid \mu_{10}^{(e)} \mid \Psi_0 \rangle \, \Psi_{1l} \, \mathrm{e}^{-iE_l t/\hbar}, \tag{7.18}$$

where the transition matrix elements $\langle \Psi_{1l} \mid \mu_{10}^{(e)} \mid \Psi_0 \rangle$ are the moments for exciting the various quasi-bound states $|\Psi_{1l}\rangle$. Let us, for the time being, ignore the coupling to the continuum, i.e., $\hat{W} = 0$. Then, the autocorrelation function, given by

$$S(t) = \langle \Phi_1(0) \mid \Phi_1(t) \rangle = \sum_l |\langle \Psi_{1l} \mid \mu_{10}^{(e)} \mid \Psi_0 \rangle|^2 \, \mathrm{e}^{-iE_l t/\hbar}, \tag{7.19}$$

represents a superposition of sinusoidal functions with periods $T_l = 2\pi\hbar/E_l$ and amplitudes which are proportional to the corresponding absorption cross sections. $|S(t)|$ is simply a constant if the photon excites

† Since the Ψ_{1l} are bound-state wavefunctions, the normalization factor $\rho = (2\pi\hbar)^{-1}$ in Equation (4.5) must be omitted in the present case.

only a single vibrational state. If, however, several bound states are simultaneously excited, the autocorrelation function may exhibit a complicated time dependence due to the superposition of several oscillatory terms with different frequencies. Since the coupling to the continuum is omitted, $S(t)$ does not diminish to zero. Inserting the expression for the autocorrelation function into Equation (4.7) the absorption cross section becomes

$$\begin{aligned}\sigma_{tot}(E) &= 2\,\mathcal{C}\,E_{photon}\,\Re\left[\int_0^\infty dt\; e^{iEt/\hbar}\,S(t)\right]\\ &= 2\pi\hbar\,\mathcal{C}\,E_{photon}\sum_l |\langle\Psi_{1l}\mid\mu_{10}^{(e)}\mid\Psi_0\rangle|^2\;\delta(E-E_l).\end{aligned}\tag{7.20}$$

In the first line of (7.20) we have exploited the symmetry relation (4.10) and the constant $\mathcal{C}$ is defined in Section 2.5 as $\mathcal{C} = \rho\pi/\hbar\epsilon_0 c$ with $\rho = (2\pi\hbar)^{-1}$. The absorption spectrum is discrete as mandatory for a bound-bound transition.

If we switch on the coupling to the continuum at $t = 0$ the excited bound states begin to decay with the consequence that the wavepacket and therefore the autocorrelation function decay too. In order to account for this we multiply, according to Equation (7.14), each term in (7.18) by

$$e^{-\Gamma^{(l)}t/2}\;e^{-i\delta E^{(l)}t/\hbar},$$

where $\Gamma^{(l)}$ and $\delta E^{(l)}$ are the total decay rate and the total shift for the lth bound state, respectively. Each state decays separately with its own decay rate. Then, the autocorrelation function becomes

$$S(t) = \sum_l |\langle\Psi_{1l}\mid\mu_{10}^{(e)}\mid\Psi_0\rangle|^2\;e^{-i(E_l+\delta E^{(l)})t/\hbar}\;e^{-\Gamma^{(l)}t/2}.\tag{7.21}$$

At short times, when all states $|\Psi_{1l}\rangle$ are still significantly populated, $S(t)$ will show a complicated time dependence. As time elapses, however, states with shorter lifetimes decay, the number of contributions in (7.21) decreases, and the behavior of $S(t)$ gradually becomes simpler. At very long times when only the state with the smallest rate is still populated $S(t)$ becomes a simple function of time and its modulus decays exponentially to zero.

Inserting (7.21) into (7.20) yields, after some straightforward algebra, the following expression for the absorption spectrum,

$$\begin{aligned}\sigma_{tot}(E) = 2\hbar\,\mathcal{C}\,E_{photon}\sum_l |\langle\Psi_{1l}\mid\mu_{10}^{(e)}\mid\Psi_0\rangle|^2\\ \times\frac{[\hbar\Gamma^{(l)}/2]}{[E-E_l-\delta E^{(l)}]^2+[\hbar\Gamma^{(l)}/2]^2}.\end{aligned}\tag{7.22}$$

Equation (7.22) is the main result of this chapter:

- The coupling to the continuum states (i.e., dissociation) broadens the discrete absorption lines and therefore the discrete spectrum turns into a continuous spectrum.
- The intensities factorize into an ordinary (multi-dimensional) FC overlap factor involving two bound states and a dynamical, energy-dependent factor which reflects the quenching of each state.
- Each absorption peak has the shape of a Lorentzian centered at $[E_l + \delta E^{(l)}]$ with full width at half maximum (FWHM) of $\Delta E_{FWHM}^{(l)} = \hbar\Gamma^{(l)} = \hbar/\tau^{(l)}$ as illustrated in Figure 7.6(a).

$\Gamma^{(l)}$ and $\tau^{(l)}$ are the total decay rate and the *lifetime* of state $|\Psi_{1l}\rangle$, respectively. The width of the Lorentzian is proportional to $\Gamma^{(l)}$ and its maximum scales with $1/\Gamma^{(l)}$. If the coupling to the continuum diminishes the Lorentzian gradually becomes narrower and at the same time the peak intensity goes to infinity.† Note, however, that the area under a Lorentzian is independent of its width. Equation (7.22) readily explains

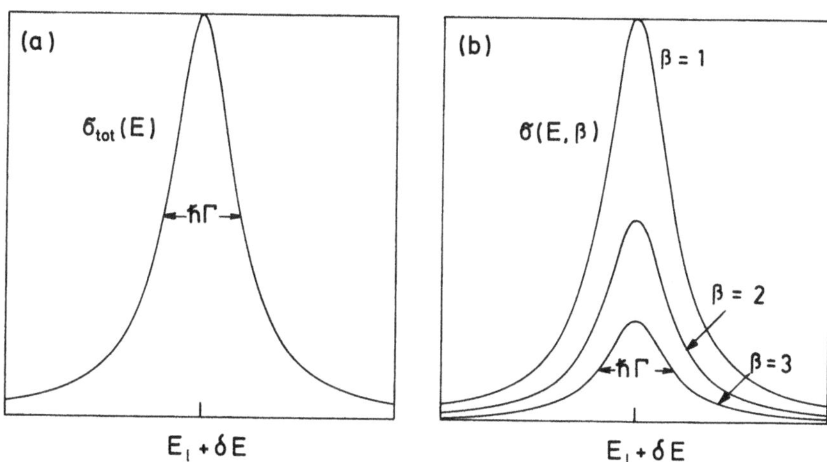

Fig. 7.6. (a) Energy dependence of a Lorentzian line-shape function with width $\hbar\Gamma$ centered at the resonance energy $(E_l + \delta E)$. (b) Partial photodissociation cross sections $\sigma(E, \beta)$ as given by (7.23). All of them have the same width $\hbar\Gamma$; the values at the maximum scale like the partial decay rates Γ_β.

† Using the following representation of the Dirac delta-function (Loudon 1983:ch.2),

$$\delta(x) = \frac{1}{\pi} \lim_{\epsilon \to 0} \frac{\epsilon}{x^2 + \epsilon^2},$$

one easily verifies that in the limit $\Gamma^{(l)} \to 0$ Equation (7.22) goes smoothly over into the discrete bound-bound spectrum (7.20).

why δE is called a shift: the maximum of the cross section occurs at an energy which is shifted by δE with respect to the zero-order energies E_l.

So far we have considered only the decay of the initial quasi-bound states and the influence on the total spectrum. Recalling that the total rate $\Gamma^{(l)}$ is the sum of all partial rates $\Gamma_\beta^{(l)}$ we can recast (7.22) as a sum of partial photodissociation cross sections, each being defined by

$$\sigma(E, \beta) = \sigma_{tot}(E) \frac{\Gamma_\beta^{(l)}}{\Gamma^{(l)}} \tag{7.23}$$

(at the lth resonance). Equation (7.23) implies that:

- All partial photodissociation cross sections $\sigma(E, \beta)$ have the same profile; they peak at the same resonance energies $E_l + \delta E^{(l)}$ and have the same width $\Delta E_{FWHM}^{(l)}$ irrespective of the final fragment state as illustrated in Figure 7.6(b).[†]
- The final state distribution, which is determined by the partial decay rates $\Gamma_\beta^{(l)}$, does not change across an absorption line.

This behavior is remarkably different from direct photodissociation where the partial cross sections can be noticeably shifted on the energy axis (see Figure 3.5 for an example for vibrational excitation). Figure 7.7 illustrates the different energy dependences of partial cross sections in direct and in indirect dissociation with two examples of rotational excitation: the photodissociation of $\mathrm{ClNO}(S_1) \rightarrow \mathrm{Cl} + \mathrm{NO}(j)$, which is a direct process, and the photodissociation of $\mathrm{HONO}(S_1) \rightarrow \mathrm{HO} + \mathrm{NO}(j)$, which evolves through a long-lived intermediate. In both cases, the degree of rotational excitation of NO is large with the rotational state distributions being maximal around $j \approx 28$. The multi-dimensional reflection principle outlined in Section 6.2.2 determines the line shape of each partial absorption cross section in the case of ClNO (Untch, Weide, and Schinke 1991b). The shift between the cross sections manifests, within the adiabatic picture, the grade of internal excitation near the transition region. As a consequence of this shift the final rotational distribution actually varies with E. The partial cross sections for HONO, on the other hand, all have Lorentzian line shapes as predicted by (7.22). All of them peak at the same energy, have the same width, and merely differ in the peak intensities. The final distribution is therefore energy-independent (Schinke, Untch, Suter, and Huber 1991).

† The Γ_β are sometimes called *partial widths* which could be erroneously interpreted to mean that the partial cross sections have different widths. That is not the case, however. All partial cross sections have the same width!

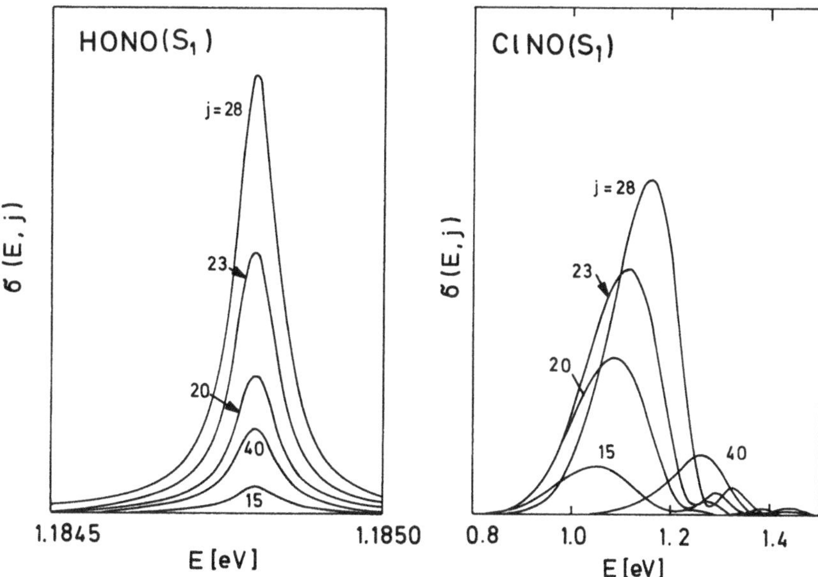

Fig. 7.7. Comparison of the different energy behavior of partial dissociation cross sections $\sigma(E, j)$ for the production of NO(j) in indirect, HONO(S_1), and in direct, ClNO(S_1), photofragmentation. Note the quite different energy scales! The results for HONO are obtained from a two-dimensional model (Schinke, Untch, Suter, and Huber 1991) and the cross sections for ClNO are taken from a three-dimensional wavepacket calculation (Untch, Weide, and Schinke 1991b).

Equation (7.22) is at the heart of spectroscopy. The positions of the absorption lines reflect the energy levels of the excited complex and the widths provide information about the lifetime and therefore about the coupling to the continuum states. The latter requires, however, that the measured widths are the true *homogeneous line widths*, i.e., unadulterated by poor resolution and/or thermal broadening, for example. Each resonance has a characteristic width. In Chapters 9 and 10 we will discuss how the final fragment distributions reflect the initial state in the complex and details of the fragmentation mechanism.

7.3.2 Example: Internal vibrational excitation

The photodissociation of methyl nitrite in the first absorption band, $CH_3ONO(S_1) \longrightarrow CH_3O + NO(n, j)$, exemplifies indirect photodissociation (Hennig et al. 1987). Figure 1.11 shows the two-dimensional potential energy surface (PES) of the S_1 electronic state as a function of the two O-N bonds. All other coordinates are frozen at the equilibrium values in the electronic ground state. Although these two modes suffice to illustrate the overall dissociation dynamics, a more realistic picture also requires the inclusion of the ONO bending angle (Nonella, Huber, Untch, and Schinke 1989). Figure 7.8 shows the same PES, however, as a function of Jacobi coordinates R and r which are defined in Figure 1.10(a).

A small potential barrier of less than 0.1 eV with respect to the bottom of the shallow potential well hinders direct dissociation of the excited $CH_3ONO(S_1)$ compound. This holds true even for total energies high above this barrier! Let us first discuss how a typical trajectory evolves on this PES after it is started near the ground-state equilibrium. At the Franck-Condon (FC) point, indicated by the heavy dot in Figure 7.8, the force in the direction of the N-O bond, $F_r = -\partial V/\partial r$, considerably

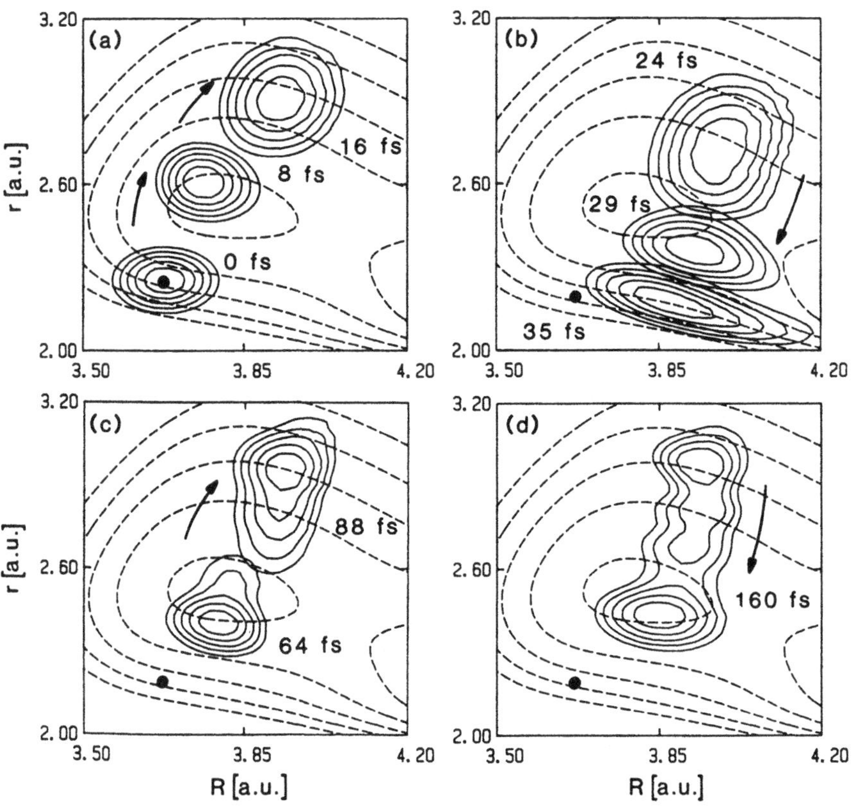

Fig. 7.8. Snapshots of the two-dimensional time-dependent wavepacket evolving on the PES of the S_1 state of CH_3ONO (indicated by the broken contours); only the inner part of $|\Phi(t)|$ is depicted. The Jacobi coordinates R and r denote the distance of CH_3O from the center-of-mass of NO and the internal separation of the NO moiety, respectively. The heavy point marks the equilibrium in the S_0 state where the evolution begins. The arrows indicate the evolution of the wavepacket. Adapted from Engel, Schinke, Hennig, and Metiu (1990).

exceeds the force in the direction of the dissociation coordinate, $F_R = -\partial V/\partial R$. Therefore, instead of dissociating the trajectory first slides down into the well. Initially, the main portion of the released energy goes into the vibrational mode of the NO moiety while the energy going into the dissociation mode does not suffice to surpass the barrier. The first step in the S_1 state is vibrational excitation of NO.

Due to coupling between R and r energy flows slowly from the vibrational to the translational mode and eventually enough energy accumulates in the dissociation coordinate to enable the trajectory to pass through the bottleneck (*internal vibrational energy redistribution*, IVR). During its "lifetime", the trajectory performs a complicated Lissajou-type motion within the well, i.e., both NO bonds are significantly excited and the coupling between them leads to a complicated internal vibration. This is a typical situation for Herzberg's type II predissociation (Herzberg 1967:ch.IV). The survival time of the trajectory within the intermediate region depends sensitively on the coupling between R and r, i.e., on the forces F_R and F_r and therefore on the shape of the PES in the well and in the barrier region. The weaker the coupling the longer is the lifetime of the excited $CH_3ONO(S_1)$ complex.

The quantum mechanical wavepacket follows essentially the classical trajectory, at least for short times (Figure 7.8). In the beginning, it remains compact and well localized. After about 30 fs it returns for the first time to the vicinity of the FC region where it originally started; this recurrence leads to the first maximum of the autocorrelation function which is shown in Figure 7.9. With increasing time the wavepacket gradually becomes more delocalized and after a few hundred femtoseconds it spreads over the entire potential well. At all times some fraction of the wavepacket, that is not shown in the figure, leaks through the bottleneck and passes out into the exit channel where it rapidly leads to dissociation.

According to the discussion in Section 7.2, the wavepacket in the inner region contains all quasi-bound stationary states $|\Psi_{1l}\rangle$ damped by $e^{-\Gamma^{(l)}t/2}$. The superposition of several states in Equation (7.21) implies a rather complicated behavior of $|S(t)|$ for intermediate times [Figure 7.9(b)]. The states with the largest decay rates $\Gamma^{(l)}$ dissociate most rapidly with the result that fewer and fewer stationary states survive in the wavepacket and the autocorrelation function becomes more regular [Figure 7.9(c)]. Eventually, the wavepacket embraces only the two states with the smallest decay rates and then the autocorrelation function exhibits a clear *quantum beat* structure with period $2\pi\hbar/\Delta E$ where ΔE is the energy separation of these two states [Figure 7.9(d)]. At even longer times, only the state with the longest lifetime is still populated

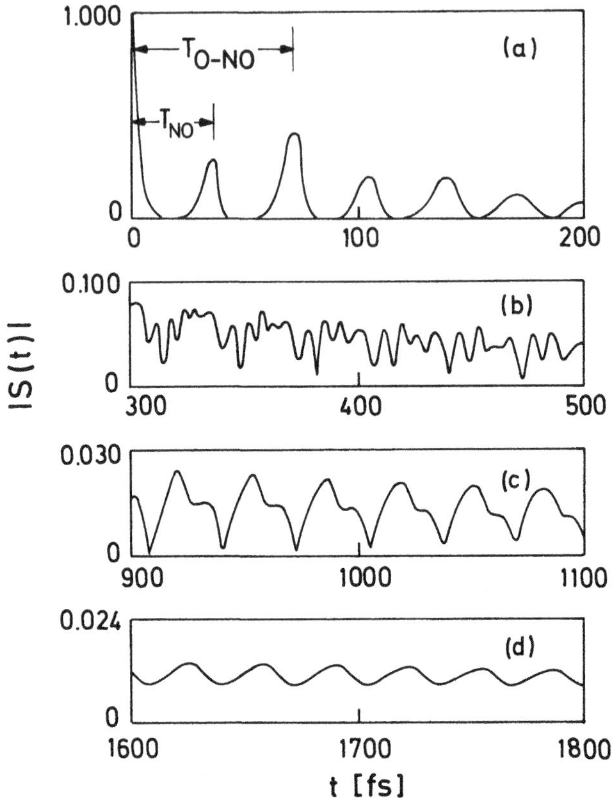

Fig. 7.9. Modulus of the autocorrelation function for the dissociation of $CH_3ONO(S_1)$. Note the different scales in (a)–(d). T_{NO} denotes the period of the NO vibration in the well and T_{O-NO} is the period of vibration along the O-NO dissociation bond, which accidentally is roughly equal to $2\,T_{NO}$. Adapted from Engel et al. (1990).

while all other states are depleted and $|S(t)|$ is then a smoothly decaying exponential.

The autocorrelation function is the fingerprint of the molecular motion on the upper PES. The stretching vibration of NO in the intermediate potential well is the dominant motion and the maxima of $S(t)$, separated by the NO vibrational stretching period $T_{NO} \approx 30$ fs, clearly illustrate this. Fourier transformation of the autocorrelation function according to (4.7) yields the absorption spectrum shown in Figure 7.10(b). It consists of a main progression with energy spacing $\Delta E \approx 2\pi\hbar/T_{NO}$. A second, much weaker progression with energy spacing $\Delta E \approx 2\pi\hbar/T_{O-NO}$ can be attributed to excitation of the CH_3O-NO dissociation bond; the corresponding period T_{O-NO} is approximately $2T_{NO}$ and therefore appears

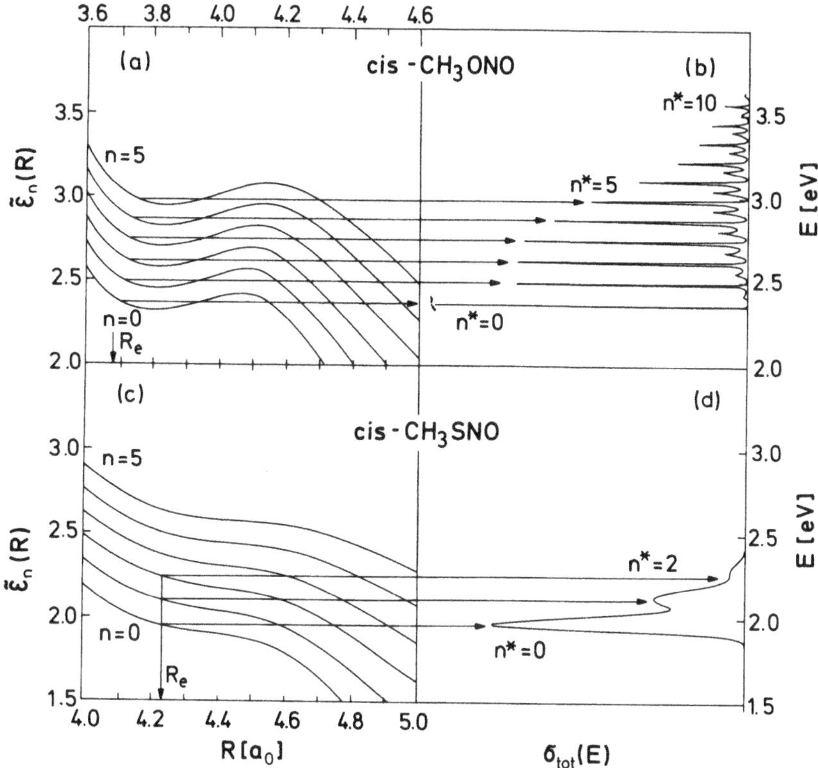

Fig. 7.10. (a) Adiabatically corrected vibrational energy curves $\tilde{\epsilon}_n(R)$, defined in Equation (7.26), as functions of the Jacobi coordinate R for the dissociation of *cis*-$CH_3ONO(S_1)$. The quantum number n represents internal vibrational excitation of the NO moiety and the arrow on the R-axis indicates the equilibrium in the electronic ground state. (b) The calculated absorption spectrum of *cis*-CH_3ONO as a function of the energy in the excited state. Normalization is such that $CH_3O + NO(r_e)$ corresponds to $E = 0$. (c) and (d) The same as in (a) and (b), respectively, but for the photodissociation of *cis*-$CH_3SNO(S_1)$ which has been discussed in Section 6.2.3. $CH_3SNO(S_1)$ exemplifies an intermediate case between direct and indirect dissociation. The broad vibrational structures can be considered as the precursors of the narrow resonances in the case of CH_3ONO. Adapted from Schinke et al. (1989).

merely as a weak undulation of the main sequence of recurrences in Figure 7.9(a).

The spectrum in Figure 7.10(b) has exactly the shape predicted by Equation (7.22). It is composed of narrow, well separated resonance lines which correspond to the metastable states within the potential trough. Their widths reflect the rate of energy transfer between the vibrational and the dissociation bond.[†] According to (7.22), the intensities are proportional to the multi-dimensional FC factors $|\langle \Psi_{1l} \mid \mu_{10}^{(e)} \mid \Psi_0 \rangle|^2$ multiplied by the lifetimes of the resonances. The first peak is therefore largest because it has the largest lifetime (i.e., the smallest width). The dependence on the lifetime, which ultimately stems from the normalization of the continuum wavefunction, can significantly distort the envelope of the calculated spectrum. In order to make comparisons with the experimental spectrum it is therefore advisable to convolute the calculated spectrum with an energy-resolution function whose width is large compared to $\hbar\Gamma$ (see Figure 7.12 and Section 7.5). A rigorous assignment of the spectrum in terms of two quantum numbers becomes readily apparent in the time-independent adiabatic picture.

7.4 Time-independent view: Resonances

Within the time-independent approach, narrow structures in absorption or scattering cross sections are interpreted as *resonances.* Resonances are unique features of the time-independent Schrödinger equation and their mathematical description is identical irrespective of whether we investigate full or half collisions. A thorough analysis of resonances for molecular systems with more than one dimension is quite complicated (Feshbach 1958, 1962, 1964; Fano 1961; Kukulin, Krasnopolsky, and Horáček 1989) and the interested reader is referred to textbooks on scattering theory (Wu and Ohmura 1962:ch.5; Taylor 1972:ch.13; Child 1974:ch.4, 1991:ch.10; Bosanac 1988) or one of the numerous review articles on this fundamental topic (Taylor 1970; Kuppermann 1981; Barrett, Robson, and Tobocman 1983; Ho 1983; Burke 1989). We confine the following discussion to a qualitative picture which, nevertheless, suffices to elucidate the relation to the time-dependent approach and to explain the subsequent examples.

7.4.1 Stationary wavefunctions and assignment

In order to calculate the absorption spectrum in the time-independent approach one solves the time-independent Schrödinger equation for a series of total energies and evaluates the overlap of the total continuum wavefunction, defined in (2.70), with the bound wavefunction of the parent molecule, $\langle \Psi_{tot}(E) \mid \mu_{10}^{(e)} \mid \Psi_0(E_i) \rangle$. Any structures in the spectrum are thus related to the energy dependence of the stationary wavefunction $\Psi_{tot}(E)$. As illustrated schematically in Figure 7.4 for the one-

---

[†] The photodissociation of $CH_3ONO(S_1)$ is an example for which the coupling operator $\hat{W}$, which initiates the decay of the quasi-bound states, cannot be clearly defined.

dimensional case, the stationary wavefunction rapidly changes its overall shape when one scans the energy across a resonance. In the vicinity of a resonance the continuum wavefunction behaves like a bound-state wavefunction; it accumulates probability in the inner region and exhibits pronounced nodal structures in the well.

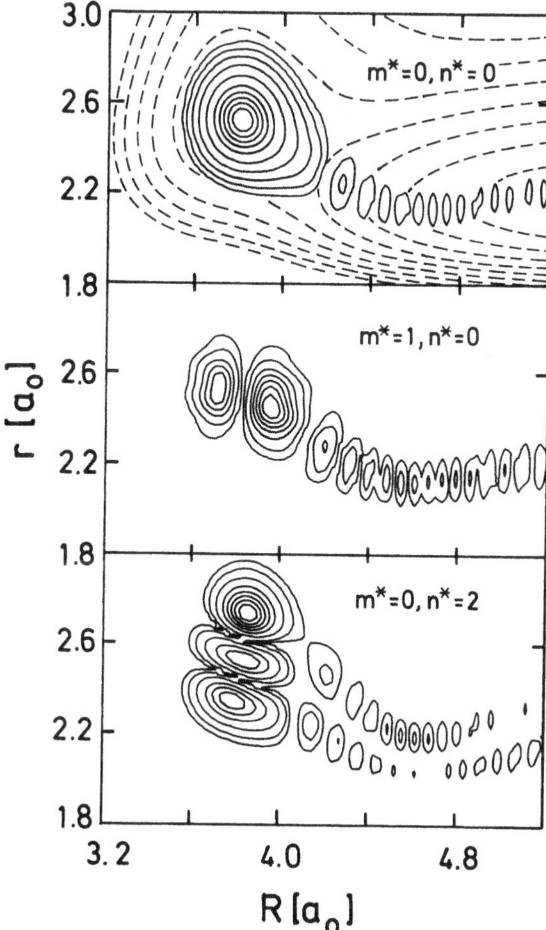

Fig. 7.11. Stationary wavefunctions $|\Psi_{tot}(E)|$ for the photodissociation of $CH_3ONO(S_1)$ for three selected resonance energies. The assignment in terms of vibrational quantum numbers m^* and n^* is described in the text. The dashed contours in the upper part represent the S_1-state PES. The magnitudes of the wavefunctions in the exit channel are very much smaller than in the inner region of the potential. In order to make the tails of the wavefunctions perceptible the contours are chosen in a nonuniform way!

Figure 7.11 shows examples of the total dissociation wavefunction $\Psi_{tot}(E)$ for the dissociation of $CH_3ONO(S_1)$ for three selected energies. The energies correspond to the first (upper part) and the third (lower part) peak of the main progression and to the first peak of the secondary progression (middle part) in the absorption spectrum shown in Figure 7.10(b). According to (4.11), $\Psi_{tot}(E)$ is the Fourier transform of the evolving wavepacket $\Phi(t)$ for a particular energy E, i.e.,

$$\Psi_{tot}(E) = \int dt \, e^{iEt/\hbar} \, \Phi(t).$$

The time-dependent wavepacket accumulates in the inner region of the PES while it oscillates back and forth in the shallow potential well as illustrated in Figure 7.8. This vibrational motion leads to an increase of the stationary wavefunction in the inner region, however, only if the energy E is in resonance with the energy of a quasi-bound level. If, on the other hand, the energy is off resonance, destructive interference of contributions belonging to different times causes cancelation of the wavefunction.

The wavefunctions shown in Figure 7.11 are true continuum wavefunctions with oscillatory tails in the exit channel; these tails manifest the continuous leakage of $\Phi(t)$ through the bottleneck. Relative to the inner part, however, the amplitudes in the exit channel are much smaller and therefore not always clearly resolved. The relative height of the wavefunction in the asymptotic region increases with decreasing lifetime, because the wavepacket escapes more rapidly from the shallow potential trough and therefore accumulates faster in the region beyond the barrier. Tuning the energy across the absorption spectrum changes the inner part of Ψ_{tot} in a systematic way, i.e., an additional node occurs if we go from one peak to the next. For energies off resonance, on the other hand, the inner part of the stationary wavefunction is much smaller than the asymptotic tail. This abrupt variation naturally leaves its fingerprints on the energy dependence of the absorption cross section.

In order to make this discussion a little more quantitative we cast $\Psi_{tot}(E)$ into a direct part, $\Psi_{dir}(E)$, which varies smoothly with energy, and an indirect part, $\Psi_{ind}(E)$, which changes rapidly in the neighborhood of a resonance. Ψ_{dir} and Ψ_{ind} represent the nonresonant and the resonant parts of Ψ_{tot}, respectively. The idea is to utilize the approximate expression for the time-dependent wavepacket to construct the indirect part by virtue of Equation (4.11). Using expansion (7.18) for the inner part of the wavepacket, damping each term according to (7.14), and taking the

Fourier transformation yields[†]

$$\Psi_{ind}(E) \propto \sum_l \frac{\langle \Psi_{1l} \mid \mu_{10}^{(e)} \mid \Psi_0 \rangle}{E - E_l - \delta E^{(l)} + i\hbar \Gamma^{(l)}/2} \Psi_{1l}, \tag{7.24}$$

where all entries have the same meaning as in Section 7.2. Expression (7.24) illustrates the energy dependence of the wavefunction in the inner region of the potential where the time-dependent wavepacket is trapped for an appreciable time. $\Psi_{ind}(E)$ is small for off-resonance energies when the denominator is comparatively large. On resonance, however, the sum over all bound states approximately reduces to only one contribution,

$$\Psi_{ind} \propto \Psi_{1l} \;/\; \Gamma^{(l)} \propto \Psi_{1l}\, \tau^{(l)}. \tag{7.25}$$

Thus, the stationary wavefunction becomes on resonance essentially a bound-state wavefunction with an amplitude which is proportional to the corresponding lifetime $\tau^{(l)}$. The larger the survival time in the well region the larger is the magnitude of the stationary wavefunction.

In view of Figure 7.11 (and similar plots for all other resonances) the peaks in the absorption spectrum can be assigned to a set of two quantum numbers (m^*, n^*), where m^* is the quantum number for excitation along the dissociation bond R and n^* is the quantum number for excitation of the N-O vibrational bond r. The asterisk indicates that these quantum numbers designate resonance, i.e., quasi-bound, states rather than true bound states. Asymptotically, n^* becomes the vibrational quantum number of the free NO molecule while m^* has no counterpart in the product channel. The main progression is built upon $m^* = 0$ and the second, much weaker progression corresponds to $m^* = 1$.

In the light of the above discussion the photodissociation of CH_3ONO proceeds in the following way,

$$CH_3ONO(S_0; 0, 0) \xrightarrow{(1)} CH_3ONO(S_1; m^*, n^*) \xrightarrow{(2)} CH_3O + NO(n).$$

The first step describes the excitation of a quasi-bound vibrational level in the excited electronic state with quantum numbers (m^*, n^*). The second step represents the dissociation of the intermediate compound due to coupling to the continuum induced by energy redistribution inside the shallow well.

7.4.2 Adiabatic picture and decay mechanism

Provided the coupling between the dissociation and the internal coordinates is not too strong, the adiabatic approximation, outlined in Section

[†] This derivation is not intended to be mathematically rigorous and it does not make a sound analysis dispensable. Nevertheless, the simple expression (7.24) describes the general energy variation of the calculated wavefunction remarkably well.

3.3.1, provides a simple picture of resonances in full as well as half collisions. The adiabatic potential curves $\epsilon_n(R)$ defined by solving the one-dimensional Schrödinger equation (3.31) are crutial. For a quantitative analysis, however, one should correct the vibrational energies by taking into account the diagonal elements of the nonadiabatic matrix $\mathbf{U}$ defined in (3.35) which can be considered as additional energy terms [see the left-hand side of Equation (3.33); Römelt 1983; Child 1991:ch.10].[†]

Figure 7.10(a) depicts the corrected vibrational energies,

$$\tilde{\epsilon}_n(R) = \epsilon_n(R) + \frac{\hbar^2}{2m} U_{nn}(R), \tag{7.26}$$

for the S_1 state of CH_3ONO. They resemble a cut through the two-dimensional PES along the minimum energy path. Each vibrationally adiabatic potential curve supports two quasi-bound levels with energies denoted by $E_{m^*n^*}$ where $m^* = 0$ and 1 indicate the number of quanta in the dissociation mode. The energies $E_{m^*n^*}$ are the adiabatic approximations of the resonance energies; they can be easily determined by one-dimensional methods. The energies of the $m^* = 1$ manifold are close to the top of the barrier and for clarity they are not indicated in Figure 7.10. The upper part of Figure 7.10 clearly elucidates the relationship between internal vibrational excitation of NO in the $CH_3ONO(S_1)$ complex on one hand and the resonance structures in the absorption spectrum on the other. It is the two-dimensional analogue of Figure 7.1. The adiabatic potential curves readily provide the correct assignment of the resonance structures.

Note the similarities as well as the distinct differences with the multi-dimensional reflection principle in direct dissociation which is illustrated for the photodissociation of $CH_3SNO(S_1)$ in the lower part of Figure 7.10. In the latter case, the very diffuse vibrational structures are due to a direct reflection of the ground-state wavefunction mediated by the adiabatic potentials. Lowering the barrier of the PES of $CH_3ONO(S_1)$ would continuously shorten the lifetime of the complex and therefore turn the relatively narrow resonances into broad features similar to those found for CH_3SNO. The autocorrelation function for the dissociation of CH_3SNO has only a single, slightly indicated recurrence in contrast to the many recurrences found for CH_3ONO (Schinke et al. 1989). The two spectra in Figure 7.10 drastically manifest the influence of small changes of the multi-dimensional PES.

The quasi-bound states can dissociate by tunneling through the potential barrier and/or by vibrational quenching. In the latter case energy

---

[†] Particularly important is the correction for the photodissociation of FNO through excitation of the S_1 state, for example (Suter et al. 1992).

is transferred from the vibrational to the dissociation mode until enough translational energy is accumulated to permit the molecule to surmount the barrier. Since the barrier is relatively broad and the reduced mass large, tunneling is unlikely for the $m^* = 0$ resonances whose energies are clearly nested inside the potential well. Therefore, nonadiabatic decay is the main fragmentation mechanism and the final vibrational state distributions of the NO product (which we will discuss in Section 9.4) support this conclusion: exciting vibrational state n^* in the S_1 complex yields NO preferentially in vibrational state $n = n^* - 1$.

The resonance widths $\Gamma(n^*)$ and the corresponding lifetimes $\tau(n^*)$ obtained from the two-dimensional calculations by Schinke et al. (1989) are listed in Table 7.1. The $n^* = 0$ resonance can decay only by tunneling and therefore it has the longest lifetime. In the present case, the energy exchange within the complex becomes more efficient with increasing NO excitation which causes the lifetime to decrease monotonically with n^*. We must emphasize, however, that such a simple dependence is not universally true. In the dissociation of ClNO via the first triplet state, for example, one finds the opposite trend (see Figure 7.14). The variation of $\Gamma(n^*)$ with internal excitation depends sensitively on the coupling between the various modes and ultimately on the shape of the multi-dimensional PES in the region of the bottleneck. The lifetimes in Table 7.1 must be compared to the internal period of the NO vibration in the complex which is approximately 30 fs.

Figure 7.12 compares the measured and the theoretical absorption spectra for $CH_3ONO(S_1)$, the latter being obtained in a three-dimensional wavepacket calculation (Untch, Weide, and Schinke 1991a). The energy spacing of the main progression reflects the vibrational frequency of NO in the S_1 complex. The ratio of the peak intensities is mainly determined by the one-dimensional Franck-Condon factors

$$f(n) = |\langle \varphi_n(r, R = R_e) \mid \varphi_r(r) \rangle|^2$$

defined in (3.42b). Here, $\varphi_n(r)$ is the wavefunction of the nth vibrational state of NO in the upper electronic state, evaluated at the equilibrium sep-

Table 7.1. Widths $\Gamma(n^*)$ and lifetimes $\tau(n^*)$ for the progression of $m^* = 0$ resonances.

n^*	0	1	2	3	4
Γ [meV]	0.29	2.6	4.5	6.1	7.2
τ [fs]	2250	250	150	110	90

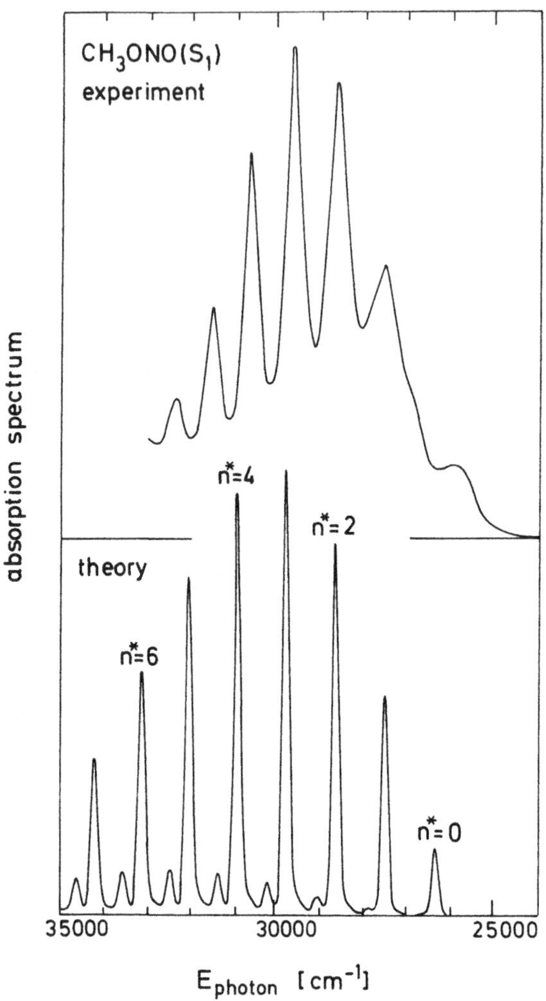

Fig. 7.12. Comparison of the measured and the calculated absorption spectra for the $S_0 \rightarrow S_1$ transition in CH_3ONO. The quantum number n^* denotes vibrational excitation of the NO moiety in the complex. The theoretical spectrum is obtained in a three-dimensional wavepacket calculation including the ONO bending angle in addition to the two N-O stretching coordinates. The spectrum is convoluted with a Gaussian function with width $\Delta E_{res} = 0.02$ eV in order roughly to mimic thermal broadening and is artificially shifted along the energy axis. Reproduced from Untch, Weide, and Schinke (1991a).

aration R_e, and $\varphi_r(r)$ is the corresponding wavefunction in the electronic ground state. Inclusion of the coordinate dependence of the transition dipole function could slightly change the envelope of the spectrum.

The widths manifest — in principle — the lifetime of the S_1 compound and therefore the rate of internal energy transfer. We must strongly emphasize, however, that the widths of the experimental spectrum do not reflect the true homogeneous lifetimes. Thermal averaging as well as the superposition of the spectra for *cis*- and *trans*-CH_3ONO inhomogeneously broaden the lines which makes a direct comparison with the calculation impossible. This problem is unfortunately rather common and significantly hampers the comparison between theory and experiment since the lifetime is one major source of information about the dissociation dynamics. Lifetimes for CH_3ONO have been estimated by Nonella et al. (1989) from measurements of the anisotropy parameter β using the quasi-diatomic model. They range from 300 fs for $n^* = 0$ to 200 fs for $n^* = 2$ and thus fall into the range of the theoretical lifetimes reported in Table 7.1.

7.4.3 Relation to resonances in full collisions

Resonances in half and in full collisions have exactly the same origin, namely the temporary excitation of quasi-bound states at short or intermediate distances irrespective of how the complex was created. In full collisions one is essentially interested in the asymptotic behavior of the stationary wavefunction $\Psi(E)$ in the limit $R \to \infty$, i.e., the scattering matrix $\mathbf{S}$ with elements S_{if} as defined in (2.59). The $\mathbf{S}$-matrix contains all the information necessary to construct scattering cross sections for a transition from state i to state f. In the case of a narrow and isolated resonance with energy E_r and width $\hbar\Gamma$ the *Breit-Wigner* expression

$$\mathbf{S}(E) \approx \mathbf{S}_{dir}(E) + \frac{\mathbf{A}}{E - E_r + i\hbar\Gamma/2} \tag{7.27}$$

accurately represents the energy dependence of the $\mathbf{S}$-matrix, where the direct term is assumed to vary only slowly with energy. The matrix $\mathbf{A}$ is independent of energy and must fulfil certain conditions in order to guarantee the unitarity of $\mathbf{S}$. Equation (7.27) is the analogue to (7.24) for the stationary wavefunction. All elements of $\mathbf{S}$ have the same resonance energy and the same width. Resonances can be considered as poles of the scattering matrix in the complex energy plane: the real part of the energy gives the resonance position and the imaginary part is related to the width (Taylor 1972:ch.13).

The observation of resonances in a scattering experiment requires high resolution of the collision energy, which, however, is rather difficult to achieve in a conventional molecular beam machine. In a photodissociation experiment, on the other hand, the frequency of the light beam determines the collision energy in the upper state and hence the necessary energy resolution is much easier to accomplish. More crucial for the detection of resonances, however, is the inevitable averaging caused by the superposition of many partial waves in a full collision as discussed in Section 1.3. As a consequence of the centrifugal potential $J(J+1)/R^2$ (see Section 11.1) the resonance energy varies with the total angular momentum and therefore the average over many partial cross sections may drastically blur the resonance structures. The effect of gradually erasing resonances if the maximum value of J is successively increased has been convincingly demonstrated by Miller (1990:fig.6) for the $H+H_2$ exchange reaction. In half collisions, on the other hand, the dipole selection rule $|\Delta \mathbf{J}| = 0$ or ± 1 (see Section 11.1) considerably restricts the possible range of total angular momenta, especially if one cools the sample in a molecular beam; the superposition of a few partial waves merely broadens the resonances without, however, erasing them.

Despite the many theoretical predictions of inelastic† resonances (mainly observed in calculations with reduced dimensionality; Manz 1989, for example) we do not know of an *unambiguous* experimental observation of resonances in atom-molecule or molecule-molecule gas-phase collisions.‡ In contrast to full atom-molecule collisions pronounced resonance-like structures are actually rather common features in bound-continuum absorption spectra (Robin 1974, 1975, 1985; Okabe 1978; Fano and Rao 1986). In fact, all sharp structures in UV absorption spectra can be considered as resonances and therefore photodissociation provides ideal opportunities to investigate resonance phenomena, such as the lifetime, the decay mechanism, and the final state distributions of the fragments, on a very detailed basis.

7.5 Resolution in the energy and in the time domain

In order to elucidate the interrelation between the energy dependence of the absorption spectrum and the time dependence of the molecular motion it is instructive to investigate the influence of degrading the spectral resolution. The comparison with a measured spectrum often makes it necessary to average the calculated spectrum over an experimental reso-

† By inelastic resonances we mean resonances which decay via energy redistribution between the internal vibrational-rotational modes or a transition from a quasi-bound to a continuum state. Elastic resonances, on the other hand, decay via tunneling through a potential barrier without the necessity of internal transitions.

‡ Incidentally we note that resonances do exist, however, in gas-surface collisions in which, as a consequence of the infinite mass of the solid, J is always zero; resonances are indeed one major source of information on the gas-surface interaction (Hoinkes 1980; Barker and Auerbach 1984). Likewise, resonances are prominent features in electron-atom or electron-molecule collisions (Schulz 1973; Domcke 1991); the extremely light mass of the electron implies that only partial waves with very low angular momentum quantum numbers contribute to the cross section.

lution function according to

$$\bar{\sigma}_{tot}(E) = \int_{-\infty}^{+\infty} dE' \, \zeta_E(E - E') \, \sigma_{tot}(E'). \tag{7.28}$$

For convenience we chose ζ_E to be a Gaussian,

$$\zeta_E(E - E') = \frac{\varepsilon}{\pi^{1/2}} \, \mathrm{e}^{-\varepsilon^2 (E-E')^2}, \tag{7.29}$$

with full width at half maximum of $\Delta E_{res} = 2(\ln 2)^{1/2}/\varepsilon$. This smears the spectrum around E on an energy scale ΔE_{res}.

Using the time-dependent expression for the total spectrum, Equation (4.7), we obtain

$$\begin{aligned} \bar{\sigma}_{tot}(E) &\propto \int_{-\infty}^{\infty} dE' \, \zeta_E(E - E') \int_{-\infty}^{\infty} dt \, \mathrm{e}^{iE't/\hbar} \, S(t) \\ &\propto \int_{-\infty}^{\infty} dt \, S(t) \int_{-\infty}^{\infty} dE' \, \zeta_E(E - E') \, \mathrm{e}^{iE't/\hbar} \\ &\propto \int_{-\infty}^{\infty} dt \, S(t) \, \mathrm{e}^{-t^2/(2\hbar\varepsilon)^2} \, \mathrm{e}^{iEt/\hbar} \\ &\propto \int_{-\infty}^{\infty} dt \, \bar{S}(t) \, \mathrm{e}^{iEt/\hbar}, \end{aligned} \tag{7.30}$$

where we have omitted all unimportant constants and exploited the fact that the Fourier transformation of a Gaussian yields another Gaussian. Averaging the spectrum in the energy domain is therefore equivalent to damping the autocorrelation function with a Gaussian function in time,

$$\bar{S}(t) = \zeta_t(t) \, S(t) = \mathrm{e}^{-t^2/(2\hbar\varepsilon)^2} \, S(t). \tag{7.31}$$

The widths of $\zeta_E(E - E')$ and $\zeta_t(t)$ are interrelated by $\Delta E_{res} \, \Delta t = 8 \, \hbar \ln 2 \approx h$ which is identical to Equation (6.12). Thus:

- Degrading the resolution in the energy domain is synonymous with discarding dynamical information, i.e., propagating the wavepacket for shorter and shorter times.
- A spectral resolution of the order of ΔE_{res} requires a resolution in the time domain of the order of at least $\Delta t \approx h/\Delta E_{res}$.

Figure 7.13 illustrates the relation between resolution in the energy and in the time domains for the dissociation of $CH_3ONO(S_1)$. The right-hand side shows the modified autocorrelation function $\bar{S}(t) = S(t) \, \zeta_t(t)$ for four "windows" Δt and the left-hand side depicts the corresponding averaged cross sections $\bar{\sigma}_{tot}(E)$. The spectrum with the lowest resolution is structureless and resembles the kind of spectrum obtained for fast and direct dissociation without any recurrence. $\bar{S}(t)$ merely shows the rapid

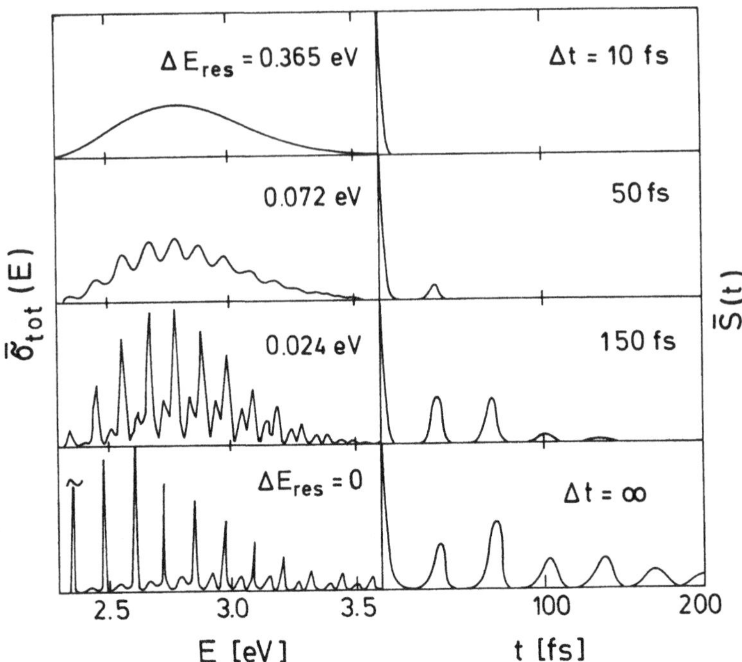

Fig. 7.13. Illustration of the interrelation between resolution in the energy domain and in the time domain. $\bar{\sigma}_{tot}(E)$ and $\bar{S}(t)$ are defined in (7.28) and (7.31), respectively. The resolutions in energy and in time fulfil the uncertainty relation $\Delta E_{res}\ \Delta t \approx h$. Adapted from Engel et al. (1990).

decay at $t = 0$ which reflects the initial motion of the wavepacket away from the Franck-Condon region; its width determines the overall width of the spectrum. For $\Delta t = 50$ fs the propagation time is long enough to allow the development of the first recurrence which has the consequence of superimposing regular undulation structures on the broad background (see also Chapter 8). The structures reflect a single vibration of NO in the complex. At even higher resolution, $\Delta t = 150$ fs, the spectrum also shows the second progression which reflects the vibration along the CH_3O-NO bond. This spectrum already contains all the structures of the spectrum with the highest resolution. A further increase of the resolution in time merely affects the intensities and the widths of the peaks without changing the overall appearance of the spectrum. Figure 7.13 clearly elucidates the relationship between the number of recurrences on one hand and the resonance width and the relative amplitudes on the other. The longer the wavepacket is trapped in the inner region, the more recurrences can be developed, and the narrower become the resonance structures in the spectrum.

So far we have deliberately discarded dynamical information by damping the autocorrelation function. Note, however, that any reduction of the lifetime of the excited complex, due either to dissociation or to a nonadiabatic transition to another electronic state, leads naturally to the same net result. Let us imagine that the barrier towards the exit channel of the $CH_3ONO(S_1)$ PES is much smaller than it is in reality so that the excited complex dissociates much faster. This would yield an autocorrelation function similar to the one shown in the second row of Figure 7.13. The corresponding spectrum would exhibit very diffuse structures superimposed on a broad background as actually found for many polyatomic molecules. The interpretation of such structures is rather cumbersome because, without accompanying calculations, it is almost impossible *unambiguously* to unravel their true dynamical origin from the measured spectrum alone. Diffuse structures are the topic of Chapter 8.

7.6 Other types of resonant internal excitation

In the preceding two sections we considered resonances induced by temporary excitation of the mode that finally becomes the vibrational mode of the diatomic fragment. In this section we feature two other types of resonant excitation of internal vibrational motion.

7.6.1 Excitation of bending motion

The photodissociation of ClNO in the first excited triplet state T_1 is a beautiful example for investigating all facets of resonances in a molecular system (Qian et al. 1991; Qian and Reisler 1991). The lower part of Figure 7.14 depicts the total absorption spectrum of ClNO in the 500-625 nm wavelength region measured at room temperature. A combined experimental and theoretical analysis assigns this band to the $S_0 \rightarrow T_1$ electronic transition (Bai et al. 1989). It shows some vague structures, but insufficient resolution as well as the overlap with the stronger $S_0 \rightarrow S_1$ electronic band at shorter wavelengths makes an unambiguous assignment rather questionable.

A much clearer picture evolves when one decomposes the total spectrum into the partial photodissociation cross sections $\sigma(\lambda, n, j)$ for absorbing a photon with wavelength λ and producing NO in a particular vibrational-rotational state with quantum numbers (n, j). Experimentally this is accomplished by measuring so-called *photofragment yield spectra.* The idea is, in principle, simple: the NO product is probed by laser-induced fluorescence (LIF). However, instead of scanning the wavelength λ_{LIF} of the probe laser (in order to determine the final rotational state distribution) one fixes λ_{LIF} to a particular transition $NO(^2\Pi, nj) \rightarrow$

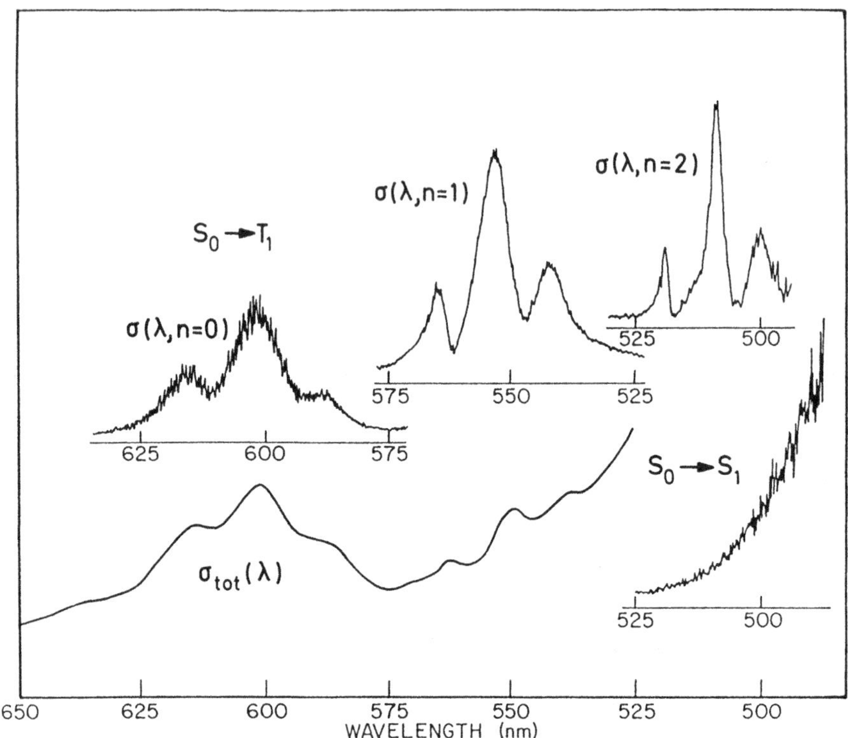

Fig. 7.14. Photofragment yield spectra for the photodissociation of ClNO through the T_1 electronic state. The lower part depicts the total absorption spectrum $\sigma_{tot}(\lambda)$ measured at room temperature. The three separate spectra in the upper part correspond to the (unnormalized) partial photodissociation cross sections $\sigma(\lambda, n, j)$ for producing NO in a particular vibrational state n as indicated. The rotational state varies between 1.5 and 4.5 in these three cases. The additional spectrum in the lower part is the $n = 0$ cross section originating from excitation in the $S_0 \rightarrow S_1$ electronic band. Recall that the sum of all partial cross sections yields the total spectrum. Adapted from Qian, Ogai, Iwata, and Reisler (1990).

NO$(^2\Sigma, n'j')$ and monitors the total fluorescence from the excited electronic state of NO as a function of the photolysis wavelength λ which was used to excite the ClNO parent molecule to the T_1 state. If properly normalized, the photofragment yield spectra correspond to the partial photodissociation spectra discussed in Section 1.4.6. In view of (7.23) and Figure 7.6 the $\sigma(\lambda, n, j)$ have the same resonance behavior as the total spectrum. The upper part of Figure 7.14 depicts three examples for various final vibrational states n; the rotational state of NO varies between 1.5 and 4.5. (Because of the electronic spin of NO, the rotational quantum number j is half-integer in this case.)

Probing NO($n = 0$) yields a LIF signal in the first band around 600 nm but no signal in the second band. On the other hand, probing NO($n = 1$) gives a signal in the second band around 550 nm but no signal in the first band. Based on this analysis and assuming a vibrationally adiabatic decay we can assign the 600 nm and the 550 nm bands to excitation of the $n^* = 0$ and 1 stretching states of NO in the T_1 complex. With this technique a third band corresponding to $n^* = 2$ becomes visible around 510 nm, which in the total spectrum is completely concealed by the $S_0 \rightarrow S_1$ electronic transition. Each vibrational band is additionally split into three resonance peaks which are attributed to excitation in the bending degree of freedom of the T_1 complex with quantum numbers $k^* = 0$, 1, and 2.

The assignment of the resonance structures in Figure 7.14 has recently been confirmed by three-dimensional wavepacket calculations employing an *ab initio* PES (Sölter et al. 1992). Figure 7.15 depicts the calculated autocorrelation function. The recurrences with the shorter period of about $T_{NO} = 19$ fs reflect the oscillations of the wavepacket along the NO stretching coordinate while the recurrence with the longer period of about $T_{bend} = 70$ fs manifests the bending motion within the T_1 complex. The essential feature of the corresponding PES is a rather flat plateau near the FC point, where the slope $\partial V/\partial R$ is almost zero. As a consequence, the Cl-NO bond breaks rather slowly. Without showing further results we note that the theoretical calculation satisfactorily reproduces most of the experimental findings.

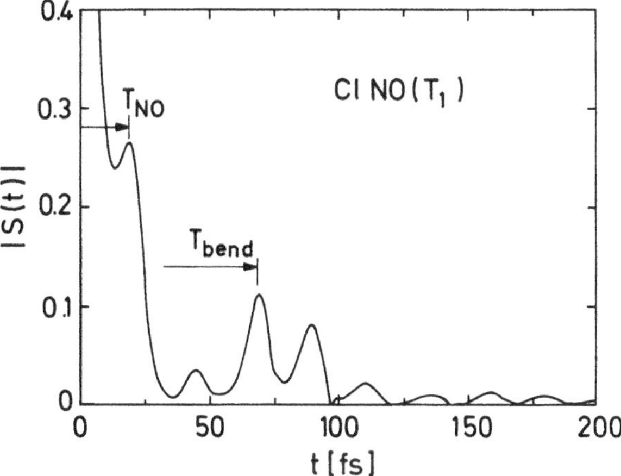

Fig. 7.15. Calculated autocorrelation function for the photodissociation of ClNO in the T_1 state. T_{NO} denotes the period of NO stretching and T_{bend} is the period for bending motion in the T_1 complex. Adapted from Sölter et al. (1992).

The photodissociation of ClNO through the T_1 state shows a lot of details which reveal a wealth of information about the dissociation dynamics. For example, the band width *decreases* with increasing NO vibrational quantum number. On the other hand, it *increases* within each vibrational band with increasing degree of bending excitation. The lifetimes range from about 15 fs to about 70 fs corresponding to roughly one and four NO stretching periods. Furthermore, there is a unique relationship between the internal excitation of the triatomic complex on one hand and the final rotational state distributions of the NO fragment on the other (Qian, Ogai, Iwata, and Reisler 1988). Excitation in the first ($k^* = 0$), the second ($k^* = 1$), and the third ($k^* = 2$) peaks within the same vibrational band yields uni-, bi-, and trimodal rotational state distributions. In other words, the final rotational state distributions "reflect" the degree of bending excitation in the complex. In Section 10.3 we will demonstrate how this "mapping" can be explained in terms of an extension of the rotational reflection principle established for direct dissociation.

The photodissociation of NH_3 through the $\tilde{A}^1A_2''$ state is an example of strong bending excitation (Ashfold, Bennett, and Dixon 1986; Dixon

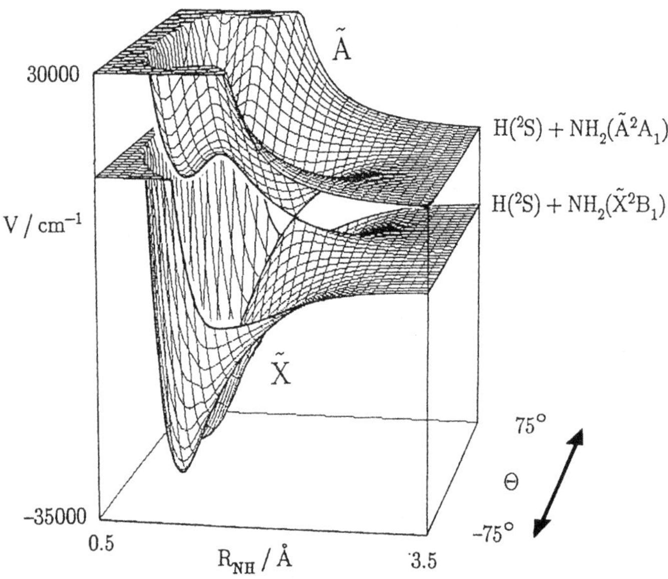

Fig. 7.16. The ground- and the excited-state potential energy surfaces of NH_3 as functions of one of the N-H bonds and the out-of-plane angle Θ. They are based on the *ab initio* calculations of Rosmus et al. (1987). The arrow indicates the oscillatory motion along the Θ-axis of the temporarily trapped wavepacket in the excited state. Reproduced from Dixon (1989).

1988). Figure 7.16 depicts the calculated two-dimensional potential energy surfaces for the $\tilde{X}^1A_1'$ and the $\tilde{A}^1A_2''$ states as functions of one of the H-N bonds, R_{NH} (the dissociation coordinate), and the corresponding out-of-plane angle Θ. They are based on the *ab initio* calculations of Rosmus et al. (1987). The upper PES has a relatively deep potential well at short distances which supports many quasi-bound bending states. A barrier hinders direct dissociation into products H and NH_2 and the evolving wavepacket first performs a large-amplitude bending motion within the potential well before the dissociation bond H-NH_2 accumulates sufficient energy to surmount the barrier. The photodissociation of NH_3 therefore qualitatively resembles the photodissociation of $CH_3ONO(S_1)$. Beyond the barrier, the $\tilde{A}^1A_2''$ state, which asymptotically correlates with $NH_2(\tilde{A}^2A_1)$, strongly couples to the $\tilde{X}^1A_1$ state and therefore the last step of the fragmentation proceeds mainly on the lower PES with the $NH_2(\tilde{X}^2B_1)$ product being strongly rotationally excited (Biesner et al. 1988, 1989; Dixon 1989; Woodbridge, Ashfold, and Leone 1991).

Because the equilibrium angles in the two electronic states differ considerably ($\Theta_e = 25°$ for the ground and $\Theta_e = 0°$ for the excited electronic states) the absorption spectrum, shown in Figure 7.17, consists of a long progression in the bending quantum number. The intensities of the room-temperature spectrum vary in a smooth way with the bending quantum number and the width is uniformly broad. Cooling the sample in a supersonic jet considerably narrows the lines and brings out details that are not visible in the room-temperature spectrum. The resonance

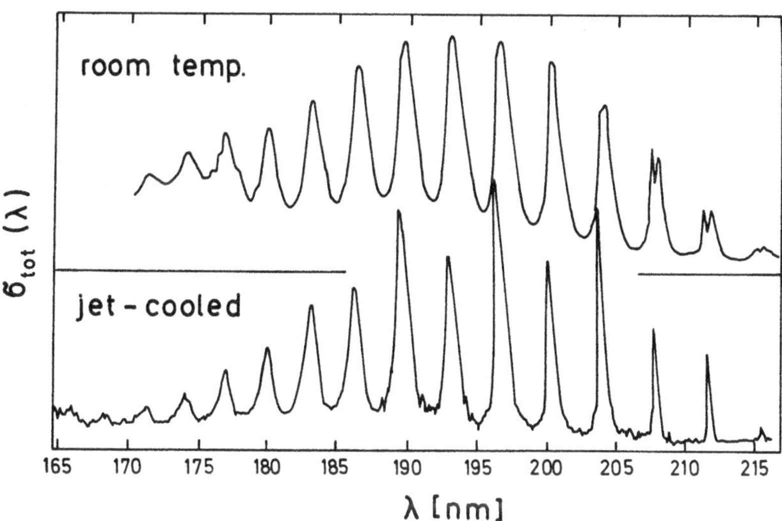

Fig. 7.17. The absorption spectra of NH_3 obtained with a room-temperature sample and in a jet-cooled molecular beam. Adapted from Vaida et al. (1987).

width, for example, changes noticeably across the spectrum and the peak intensities show some systematic fluctuations. Figure 7.17 intriguingly illustrates the influence of thermal broadening on the absorption spectrum and the need for high-resolution spectra if one wants to compare with *ab initio* theory. Other examples of high excitation of the bending degree of freedom are HCN (Macpherson and Simons 1978) and HCO (Loison, Kable, Houston, and Burak 1991).

7.6.2 Excitation of symmetric and anti-symmetric stretch motion

The photodissociation of symmetric triatomic molecules of the type ABA is particularly interesting because they can break apart into two identical ways: ABA $\rightarrow$ AB + A and ABA $\rightarrow$ A + BA. Figure 7.18(a) shows a typical PES as a function of the two equivalent bond distances. It represents qualitatively the system IHI which we will discuss in some detail below. We consider only the case of a collinear molecule as illustrated in Figure 2.1. The potential is symmetric with respect to the C_{2v}-symmetry line $R_{\mathrm{IH}} = R_{\mathrm{HI}}$ and has a comparatively low barrier at short distances. The minimum energy path smoothly connects the two product channels via the saddle point. A trajectory that starts somewhere in the inner region can exit in either of the two product channels. However, the branching ratio $\sigma_{\mathrm{IH+I}}/\sigma_{\mathrm{I+HI}}$ obtained by averaging over many trajectories or from the quantum mechanical wavepacket must be exactly unity.

The corresponding exchange reaction of the type A+BA $\rightarrow$ AB+A has been extensively studied in the last twenty years or so, quantum mechanically as well as classically, and detailed insight into chemical reactivity has been gained (Baer 1985a; Clary, 1986b). One of the most appealing aspects is the possibility of resonances, i.e., quasi-bound states embedded in the region of the *transition state*, even if the potential does not have an intermediate well that could trap the system. Despite the numerous theoretical predictions no scattering experiment has yet *unambiguously* confirmed the existence of such reactive resonances.

Photodetachment spectroscopy of negative ions like IHI$^-$ and similar systems, studied by Neumark and coworkers (Metz, Kitsopoulos, Weaver, and Neumark 1988; Weaver, Metz, Bradforth, and Neumark 1988; Metz et al. 1990; Bradforth et al. 1990; Neumark 1990; Weaver and Neumark 1991) has provided the first conclusive manifestation of reactive resonances for a purely repulsive PES. The idea of the experiment goes as follows: a photon with frequency ω detaches the electron from the negative ion producing e$^-$ and IHI. If the PES for the neutral molecule is dissociative, the IHI complex subsequently breaks apart into I and HI.

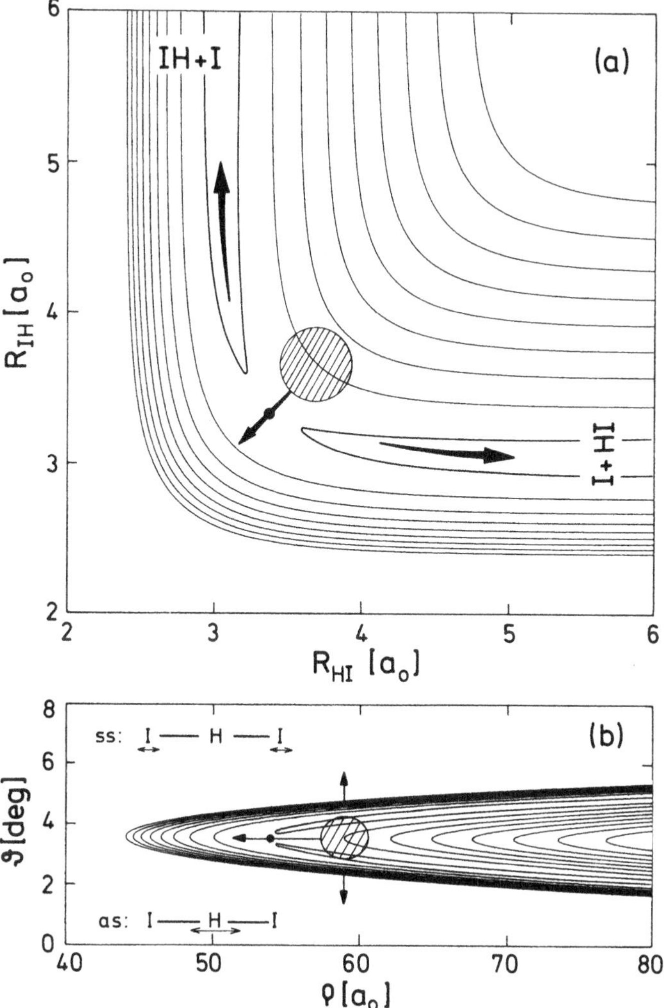

Fig. 7.18. (a) Model potential energy surface for collinear IHI as adapted from Manz and Römelt (1981). $R_{\rm IH}$ and $R_{\rm HI}$ are the two I-H bond distances. The heavy point marks the saddle point and the shaded area indicates *schematically* the Franck-Condon region in the photodetachment experiment. The arrow along the symmetric stretch coordinate ($R_{\rm HI} = R_{\rm IH}$) illustrates the early motion of the wavepacket and the two heavy arrows manifest dissociation into the two identical product channels. (b) The same PES as in (a) but represented in terms of hyperspherical coordinates (ϱ, ϑ) defined in (7.33). The horizontal and the vertical arrows illustrate symmetric and anti-symmetric stretch motions, respectively, as indicated by the two insets.

Energy conservation dictates that

$$E_{tot} = E_i + \hbar\omega = E_{electron} + E_{\rm IHI}, \tag{7.32}$$

where E_i is the initial energy of the parent ion, $E_{electron}$ is the kinetic energy of the electron, and E_{IHI} denotes the total energy of the neutral IHI complex.

Measuring the electron intensity as a function of the kinetic energy $E_{electron}$ yields the photodetachment spectrum. An example for IHI$^-$ is shown in Figure 7.19. It exhibits three well resolved peaks which are attributed to resonant excitation of the anti-symmetric (as) stretch mode of the neutral IHI complex with quantum numbers $v_{as}^* = 0$, 2, and 4.[†] The advantage of photodetachment spectroscopy in contrast to a normal scattering experiment is twofold. First, in view of (7.32) the energy of the neutral system can be continuously scanned by measuring $E_{electron}$. Second, preparing the IHI complex through the absorption of a photon, which obeys the dipole selection rule $\Delta \mathbf{J} = 0$ and ± 1, avoids the summation over many partial waves, which is inevitable in a full collision.

In the case that the two outer atoms are heavy and the central atom is light the two bond coordinates R_{IH} and R_{HI} are not adequate for

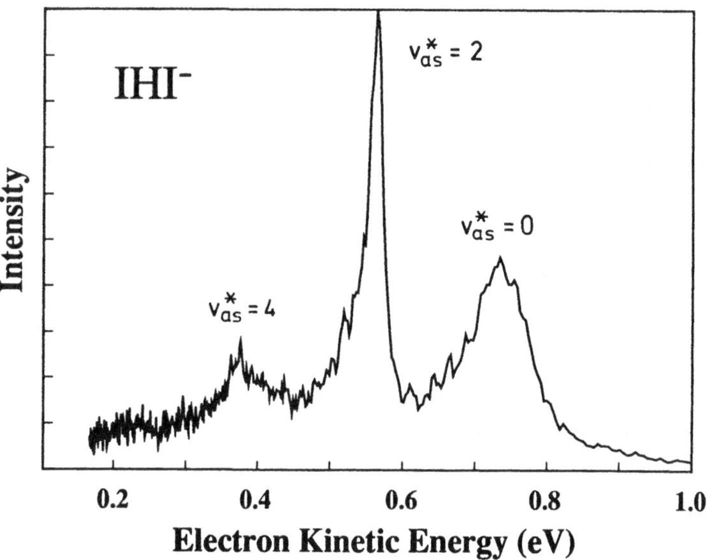

Fig. 7.19. Photodetachment spectrum for IHI$^-$; the wavelength of the excitation laser is 266 nm. v_{as}^* denotes the number of quanta in the anti-symmetric stretch mode in the transition region. The widths of the peaks do not represent the true lifetimes of the meta-stable states. Adapted from Neumark (1990)

† It should be noted, however, that contributions from electronically excited states of the neutral molecule cannot be ruled out as has been pointed out by Yamashita and Morokuma (1990) for ClHCl.

visualizing the dissociation dynamics because the corresponding Hamiltonian contains a large kinetic coupling term, i.e., a kinetic energy term of the form $\partial^2/\partial R_{\mathrm{IH}}\partial R_{\mathrm{HI}}$ [see Equation (8.1)]. The dissociation of IHI and similar systems is more conveniently described by so-called *hyperspherical coordinates* (ϱ, ϑ) [see Manz (1985) and Child (1991:ch.10), for example]. For a general collinear A + BC system they are defined by

$$\varrho = (\bar{R}^2 + \bar{r}^2)^{1/2} \quad , \quad 0 < \varrho < \infty \tag{7.33a}$$

$$\vartheta = \arctan(\bar{r}/\bar{R}) \quad , \quad 0 < \vartheta < \vartheta_{max}, \tag{7.33b}$$

where $\bar{R} = (m/\mu)^{1/2} R$ and $\bar{r} = r$ are *mass-scaled Jacobi coordinates* and $\vartheta_{max} = \arctan[(m_{\mathrm{B}} M / m_{\mathrm{A}} m_{\mathrm{C}})]^{1/2}$ is the so-called *skewing angle.* The reduced masses m and μ are defined in (2.38) and $M = m_{\mathrm{A}} + m_{\mathrm{B}} + m_{\mathrm{C}}$. The hyperspherical coordinates are simply the polar coordinates in the $(\bar{R}, \bar{r})$-plane. The Jacobi coordinates (R, r) defined in Figure 2.1 are not appropriate for symmetric molecules because they cannot simultaneously describe both product channels, A + BC and AB + C. The set of Jacobi coordinates appropriate for one dissociation channel is inappropriate for the other one and *vice versa.* The hyperspherical coordinates, on the other hand, describe both channels equally well and they can be used at short distances as well as in the asymptotic regions. The two-dimensional Hamiltonian for the linear triatomic molecule is given in hyperspherical coordinates by

$$\hat{H}(\varrho, \vartheta) = -\frac{\hbar^2}{2\mu}\left(\frac{\partial^2}{\partial \varrho^2} + \frac{1}{\varrho}\frac{\partial}{\partial \varrho} + \frac{1}{\varrho^2}\frac{\partial^2}{\partial \vartheta^2}\right) + V(\varrho, \vartheta). \tag{7.34}$$

Figure 7.18(b) depicts the same PES as in (a) in terms of hyperspherical coordinates. Because of the extreme mass ratio $m_{\mathrm{I}}/m_{\mathrm{H}}$ the skewing angle is very small, $\vartheta_{max} = 7.2°$. The line $\vartheta = \vartheta_{max}/2$ corresponds to symmetric stretch motion of both iodine atoms with respect to the stationary hydrogen atom whereas motion along the hyperspherical angle ϑ describes anti-symmetric stretch vibration of the hydrogen atom between the two stationary iodine atoms. As a consequence of the extremely small skewing angle, the fast anti-symmetric stretching motion of hydrogen decouples almost perfectly from the slow vibration of the iodine atoms. The large difference in the corresponding frequencies strongly favors an adiabatic separation.

Let us imagine the motion of a quantum mechanical wavepacket starting in the FC region at the outer slope of the saddle point (Engel 1991a). It moves first slowly along the symmetric stretch coordinate towards the saddle point, is reflected at the inner branch of the potential, and then recurs to its origin. A small fraction will dissociate while the major portion begins another round. At the same time, the wavepacket oscillates along the anti-symmetric stretch coordinate as well. Because the initial wavepacket is symmetric, $\Phi(t)$ remains symmetric for all times with the result that the center of the wavepacket stays on the line $\vartheta = \vartheta_{max}/2$; merely its angular breadth oscillates.

The autocorrelation function $S(t)$ shown in Figure 7.20 reflects these two types of motion: the wide oscillations with period T_{ss} represent the slow symmetric stretch vibration and the rapid oscillations with period T_{as} represent the fast anti-symmetric stretch vibration. Because the two exit channels are so exceedingly narrow leakage into the product channels is rather weak. Although no barrier hinders the dissociation the system is trapped in the inner region for a long time.

The spectrum obtained by Fourier transformation of $S(t)$ exhibits three bands separated by $2\pi\hbar/T_{as}$ in reasonable agreement with the measured spectrum. They represent the eigenstates of the anti-symmetric stretch mode ν_{as} with quantum numbers $v_{as}^* = 0, 2$, and 4. Because of the symmetry of the initial IHI^- state levels with odd quantum numbers are not excited. Each anti-symmetric stretch band is split into several lines separated by $2\pi\hbar/T_{ss}$ which represent the eigenstates of the symmetric stretch mode ν_{ss}. The energy resolution in Figure 7.19 is not sufficient, however, to resolve the symmetric stretch progression, but recent, more refined measurements clearly verify their existence (Waller, Kitsopoulos, and Neumark 1990).

Let us add that, as a consequence of the very different time scales involved, the time-independent adiabatic picture based upon hyperspherical coordinates offers a simple explanation and assignment of the resonances as well (Römelt 1983; Manz et al. 1984; Child 1991:ch.10; Vien, Richard-Viard, and Kubach 1991), in the same way as for $CH_3ONO(S_1)$.

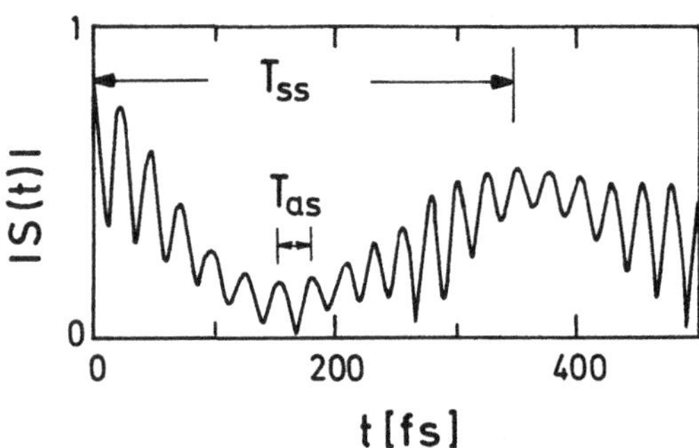

Fig. 7.20. Autocorrelation function for the dissociation of IHI. The wide oscillation reflects the symmetric stretch motion of the two iodine atoms with respect to the stationary hydrogen atom with period T_{ss} whereas the fast oscillations manifest the anti-symmetric stretch motion of hydrogen between the two iodine atoms with period T_{as}. Adapted from Engel (1991a).

For the three-dimensional modeling of photoelectron spectra of IHI and similar systems see Schatz (1989a,b, 1990), Bowman and Gazdy (1989), Gazdy and Bowman (1989), Zhang and Miller (1990), and Zhang, Miller, Weaver, and Neumark (1991).

7.7 Epilogue

Indirect photodissociation involves two more or less separate steps: the absorption of the photon and the fragmentation of the excited complex. Resonances, which mirror the quasi-bound states of the intermediate complex in the upper electronic state, are the main features. They have an inherently quantum mechanical origin. If we consider — in very general terms — the inner region, before the fragments have obtained their identities, as the transition state, then the resolution of resonance structures in the absorption spectrum manifests *transition state spectroscopy* in the original sense of the word (Foth, Polanyi, and Telle 1982; Brooks 1988).

In the time-dependent picture, resonances show up as repeated recurrences of the evolving wavepacket. Resonances and recurrences reveal, in different ways, the same dynamical effect, namely the temporary excitation of internal motion within the complex. In the context of classical mechanics, the existence of quantum mechanical resonances is synonymous with trapped trajectories performing complicated Lissajou-type motion before they finally dissociate. The larger the lifetime, the more frequently the wavepacket recurs to its starting position, and the narrower are the resonances.

Which approach, the time-independent or the time-dependent, is computationally advantageous? If we are satisfied with a low resolution spectrum and if we are not interested in the final fragment distributions, the time-dependent approach has certain advantages because propagating the wavepacket for only a short time can yield a satisfactory survey of the spectrum. If, however, we want to determine the real line shape, the lifetimes, and the final state distributions, the wavepacket must be propagated for at least a time $\Delta t \approx h/\Delta E_{res}$. If the desired energy resolution ΔE_{res} is small then the time-independent formalism may become more opportune. Solving the time-independent Schrödinger equation for many energies can actually be less time consuming than propagating a single wavepacket over a long period. Due to the accumulation of numerical errors the propagation becomes questionable after a certain time anyhow.†

† This problem can be circumvented, however, if we start the evolution of the wavepacket in the dissociative state with an (approximate) eigenstate of the Hamiltonian. According to (7.21), $|S(t)|^2$ is then an exponentially decaying function from which we can easily extract the dissociation rate $\Gamma^{(l)}$.

Incidentally we note that in addition to the methods pursued in this monograph, i.e., the exact solution of Schrödinger's equation, there are other methods available for determining resonance energies and widths, namely the method of *complex scaling* (Reinhardt 1982; Junker 1982; Ho 1983; Moiseyev 1984) and the *stabilization procedure* (Taylor 1970; Meier, Cederbaum, and Domcke 1980; Lefebvre 1985).

The assignment, i.e., the labeling of each resonance by a set of quantum numbers was quite obvious in the cases discussed in this chapter. However, this is not always so, especially if more than two, strongly coupled degrees of freedom are involved. Ultimately, only the stationary wavefunctions allow, by virtue of their nodal structure, an unambiguous assignment. Many structured absorption spectra for polyatomic molecules are unassigned and many others are probably erroneously assigned because it has been assumed, without justification by *ab initio* calculations, that the PES in the upper state has the same qualitative shape as in the electronic ground state. CH_3ONO, H_2O, and other examples (see Figures 1.11 and 1.12) demonstrate that this is mostly not the case. Potential energy surfaces of electronically excited states can be rather complicated and without knowing these surfaces the assignment of spectra becomes a matter of luck!

Resonances occur whenever the PES has a well, deep enough to trap the molecule for several internal periods. The barrier dividing the inner region from the product channel need not necessarily be large as the example of $CH_3ONO(S_1)$ and similar systems indicate. Figures 1.2 and 1.3 illustrate some typical situations in which resonances are prominent: electronic predissociation [Figure 1.2(a)], vibrational predissociation [Figure 1.2(b)], and unimolecular decay (Figure 1.3). We must emphasize, however, that under certain conditions even potentials *without* an intermediate well may trap the molecule and support resonances as the examples of IHI and $ClNO(T_1)$ prove. Weakly bound van der Waals molecules represent a special class of very long-lived resonances; they will be discussed in Chapter 12.

The interpretation of final state distributions following the decay of metastable states is a very interesting topic. If the intermediate complex lives longer than an internal period, the memory on the parent molecule in the electronic ground state will be essentially erased and the product state distributions will no longer reflect the initial wavefunction. As we will show in Section 10.3, they instead reflect the wavefunction in the transition state of the upper electronic state and the dynamics in the exit channel following the transition region.

Finally, the transition from direct to indirect dissociation is continuous as illustrated in Figure 7.21 for a one-dimensional model. As the upper-

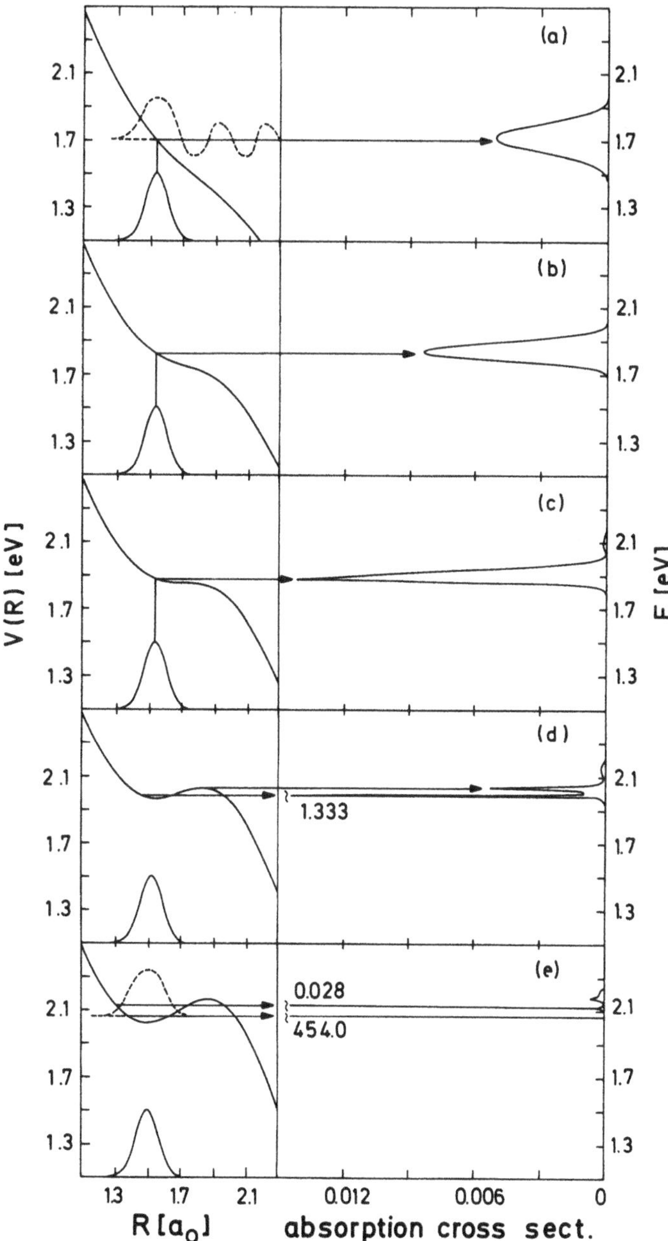

Fig. 7.21. One-dimensional illustration of the transition from direct, (a) and (b), to indirect, (d) and (e), photodissociation. While the spectra in (a) and (b) reflect essentially the initial wavefunction in the electronic ground state, the spectra in (d) and (e) mirror the resonance states of the upper electronic state. The spectrum in (c) illustrates an intermediate case. According to (2.72), the integrated cross sections are the same in each case. Reproduced from Schinke et al. (1989).

state potential flattens within the Franck-Condon region the absorption spectrum gradually becomes narrower. As soon as the flat plateau turns into a shallow well pronounced resonances appear in the spectrum and with increasing well depth more and more resonances appear. The systems $CH_3ONO(S_2)$, $CH_3SNO(S_1)$, and $CH_3ONO(S_1)$ illustrate this behavior very well. The PES for $CH_3ONO(S_2)$ is steep in the FC region (see the upper part of Figure 1.11) and the corresponding absorption spectrum is consequently broad and structureless. The potential for $CH_3SNO(S_1)$ is rather flat near the absorption point [see Figure 7.10(c) for the adiabatic potential curves] and the absorption spectrum consists of three relatively broad vibrational bands [Figures 6.5 and 7.10(d)]. The PES of $CH_3ONO(S_1)$, finally, exhibits a pronounced potential well which gives rise to a structured absorption spectrum with well defined resonances. The broad vibrational bands for $CH_3SNO(S_1)$ are the precursors for the resonances observed for $CH_3ONO(S_1)$.

Resonances are exceedingly sensitive to subtle features of the multidimensional PES and an interesting question for future applications is to what extent *ab initio* theory is able to reproduce all the details of indirect photodissociation.

8
Diffuse structures and unstable periodic orbits

Direct and indirect processes represent the two major classes of photodissociation. In the direct case, the fragmentation proceeds too fast for the molecule to develop a complete internal vibration in the upper electronic state before it breaks apart. The wavepacket leaves the FC region and never returns to its starting place with the consequence that the autocorrelation function decays rapidly to zero. The resulting absorption spectrum is broad and without any vibrational structures. In indirect photodissociation, on the other hand, the excited complex in the upper electronic state lives for a sufficiently long time to allow the development of internal vibration. The wavepacket oscillates in the inner region, frequently recurs to its place of birth, and continuously leaks out into the exit channel. The autocorrelation function exhibits many recurrences and decays slowly to zero. The resulting absorption spectrum is composed of narrow lines which reflect the quasi-bound (resonance) states of the complex and their widths reflect the coupling to the continuum.

The transition from direct to indirect photodissociation proceeds continuously (see Figure 7.21) and therefore there are examples which simultaneously show characteristics of direct as well as indirect processes: the main part of the wavepacket (or the majority of trajectories, if we think in terms of classical mechanics) dissociates rapidly while only a minor portion returns to its origin. The autocorrelation function exhibits the main peak at $t = 0$ and, in addition, one or two recurrences with comparatively small amplitudes. The corresponding absorption spectrum consists of a broad background with superimposed undulations, so-called *diffuse structures.* The broad background indicates direct dissociation whereas the structures reflect some kind of short-time trapping.

Figure 8.1 illustrates two examples, the photodissociation of H_2O in the first and in the second absorption bands. Both spectra exhibit very diffuse

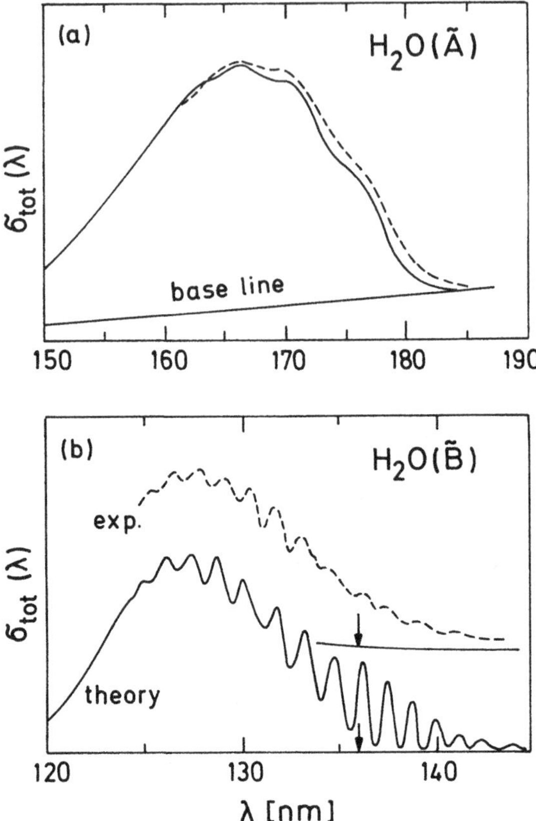

Fig. 8.1. (a) Absorption spectrum of H_2O in the first continuum. The experimental spectrum (broken line) is taken from Wang, Felps, and McGlynn (1977) and the theoretical spectrum (solid line) has been calculated by Engel, Schinke, and Staemmler (1988) using the three-dimensional *ab initio* PES of Staemmler and Palma (1985). (b) Absorption spectrum of H_2O in the second continuum. The experimental result (broken line) is redrawn from Wang, Felps, and McGlynn (1977) and the theoretical curve (solid line) has been calculated by Weide, Kühl, and Schinke (1989) employing the two-dimensional *ab initio* PES of Theodorakopulos, Petsalakis, and Buenker (1985). The experimental and the calculated spectra are vertically displaced for clarity. The arrows mark the $H + OH(^2\Sigma)$ dissociation threshold in the $\tilde{B}$ state. Reproduced from Schinke, Weide, Heumann, and Engel (1991).

structures that in the past had been attributed to excitation of bending motion in the upper electronic state. Why excitation of the bending mode and not excitation of a stretching mode? Figure 8.1 illustrates *the* major problem with diffuse structures, namely their assignment in terms of excitation of the internal modes. As we will demonstrate in this chapter the underlying molecular dynamics is completely different for the two absorption bands of H_2O.

The term "diffuse structures" is not well defined in the literature. It is used whenever structures in the spectrum cannot be unambiguously assigned. In the context of this chapter we identify diffuse structures with very broad resonances. The excited complex is so short-lived that the corresponding autocorrelation function exhibits one or at most two recurrences.

8.1 Large-amplitude symmetric and anti-symmetric stretch motion

Symmetric triatomic molecules like H_2O, O_3, and CO_2, for example, are good candidates for the discussion of diffuse vibrational structures in absorption spectra. We will first discuss a simple model system and then illustrate the general predictions by realistic examples.

8.1.1 A collinear model system

Figure 8.2 depicts a typical potential energy surface (PES) for a symmetric molecule ABA with intramolecular bond distances R_1 and R_2; the ABA bond angle is assumed to be 180° (collinear configuration). The PES is symmetric with respect to the line defined by $R_1 = R_2$; it has a saddle point at short distances and decreases monotonically from the saddle point out into the two identical product channels A + BA and AB + A (see also Figure 7.18). The shaded area indicates the Franck-Condon (FC) region accessed via photon absorption and the two arrows illustrate the main dissociation paths for the quantum mechanical wavepacket or, equivalently, a swarm of classical trajectories. Because no barrier obstructs dissociation, the majority of trajectories immediately evanesce in either one of the two product channels without ever returning to the vicinity of the FC point.

The model potential displayed in Figure 8.2 had originally been used by Kulander and Light (1980) to study, within the time-independent R-matrix formalism, the photodissociation of linear symmetric molecules like CO_2. It will become apparent below that in this and similar cases the time-dependent approach, which we shall pursue in this chapter, has some advantages over the time-independent picture. The motion of the ABA molecule can be treated either in terms of the hyperspherical coordinates defined in (7.33) or directly in terms of the bond distances R_1 and R_2. The Hamiltonian for the linear molecule expressed in bond distances

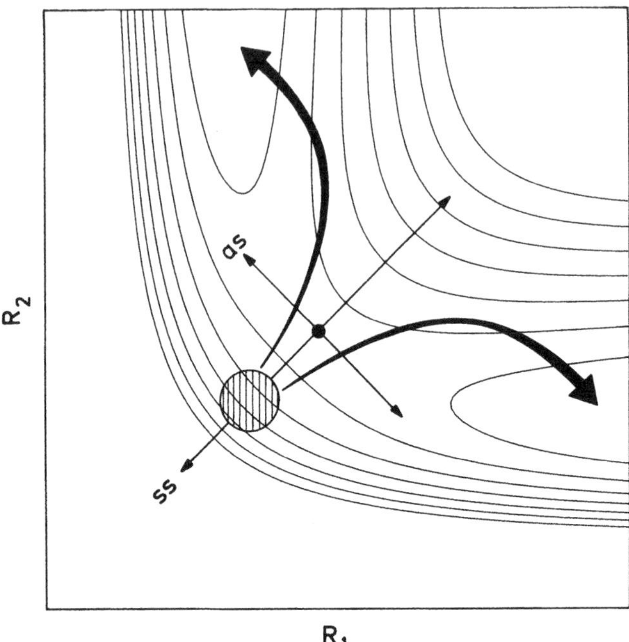

Fig. 8.2. Typical potential energy surface for a symmetric triatomic molecule ABA. The potential energy surface of H_2O in the first excited electronic state for a fixed bending angle has a similar overall shape. The two thin arrows illustrate the symmetric and the anti-symmetric stretch coordinates usually employed to characterize the bound motion in the electronic ground state. The two heavy arrows indicate the dissociation path of the major part of the wavepacket or a swarm of classical trajectories originating in the FC region which is represented by the shaded circle. Reproduced from Schinke, Weide, Heumann, and Engel (1991).

follows straightforwardly from Equation (2.39),

$$\hat{H}(R_1, R_2) = -\frac{\hbar^2}{2\mu}\frac{\partial^2}{\partial R_1^2} - \frac{\hbar^2}{2\mu}\frac{\partial^2}{\partial R_2^2} + \frac{\hbar^2\eta}{\mu}\frac{\partial^2}{\partial R_1 \partial R_2} + V(R_1, R_2), \quad (8.1)$$

where $\mu = m_A m_B/(m_A + m_B)$ and $\eta = m_A/(m_A + m_B)$. In contrast to Jacobi or hyperspherical coordinates, the kinetic energy operator is not diagonal! However, application of $\partial^2/\partial R_1 \partial R_2$ does not cause any numerical difficulties if the time-dependent wavepacket is discretized and represented by a two-dimensional Fourier-series as sketched in Section 4.2.2. Equation (4.22) allows the straightforward evaluation of all first- and second-order derivatives, even if they contain mixed terms. In the following, we examine some illuminating model calculations, with masses appropriate for CO_2, which are relevant for other symmetric triatoms as well.

Figure 8.3(a) depicts a calculated absorption spectrum (Kulander and Light 1980; Schinke and Engel 1990). It consists of a broad background with superimposed finer structures that look rather irregular. The broad background is reminiscent of fast dissociation while the diffuse structures indicate some kind of short-time trapping. The PES in the upper electronic state is repulsive and there is no potential well that could trap the wavepacket or the trajectories as they rush down the potential slope into either of the exit channels. In order to magnify the low and the high energy portions of the spectrum in (a) Schinke and Engel (1990) calculated the spectra for two slightly different equilibrium separations in the electronic ground state. They are shown in Figures 8.3(b) and (c), respectively. Note that this alteration does not affect the general dynamics in the upper state. For example, possible resonances and their widths are the same in all three cases in Figure 8.3.

The spectrum in Figure 8.3(c) looks particularly bizarre and an obvious assignment, like for the examples discussed in Chapter 7, is not apparent. The primary questions are:

- What kind of internal vibrational motion causes the superimposed structures in the spectrum and how can we characterize and assign the structures?

A much simpler picture emerges in the time domain. The corresponding autocorrelation function, depicted in Figure 8.4, exhibits three well resolved recurrences with very small amplitudes. The recurrence times T_1, T_2, and T_3 are incommensurable which indicates that they reflect different types of molecular motion. Since the recurrences are well separated we can write $S(t)$ as a sum $S(t) = \sum_i S_i(t)$, $i = 0, \ldots, 3$, with S_0 representing the main peak at $t = 0$. The Fourier transformation is linear so that the absorption cross section also splits into four individual terms,

$$\sigma_{tot}(E) \propto \sum_{i=0}^{3} \tilde{\sigma}_i(E) \propto \sum_{i=0}^{3} \int_{-\infty}^{+\infty} dt\ S_i(t)\ \mathrm{e}^{iEt/\hbar}. \tag{8.2}$$

Let us, for simplicity, represent each $S_i(t)$ by

$$S_i(t) = a_i \left[\mathrm{e}^{-\alpha_i(t-T_i)^2} + \mathrm{e}^{-\alpha_i(t+T_i)^2}\right] \mathrm{e}^{-i\beta_i t/\hbar}, \tag{8.3}$$

where a_i, α_i, and β_i are real constants, T_i $(i = 1, 2, 3)$ are the three recurrence times, and $T_0 = 0$ [see Figure 8.5(a)]. Equation (8.3) is written in this particular form in order to satisfy the symmetry relation (4.10).

Inserting (8.3) into (8.2) readily yields

$$\tilde{\sigma}_i(E) \propto a_i \cos\left[(E - \beta_i)\frac{T_i}{\hbar}\right] \mathrm{e}^{-[(E-\beta_i)/\hbar]^2/(4\alpha_i)}. \tag{8.4}$$

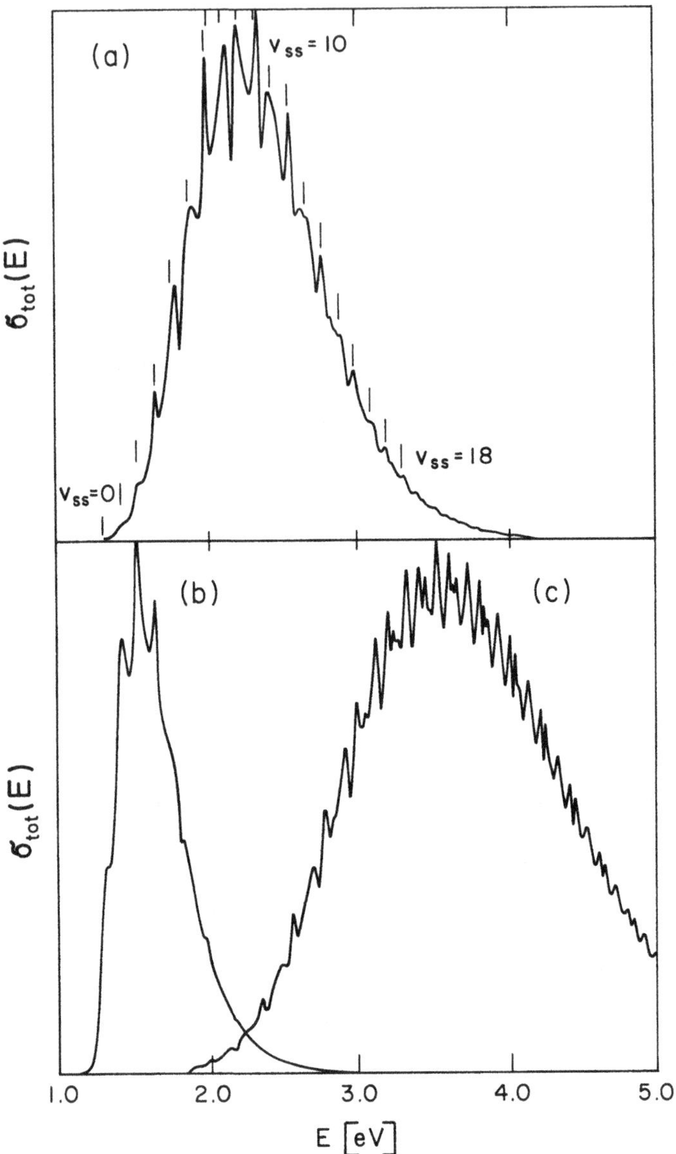

Fig. 8.3. (a) Absorption cross section for the model CO_2 system studied by Kulander and Light (1980). The vertical lines indicate the energies of the symmetric stretch mode on top of the saddle of the upper electronic state and v_{ss} denotes the corresponding quantum number. (b) and (c) Absorption spectra for equilibrium bond distances which are slightly longer respectively shorter than those with which the spectrum in (a) has been calculated. The dynamics in the upper state, however, is unaffected by this shift. Reproduced from Schinke and Engel (1990).

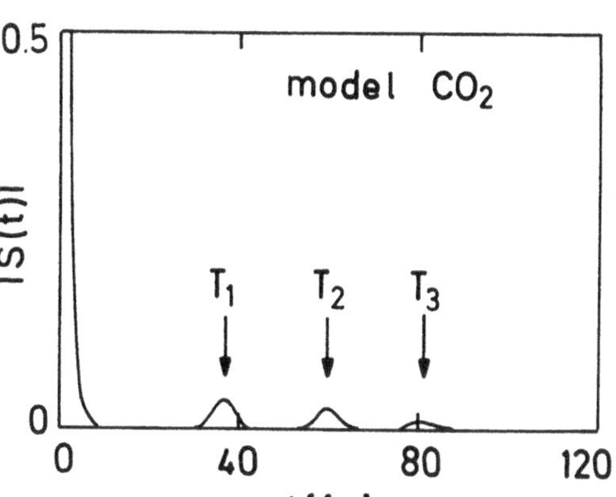

Fig. 8.4. The autocorrelation function $S(t)$ corresponding to the absorption spectrum shown in Figure 8.3(a). The arrows mark the periods of the three unstable periodic orbits discussed in Section 8.1.2. Reproduced from Schinke and Engel (1990).

The sinusoidal factor follows from the translation property of the Fourier transformation (Butkov 1968:ch.4). Thus, each "spectrum" $\tilde{\sigma}_i(E)$ is a Gaussian in energy centered at $E = \beta_i$ and modulated by a sinusoidal function with wavelength

$$\Delta E = \frac{2\pi\hbar}{T_i} \tag{8.5}$$

as schematically illustrated in Figure 8.5(b). Except for $\tilde{\sigma}_0$, the $\tilde{\sigma}_i$ can be both positive *and* negative and therefore they are not intrinsic cross sections; only the sum of all $\tilde{\sigma}_i$ has a physical meaning.

According to this simplified analysis the absorption cross sections in Figure 8.3 represent the sum of three oscillatory terms with wavelengths $2\pi\hbar/T_i$ superimposed on the structureless background $\tilde{\sigma}_0(E)$. Since the recurrence times are incommensurable, the oscillations are uncorrelated which explains the somewhat erratic appearance of the diffuse structures, especially in Figure 8.3(c). The spectrum would show a much simpler energy dependence if $S(t)$ were composed of only a single recurrence instead of three (see Section 8.2 for an example).

The analysis presented so far explains the energy dependence of the spectrum in terms of the recurrences of the wavepacket. It does not, however, elucidate what kind of molecular motion actually causes the recurrences. In this particular example, the evolving wavepacket by itself does not clearly reveal the origin of the recurrences. The major part of the wavepacket follows the shortest route to dissociation and buries the much smaller portion, which returns at least once to the starting position.

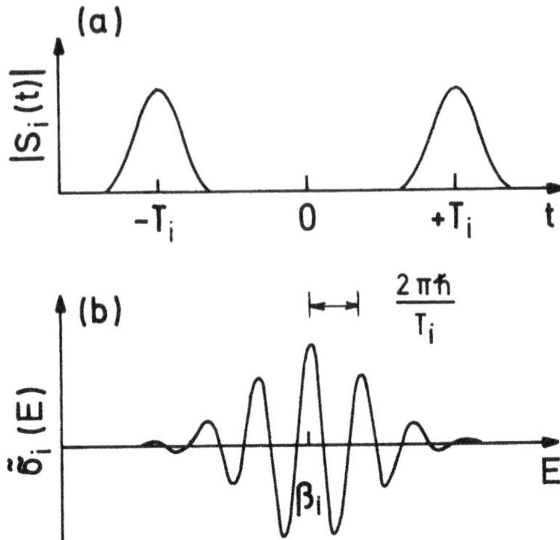

Fig. 8.5. (a) Schematic illustration of $S_i(t)$ defined in Equation (8.3). (b) The corresponding "cross section" $\tilde{\sigma}_i(E)$ defined in (8.4).

According to Section 4.1.1 the wavepacket is a superposition of stationary wavefunctions corresponding to a relatively wide range of energies. This and the superposition of three apparently different types of internal vibrations additionally obscures details of the underlying molecular motion that causes the recurrences. A particularly clear picture emerges, however, if we analyze the fragmentation dynamics in terms of classical trajectories.

8.1.2 Unstable periodic orbits

Following individual trajectories, which start with randomly selected initial conditions in the FC region, evinces that some of them behave differently than the majority of trajectories. While most trajectories dissociate directly as indicated by the heavy arrows in Figure 8.2, these distinctive trajectories are reflected, sometimes in an unexpected way, back to the proximity of the FC point before they finally lead to fragmentation. A more systematic search brings to light the existence of so-called *unstable periodic orbits* which, loosely speaking, guide these "indirect" trajectories. They provide an intriguing explanation of the recurrences in the autocorrelation function and thus of the diffuse structures in the spectrum.

Figure 8.6 depicts the four simplest types of unstable periodic orbits for the model CO_2 system. Despite the fact that the total energy is well above the dissociation threshold and despite the lack of an intermediate potential well, that could possibly trap the molecule at short distances, these trajectories live forever without dissociating! They are perfectly periodic in the four-dimensional phase-space. The periods depend only weakly on the total energy. These trajectories are, however, highly unstable and fragile: the slightest distortion destroys the periodicity with the consequence that they promptly dissociate.

The simplest periodic orbit [Figure 8.6(a)] is totally symmetric, i.e., $R_1 = R_2$ and $P_1 = P_2$ at all times, where P_1 and P_2 are the classical momenta corresponding to R_1 and R_2, respectively. It oscillates, like a tightrope dancer, on top of the potential ridge without loosing its balance and falling into either of the two product channels. This periodic orbit naturally follows from symmetry considerations and it can be surmised without any calculation. Its period, T_1, agrees exceedingly well with the first recurrence time in Figure 8.4 and this coincidence undoubtedly

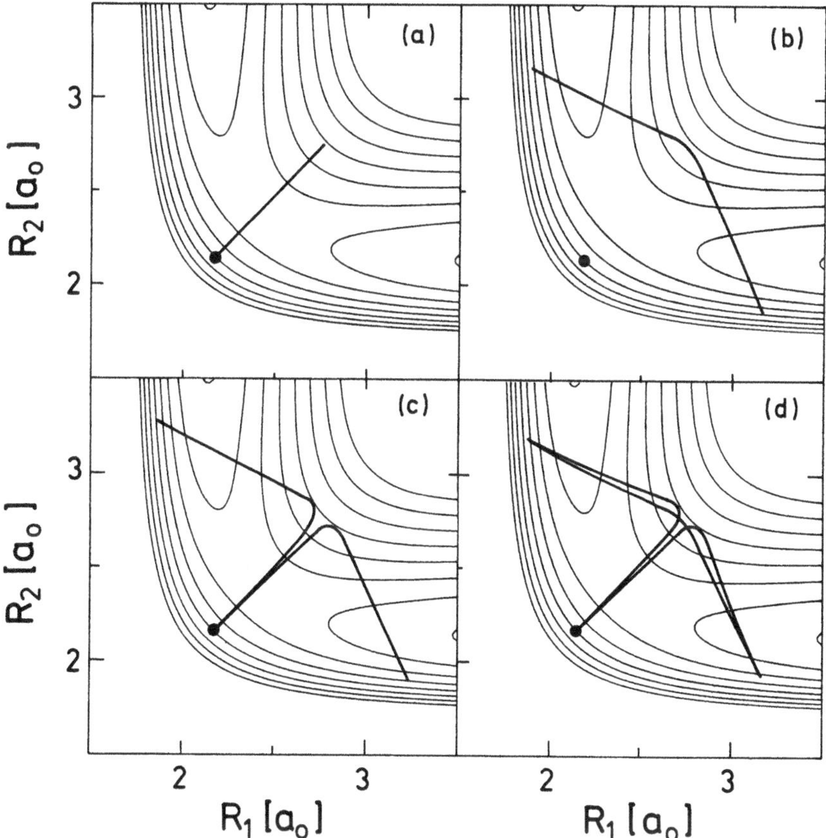

Fig. 8.6. The four simplest types of unstable periodic orbits for the model CO_2 system. The heavy dot marks the equilibrium in the electronic ground state. Adapted from Schinke and Engel (1990).

corroborates that the first and largest recurrence is caused by symmetric stretch motion.

The quantum mechanical picture is the following. While the main part of the quantum mechanical wavepacket slides down into one of the product channels and dissociates at once, a small part of it stays on top of the potential ridge, swings along the symmetric stretch line, and recurs to its origin. Due to the strong coupling to the mode that is "perpendicular" to symmetric stretch motion (i.e., dissociation) this portion of the wavepacket diminishes rapidly with the consequence that only its first recurrence leads to an appreciable increase of the autocorrelation function. Heller (1978b) was the first to recognize this type of periodic orbit and its importance for the absorption spectrum of symmetric triatomic molecules.

Figure 8.6(b) features another type of periodic orbit which, roughly speaking, is orthogonal to the symmetric stretch periodic orbit. It has been extensively investigated in reactive collisions in order to explain resonances of the kind we discussed in Section 7.6.2 (Pollak, Child, and Pechukas 1980; Pollak 1981; Pollak and Child 1981; Marston and Wyatt 1984a,b, 1985; Pollak 1985; Child 1991:ch.10). Because the two outer atoms oscillate in opposite directions, i.e., when one of the intramolecular bonds is large the other one is small and *vice versa*, we will call this trajectory an *anti-symmetric stretch* or *hyperspherical periodic orbit*. Anti-symmetric stretch motion corresponds roughly to motion along the hyperspherical angle [see Figure 7.18(b)]. While the symmetric stretch periodic orbit is rather obvious, the existence of the anti-symmetric stretch periodic orbit is at first glance unexpected and very astounding. At its turning points the orbit is reflected onto itself and oscillates between the two sides of the PES. Although it stretches far into the product channels it does not lead to dissociation.†

The period of the anti-symmetric stretch periodic trajectory does not correspond, however, to any of the three recurrences we see in Figure 8.4. This is not at all surprising; in order to come back to the FC region, which in this case is considerably displaced from the anti-symmetric stretch orbit, the trajectory must necessarily couple to the symmetric stretch mode. If we were to launch the wavepacket at the outer slope of the saddle point, the anti-symmetric stretch periodic orbit would support recurrences by itself without coupling to the symmetric stretch mode. An example is the dissociation of IHI discussed in Section 7.6.2.

† The existence of the anti-symmetric stretch periodic orbit is more evident, however, in terms of either hyperspherical coordinates or mass-scaled Jacobi coordinates (Kulander, Cerjan, and Orel 1991). In the present case, the relatively large kinetic coupling term in Equation (8.1) has the consequence that it does not perpendicularly intersect the potential contours at the turning points.

Figures 8.6(c) and (d) illustrate two additional types of periodic orbits that actually combine symmetric and anti-symmetric stretch motion. Both trajectories start their journey in the FC region, but are slightly inclined to the symmetry line. At the outer slope of the saddle point they couple to the anti-symmetric mode and are reflected into one of the product channels without, however, dissociating. At the next turning point, the orbit in Figure 8.6(c) is reflected back onto itself and returns to its starting point on exactly the same path on which it came in the first place. The other periodic orbit, Figure 8.6(d), continues its journey differently. At the second turning point it is first thrown across the saddle point into the other channel before it finds the way home. The delay times after which they return to their origin, T_2 and T_3, agree extremely well with the remaining two recurrences of Figure 8.4.

The classical picture of how the periodic orbits influence the dissociation dynamics goes as follows: The majority of trajectories that we start randomly in the FC region immediately dissociate. Some of them, however, set out very close to a periodic orbit and therefore they stay in its vicinity for at least one full round. If a trajectory were accidentally to begin *exactly* on a periodic orbit, it would follow this orbit forever, or until numerical inaccuracies accumulate and ultimately throw it out of the periodic orbit:

- The periodic orbits serve as "guide ropes" for those classical trajectories which start in their neighborhood.

Because the periodic orbits are highly unstable in the present example (i.e., the slightest distortion destroys the periodicity and leads to dissociation) the trapped trajectories recur only once; after returning to their origin they start again, with a new set of initial conditions, and most probably will dissociate in the second round.

Knowing that the quantum mechanical wavepacket follows — at least for short times — the route of a swarm of classical trajectories, it is plausible to surmise that the periodic orbits also influence the motion of the wavepacket:

- The periodic orbits are — figuratively speaking — the "backbones" for the quantum mechanical wavepacket evolving on the upper-state PES.

Because of the superposition of three distinct types of periodic motion with different periods the wavepacket by itself does not reveal a clear picture in the present case, i.e., the "classical skeleton" is hardly visible through the "quantum mechanical flesh". The perfect agreement between the recurrence times of the quantum mechanical wavepacket and the periods of the classical periodic orbits, however, provides convincing evidence that the structures in the absorption spectrum are ultimately the consequence of the three generic unstable periodic orbits. This correlation is of course not always unambiguous, especially if the behavior of the autocorrelation function is more complex and if more periodic orbits exist (see Figure 8.8, for example).

The stability or robustness of the periodic orbits determines the amplitudes of the recurrences and hence the diffuseness of the spectrum. The less stable the orbit, the less pronounced are the diffuse structures superimposed on the background. With increasing stability the wavepacket might recur more than once with the consequence that the structures become narrower and more like resonances of the type discussed in Chapter 7. Let us consider the symmetric stretch orbit. The shape of the potential along the minimum energy path (perpendicular to the symmetric stretch coordinate) governs the stability of the symmetric stretch periodic orbit. The flatter the saddle between the two product channels, the larger is the trapping probability and hence the amplitude of the recurrence. A marble rolling roughly along the symmetric stretch coordinate maintains its balance more easily if the saddle is flatter. A shallow potential minimum in the region of the saddle would stabilize the symmetric stretch motion and thus it would ultimately lead to narrow, i.e., longer-lived resonances in the same way as described for $CH_3ONO(S_1)$ in Sections 7.3 and 7.4, for example.

The importance and usefulness of periodic orbits for molecular motion have been widely recognized and illustrated with many beautiful examples in the last decade (Gutzwiller 1990:ch.6). These include:

1) Quantization of nonseparable Hamiltonians and the calculation of spectral densities (Noid, Koszykowski, and Marcus 1981; Tabor 1981, 1989; Stechel and Heller 1984; Gutzwiller 1990:chs.16,17; Child 1991: ch.7).
2) Resonances in full and in half collisions (Pollak, Child, and Pechukas 1980; Pollak 1981; Pollak and Child 1981; Marston and Wyatt 1984a,b, 1985; Schinke, Weide, Heumann, and Engel 1991).
3) The analysis of quantum mechanical eigenfunctions in energy regimes where the classical motion is chaotic (Heller 1984, 1986; Founargiotakis, Farantos, Contopoulos, and Polymilis 1989).

The semiclassical theories of Gutzwiller (1967, 1971, 1980), Balian and Bloch (1972), and Berry and Tabor (1976, 1977), which are based on Feynman's path-integral formulation of quantum mechanics (Feynman and Hibbs 1965), provide a solid theoretical justification.

When conventional spectroscopic pictures and means, like normal and local modes or adiabatic separation, are useless for the interpretation of absorption spectra because either the displacements from equilibrium are too large or the coupling between the degrees of freedom is too strong, unstable periodic orbits might provide the desired link between molecular motion on one hand and structures in the absorption spectra on the other. This applies to bound-bound transitions in an energy region where the classical Hamiltonian shows chaos, photodissociation, or resonances in full collisions.

The highly excited hydrogen atom in a strong magnetic field is probably the most beautiful example (Wintgen et al. 1986; Holle et al. 1986, 1987; Main, Wiebusch, Holle, and Welge 1986; Main, Holle, Wiebusch, and Welge 1987; Wintgen and Friedrich 1987; Du and Delos 1988a,b; Eichmann, Richter, Wintgen, and Sandner 1988; for an overview see Friedrich and Wintgen 1989). Figure 4.4(a) shows part of a high resolution spectrum and Figure 4.4(b) depicts the corresponding autocorrelation function. While the spectrum is seemingly unassignable, its Fourier transform has a much simpler behavior. The main "resonances" in the autocorrelation function can be unambiguously related to unstable periodic orbits of the electron in the Coulomb field of the proton and the external magnetic field. Molecular examples, in which unstable periodic orbits play a crucial role, include the IR photodissociation of H_3^+ (Carrington and Kennedy 1984; Gómez Llorente and Pollak 1987, 1988; Taylor and Zakrzewski 1988; Gómez Llorente, Zakrzewski, Taylor, and Kulander 1988, 1989; Berblinger, Pollak, and Schlier 1988; Brass, Tennyson, Pollak 1990), the UV photodissociation of H_2O via the $\tilde{B}$ state (see the next section), and Na_3 (Gómez Llorente and Taylor 1989).

8.1.3 Examples

The photodissociation of H_2O in the first absorption band ($\tilde{X}^1A_1 \rightarrow \tilde{A}^1B_1$; see Figure 1.5) exemplifies large-amplitude symmetric stretch motion in the upper electronic state. Both the measured as well as the calculated spectra exhibit very diffuse structures superimposed on a broad background [Figure 8.1(a)]. The corresponding PES (Staemmler and Palma 1985) for fixed HOH bending angle has the same overall shape as the model potential shown in Figure 8.2. The calculated autocorrelation function exhibits a single, very weak recurrence that, according to Henriksen, Zhang, and Imre (1988), is caused by symmetric stretch motion. The other types of periodic orbits discussed in Figure 8.6 for the model system are too unstable in this case to support significant recurrences.

A time-independent adiabatic approximation, based on the *local* separability of symmetric and anti-symmetric stretch motion in the region of the saddle point, provides a complementary picture (Pack 1976). Within the adiabatic limit the eigenenergies of the symmetric stretch motion on top of the potential ridge are defined through the one-dimensional

Schrödinger equation[†]

$$\left[-\frac{\hbar^2}{2\tilde{\mu}}\frac{\partial^2}{\partial R_+^2} + V(R_+, R_- = 0) + \epsilon_{v_{ss}}\right]\varphi_{v_{ss}}(R_+) = 0, \tag{8.6}$$

where $R_+ = R_1 + R_2$ and $R_- = R_1 - R_2$ and R_1 and R_2 are the two H–O bond separations. The mass $\tilde{\mu}$ is given by $\tilde{\mu} = \mu/(2+\eta)$ with μ and η being defined below (8.1).

Figure 8.7(a) depicts a cut of the $\tilde{A}$-state PES along the symmetric stretch coordinate together with the corresponding eigenenergies obtained from Equation (8.6) and Figure 8.7(b) shows, on the same energy scale, the absorption spectrum calculated for a fixed HOH bending angle of 104°. Figure 8.7 clearly manifests that:

- The diffuse structures in the absorption spectrum of H_2O in the first absorption band reflect the eigenenergies of the symmetric stretch mode in the upper electronic state.

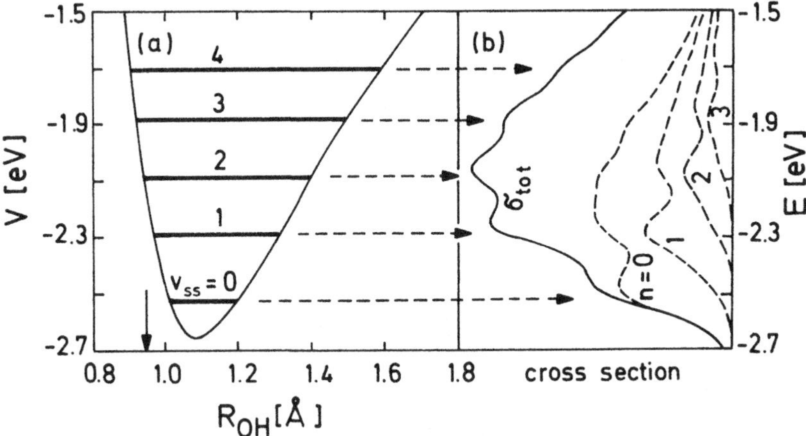

Fig. 8.7. (a) Cut through the potential energy surface of H_2O in the first excited electronic state along the symmetric stretch coordinate. The horizontal lines indicate the symmetric stretch energies obtained from solving (8.6). The HOH bending angle is 104° and the energy normalization is such that $E = 0$ corresponds to three ground-state atoms. The arrow on the abscissa marks the equilibrium bond distance in the electronic ground state. (b) The total absorption spectrum (for fixed bending angle) as a function of energy. The dashed curves represent the partial photodissociation cross sections for producing OH in vibrational state n.

[†] Equation (8.6) follows by transforming (8.1) to the symmetry-adapted coordinates R_+ and R_-, setting R_- to zero, and omitting the kinetic energy term in R_-.

- The relative intensities of the broad resonance structures manifest the one-dimensional FC factors of the symmetric stretch wavefunctions in the ground and in the upper electronic states.

Since the equilibrium distances in the two electronic states are noticeably different, the maximum occurs near the symmetric stretch quantum number $v_{ss} = 2$. The same general conclusions are also valid for the model CO_2 system discussed above, as the good agreement between the main peaks, caused by symmetric stretch motion, and the corresponding eigenenergies, indicated by the vertical lines in Figure 8.3(a), emphasizes.[†]

So far we considered only the total absorption spectrum $\sigma_{tot}(E)$; but one must not forget that the total spectrum is merely the sum of the partial photodissociation cross sections $\sigma(E, n)$ for producing the OH fragment in a particular vibrational state n. Each partial cross section is structured [Figure 8.7(b)] and the superposition of all of them leads to the structures in the total spectrum. Actually, the modulation of the partial spectra is even more pronounced than the modulation of the total spectrum; the summation of several $\sigma(E, n)$ naturally blurs the diffuse structures. Averaging over the bending angle according to the sudden formula (3.48) as well as thermal broadening additionally smears them out as comparison of Figures 8.1 and 8.7 clearly shows. In the light of Figure 8.7, we may interpret the diffuse structures as very broad resonances due to temporary excitation of symmetric stretch motion. The diffuseness is the consequence of coupling to the dissociation channel.

Despite their vagueness the vibrational structures in the measured spectrum were firmly attributed to bending excitation in the upper electronic state, probably because the energy spacing of about 1850 cm^{-1} corresponds roughly to the bending frequency in the electronic ground state. The theoretical analysis based on an *ab initio* PES and exact dynamical calculations unambiguously proves this assignment to be wrong! The vibrational structures are due to symmetric stretch excitation, as predicted by Pack (1976) and Heller (1978b) for general symmetric molecules. Bending motion is inactive in the $\tilde{A}$ state of water; it merely broadens the diffuse structures.

[†] The success of this simple adiabatic picture rests upon the fact that, because of symmetry considerations, the nonadiabatic coupling elements $\langle \varphi_{v_{ss}}(s;a) \mid \partial/\partial a \mid \varphi_{v'_{ss}}(s;a) \rangle$ [see Equation (3.36)], which couple symmetric and anti-symmetric stretch motion, are exactly zero on top of the potential saddle. Here, a represents the reaction, i.e., anti-symmetric stretch, coordinate, s is the symmetric stretch coordinate, and the $\varphi_{v_{ss}}(s;a)$ are the symmetric stretch wavefunctions. Away from the saddle point, however, nonadiabatic coupling between symmetric and anti-symmetric stretch motion is substantial and the adiabatic approximation breaks down.

The absorption spectrum of O_3 in the Hartley band [Figure 8.8(a)] qualitatively resembles the model spectrum in Figure 8.3. Johnson and Kinsey (1989a,b) determined the autocorrelation function by Fourier transformation of the experimental spectrum [Figure 8.8(b)] and related the larger recurrences to unstable periodic orbits which they calculated using the *ab initio* PES of Sheppard and Walker (1983). Orbit B combines symmetric and anti-symmetric stretch motion in the same way as we discussed above for the model CO_2 system. The calculated potential has two shallow wells slightly displaced from the symmetry line which might explain the existence of the complicated orbits C and D. According to this analysis the recurrence due to symmetric stretch motion (orbit A) is comparatively small; the strongest peaks are probably the result of motion along orbit D (see also Farantos and Taylor 1991).

Le Quéré and Leforestier (1990, 1991) calculated the autocorrelation function *directly* using the same PES and found fair agreement in regard of the recurrence times while the amplitudes were in remarkable disagreement. This may be due to either deficiencies of the calculated PES or the neglect of nonzero total angular momentum states in the theory. If the recurrences in the autocorrelation function are rescaled, the quantum mechanically calculated spectrum agrees well with experiment.

Despite the apparent difficulties in *quantitatively* reproducing a structured spectrum like that of O_3 we must emphasize, however, that unstable periodic orbits are probably the only means of interpreting these struc-

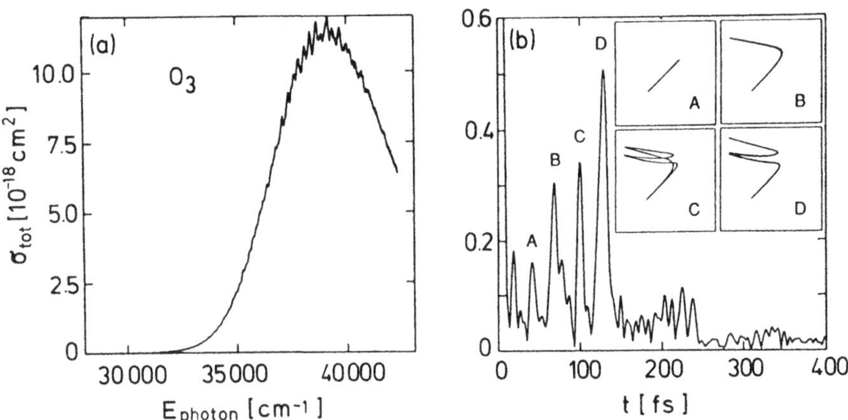

Fig. 8.8. (a) Absorption spectrum of O_3 in the Hartley band measured by Freeman, Yoshino, Esmond, and Parkinson (1984). (b) Fourier transformation of the measured spectrum (multiplied by 100). The inset depicts unstable periodic orbits which are possibly responsible for the four most prominent recurrences denoted by A–D. Adapted from Johnson and Kinsey (1989a).

tures and revealing the underlying molecular motion. Necessary prerequisites are, however, accurate potential energy surfaces and exact dynamics calculations (Banichevich, Peyerimhoff, and Grein 1990; Yamashita and Morokuma 1991).

8.2 Large-amplitude bending motion

In this section we describe another kind of large-amplitude motion which leads — again guided by an unstable periodic orbit — to very diffuse structures. The absorption spectrum of H_2O in the second band ($\tilde{X}^1A_1 \rightarrow \tilde{B}^1A_1$; see Figures 1.6 and 8.1) is composed of a broad background, like in the first band, with a superimposed, rather regular modulation. The result of a two-dimensional model calculation, in which one of the O-H bonds is fixed, reproduces the experimental spectrum quite satisfactorily. Deviations at the onset of the spectrum are probably due to the neglect of electronic nonadiabatic coupling to the electronic ground state. The two-dimensional Hamiltonian including only the H-OH dissociation bond and the rotation of the OH fragment (rigid-rotor model) is given by Equation (3.16) and the upper part of Figure 1.12 shows the corresponding PES. Figure 8.9 shows the same PES in a different representation. It has a deep well at the linear H–O–H configuration with a depth of about 3.3 eV relative to the $H(^2S) + OH(^2\Sigma, r_e)$ dissociation limit. This well is the result of an avoided crossing with the electronic ground state. It influences the overall dissociation dynamics and explains the undulations of the absorption spectrum as well as the strong rotational excitation of $OH(^2\Sigma)$ which we will discuss in Section 10.2.

8.2.1 *Breakdown of adiabatic separability*

The photodissociation of water through the $\tilde{B}$ state provides an illustrative example for the failure of traditional spectroscopic models when the molecule has large-amplitude motion and when the coupling between the relevant modes is strong. According to conventional ideas, the diffuse structures in the second continuum of H_2O are attributed to the excitation of high bending states as illustrated schematically in Figure 8.10 (Wang, Felps, and McGlynn 1977). The photon promotes H_2O from a bent ground state to a linear excited electronic state. The large difference in the two equilibrium angles, $\alpha_e = 104°$ in the lower state compared to $\alpha_e = 180°$ in the upper state, causes excitation of high bending mode overtones, in the same way as described for NH_3, for example. This picture, which is strictly based on the adiabatic separation of the bending and the stretching degrees of freedom, is realistic for NH_3, but it is essentially wrong in the present case (Weide and Schinke, 1989). It neglects

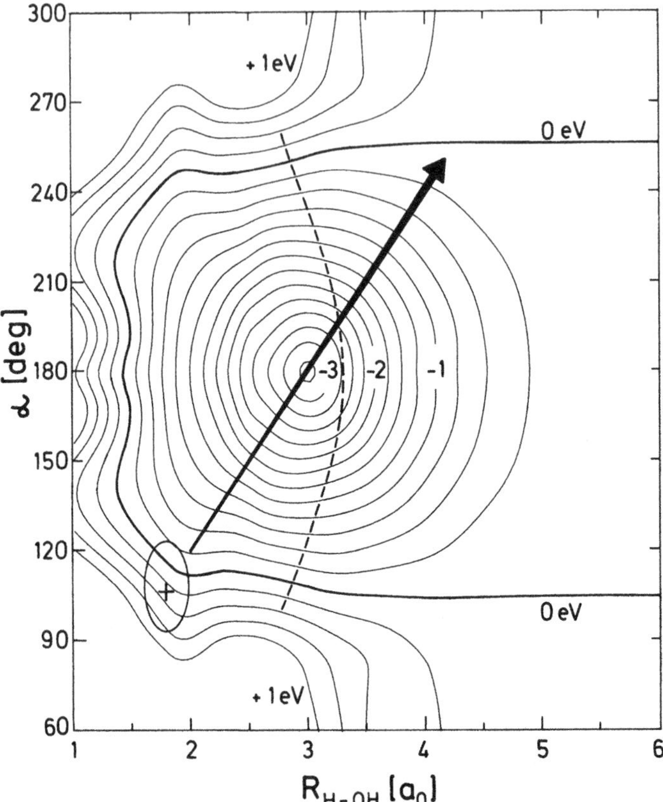

Fig. 8.9. Contour plot of the potential energy surface of H_2O in the $\tilde{B}^1A_1$ state as a function of the H-OH dissociation bond R_{H-OH} and the HOH bending angle α; the other O-H bond is frozen at the equilibrium value in the ground electronic state. The energy normalization is such that $E = 0$ corresponds to $H(^2S) + OH(^2\Sigma, r_e)$. This potential is based on the *ab initio* calculations of Theodorakopulos, Petsalakis, and Buenker (1985). The structures at short H-OH distances are artifacts of the fitting procedure. The cross marks the equilibrium in the ground state and the ellipse indicates the breadth of the ground-state wavefunction. The heavy arrow illustrates the main dissociation path and the dashed line represents an unstable periodic orbit with a total energy of 0.5 eV *above* the dissociation threshold.

the fact that in addition to the bending angle the H-OH dissociation bond also changes substantially, from $1.8a_0$ in the ground state to about $3a_0$ in the excited state as Figure 8.9 clearly illustrates. As a consequence, the wavepacket is launched on the $\tilde{B}$-state PES far away from equilibrium. The maximum of the absorption spectrum corresponds to energies of 0.5–0.75 eV in the continuum, i.e., *above* the dissociation threshold of the $\tilde{B}$ state [see Figure 8.1(b)]. Thus, the $\tilde{X} \to \tilde{B}$ transition in water is a bound-continuum rather than a bound-bound transition!

In the relevant energy regime the bending and the H-OH stretching degrees of freedom do not adiabatically separate, like they do, for example, deep inside the potential well. For energies slightly below the dissociation limit, the (classical) motion behaves chaotically whereas above the threshold H_2O dissociates with strong coupling between bending and stretching motion. Simple one-dimensional bending wavefunctions of the kind depicted schematically in Figure 8.10 do not exist in this energy regime! We strongly emphasize this point because one often finds that structures in bound-continuum spectra are labeled by a set of quantum numbers. However, one should always bear in mind that this kind of assignment depends ultimately upon adiabatic separability, at least approximately. A prototype for the validity of such an assignment, although the vibrational structures are very diffuse and although the spectrum is a continuum spectrum, is the photodissociation of $CH_3SNO(S_1)$ (see Figure 6.5). If, however, the necessary conditions are not met, which is often the

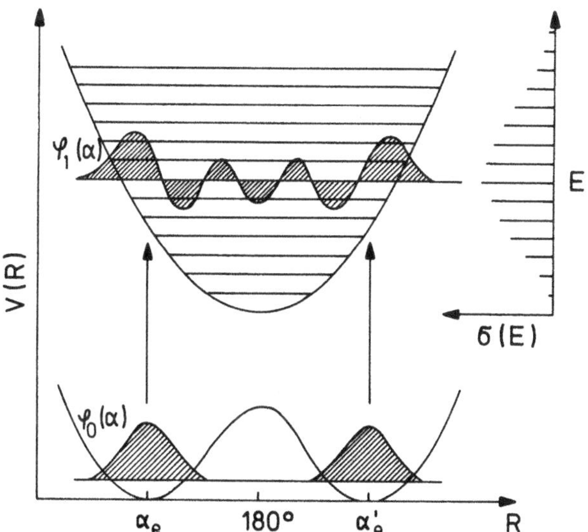

Fig. 8.10. One-dimensional illustration of the excitation of high bending states in a bent-linear transition as in the case of NH_3, for example (Figure 7.16). The potentials and hence the bending wavefunctions are symmetric with respect to the linear configuration. The horizontal lines indicate the bending energies in the upper state. Because the equilibrium angles in the ground and in the excited electronic states are so different, the distribution of Franck-Condon factors $|\langle\varphi_1(\alpha) \mid \varphi_0(\alpha)\rangle|^2$ peaks at highly excited bending states. φ_0 and φ_1 are the bending wavefunctions in the ground and the excited electronic state, respectively.

case at energies high up in the continuum, an erroneous assignment may entail severe misinterpretations of the underlying molecular dynamics.

Large-amplitude bending motion indeed causes the diffuse structures in the $H_2O(\tilde{B})$ spectrum, but in a different way than suggested by Figure 8.10.

8.2.2 Dissociation guided by an unstable periodic orbit

Let us imagine a swarm of classical trajectories starting randomly in the transition region. They immediately experience an extremely strong force in the direction of the potential minimum. Both the torque $-\partial V/\partial\alpha$, which opens the bending angle, and the radial force $-\partial V/\partial R_{\text{H-OH}}$, which first stretches and finally cuts the bond between H and OH, are large in the Franck-Condon region and along the entire dissociation path. The vast majority of trajectories traverse the potential well (as indicated by the heavy arrow in Figure 8.9) and dissociate without any significant delay, producing $OH(^2\Sigma)$ in highly excited rotational states (Weide and Schinke 1987; Dunne, Guo, and Murrell 1987). The momentum gained in the first half of the fragmentation, from $\alpha = 104°$ to $180°$, suffices for the trajectories to climb up the outer wall of the potential and to fragment. A small fraction of trajectories, however, do not succeed in dissociating at the first attempt but recur, like a boomerang, to the starting position. The second try with new initial conditions, however, is then successful (Weide and Schinke 1989).

The quantum mechanical wavepacket closely follows the main classical route. It slides down the steep slope, traverses the well region, and travels toward infinity. A small portion of the wavepacket, however, stays behind and gives rise to a small-amplitude recurrence after about 40–50 fs. Fourier transformation of the autocorrelation function yields a broad background, which represents the direct part of the dissociation, and the superimposed undulations, which are ultimately caused by the temporarily trapped trajectories (Weide, Kühl, and Schinke 1989). A purely classical description describes the background very well (see Figure 5.4), but naturally fails to reproduce the undulations, which have an inherently quantum mechanical origin.

As in the cases discussed in Section 8.1 an unstable periodic orbit, illustrated by the broken line in Figure 8.9, guides the indirect classical trajectories and likewise the temporarily trapped part of the quantum mechanical wavepacket. It essentially represents large-amplitude bending motion, that is strongly coupled, however, to the stretching coordinate $R_{\text{H-OH}}$.

This type of periodic orbit extends, without significantly changing its generic shape, from negative energies deep inside the potential well, through the (classically) chaotic energy regime around the dissociation threshold, and high up into the continuum. As the total energy rises from about -3 eV to more than $+1$ eV (measured with respect to the dissociation threshold) the period increases by roughly a factor of three, the coupling to the H-OH stretching mode (i.e., the "curvature" of the orbit) becomes stronger, and at the same time the stability decreases rapidly. The periodic orbit is quite robust deep in the well whereas high above the dissociation threshold it becomes extremely fragile with the consequence that the slightest distortion destroys it. The existence of this type of bending periodic orbit is certainly not unexpected for energies inside the well; its persistence high up into the continuum, however, comes as a surprise.

Figure 8.11 illustrates how the periodic orbit influences the dissociation. It shows snapshots of the *classical coordinate distribution* $P_{cl}(R_{H-OH}, \alpha; t)$ serving as the classical analogue of the quantum mechanical distribution $|\Phi(R_{H-OH}, \alpha; t)|^2$. P_{cl} represents a time-resolved average over many individual trajectories. For clarity of presentation, the distributions in Figure 8.11 include only indirect trajectories, i.e., those which perform at least one bending oscillation before they dissociate. Incorporating all trajectories would have the undesired effect that the direct trajectories conceal the interesting portion of the "classical wavepacket". This discrimination of direct and indirect processes is only possible within the classical approach where each trajectory represents an individual event. As the trajectories slide down towards the bottom of the potential well they perceive the existence of the periodic orbit and instead of dissociating they try to stay in its proximity. However, the periodic orbit is highly unstable with the result that all trajectories, except those which come extremely close to it, are forced to let the lifeline go and to dissociate after only one cycle. If the periodic orbit were more stable a larger fraction of all trajectories would recur, probably for several times, and the structures in the spectrum would become narrower and narrower and resemble resonances as described in Chapter 7 (see, for example, Segev and Shapiro 1982).

The individual peaks of the absorption spectrum do *not* correspond to prominent quasi-bound quantum states with a well defined number of quanta in one or other mode, like in the photodissociation of $CH_3ONO(S_1)$, for example (see Figure 7.11). In order to develop a distinct nodal structure, which varies systematically from peak to peak, the complex must live sufficiently long. The photodissociation of H_2O, however, does not fulfil this requirement and therefore we cannot label the individual peaks in Figure 8.1(b) by a sequence of bending quantum

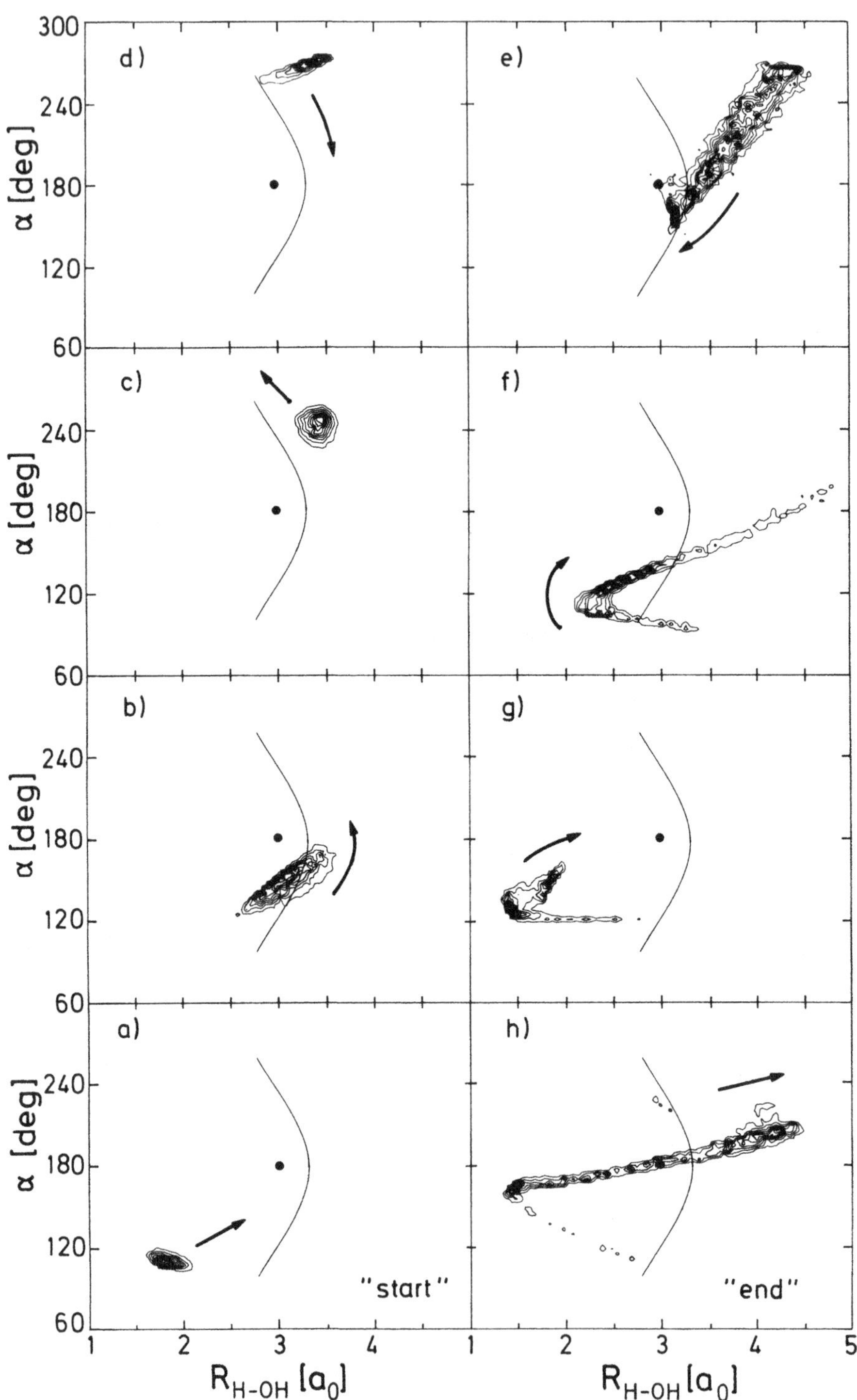

Fig. 8.11. Time-resolved coordinate distribution $P_{cl}(R_{\mathrm{H-OH}}, \alpha; t)$ obtained by averaging over a large number of classical trajectories. In order to reveal how (*cont.*)

numbers $k^* = \ldots, 10, 11, 12, \ldots$ as done by Wang, Felps, and McGlynn (1977).†

The excited complex breaks apart very rapidly and only a minor fraction performs, on the average, one single internal vibration. Therefore, the total stationary wavefunction does not exhibit a clear change of its nodal structure when the energy is tuned from one peak to another (Weide and Schinke 1989). In the light of Section 7.4.1 we can argue that the direct part of the total wavefunction, Ψ_{dir}, dominates and therefore obscures the more interesting indirect part, Ψ_{ind}. The superposition of the direct and the indirect parts makes it difficult to analyze diffuse structures in the time-independent approach. In contrast, the time-dependent theory allows, by means of the autocorrelation function, the separation of the direct and resonant contributions and it is therefore much better suited to examine diffuse structures.

The photodissociation of $H_2O(\tilde{B})$ resembles in several respects the highly excited hydrogen atom in the strong magnetic field. Averaging the high resolution spectrum shown in Figure 4.4, i.e., throwing away dynamical information, yields a regular modulation known as *quasi-Landau resonances* which extend far into the ionization continuum. Recall that averaging in the energy domain corresponds to damping the autocorrelation function in the time domain as we demonstrated in Section 7.5. If the damping is so strong that only the first recurrence with the shortest period in Figure 4.4(b) survives, the resulting spectrum is continuous with a regular modulation like the $H_2O(\tilde{B})$ spectrum (Reinhardt 1983; Holle et al. 1986). In the latter case, dissociation rather than incoherent

---

Fig. 8.11. (*cont.*) the unstable periodic orbit, represented by the solid line, influences the dissociation dynamics all direct trajectories, which fragment immediately without any recurrence, are discarded. The times range from 0 fs in (a) to 50.8 fs in (h). The arrows schematically indicate the evolution of the "classical" wavepacket and the heavy dot marks the equilibrium of the $\tilde{B}$-state potential energy surface. Adapted from Weide, Kühl, and Schinke (1989).

---

† In reality the photodissociation of H_2O through excitation in the second absorption band is much more complicated than discussed in this section. As illustrated in Figure 1.5, strong nonadiabatic coupling with either the $\tilde{A}$ or the $\tilde{X}$ state causes only less than 10% of the OH products to be produced in the excited electronic state $^2\Sigma$ whereas more than 90% are created in the electric ground state. The strong nonadiabatic effect presumably influences the motion of the wavepacket and therefore the absorption spectrum. Furthermore, the freezing of the other O-H bond seems to be arguable. Nevertheless, we are confident that the two-dimensional model including only one electronic state provides a realistic description of the general intramolecular dynamics and a faithful explanation of the diffuse structures.

averaging cuts off the autocorrelation function. Another example is the IR photodissociation of H_3^+. The high resolution spectrum is extremely dense and apparently not assignable (Carrington and Kennedy 1984). The low resolution spectrum, on the other hand, exhibits a regular undulation, like in the case of H_2O, which can be attributed to a simple periodic orbit, the "horseshoe orbit" (Gómez Llorente and Pollak 1988; Gómez Llorente et al. 1988).

8.3 Epilogue

Many UV absorption spectra of polyatomic molecules exhibit diffuse structures of the kind discussed in this chapter. The books of Robin (1974, 1975, 1985) and Okabe (1978) contain many examples. We may consider them as very broad resonances high above the dissociation threshold of the respective electronic state or very short-lived metastable states embedded in the continuum. The broadness reflects the coupling to the dissociation channel. The "lifetime" is so short that the autocorrelation function has only one single recurrence with small amplitude. Despite their diffuseness and the sometimes uninteresting overall behavior, diffuse structures can hide very interesting molecular motion. The dynamical information, however, is difficult to deduce from the spectra without knowing the underlying PES and without exact dynamical calculations. In view of the notorious lack of reliable PESs for excited electronic states we conjecture that in many cases the diffuse structures have been erroneously assigned. The absorption spectra of H_2O in the first two continua and of H_2S (Weide, Staemmler, and Schinke 1990; Schinke, Weide, Heumann, and Engel 1991; see also Section 15.3.2) support our point. An assignment based on the comparison of the frequencies in the upper state with the frequencies in the ground state is dangerous because the corresponding potentials can be, and often really are, drastically dissimilar.

However, even if we know the corresponding PES and perform exact dynamical calculations, the correct interpretation of diffuse structures might still turn out to be rather involved and not at all trivial if an appropriate adiabatic, i.e., zeroth-order, picture does not hold. Conventional assignment rests upon adiabatic separability which cannot necessarily be expected in the energy regime high up in the continuum. At high energies the various degrees of freedom must be expected to be strongly coupled with the consequence that the stationary wavefunctions do not exhibit a simple nodal structure.

Because of the very short "lifetime" the time-dependent approach provides the more illuminating explanation of diffuse structures. It allows the separation of the indirect part of the dissociation from the less interesting direct one, which dominates the overall fragmentation dynamics.

This is inherently impossible in the time-independent approach because the wavefunction contains the entire history of the wavepacket. The real understanding, however, is provided by classical mechanics. Plotting individual trajectories easily shows the type of internal motion leading to the recurrences which subsequently cause the diffuse structures in the energy domain. The next obvious step, finding the underlying periodic orbits, is rather straightforward.

There are usually more than one periodic orbits for a given system. Which one actually guides the dissociating trajectories depends crucially on the overall fragmentation dynamics. Let us consider the dissociation of $H_2O(\tilde{B})$. The most obvious periodic orbit is that which describes pure H-OH stretching for fixed bending angle $\alpha = 180°$. However, since it is roughly perpendicular to the main dissociation path, this orbit does not noticeably affect the breakup.

Periodic orbits also explain the long-lived resonances in the photodissociation of $CH_3ONO(S_1)$, for example, which we amply discussed in Chapter 7. But the existence of periodic orbits in such cases really does not come as a surprise because the potential barrier, independent of its height, stabilizes the periodic motion. If the adiabatic approximation is reasonably trustworthy the periodic orbits do not reveal any additional or new information. Finally, it is important to realize that, in general, the periodic orbits do not provide an assignment in the usual sense, i.e., labeling each peak in the spectrum by a set of quantum numbers. Because of the short lifetime of the excited complex, the stationary wavefunctions do not exhibit a distinct nodal structure as they do in truly indirect processes (see Figure 7.11 for examples).

9
Vibrational excitation

The final vibrational state distribution of the photofragment manifests the change of its bond length along the dissociation path [see Simons and Yarwood (1963) and Mitchell and Simons (1967) for early references]. Let us consider the dissociation of a linear triatomic molecule ABC described by Jacobi coordinates R and r as defined in Figure 2.1. In classical mechanics, it is the force

$$F_r = -\frac{\partial V(R,r)}{\partial r} = -\frac{dv_{\rm BC}(r)}{dr} - \frac{\partial V_I(R,r)}{\partial r}, \tag{9.1}$$

that controls the degree of internal vibrational excitation, where $V(R,r)$ is the full potential energy surface (PES) in the upper electronic state; $v_{\rm BC}(r)$ represents the vibrational potential of the free oscillator and $V_I(R,r)$ is the interaction potential (2.41), which provides the coupling between R and r. In the close-coupling approach, the energy transfer between R and r is governed by the coupling matrix elements

$$V_{nn'}(R) = \langle \varphi_n(r) \mid V_I(R,r) \mid \varphi_{n'}(r) \rangle, \tag{9.2}$$

where the $\varphi_n(r)$ are the vibrational wavefunctions of the product molecule (see Section 3.1). Both points of view are of course equivalent: the stronger the r dependence of V_I the larger are the nondiagonal elements of the potential matrix. For clarity of the presentation we distinguish the *elastic* case ($\partial V_I/\partial r = 0$) and the *inelastic* cases ($\partial V_I/\partial r \neq 0$).

The main purpose of this chapter is to emphasize the intimate relation between the topology of the dissociative PES and the vibrational excitation of the fragment molecule. In Sections 9.1 and 9.2 we consider exclusively direct processes. The photodissociation of symmetric molecules with two equivalent product channels is the topic of Section 9.3. Finally, vibrational state distributions following the decay of a long-lived intermediate complex will be discussed in Section 9.4. The theory and most of the applications are based on the two-dimensional model outlined in Section 2.4, with fixed bending angle.

9.1 The elastic case: Franck-Condon mapping

The photodissociation dynamics becomes particularly simple if the translational motion (associated with R) and the vibrational motion (described by r) exactly decouple, i.e., if the full PES separates as

$$V(R,r) = V_I(R) + v_{\mathrm{BC}}(r). \tag{9.3}$$

If the interaction potential V_I depends only on R, energy *cannot* flow between the translational and the vibrational modes; the full or half collision is elastic (no final state interaction). The resulting quantum mechanical or classical equations of motion separate in two uncoupled blocks and the motions in R and r evolve independently of each other.

Figure 9.1 illustrates a representative example of an elastic PES; the dependence on the vibrational coordinate r is independent of the dissociation coordinate R with the consequence that the minimum energy path runs parallel to the R-axis. If the equilibrium bond distances in the lower and in the upper states are roughly equal, the fragment will not experience any significant restoring force along the entire dissociation path and therefore it will be born predominantly in its vibrational ground state [case (a) in Figure 9.1]. On the other hand, if the FC point is significantly displaced from the minimum energy path of the upper state PES the molecule is subject to a strong restoring force and finally leaves the interaction zone with a substantial amount of vibrational energy [case (b) in Figure 9.1].

9.1.1 Franck-Condon distribution

Let us make this more quantitative using the time-independent quantum mechanical theory outlined in Section 3.1. Because the interaction potential is independent of r the potential matrix $\mathbf{V}$ defined in (9.2) is diagonal, i.e., different vibrational fragment states do not mutually couple. As a result, the matrix of radial wavefunctions $\chi_{n'}(R,r;E,n)$, which solve the coupled equations (3.5), is diagonal as well, i.e., $\chi_{n'}(E,n) \propto \delta_{nn'}$. If we assume, in order to simplify the subsequent discussion, that the nuclear wavefunction in the ground electronic state factorizes as $\varphi_R(R)\, \varphi_r(r)$ [see Equation (3.40)] the dissociation amplitudes $t(E,n)$ in Equation (3.14) reduce to a single term for each final state n and the unnormalized final state distribution becomes

$$P(E,n) = \bar{g}(E,n)\, \bar{f}(n). \tag{9.4}$$

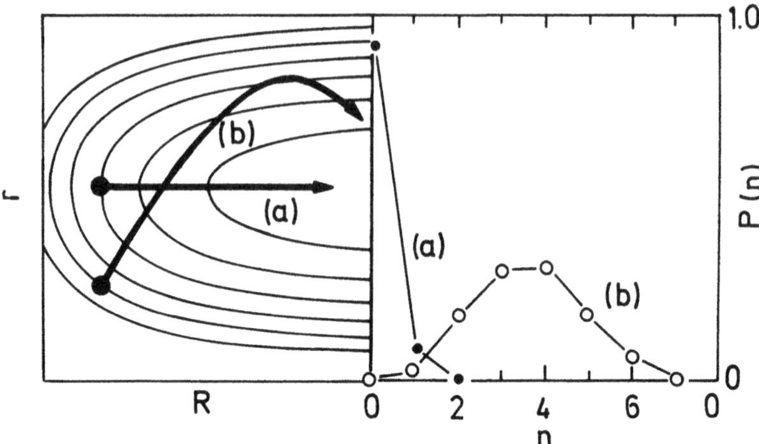

Fig. 9.1. Left-hand side: Representation of an elastic potential energy surface. It has the general form (6.35) with coupling strength parameter $\epsilon = 0$. In case (a), the equilibrium bond distance in the electronic ground state equals the equilibrium separation of the free BC fragment. The heavy arrows schematically indicate two representative trajectories starting at the respective FC points. Right-hand side: The corresponding final state distributions.

The $\bar{f}(n)$ are one-dimensional FC factors of the form

$$\bar{f}(n) = |\langle\varphi_n^{\mathrm{BC}} \mid \varphi_r\rangle|^2 = \left|\int_0^\infty dr\, \varphi_n^{\mathrm{BC}}(r)\, \varphi_r(r)\right|^2 \qquad (9.5\mathrm{a})$$

and the

$$\bar{g}(E,n) = |\langle\chi_n(E,n) \mid \varphi_R\rangle|^2 = \left|\int_0^\infty dR\, \chi_n^*(R;E,n)\, \varphi_R(R)\right|^2 \qquad (9.5\mathrm{b})$$

are one-dimensional overlap integrals in the radial coordinate R. φ_n^{BC} represents a vibrational wavefunction of the fragment and χ_n is a continuum wavefunction solving (3.5) with wavenumber k_n.[†] The *vibrational Franck-Condon factors* $\bar{f}(n)$ represent the squared moments of the wavefunction $\varphi_r(r)$ in the electronic ground state in terms of the wavefunctions $\varphi_n^{\mathrm{BC}}(r)$ of the free fragment.

If the radial overlap integrals $\bar{g}(E,n)$ are considered as functions of the translational energy $E_{trans} = E - \epsilon_n$, where the ϵ_n are the vibrational energies of the product molecule, they do not depend on the vibrational quantum number n because the radial functions $\chi_n(E_{trans}, n)$ all solve

† Equations (9.4) and (9.5) are analogous to the adiabatic expressions (3.41) and (3.42). In the adiabatic approximation the variation of the internal vibrational potential with R is approximately taken into account and it is therefore more general than the elastic limit discussed here. For a truly elastic PES both sets of equations become identical and exact.

the Schrödinger equation (3.5) with the *same* potential $V_I(R)$ and the *same* wavenumber $k^2 = 2mE_{trans}/\hbar^2$ defined in (2.52) [see Figure 9.2(a)]. If they are considered as functions of the total energy $E = E_{trans} + \epsilon_n$, however, they depend on E as well as n as illustrated in Figure 9.2(b). The shift between the individual cross sections manifests the energy separation between the corresponding oscillator states.

The interplay of the two factors $\bar{f}(n)$ and $\bar{g}(E,n)$ leads to an interesting difference in the energy dependence of the final vibrational state distributions for the two cases of Figure 9.1. This is illustrated in Figure 9.3. For case (a), the equilibrium bond lengths in the lower and in the upper electronic state are very similar so that the distribution of FC factors $\bar{f}(n)$ peaks at $n = 0$ and rapidly decays to zero [see Figure 3.4(a)]. The resulting final state distributions $P(E,n)$, depicted in Figure 9.3(a), consequently show almost no energy dependence. The ground state is always the dominant channel and with increasing energy the distributions merely become slightly broader. Quite a different behavior, on the other hand, is found for case (b). Because the two equilibrium distances differ considerably, the distribution of FC factors is rather broad with a maximum at $n = 3$. In this case the dependence of the $\bar{g}(E,n)$ on the vibrational quantum number n induces a gradual shift of the final state distribution to higher and higher quantum numbers (Freed and Band 1977). Remember that the coupling between translation and vibration is zero in both cases!

This rather different energy dependence for the two cases can be directly deduced from Figure 9.1. Increasing the energy in case (a) merely

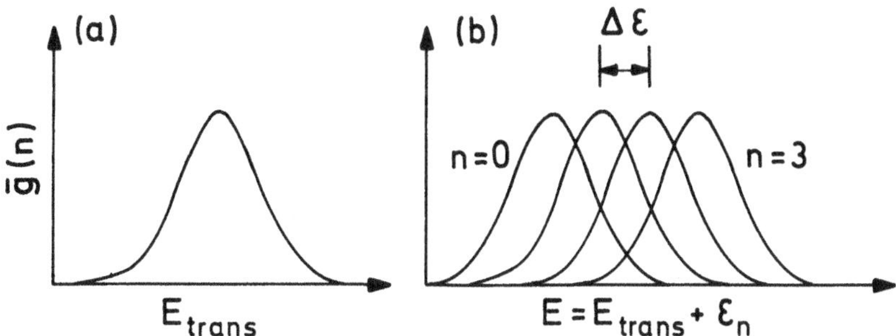

Fig. 9.2. (a) One-dimensional overlap integrals $\bar{g}(E_{trans})$ defined in (9.5b) as a function of the translational energy. They are independent of n. (b) The same overlap integrals plotted as functions of the total energy $E = E_{trans} + \epsilon_n$, where the ϵ_n are the vibrational energies of the fragment. $\Delta\epsilon$ is the vibrational energy spacing between adjacent channels. Multiplication of the $\bar{g}(E,n)$ with the appropriate FC factors $\bar{f}(n)$ yields the partial photodissociation cross sections.

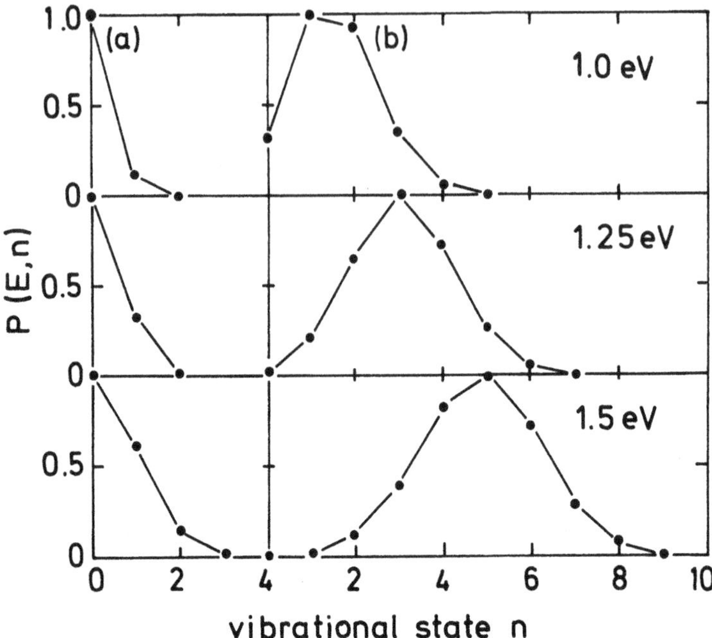

Fig. 9.3. Energy dependence of final vibrational state distributions for the cases (a) and (b) illustrated in Figure 9.1. The total energy is normalized such that $E = 0$ corresponds to $\mathrm{A} + \mathrm{BC}(r_e)$.

moves the transition point to smaller R values without changing the position along the vibrational coordinate r. It affects only the translational but not the vibrational motion. In case (b), on the other hand, the molecule starts its journey on the upper-state PES at smaller dissociation bonds R and, at the same time, at ever shorter vibrational bond distances r if the energy increases, i.e., the transition region moves gradually along the backward continuation of the heavy arrow in Figure 9.1. Thus, as the total energy increases the initial amount of vibrational energy of the oscillator grows as well and because there is no coupling between R and r it resides completely within the vibrational bond. As a consequence, the final distribution shifts gradually to larger quantum numbers.

In general, final vibrational state distributions following direct dissociation can be explained as a mapping of the initial coordinate distribution in the electronic ground state onto the final quantum number axis of the fragment as outlined in Section 6.4. Within the elastic limit this mapping is essentially governed by the FC factors $\bar{f}(n)$, modified however by the n dependence of the radial overlap integrals $\bar{g}(E,n)$. Therefore, we introduce the name *Franck-Condon mapping* (Simons and Tasker 1973, 1974) in order to distinguish it from dynamical mapping which we will discuss in Section 9.2.

9.1.2 Examples

Purely elastic potentials are very unlikely in reality. But, nevertheless, there are a lot of cases in which the translational-vibrational coupling is actually so weak that the FC limit is applicable. The upper portion of Figure 1.11 depicts the two-dimensional PES for the S_2 state of CH_3ONO as calculated by Suter, Brühlmann, and Huber (1990). Unlike the dissociation in the S_1 state (see Chapter 7) the fragmentation of $CH_3ONO(S_2) \rightarrow CH_3O + NO$ is fast and direct. The internal NO potential changes only very slightly along the dissociation path and therefore inelastic effects are indeed marginal. Since the N-O bond distances in the S_0 state and for the free radical are very similar it does not come as a surprise that NO is generated with little vibrational excitation. The measured ratio $\sigma(n=0) : \sigma(n=1)$ of 0.76 : 0.24 in the 248 nm photolysis confirms this prediction (Suter, Brühlmann, and Huber 1990).

The photodissociation of HONO $\rightarrow$ HO + NO via the $\tilde{A}$ state produces OH radicals predominantly in the lowest vibrational state (Vasudev, Zare, and Dixon 1984). Although the lifetime of the excited complex amounts to several NO and OH vibrational periods, there is no appreciable restoring force which could change the O-H bond distance during the fragmentation. The calculated two-dimensional PES of Suter and Huber (1989) clearly elucidates this particular aspect of the dissociation dynamics of HONO which is otherwise more complex.†

Other examples with very little final state interaction, concerning the vibration of the fragment, are ICN (for a review see Jackson and Okabe 1986:tab.2) and H_2O_2 (Ondrey, Van Veen, and Bersohn 1983; Jacobs, Kleinermanns, Kuge, and Wolfrum 1983). In the photodissociation of H_2CO , the CO fragment experiences very little vibrational excitation (Bamford et al. 1985) whereas the vibrational distribution of H_2 is inverted with a peak at $n = 1$ (Debarre et al. 1985). The FC limit, with the wavefunction of the parent molecule being replaced by the wavefunction at the transition state, reproduces the CO vibrational distribution remarkably well which indicates that inelastic effects in the exit channel are presumably very weak, as far as the C-O bond is concerned. In contrast, the same model fails to describe the vibrational distribution of H_2 which lead Debarre et al. (1985) to speculate that final state interaction in the exit channel is not negligible as far as the H-H bond is concerned.

† Like in the photodissociation of $CH_3ONO(S_1)$ (Section 7.3) the terminal NO moiety is first vibrationally excited and in order to initiate fragmentation energy must flow from the N-O vibrational mode to the OH-NO dissociation bond. The OH radical merely plays the role of a spectator and remains rotationally and vibrationally unexcited throughout the bond rupture. Vibrational excitation of NO behaves quite differently as we will discuss in Section 9.4

9.2 The inelastic case: Dynamical mapping

If the interaction potential V_I depends simultaneously on R and r, the translational and the vibrational modes are coupled and energy can flow between them. Figure 9.4 depicts a typical potential energy surface $V(R, r)$ of the form (6.35) where the parameter ϵ controls the coupling between R and r. The curvature of the minimum energy path, which increases with ϵ, shows the coupling. Owing to the r dependence of V_I the potential matrix $\mathbf{V}$ defined in (9.2) is no longer diagonal and therefore all vibrational channels are mutually coupled and the sum in (3.14) for the dissociation amplitude extends over all "intermediate" channels n'. This makes the analysis of final state distributions within the close-coupling approach rather complicated, except for very weak coupling. As emphasized in Chapter 6, a purely classical analysis, however, provides a clear understanding and is, in addition, trustworthy.

9.2.1 Energy redistribution

A trajectory that we launch significantly displaced from the minimum energy path [case (a) in Figure 9.4] is subject to a significant restoring force and leads to strong vibrational excitation, despite the fact that the equilibrium bond distances in the parent molecule and for the free fragment do not differ much. Figure 6.9 shows vibrational state distributions for several values of the coupling parameter ϵ; increasing ϵ, i.e., the curvature of the minimum energy path, gradually shifts the peak of the distribution

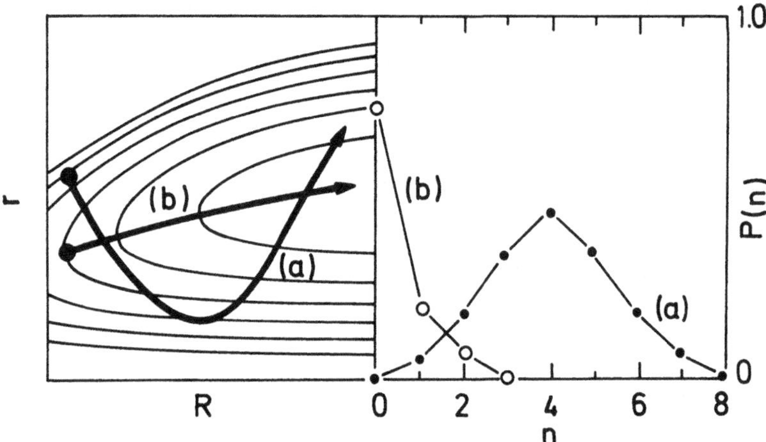

Fig. 9.4. Left-hand side: Representation of an inelastic potential energy surface of the form (6.35) with $\epsilon = 0.2$. The heavy arrows represent two characteristic trajectories starting at the respective FC points. Right-hand side: The corresponding final vibrational state distributions.

to higher quantum numbers without changing the overall shape. Raising the total energy has the same qualitative effect (Figure 9.5): the higher the total energy, the further away from the minimum energy path start the trajectories and consequently the stronger is the degree of vibrational excitation. As opposed to case (a), a trajectory starting on or in the proximity of the minimum energy path [case (b) in Figure 9.4] slides down the valley without being significantly excited, despite the relatively large change in the bond distance! The result is a final state distribution peaking at $n = 0$ which is only marginally energy-dependent [Figure 9.5(a)].

The vibrational reflection principle outlined in Section 6.4 provides a simple, but yet *quantitative* explanation of final vibrational state distributions and their variation with the coupling strength and the total energy. The central quantity is the vibrational excitation function $N(r_0)$. It comprehensively manifests the dynamical details of the fragmentation process in the upper electronic state. Usually, one needs only very few trajectories to construct $N(r_0)$ which makes the simple classical theory outlined in Section 6.4 very efficient for calculating and understanding final state distributions. This is particularly beneficial for fitting experimental data.

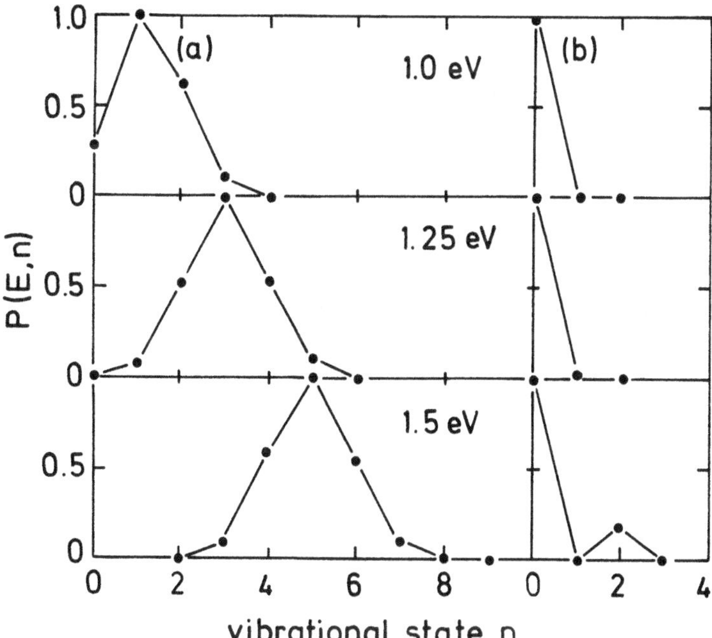

Fig. 9.5. Energy dependence of the vibrational state distributions for the two cases illustrated in Figure 9.4. Energy normalization is such that $E = 0$ corresponds to $A + BC(r_e)$.

Final vibrational state distributions following direct photodissociation can be considered as a mapping of the initial coordinate distribution onto the final quantum number axis. As opposed to the elastic limit, the mapping in the inelastic case is a *dynamical mapping* with the final state distributions reflecting the translational-vibrational coupling in the exit channel. While in the elastic limit the vibrational energy of the fragment remains constant throughout the breakup, in the inelastic case it varies along the trajectory. Thus, during the fragmentation energy flows either into or out of the vibrational mode. Despite the similarity of the distributions in Figures 9.3 and 9.5, they represent quite different dynamical situations!

9.2.2 The photodissociation of CH_3I *and* CF_3I

The photodissociation of trifluoromethyl iodide, $CF_3I \rightarrow CF_3 + I/I^*$, which was briefly discussed in Section 6.4, seems to illustrate case (a) of Figure 9.4 while the photodissociation of methyl iodide, $CH_3I \rightarrow CH_3 + I/I^*$, appears more to represent case (b). In both examples, the ν_2 "umbrella" mode, in which the C atom oscillates relative to the H_3-respectively F_3-plane, is predominantly excited. Following Shapiro and Bersohn (1980) the dissociation of CH_3I and CF_3I may be approximately treated in a two-dimensional, pseudo-linear model in which the vibrational coordinate r describes the displacement of the C atom from the H_3-/F_3-plane and the dissociation coordinate R is the distance from iodine to the center-of-mass of CH_3/CF_3 (see Figure 9.6).[†]

While CF_3 is pyramidal both within the parent molecule ($r_e = 0.48$ Å) and as free radical ($r_e = 0.40$ Å), CH_3 is pyramidal within CH_3I ($r_e = 0.33$ Å) but planar ($r_e = 0$) asymptotically. Thus, if the coupling between R and r were negligibly small, the FC limit would predict that the dissociation of CH_3I yields, as a consequence of the large change in bond lengths, a substantially inverted vibrational distribution, similar to that in Figure 9.3(b).[‡] In contrast to CH_3I, the FC limit would predict

† The photodissociation of CH_3I has received considerable interest in the last decade, both experimentally and theoretically. Extensive lists of references can be found in the articles of Guo and Schatz (1990b) and Amatatsu, Morokuma, and Yabushita (1991). There are many details which make this system a prototype for the fragmentation of a small polyatomic molecule in two electronic states. Here we consider only the vibrational distribution in the main dissociation channel, $CH_3 + I^*$. A discussion of nonadiabatic effects is deferred to Chapter 15.

‡ Older experiments indeed yielded inverted vibrational distributions with peaks around $n = 2$–3 (Sparks, Shobatake, Carlson, and Lee 1981; van Veen, Baller, de Vries, and van Veen 1984) which inevitably influenced the earliest theoretical attempts to model the photodissociation of CH_3I (Shapiro and Bersohn 1980; Lee and Heller 1982; Gray and Child 1984).

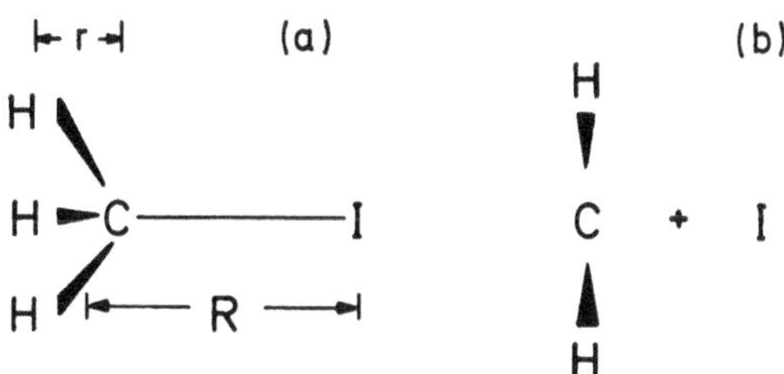

Fig. 9.6. (a) The geometry of CH_3I in the electronic ground state. The view is perpendicular to the C-I bond and parallel to the H_3-plane. R is the distance from I to the center of mass of CH_3 and r is the displacement of C from the H_3-plane. (b) The CH_3 + I system after the dissociation with CH_3 being planar.

that CF_3 is preferentially produced in the vibrational ground state if the final state interaction were negligibly small.

The most recent measurements for both systems, however, show just the opposite trend: CH_3 is preferentially produced in the vibrational ground state [Hall, Sears, and Frye 1989; Zhu, Continetti,Zhao, Balko, Hintsa, and Lee cited by Guo and Schatz (1990b); Suzuki, Kanamori, and Hirota 1991; see Figure 9.7(a)] while for CF_3 a highly inverted distribution is obtained (Felder 1990; Figure 6.10). The discrepancy from the simple FC picture clearly manifests the influence of final state interaction. Calculated potential energy surfaces became available for CH_3I only very recently (Amatatsu, Morokuma, and Yabushita 1991) thus all previous theoretical studies were restricted to the fitting of the available experimental data.

Case (a) of Figure 9.4 illustrates a possible dissociation path for CF_3I. By changing the coupling parameter ϵ in the potential expression (6.35) we can arbitrarily shift the peak of the final state distribution as demonstrated in Figure 6.9 (Hennig, Engel, and Schinke 1986; Untch, Hennig, and Schinke 1988). A value of $\epsilon = 0.25$ gives reasonable agreement with the measured distribution (see Figure 6.10) concerning the peak position as well as the width. Thus, we may conclude that:

- In the photodissociation of CF_3I, the inelasticity of the PES, i.e., the force $-\partial V_I/\partial r$ causes, relative to the pure FC limit, significant excitation of the CF_3 fragment.

Employing the two-state model of Shapiro (1986) Guo and Schatz (1990b) fitted the most recent experimental data for the photodissociation of CH_3I. Figure 9.8 depicts their fitted two-dimensional PES. The evolving quantum mechanical wavepacket follows roughly the minimum energy path and leads to CH_3 products preferentially in the lower vibrational states, as qualitatively predicted in Figure 9.4.

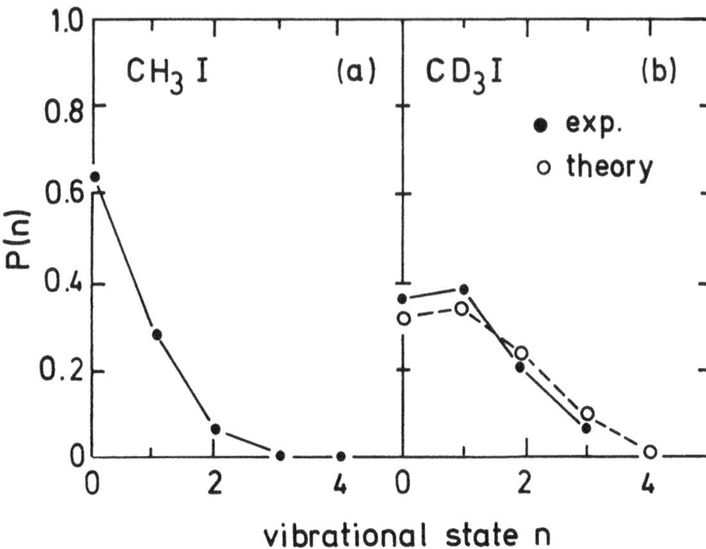

Fig. 9.7. (a) Final vibrational state distribution in the umbrella mode of CH_3 following the dissociation of CH_3I at 248 nm; iodine is produced in the excited electronic state. The experimental and the theoretical distributions virtually coincide. (b) The same as in (a) but for CD_3I. The experimental data are those of Zhu, Continetti, Zhao, Balko, Hintsa, and Lee cited by Guo and Schatz (1990b). Adapted from Guo and Schatz (1990b).

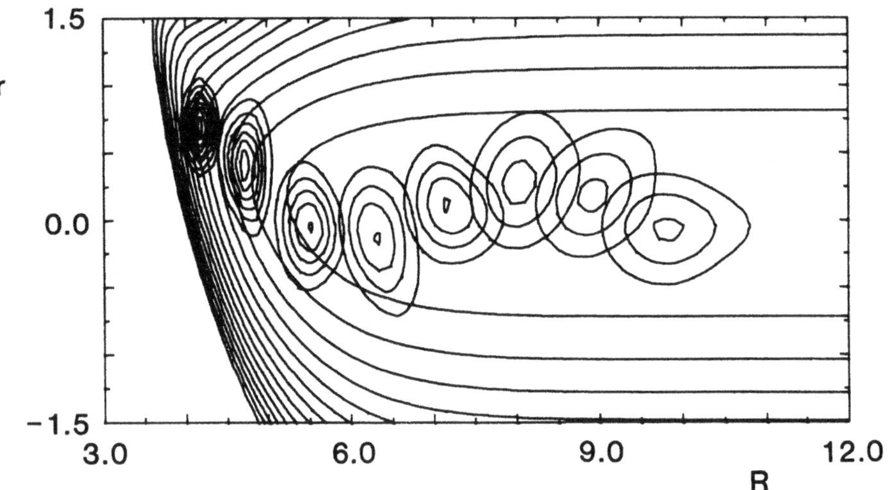

Fig. 9.8. Two-dimensional potential energy surface for the photodissociation of CH_3I fitted to reproduce the latest experimental data. The coordinates are given in atomic units. Adapted from Guo and Schatz (1990b).

- In the photodissociation of CH_3I, the inelastic force $-\partial V_I/\partial r$ partly compensates the strong restoring force $-dv_{BC}/dr$ with the result that, relative to the pure FC limit, CH_3 is generated with comparatively weak vibrational excitation during the fragmentation; the final state interaction *de-excites* rather than excites the umbrella vibration of CH_3.

The *ab initio* calculations of Amatatsu, Morokuma, and Yabushita (1991) qualitatively confirm this picture. The vibrational state distributions obtained with their potential in classical trajectory calculations also agree satisfactorily with the experimental data.

These two examples, CF_3I and CH_3I, clearly underscore the influence of the translational-vibrational coupling and the need for knowing, at least qualitatively, the topology of the dissociative PES. Necessary prerequisites for adjusting the PES are experimental data of high quality.

9.3 Symmetric triatomic molecules

The photodissociation of symmetric molecules such as H_2O, H_2S, and O_3 illustrates particularly clearly the close relation between the change of the molecular configuration along the dissociation path and the final vibrational state distribution. Figure 9.9 shows the two-dimensional PES

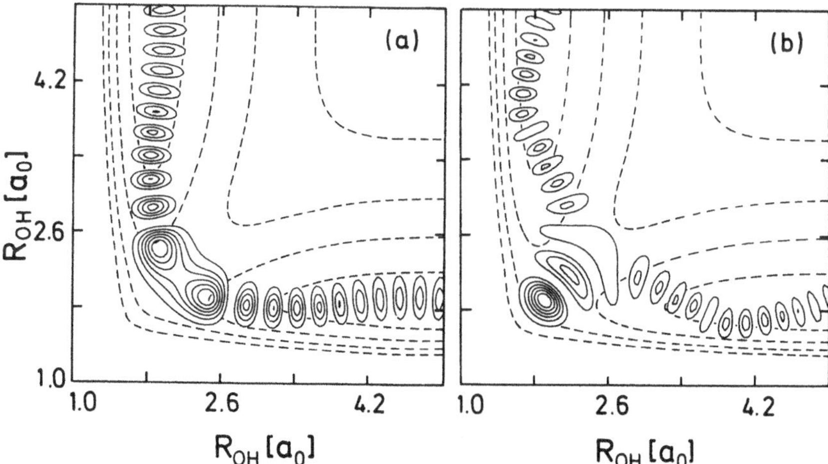

Fig. 9.9. Contour plot of the potential energy surface of H_2O in the $\tilde{A}^1B_1$ state; the bending angle is fixed at 104°. Superimposed are the total stationary wavefunctions $\Psi_{tot}(E)$ defined in (2.70). The total energies are -2.6 eV and -2.0 eV corresponding to wavelengths of $\lambda = 180$ nm and 165 nm, respectively. Energy normalization is such that $E = 0$ corresponds to three ground-state atoms.

of H_2O in the $\tilde{A}^1B_1$ state for a fixed HOH bending angle. The FC point ($R_{OH} = 1.81a_\circ$) lies on the inner slope of the potential, significantly displaced from the saddle point ($R_{OH} = 2.06a_\circ$). A classical trajectory initiating near the equilibrium of the electronic ground state thus first experiences a force toward the saddle point before it is reflected in either of the two dissociation channels (see also Figure 8.2).

The acceleration along the symmetric stretch coordinate leads to appreciable vibrational excitation of $OH(^2\Pi)$ (Figure 9.10), despite the fact that the OH equilibrium separations in $H_2O(\tilde{X})$ and the free OH radical are almost identical. During the first few femtoseconds of the dissociation both O-H bonds stretch simultaneously before one of them breaks leaving the surviving OH fragment vibrationally excited [Figure 9.9(b)]. With increasing energy the dissociation begins with progressively more compressed O-H bonds further up on the inner potential slope. This boosts the initial repulsion with the result that OH becomes even more excited. For lower energies, on the other hand, the upper PES is accessed closer to the saddle point and the wavepacket slides directly down into the two fragment channels yielding OH preferentially in the vibrational ground state [Figure 9.9(a)]. The stationary total dissociation wavefunctions $\Psi_{tot}(E)$ shown in Figure 9.9 for two representative total energies clearly illustrate the dissociation path and readily explain the observed energy dependence of the final state distribution.

Up to now OH vibrational state populations have been measured for only two wavelengths. In accordance with the above considerations, absorption in the far red wing of the absorption spectrum, λ =193 nm,

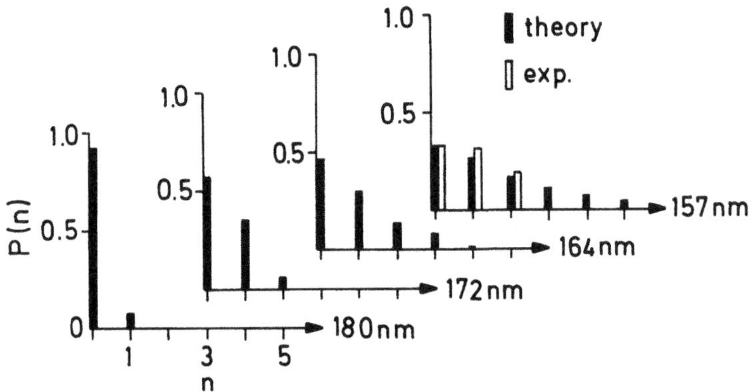

Fig. 9.10. Vibrational state distributions of $OH(^2\Pi)$ following the dissociation of H_2O in the first absorption band. The theoretical distributions are adapted from Engel, Schinke, and Staemmler (1988) and the experimental results for 157 nm are those of Andresen et al. (1984). Theory and experiment are normalized so that both have the same probability for $n = 0$.

yields exclusively OH($n = 0$) (Häusler, Andresen, and Schinke 1987). In the 157 nm photolysis Andresen, Ondrey, Titze, and Rothe (1984) measured a ratio of $1 : 1 : 0.58$ for $\sigma(n = 0) : \sigma(n = 1) : \sigma(n = 2)$ in good agreement with theory.† For even shorter wavelengths the distribution becomes inverted with a peak at $n = 1$. Although the measurements are rather limited they clearly confirm the strong energy dependence of the vibrational excitation predicted by the simple "geometrical" considerations in the above discussion.

Vibrational product distributions following the direct fragmentation of symmetric molecules depend sensitively on the shape of the dissociative PES in the region of the saddle point. Figure 9.11 depicts vibrational distributions of SH measured in the photodissociation of H_2S in the first absorption band (Xie et al. 1990; similar results were also obtained by van Veen, Mohamed, Baller, and de Vries 1983 and Xu, Koplitz, and Wittig 1987). The corresponding absorption spectrum peaks around 195 nm. Although the overall dissociation dynamics of H_2S is similar to the photolysis of H_2O, the vibrational state distributions manifest obvious differences. For all wavelengths probed in the experiment, SH($n = 0$) is by far the dominant channel. Higher excited levels gradually become populated with decreasing wavelength, but the sum of the population over all excited states never exceeds 40%. Figures 9.10 and 9.11 clearly underscore the different energy dependences in the two cases.

While $H_2O(\tilde{A})$ dissociates on a single PES, the dissociation of H_2S involves two excited electronic states. *Ab initio* calculations yield two adiabatic potential energy surfaces which are very close in energy at and in the proximity of the C_{2v} symmetry line. The lower one is dissociative with an overall shape like the $\tilde{A}$-state PES of H_2O and the upper one is binding (Theodorakopoulos and Petsalakis 1991; Heumann, Düren, and Schinke 1991). If both H-S bond distances are equally long, the two states have $^1A_2$ and $^1B_1$ electronic symmetry, respectively. The corresponding states in H_2O are much further apart and therefore the dissociation in the first band proceeds on a single PES.

As a consequence of the strong interaction of the two electronic states the shape of the dissociative PES in the saddle point region differs noticeably from the dissociative PES for $H_2O(\tilde{A})$ [see Figure 15.6(a)]. The saddle is considerably narrower than for H_2O. Although the differences are rather subtle, they explain qualitatively the dissimilarity of the final vibrational state distributions. Trajectories starting near the tran-

† The original value of 0.15 for $n = 2$ reported by Andresen et al. (1984) had been later corrected by taking into account predissociation in the $^2\Sigma$ state of OH (Andresen and Schinke 1987).

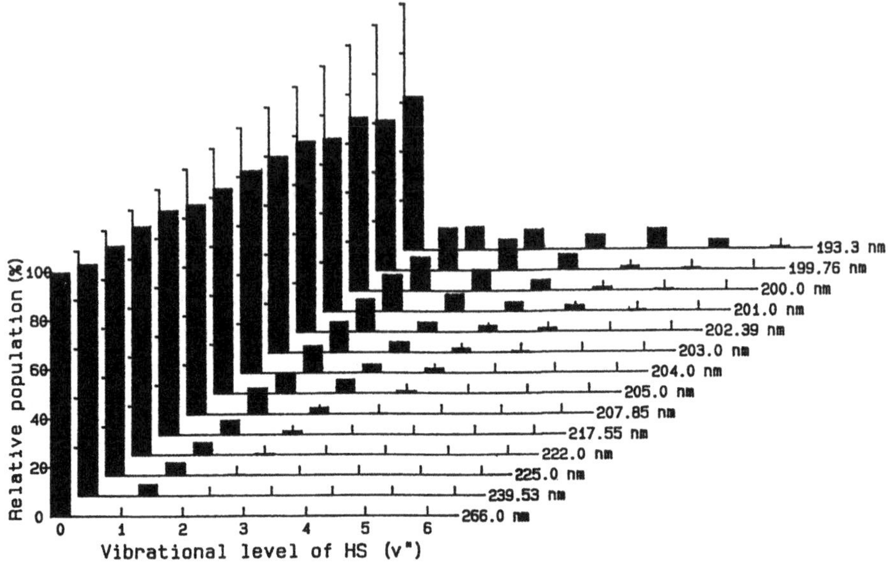

Fig. 9.11. Measured vibrational state distributions of SH following the photodissociation of H_2S in the first absorption band. Reproduced from Xie et al. (1990).

sition region are rather promptly deflected into one of the exit channels before both H-S bonds can simultaneously elongate. Thus, HS is generated primarily in the vibrational ground state. Dixon, Marston, and Balint-Kurti (1990) fitted the experimental distributions and obtained a two-dimensional PES similar to the *ab initio* PES. The photodissociation of H_2S is much more involved than the dissociation of H_2O in the first continuum; a more detailed discussion will be deferred to Section 15.3.2.

Incidentally we note that Levene, Nieh, and Valentini (1987) measured vibrational state distributions for O_2 following the photodissociation of O_3 in the Chappuis band which qualitatively resemble those shown in Figure 9.11 for H_2S. In order to interpret their results the authors surmised a model PES that actually has the same overall topology in the saddle point region as the one calculated for H_2S.

The photofragments originating from a symmetric parent molecule are often generated with a relatively high degree of vibrational excitation. Representative examples are SO_2 (Effenhauser, Felder, and Huber 1990), CO_2 (Lee and Judge 1973), and CS_2 (Yang, Freedman, Kawasaki, and Bersohn 1980). The comparatively low grade of excitation found for H_2S seems to be more the exception.

9.4 Adiabatic and nonadiabatic decay

While all examples discussed in the foregoing sections belong to the category of direct photodissociation we consider in this section three cases of indirect bond fission and the subsequent vibrational state distributions. We will focus our attention on the question, how do the final distributions "reflect" the fragmentation mechanism?

The photodissociation of ClNO via the T_1 state has been discussed in Section 7.6 and Figure 7.14 shows the measured photofragment yield spectra, i.e., the relative partial absorption cross sections for producing NO in vibrational state n (Qian, Ogai, Iwata, and Reisler 1990). The three main bands in the absorption spectrum correspond to internal vibrational excitation of the NO moiety with quantum numbers $n^* = 0, 1$, and 2. Each vibrational band splits into three peaks corresponding to excitation of the bending degree of freedom with quantum numbers $k^* = 0, 1$ and 2.

Excitation of ClNO(T_1) in any one of the three vibrational bands yields exclusively NO products in vibrational state $n = n^*$ (Qian et al. 1990). The left-hand side of Figure 9.12 depicts the results of a three-dimensional wavepacket calculation including all three degrees of freedom and using an *ab initio* PES (Sölter et al. 1992). This calculation reproduces the absorption spectrum and the final vibrational and rotational distributions of NO in good agreement with experiment.

It is immediately apparent that the excited ClNO(T_1) complex dissociates adiabatically as illustrated schematically in Figure 9.13(a): the quasi-bound state generated by the light pulse decays via the corresponding vibrationally adiabatic potential curve $\epsilon_n(R)$ defined in (3.31) (or adiabatic PES if we include also the bending degree of freedom) without significant coupling to the adjacent vibrational states $(n^* - 1)$ or $(n^* + 1)$. The initial degree of vibrational excitation in the FC region is retained during the entire fragmentation step. Therefore, we can *schematically* represent the fragmentation of ClNO(T_1) by

$$\text{ClNO}(T_1, n^*) \longrightarrow \text{Cl} + \text{NO}(n = n^*).$$

As estimated from the widths of the partial absorption cross sections, the lifetime varies from 15 fs to 70 fs which corresponds, on the average, to merely one to four vibrational periods of NO within the T_1 complex. The photodissociation of ClNO(T_1) is thus a hybrid of direct and indirect dissociation.

The photodissociation of CH_3ONO via the S_1 state has been discussed in detail in Sections 7.3 and 7.4 and Figure 7.12 shows the comparison of the experimental and the calculated absorption spectra. The photon

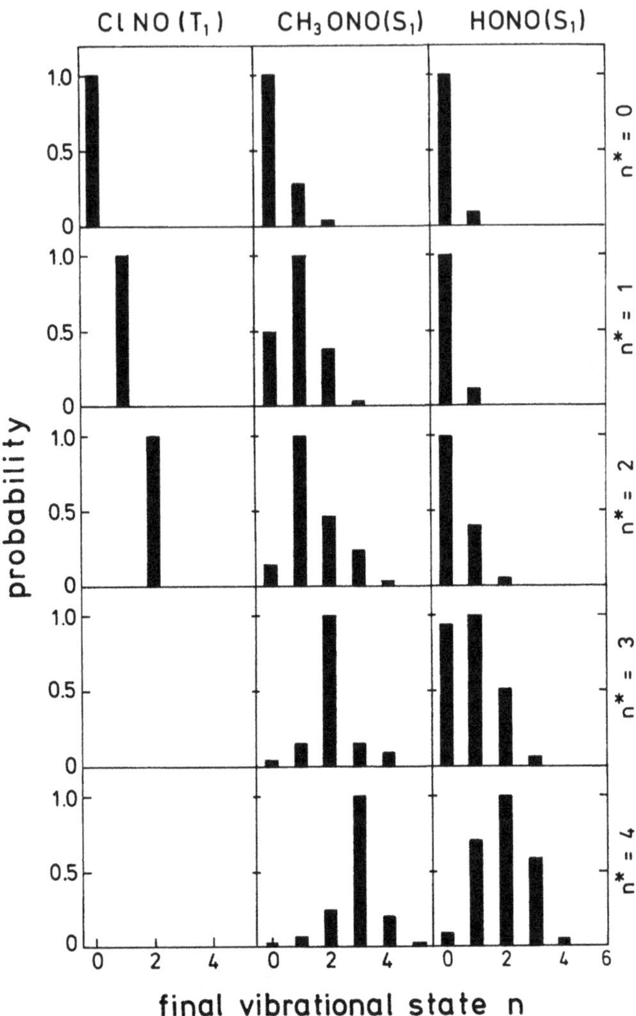

Fig. 9.12. Calculated vibrational state distributions of NO following the dissociation of $ClNO(T_1)$ (Sölter et al. 1992), $CH_3ONO(S_1)$ (Untch and Schinke 1992), and $HONO(S_1)$ (Suter, Huber, Untch, and Schinke 1992), respectively, in various vibrational bands indicated by $n^* = 0, 1, 2, \ldots$. All calculations include the N-O vibrational bond, the X-NO dissociation bond, and the XNO bending angle with X=Cl, CH_3O, and HO.

excites resonantly vibrational states of NO in the complex with lifetimes corresponding to several internal NO periods. The photodissociation of $CH_3ONO(S_1)$ is thus a truly indirect process. The middle part of Figure 9.12 depicts the vibrational state distributions for excitation in the various absorption peaks. They were obtained by a three-dimensional wavepacket calculation including the O-N and the N-O bonds as well

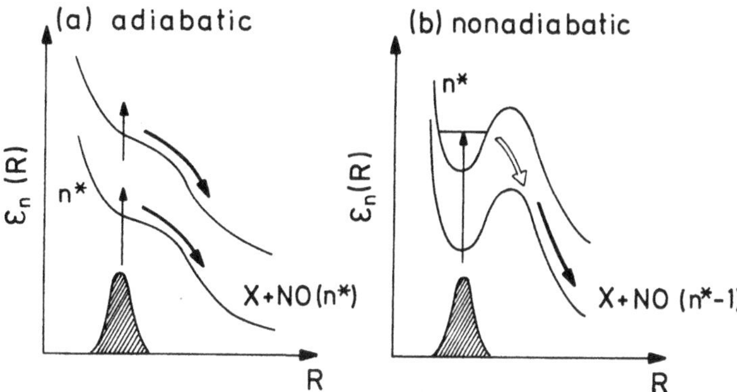

Fig. 9.13. Schematic illustration of vibrationally adiabatic and nonadiabatic dissociation of XNO with X = Cl, CH_3O, or HO, respectively.

as the ONO bending angle (Untch and Schinke 1992). As opposed to $ClNO(T_1)$ the photodissociation of CH_3ONO yields noticeably broader final vibrational state distributions with a clear overall propensity for the final product channel $n = n^* - 1$:

- The photodissociation of $CH_3ONO(S_1)$ reveals an intriguing *quantum state selectivity*: excitation of $NO(n^*)$ in the complex preferentially yields NO fragments in vibrational state $n^* - 1$.

The only exception is the decay of the $n^* = 1$ resonance.

This propensity points to a vibrationally nonadiabatic decay mechanism as illustrated in Figure 9.13(b). The vibrationally excited quasi-bound state generated by the photon can decay either by tunneling through the barrier or by a nonadiabatic transition to the next lower vibrational state whose potential barrier is lower than the total energy. If the nonadiabatic coupling between the vibrational states is not too strong beyond the barrier, the final fragmentation will proceed predominately in an adiabatic way. Tunneling is rather unlikely for a heavy molecule such as CH_3ONO and therefore the excited complex decays mainly via the second mechanism. Thus, we may *schematically* represent the dissociation of $CH_3ONO(S_1)$ by

$$\begin{aligned} CH_3ONO(S_1, n^*) &\longrightarrow CH_3ONO(S_1, n^* - 1) \\ &\longrightarrow CH_3O + NO(n = n^* - 1). \end{aligned}$$

Depending on the strength of the vibrational-translational coupling in the barrier region and in the exit channel subsequent transitions to lower or higher vibrational states become feasible and explain the small but not negligible population of levels adjacent to $(n^* - 1)$. The fall off towards higher states is not dictated by energy conservation but follows from dynamical constraints. The total energy would suffice to fill even higher vibrational states. For very weak coupling, like in van der Waals molecules, the fragment will almost exclusively populate state $(n^* - 1)$, i.e., the propensity rule turns into a selection rule (see Section 12.3.1).

The experimental investigation of Benoist d'Azy et al. (1985) clearly confirms the $n^* \to n^* - 1$ propensity. However, the measured NO vibrational state distributions are — like those obtained in classical trajectory calculations (Nonella et al. 1989) — significantly broader than the quantum mechanical distributions. We must strongly emphasize at this point that the inclusion of the bending degree of freedom crucially affects the final vibrational distributions. Two-dimensional calculations with the bending angle fixed at the ground-state equilibrium angle yield vibrational distributions that are significantly shifted to lower states and which are therefore in poor agreement with experiment.

The photodissociation of HONO via the S_1 state proceeds in a way very similar to the dissociation of $CH_3ONO(S_1)$ (Hennig et al. 1989). The absorption spectrum exhibits a main progression which corresponds to excitation of the NO stretching mode. Earlier experimental investigations focussed attention primarily on the OH fragment which is produced both vibrationally and rotationally "cold" (Vasudev, Zare, and Dixon 1983, 1984). Because of technical problems related to the generation of HONO, the NO fragment is more difficult to probe (Dixon and Rieley 1989) and therefore experimental information is comparatively sparse. As for CH_3ONO a three-dimensional PES has been calculated including the two O-N bonds as well as the ONO bending angle (Suter, Huber, Untch, and Schinke 1992). It qualitatively resembles the corresponding PES for $CH_3ONO(S_1)$ shown in Figure 1.11. However, the intermediate potential well in which the excited complex is trapped is about 4–5 times deeper than the well for CH_3ONO.[†]

The right-hand part of Figure 9.12 depicts the final vibrational state distributions of NO obtained in a three-dimensional wavepacket calculation. They reveal a weak propensity for a final vibrational state $n = n^* - 2$. Excitation of vibrational state n^* in the $HONO(S_1)$ complex leads preferentially to NO products in vibrational state $(n^* - 2)$. Because the intermediate potential well is rather deep, 0.362 eV, the breakup of

[†] Since the dynamical calculations using this 3D PES (Untch 1992) do not satisfactorily reproduce the experimental absorption spectrum (King and Moule 1962) and the measured lifetimes in the S_1 state (Shan, Wategaonkar, and Vasudev 1989) we must conclude that the presently available *ab initio* PES is not sufficiently accurate to account for all details of the photodissociation of HONO. Nevertheless, we present the calculated vibrational state distributions of NO in Figure 9.12, irrespective whether they reflect reality or no. The purpose of this section is to underline the influence of translational-vibrational coupling on the final vibrational state distributions and this can be done with a slightly deficient PES.

the complex requires more than one vibrational quantum jump of NO to a lower vibrational state in order to surpass the barrier. The calculations are too involved, however, to reveal whether a sequential process, $n^* \to n^* - 1 \to n^* - 2$ or one quantum jump $n^* \to n^* - 2$ dominates the decay. The vibrational-translational coupling is substantially stronger for HONO than for CH_3ONO.

Dixon and Rieley (1989) measured a ratio of $\sigma(n = 1) : \sigma(n = 2) : \sigma(n = 3)$ of $2 : 1 : 0.01$ for excitation in the $n^* = 2$ band. The calculated ratio shows the same overall trend, i.e., a decrease with n but without the result for $n = 0$, which experimentally could not be determined for technical reasons, the comparison between theory and experiment remains incomplete.

Figure 9.12 strikingly illustrates the variety of possible final vibrational state distributions following the decay of trapped intermediate states. Although the overall topology of the multi-dimensional potential energy surfaces, especially for CH_3ONO and HONO, is similar, the detailed dissociation dynamics differs remarkably. The coupling between the translational and the vibrational modes crucially controls the efficiency of internal vibrational energy transfer and thus the degree of excitation of the fragment molecule. The curvature of the minimum energy path in a (R, r) representation illustrates this coupling. It is weakest for ClNO and strongest for HONO. The bending degree of freedom is also important, at least for CH_3ONO and HONO and must not be ignored.

The participation of three degrees of freedom makes a simple analysis impossible, even if the corresponding PES is known from *ab initio* calculations. The motion of a wavepacket or a swarm of classical trajectories is difficult to monitor and to imagine in three dimensions except for simple cases with weak coupling. In this light, the above discussion and the schematic illustration in Figure 9.13 must not be overinterpreted; they are *qualitative* and inevitably influenced by the results of the measurements or the calculations. Nevertheless, the three examples discussed in this section manifest a clear-cut dependence on the "initial" state of the excited complex. More examples of state-specific molecular fragmentation will be given in the next two chapters.

10
Rotational excitation I

Rotational state distributions of the photofragments provide a wealth of information on the dissociation dynamics. The total available energy often exceeds 1 eV which suffices to make many rotational states energetically accessible. Since the torque imparted to the rotor is typically large, it is not unusual that 50 or even more rotational states become populated during dissociation. Modern detection methods, on the other hand, make it feasible to resolve even the most detailed aspects of rotational excitation: *scalar properties* , such as the final state distribution as well as *vector properties*, such as the orientation of the angular momentum vector of the fragment with respect to any axis of reference. In the same way as the vibrational distribution reflects the change of the bond length of the fragment molecule, the rotational distribution elucidates the change of the bond angle along the reaction path.

In this chapter we discuss only the scalar aspect of rotational excitation, i.e., the forces which promote rotational excitation and how they show up in the final state distributions. The simple model of a triatomic molecule with total angular momentum $\mathbf{J} = 0$, outlined in Section 3.2, is adequate for this purpose without concealing the main dynamical effects with too many indices and angular momentum coupling elements. The vector properties and some more involved topics will be discussed in Chapter 11.

Sources of rotational excitation

Three major sources contribute to the rotational excitation of the fragment molecule:

1) Overall rotation of the parent molecule in the electronic ground state.
2) Bending or torsional vibration within the parent molecule.
3) The torque generated by the anisotropy of the upper-state potential energy surface, i.e., the so-called *exit channel dynamics.*

Rotational excitation as a consequence of overall rotation of the parent molecule *before* the photon is absorbed does not reveal much dynamical information about the fragmentation process. It generally increases with the magnitude of the total angular momentum **J** and thus increases with the temperature of the molecular sample. In order to minimize the thermal effect and to isolate the dynamical aspects of photodissociation, experiments are preferably performed in a supersonic molecular beam whose rotational temperature is less than 50 K or so. Broadening of final rotational state distributions as a result of initial rotation of the parent molecule will be discussed at the end of this chapter.

Internal bending or torsional vibration in the electronic ground state can be considered as *hindered* or *frustrated rotation* of the fragments within the restraint of the parent molecule. Let us, for simplicity, consider the dissociation of a triatomic molecule ABC $\rightarrow$ A + BC. After the photon has cut the A-BC bond the diatom rotates freely around its own center-of-mass and the original bending vibration, associated with the ABC bond angle, is converted into rotational motion (sudden release). The final rotational state distribution then reflects the initial bending wavefunction in the electronic ground state, provided subsequent excitation or de-excitation in the exit channel is negligibly small. This kind of mapping is easily described within the elastic or the Franck-Condon limit (see Section 10.1).

In many cases, however, rotational excitation is mainly generated in the excited electronic state. As the complex breaks apart the torque induced by the anisotropy of the potential energy surface (PES) in the excited electronic state propels rotation of the fragment molecule. The photodissociation of H_2O in the first and in the second absorption bands illustrates most intriguingly the influence of the exit channel dynamics (see Figure 1.12 for the corresponding potentials). In the case of $H_2O(\tilde{A})$, the torque imparted to $OH(^2\Pi)$ is very small and the resulting rotational state distribution shown in Figure 1.8 reflects mainly the zero-point bending motion of $H_2O(\tilde{X})$. This contrasts the dissociation via the $\tilde{B}$ state where the final state interaction is extremely strong with the consequence that $OH(^2\Sigma)$ is born with extremely high rotational excitation.

What are the right coordinates?

Before we start a more quantitative discussion it is worthwhile thinking about which coordinates to use for interpretation of rotational excitation. Let us consider the photodissociation of a triatomic molecule, ClNO $\rightarrow$ Cl + NO(j), for example. Researchers with a background in electronic structure calculations and spectroscopy normally use *bond coordinates*, i.e., the two bond distances R_{ClN} and R_{NO} and the ClNO bond angle α [see Figure 10.1(a)]. These are the appropriate coordinates for generating the multi-dimensional PES by *ab initio* calculations and setting up a normal mode analysis. Figure 10.2(a) depicts the calculated PES of the S_1 state of ClNO as a function of R_{ClN} and α. The N-O bond distance is fixed.

The bond coordinates are not appropriate, however, for describing the dissociation into the fragments Cl and NO(j). The kinetic energy expression does not separate in R_{ClN} and α but contains a relatively large kinetic coupling term which makes it difficult to imagine a classical trajectory or the route of a wavepacket as it evolves on the PES $V(R_{\text{ClN}}, \alpha)$. If we were erroneously to ignore the kinetic coupling between R_{ClN} and α, the trajectory would follow essentially the minimum energy path, indicated by the heavy arrow in Figure 10.2(a), without significant action in the angular direction, i.e., without noticeable rotational excitation of NO. The angle α would remain almost constant. This picture does not correctly describe the true dissociation process!

In full as well as half collisions one uses Jacobi or scattering coordinates R and γ defined in Figure 10.1(b) where the vector $\mathbf{R}$ points to the center-of-mass of NO and γ is the orientation angle of $\mathbf{r}$ with respect to $\mathbf{R}$, i.e., $\cos\gamma = \mathbf{R} \cdot \mathbf{r}/Rr$. Figure 10.2(b) shows the ClNO(S_1) PES represented in R and γ. The distortion relative to Figure 10.2(a) reflects the transformation between the two sets of coordinates (R_{ClN}, α) and (R, γ). The Jacobi coordinates have the advantage that the kinetic energy separates into two terms, one representing the motion along R and the other one describing the motion in the angle γ [see Equation (3.16)]. The absence of kinetic coupling makes it comparatively simple to surmise the route of a dissociative trajectory on the PES $V(R, \gamma)$. The minimum energy path is considerably tilted and thus a trajectory sliding down the hill experiences a substantial torque which strongly excites rotation of NO in full accord with experiment (see Figures 6.7 and 6.8). The classical equations of motion demonstrate the advantage of the Jacobi coordinates

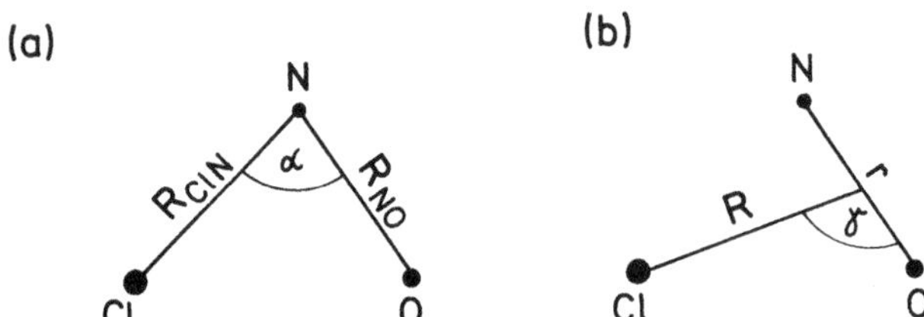

Fig. 10.1. Bond coordinates R_{ClN} and α (a) and Jacobi coordinates R and γ (b) for a triatomic molecule, ClNO for example.

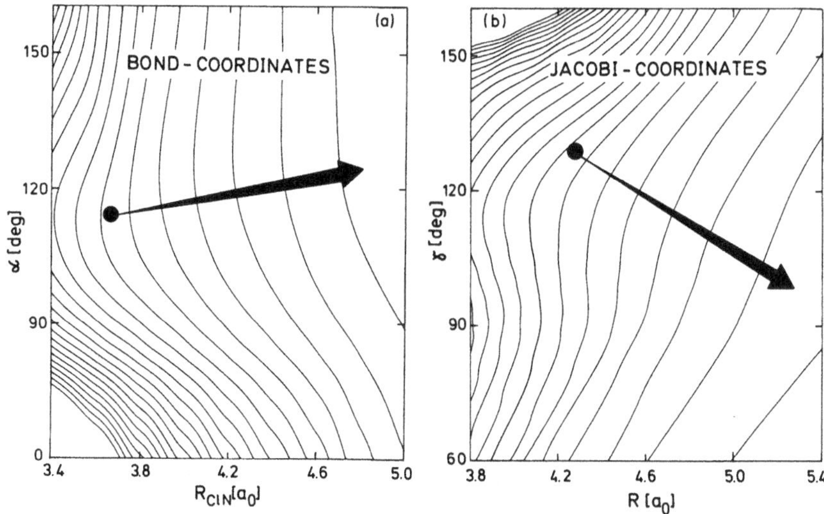

Fig. 10.2. (a) The potential energy surface of ClNO(S_1) for fixed N-O bond distance represented in terms of bond coordinates R_{ClN} and α calculated by Schinke, Nonella, Suter, and Huber (1990). The heavy arrow indicates the dissociation path if one were erroneously to neglect the strong kinetic coupling. (b) The same PES as in (a) but represented in Jacobi coordinates R and γ. The heavy arrow illustrates the true dissociation path.

in a clear way. According to Equation (5.4f), it is the torque

$$F_\gamma = -\frac{\partial V(R,\gamma)}{\partial \gamma} \tag{10.1}$$

which controls the change of the molecular angular momentum.

As in Chapter 9 we discuss first the elastic limit (no exit channel excitation) in Section 10.1 and subsequently the more interesting inelastic case in Section 10.2. In Section 10.3 we consider the decay of long-lived resonance states and the impact of exit channel dynamics on the product distributions. A simple approximation, the so-called impulsive model, which is frequently employed to analyze experimental distributions in the absence of a PES, is discussed critically in Section 10.4. The chapter ends with a more qualitative assessment of thermal broadening of rotational state distributions in Section 10.5

10.1 The elastic case: Franck-Condon mapping

For simplicity we consider the photodissociation of a triatomic molecule, ABC $\rightarrow$ A + BC(j), as described in Section 3.2. Let us assume that the PES in the excited electronic state is isotropic, i.e., it depends only on the Jacobi coordinate R, but not on the orientation angle γ. Then, the radial and the angular motions separate completely and energy cannot flow between the two modes.

10.1.1 Rotational Franck-Condon factors

In analogy to Section 9.1 and Equations (9.4) and (9.5) the (unnormalized) final rotational state distribution becomes

$$P(E,j) = \bar{f}(j)\,\bar{g}(E,j) \tag{10.2}$$

with the $\bar{f}(j)$ being one-dimensional FC-type factors,

$$\bar{f}(j) = |\langle Y_{j0} \mid \varphi_\gamma \rangle|^2 = \left| 2\pi \int_0^\pi d\gamma \, \sin\gamma \, Y_{j0}(\gamma,0)\, \varphi_\gamma(\gamma) \right|^2 \tag{10.3a}$$

and the

$$\bar{g}(E,j) = |\langle \chi_j(E,j) \mid \varphi_R \rangle|^2 = \left| \int_0^\infty dR \, \chi_j^*(R;E,j)\, \varphi_R(R) \right|^2 \tag{10.3b}$$

are one-dimensional overlap integrals in the radial coordinate R. In deriving (10.3) we assumed that the ground-state wavefunction factorizes as $\varphi_R(R)\,\varphi_\gamma(\gamma)$. The spherical harmonics Y_{j0} are the eigenfunctions of the free rotor with quantization axis in the molecular plane, which for $\mathbf{J} = 0$ remains fixed in space for all times. The continuum functions $\chi_j(R;E,j)$ are solutions of the single-channel version of Equation (3.20) in the limit that the coupling matrix $\mathbf{V}$ is diagonal. In analogy to the well-known vibrational FC factors $\bar{f}(n)$, Equation (9.5a), we call the $\bar{f}(j)$ *rotational* FC *factors*. They are the squared moments of the bending wavefunction $\varphi_\gamma(\gamma)$ in the electronic ground state in terms of the rotor wavefunctions $Y_{j0}(\gamma,0)$ of the free fragment.

The one-dimensional Schrödinger equations (3.20) for the radial wavefunctions $\chi_j(R;E,j)$ depend on the rotational quantum number j only through the wavenumbers k_j, Equation (3.21), and the centrifugal potential.[†] The rotational energy $B_{rot}j(j+1)$ is usually small compared to the total energy E and the centrifugal potential $\hbar^2 j(j+1)/2mR^2$ is likewise small compared to the interaction potential $V(R)$. As a result, the radial wavefunctions and hence the overlap integrals $\bar{g}(E,j)$ are approximately independent of the particular rotational channel and therefore the final rotational state distribution reduces to

$$P(E,j) \propto \bar{f}(j) = |\langle Y_{j0}(\gamma,0) \mid \varphi_\gamma(\gamma) \rangle|^2. \tag{10.4}$$

As opposed to vibrational excitation, the overlap factors $\bar{g}(E,j)$ do *not* significantly affect the final state distribution. Thus:

[†] In the elastic limit the potential is independent of the angle γ so that, according to Equation (3.22), $V_{jj'}(R) = \delta_{jj'}V(R)$ independent of j.

- The final rotational state distribution in the elastic limit is approximately proportional to the rotational FC factors $\bar{f}(j)$.

In order to discriminate this special case from dynamical mapping we introduce the term FC *mapping.* As an immediate consequence we note that:

- The final rotational state distribution in the FC limit is nearly independent of the total energy and hence independent of the photolysis wavelength.

The projection integrals $\langle Y_{j0} \mid \varphi_\gamma \rangle$ can be interpreted as the (discrete) angular momentum representation of the initial bending wavefunction in the electronic ground state. Employing the semiclassical limit for the spherical harmonics,

$$Y_{j0}(\gamma, 0) \approx \frac{1}{2i\pi(\sin\gamma)^{1/2}} \left[e^{i(j\gamma+\pi/4)} - e^{-i(j\gamma+\pi/4)} \right], \tag{10.5}$$

valid for large j (Ford and Wheeler 1959; Child 1991:ch.8; the argument $j+1/2$ has been replaced by j on the right-hand side of Equation (10.5)), and utilizing that the bending wavefunction is well localized around the equilibrium angle γ_e, we can approximately write (Schinke, Vander Wal, Scott, and Crim 1991)

$$\langle Y_{j0} \mid \varphi_\gamma^{(k)} \rangle \approx \Im \left[e^{i(j\gamma_e+\pi/4)} \int_{-\infty}^{+\infty} d\gamma' \varphi_\gamma^{(k)}(\gamma')\, e^{ij\gamma'} \right], \tag{10.6}$$

where $\gamma' = \gamma - \gamma_e$ and $\Im[\cdots]$ denotes the imaginary part of the expression in brackets. In what follows k denotes the particular bending quantum number of the initial state of the parent molecule. The integral on the right-hand side is just the Fourier transformation of the bending wavefunction $\varphi_\gamma^{(k)}(\gamma)$. If we furthermore assume that the bending vibration is harmonic, the Fourier integral can be evaluated analytically and we obtain (see also Morse and Freed 1980)

$$\begin{aligned} |\langle Y_{j0} \mid \varphi_\gamma^{(k)} \rangle|^2 &\approx \sin^2 \left[j\gamma_e + (-1)^k \frac{\pi}{4} \right] \left[\varphi_\gamma^{(k)} \left(\gamma' = \frac{j\hbar}{\tilde{m}\omega} \right) \right]^2 \\ &\approx \frac{1}{2^k k!} \sin^2 \left[j\gamma_e + (-1)^k \frac{\pi}{4} \right] H_k^2 \left[j(\hbar/\tilde{m}\omega)^{1/2} \right] e^{-j^2\hbar/\tilde{m}\omega}, \end{aligned} \tag{10.7}$$

where H_k is the Hermite polynomial of order k; ω is the bending frequency in the electronic ground state (with units 1/time) and $\tilde{m}$ is defined by

$$\frac{1}{\tilde{m}} = \frac{1}{\mu r_e^2} + \frac{1}{m R_e^2}, \tag{10.8}$$

where R_e and r_e are the equilibrium distances in the ground state and m and μ are the reduced masses defined in (2.38). Unimportant normalization constants have been omitted in (10.7). $\varphi_\gamma^{(k)}(j\hbar/\tilde{m}\omega)$ is the bending wavefunction with the displacement in the angle $(\gamma - \gamma_e)$ replaced by the argument $j\hbar/\tilde{m}\omega$. Equation (10.7) is valid for the ground $(k = 0)$ as well as excited $(k > 0)$ bending states. In the above derivation we made use of the simple relation between the wavefunctions in the coordinate representation and in the momentum representation of the harmonic oscillator (Cohen-Tannoudji, Diu, and Laloë 1977:ch.V). Thus:

- Within the harmonic approximation for the bending motion the rotational FC factors are proportional to the square of the bending wavefunction with argument $(\gamma - \gamma_e) = j\hbar/\tilde{m}\omega$, modulated by a sinusoidal factor with wavelength $\Delta j \approx \pi/\gamma_e$.

The sinusoidal function is the result of the displacement of the equilibrium angle γ_e from the linear configuration. If $\gamma_e > \pi/2$ we must use $\pi - \gamma_e$ instead of γ_e.

Let us consider a realistic example. Figure 10.3 shows the distributions of FC factors $\bar{f}(j)$, calculated exactly by using Equation (10.3a), for the three lowest bending states of H_2O in the electronic ground state. As predicted by (10.7), they *qualitatively* reflect the coordinate dependence of the bending wavefunctions, shown in the insets, modulated by fast oscillations with approximate wavelength $\Delta j \approx 2.4$.[†] The FC distribution of the lowest bending state is, apart from the oscillations, a Gaussian centered at $j = 0$ whose width is inversely proportional to the width of the bending wavefunction; the narrower the angular wavefunction the broader is, according to the uncertainty relation, the rotational angular momentum distribution. The distribution for the first excited state has a minimum at $j = 0$ and maxima at $j = \pm 4$, and so on.

In order to elucidate the general behavior we also show in Figure 10.3 what we will call the harmonic oscillator (HO) approximation, i.e., expression (10.7) without the sinusoidal factor. It represents the momentum distribution of the harmonic oscillator in the kth bending vibrational state. Suppression of the fast oscillations has the advantage of elucidating more clearly the wide oscillations for $k > 0$, which reflect the nodal structure of the excited bending wavefunctions. The superimposed fast oscillations, on the other hand, reflect the shift of the equilibrium angle γ_e away from zero. They are absent for a linear molecule and most pronounced for $\gamma_e \approx \pi/2$, as for H_2S and H_2O, for example.

[†] The FC distributions are symmetric with respect to $j = 0$. Since we consider in this chapter only the magnitude of the rotational angular momentum vector, it suffices to show merely one half of the full distribution. Because of the anharmonicity of the ground-state potential of H_2O, the bending wavefunctions in the insets of Figure 10.3 are, of course, *not* symmetric about $\alpha_e = 104°$.

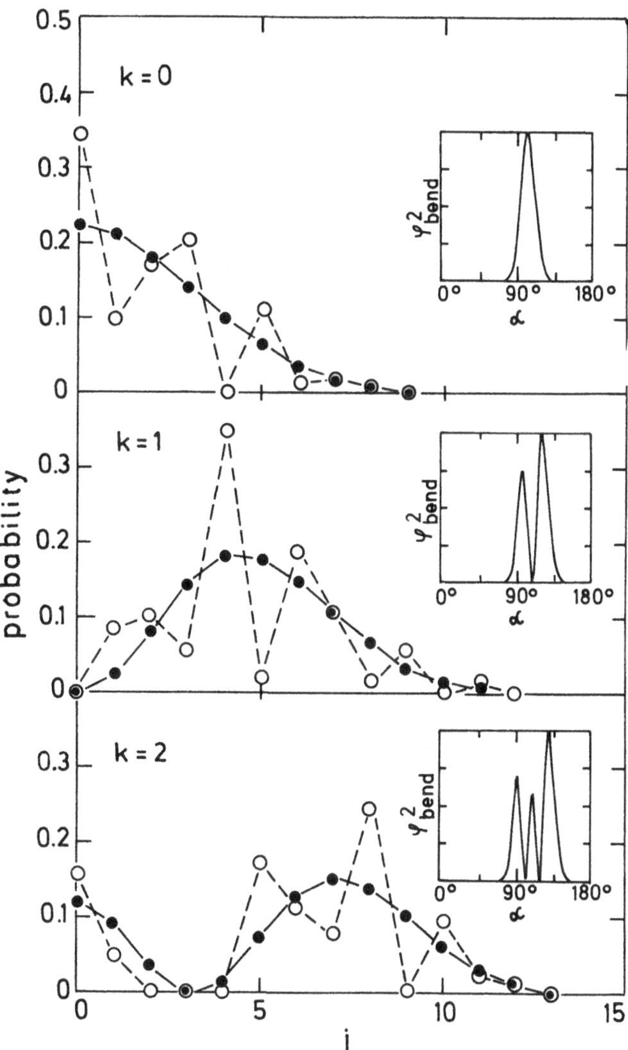

Fig. 10.3. Rotational FC factors $\bar{f}(j)$ as defined in (10.3a) for the first three bending states of H_2O in the electronic ground state (open circles). The filled circles represent the results of the harmonic oscillator approximation, i.e., the right-hand side of (10.7) *without* the sinusoidal term. k is the bending quantum number. The insets depict the squares of the corresponding bending wavefunctions as functions of the HOH bending angle α. Because of the heavy central atom the bending angle α and the Jacobi angle γ are almost identical for H_2O. Adapted from Schinke, Vander Wal, Scott, and Crim (1991).

10.1.2 *Example: Photodissociation of* $H_2O(\tilde{A})$

The photodissociation of H_2O via the $\tilde{A}^1B_1$ state (the first absorption band) provides a perfect example for FC mapping. Figure 10.4 depicts the corresponding PES as function of one of the O-H bonds and the HOH bending angle with the other O-H bond being fixed at the equilibrium value in $H_2O(\tilde{X})$. Because the center-of-mass of OH nearly coincides with the oxygen atom the bond coordinates (R_{OH}, α) are almost equivalent to the Jacobi coordinates (R, γ).

The $\tilde{A}$-state potential is, in general, quite anisotropic, especially for small angles where the two hydrogen atoms come close and repel each other. But near 104°, where the wavepacket in the photodissociation of the lowest bending state starts, it is almost isotropic ($\partial V/\partial\alpha \approx 0$) with the result that no torque excites OH while the molecule breaks apart. As a consequence, the photodissociation of H_2O via the $\tilde{A}$ state yields

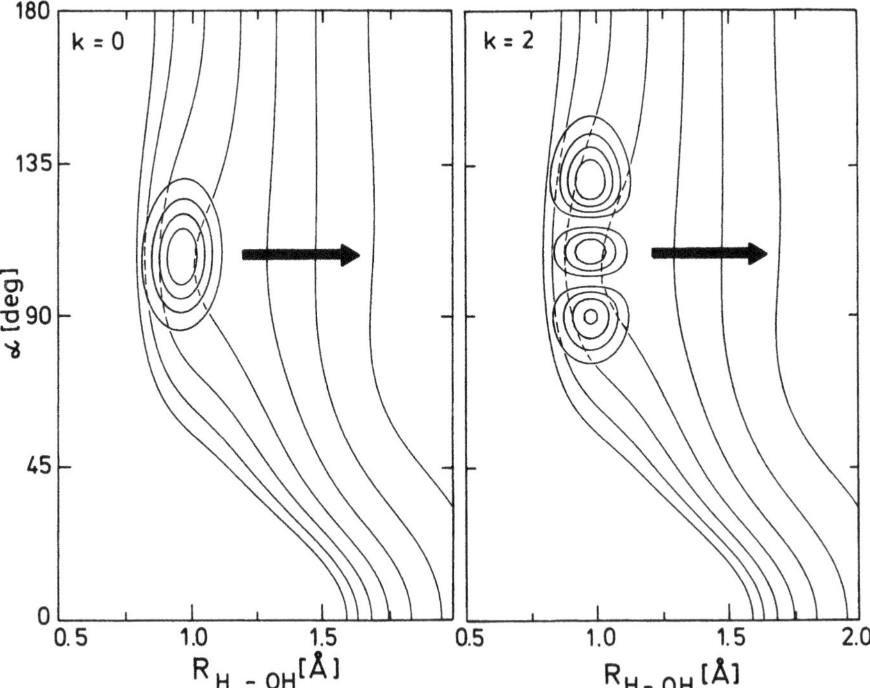

Fig. 10.4. Two-dimensional plot of the $\tilde{A}$-state PES of H_2O. The energies of the contours decrease with increasing H-OH bond distance R_{H-OH}. One of the O-H bonds is frozen at its equilibrium value; α is the HOH bending angle. Superimposed are the $k = 0$ and the $k = 2$ bending wavefunctions of the electronic ground state with the heavy arrows indicating the main dissociation path.

rotationally "cold" OH radicals as indeed measured by Andresen, Ondrey, Titze, and Rothe (1984; see Figure 1.8) and calculated by Schinke, Engel, and Staemmler (1985). The rotational distribution initially generated in the FC transition remains almost unchanged during the dissociation and eventually it becomes the final rotational distribution of the OH product.

Figure 10.5 depicts measured rotational state distributions of OH following the dissociation of H_2O with zero, one, and two quanta of bending vibration in comparison to the corresponding FC distributions (Schinke, Vander Wal, Scott, and Crim 1991). For clarity of the comparison we have suppressed the fast oscillations [HO model, Equation (10.7) without the sinusoidal factor]; the finer structures are not pertinent at this point of the discussion.† In Section 11.3 we will briefly describe how these experiments are performed. The total angular momentum is zero in all three cases.‡ As argued above, the dissociation of the lowest bending state samples only the isotropic portion of the $\tilde{A}$-state PES around $\gamma_e = 104°$ with the consequence that the final distribution clearly resembles the FC distribution (compare the top of Figure 10.3 with the top of Figure 10.5).

With increasing bending quantum number, however, the breadth of the bending wavefunction becomes larger so that the evolving wavepacket samples more and more the anisotropic portion of the PES. The effect is a slight but nevertheless noticeable distortion of the original FC distribution along the fragmentation path which gradually becomes more prominent with increasing k. However, the torque is not strong enough to erase the reflection-type structures completely. As opposed to most of other molecules which we have investigated up to now, the final state interaction in the photodissociation of $H_2O(\tilde{A})$ *de-excites* rather than *excites* the OH fragment, relative to the FC limit. Figure 10.5 distinctly shows the influence of the anisotropy of the dissociative PES on the final rotational state distributions. In passing we note that exact dynamical calculations including the full anisotropy of the PES *quantitatively* repro-

---

† The pronounced oscillations of the FC distributions in Figure 10.3 are not realistic for the dissociation of H_2O into $H(^2S)$ and $OH(^2\Pi)$. A correct treatment must also include the electronic degrees of freedom of both products which would damp and modify these oscillations. Truly state-resolved measurements yield oscillatory rotational state distributions for $OH(^2\Pi)$, but their description requires a more sophisticated FC theory than the one presented in this section (Balint-Kurti 1986; Schinke et al. 1985). See also Section 11.3.

‡ The initial state of $H_2O(\tilde{X})$ contains, in addition to bending excitation, some quanta of stretching excitation; this does not, however, noticeably affect the present discussion of final rotational state distributions because stretching and bending vibrations separate to a very good approximation.

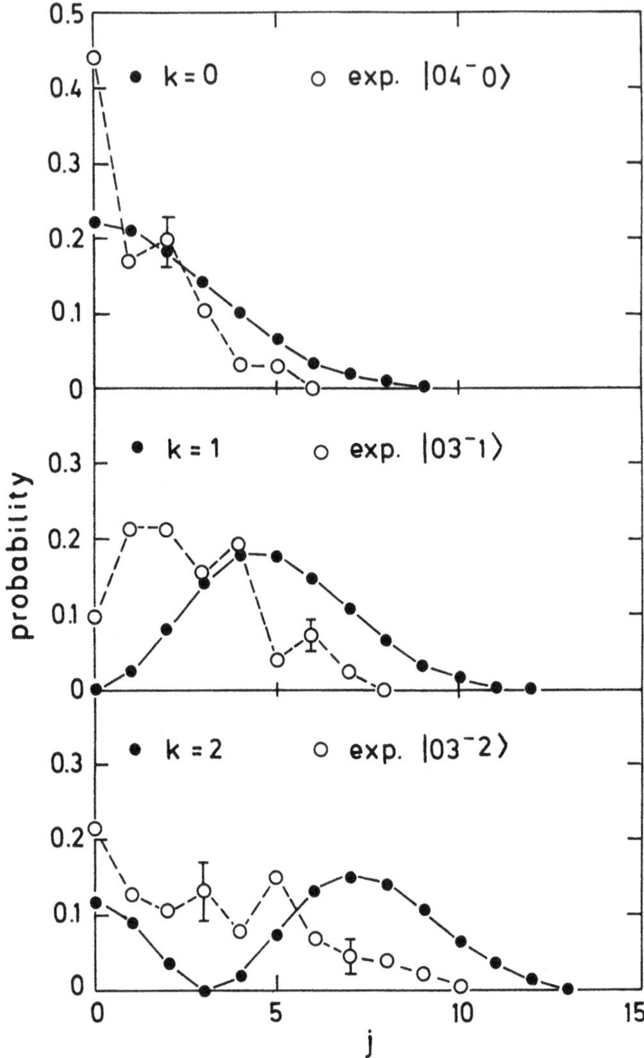

Fig. 10.5. Measured rotational state distributions of OH following the dissociation of the three lowest bending states of H_2O (open circles). In addition to the bending quanta $H_2O(\tilde{X})$ also contains 4 respectively 3 quanta of OH stretching excitation. The local mode nomenclature $|nm^{\pm}k\rangle$ is explained in Section 13.2. The total angular momentum is zero in all cases. The filled circles represent the harmonic oscillator approximation defined in the text. Reproduced from Schinke, Vander Wal, Scott, and Crim (1991).

duce the experimental results for $k = 0$ as well as $k = 1$ and 2 (Schinke, Vander Wal, Scott, and Crim 1991).

Before concluding this section it is instructive to draw attention to the following point. It is common practice to compare measured rotational state distributions to the Boltzmann distribution

$$P_T(j) \propto (2j+1)\, \mathrm{e}^{-B_{rot}j(j+1)/k_BT}, \qquad (10.9)$$

where k_B is Boltzmann's constant. If the plot $\ln[P(j)/(2j+1)]$ versus the rotational energy $B_{rot}j(j+1)$ is linear, one extracts a "temperature" T which is then compared to the temperature of the molecular sample. In the photodissociation of H_2O, performed in a molecular beam, this analysis yields "temperatures" ranging from 210 K to 475 K depending on the particular electronic state of OH (Andresen et al. 1984). What is the origin of this "temperature" and how does it relate to the temperature $T_{H_2O} \approx 50$ K of H_2O before the photolysis?

According to (10.7), the FC distribution for the dissociation of the lowest bending state is a Gaussian in j, i.e., $\mathrm{e}^{-\alpha j^2}$. For large enough values of the angular momentum quantum number (which is already fulfilled for $j \geq 3$–4 or so) this Gaussian yields to a very good approximation a linear Boltzmann plot with a "temperature" that is related to the parameters of the bending oscillator, namely the moment of inertia $\tilde{m}$ and the frequency ω. The photodissociation of H_2O is fast and direct and there is absolutely no reason for a temperature to develop in the sense of thermodynamics:

- The linear Boltzmann plot obtained in the photodissociation of H_2O in the first continuum merely reflects the rotational FC distribution for the zero-point bending motion in the electronic ground state. It has nothing to do with energy randomization in a long-lived intermediate complex!

Inserting the appropriate parameters yields a "temperature" of approximately 400 K in reasonable accord with the average of the two measured values. A linear Boltzmann plot of the final rotational state distribution usually indicates that the torque in the exit channel is weak.

The FC or elastic limit should be considered as a zeroth-order approach (Morse, Freed, and Band 1979; Beswick and Gelbart 1980; Morse and Freed 1981, 1983). All deviations from measured distributions must be attributed to rotational-translational coupling in the exit channel. Considering the many experimental results on rotational state distributions we must conclude, however, that negligibly weak final state interaction as in the case of $H_2O(\tilde{A})$ is the exception. Other examples which might fall into this category are H_2S (Hawkins and Houston 1982; Weiner, Levene, Valentini, and Baronavski 1989), OH following the dissociation of HONO (Vasudev, Zare, and Dixon 1984) and CH_3I (Suzuki, Kanamori, and Hirota 1991).

10.2 The inelastic case: Dynamical mapping

The photodissociation of $H_2O(\tilde{A})$ represents a very special example of rotational excitation: the initial FC distribution, which reflects solely the bending motion in the parent molecule, remains unchanged because the torque $-\partial V/\partial\gamma$ is essentially zero along the reaction path. In most other cases, however, the coupling between translation and rotation is substantial and the final state distribution of the fragment reflects the net result of this coupling.

Figure 10.6 illustrates, for example, the evolution of rotational excitation in the dissociation of ClNO via the S_1 state. It shows time-resolved rotational state distributions defined as

$$P(j;t) = \int_0^\infty dR \left| \langle \Phi(R,\gamma;t) \mid Y_{j0}(\gamma,0) \rangle \right|^2, \tag{10.10}$$

where $\Phi(R,\gamma;t)$ is the two-dimensional wavepacket evolving on the S_1 PES depicted in Figure 10.2(b). The distribution at the starting point, $t = 0$, is simply the FC distribution; it has the overall shape of a Gaussian with superimposed oscillations, just as predicted by (10.7). The overall width is significantly larger than the width for H_2O because the corresponding bending wavefunction is considerably narrower. As the molecule dissociates, NO experiences a strong torque with the consequence that the peak of the distribution gradually shifts to higher states. Simultaneously, the pronounced oscillations are damped and quickly disappear.

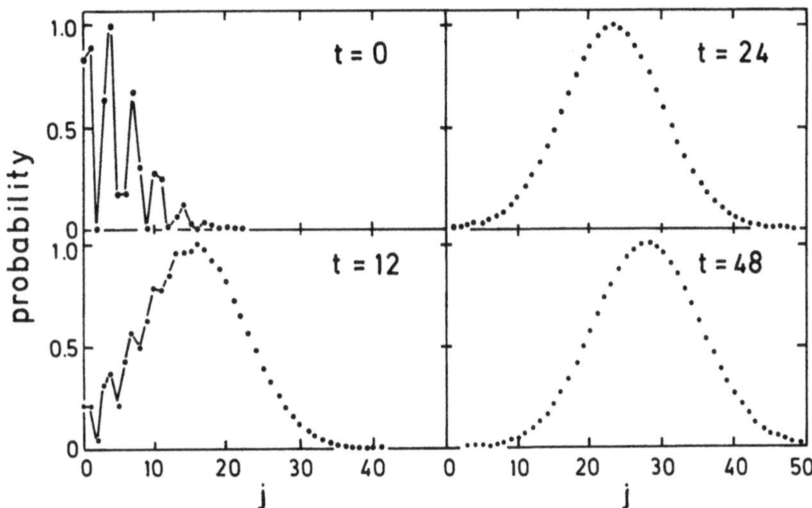

Fig. 10.6. Time-dependent rotational state distribution, defined in Equation (10.10), in the photodissociation of ClNO through the S_1 state. The times are given in femtoseconds. Adapted from Untch, Weide, and Schinke (1991b).

After about 50 fs $\Phi(t)$ has reached the asymptotic region where the torque is essentially zero and the distribution does not alter any further, i.e., the dissociation is over. Figure 10.6 illustrates rotational excitation in the angular momentum representation, whereas Figure 10.2(b) manifests rotational excitation in the coordinate picture.

In this section we focus our attention on two particular examples, the photodissociation of H_2O_2 and the photodissociation of H_2O via the $\tilde{B}^1A_1$ state. In the first case, the extent of rotational excitation is moderate, i.e., only a few percent of the total available energy flow into rotation. The dissociation of $H_2O(\tilde{B})$, on the other hand, represents an example of extremely strong rotational excitation.

10.2.1 *Photodissociation of* H_2O_2

The photodissociation

$$H_2O_2(\tilde{X}) + \hbar\omega \longrightarrow H_2O_2(\tilde{A},\tilde{B}) \longrightarrow 2\ OH(j)$$

in the 193 nm band has been extensively studied in recent years and a wealth of very detailed data has been accumulated by several research groups (Ondrey, van Veen, and Bersohn 1983; Docker, Hodgson, and Simons, 1986a,b; Gericke, Klee, Comes, and Dixon 1986; Jacobs et al. 1987; Grunewald, Gericke, and Comes 1988; Brouard, Martinez, O'Mahony, and Simons 1990; Brouard, Martinez, Milne, Simons, and Wang 1990). It is particularly interesting because torsional motion (associated with the angle φ; see Figure 10.7 for definition of the coordinates) rather than bending motion (associated with the angles α_1 and α_2) generates the rotational excitation of both OH fragments. After the rupture of the O-O bond the two OH radicals rotate like cartwheels in opposite directions around the O-O axis. In this section we will primarily consider the final

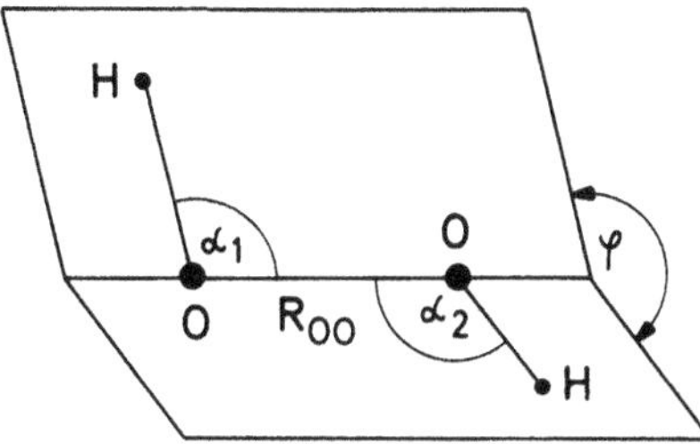

Fig. 10.7. The bond coordinates used to describe the photodissociation of H_2O_2. φ is the torsional angle. The dynamical calculations, however, have been performed in Jacobi coordinates (Schinke and Staemmler 1988). The intermolecular vector $\mathbf{R}$ joins the centers-of-mass of the two rotamers.

rotational state distributions of the OH fragments. Additional aspects will be addressed in subsequent sections.

The photodissociation of H_2O_2 near 193 nm evolves via two electronic states, $\tilde{A}^1A$ and $\tilde{B}^1B$, whose potential energy surfaces are both strongly repulsive in the O-O bond coordinate. At the FC point, the $\tilde{A}$ state is the lower one with the $\tilde{B}$ state being about 1.3 eV higher in energy. Detailed experimental (Grunewald, Gericke, and Comes 1987) and theoretical (Schinke and Staemmler 1988) investigations yield a ratio $\sigma_A : \sigma_B$ of roughly 3 : 1 for 193 nm and the final state distributions obtained separately for both electronic states must be averaged accordingly.

Figure 10.8 depicts the φ dependence of the potentials for the three lowest states. The equilibrium angle in the electronic ground state, $\tilde{X}$, is $\varphi_e = 110°$ as compared to $\varphi_e = 180°$ (*trans*-configuration) for the $\tilde{A}$ state and $\varphi_e = 0°$ respectively 360° (*cis*-configuration) for the $\tilde{B}$ state. The large displacements of the equilibrium angles imply relatively strong anisotropies and large torques $-\partial V/\partial\varphi$ at the FC point with the consequence that the two cartwheels immediately start to spin around the common axis after the molecule is promoted to the upper electronic states. However, the direction of rotation differs in the two excited states. While in the $\tilde{A}$ state the radicals rotate into the *trans*-configuration, the opposite motion takes place in the $\tilde{B}$ state. The φ dependences of the potential energy surfaces generate considerable rotational excitation of the two OH rotamers with the corresponding angular momentum vectors $\mathbf{j}_1$ and $\mathbf{j}_2$ pointing in opposite directions.

The photodissociation of H_2O_2 represents an instructive example of the rotational reflection principle, which was outlined in detail in Section 6.3. Figure 10.9 depicts the rotational excitation functions $J_A(\varphi_0)$ and $J_B(\varphi_0)$ obtained in five-dimensional classical trajectory calculations including the O-O bond distance, the two polar angles α_i and the two azimuthal angles φ_i of each OH rotamer ($i = 1, 2$) (Schinke and Staemmler 1988). φ_0 is the initial torsional angle of the trajectory. The excitation functions have the same qualitative behavior although the φ dependences of the $\tilde{A}$- and the $\tilde{B}$-state PES differ remarkably.

The right-hand side of Figure 10.9 shows the resulting final rotational state distributions obtained by Monte Carlo averaging; they are highly inverted with peaks at j=11 and 10, respectively. The distributions essentially reflect the torsional wavefunction in the electronic ground state *and* the anisotropy of the excited potential energy surfaces as has been explained in Section 6.3. In contrast to the elastic case, the mapping is mediated by the dynamics in the two excited electronic states. The lower panel in Figure 10.10 compares the theoretical to the experimental distributions, the latter being measured in a molecular beam. The excellent

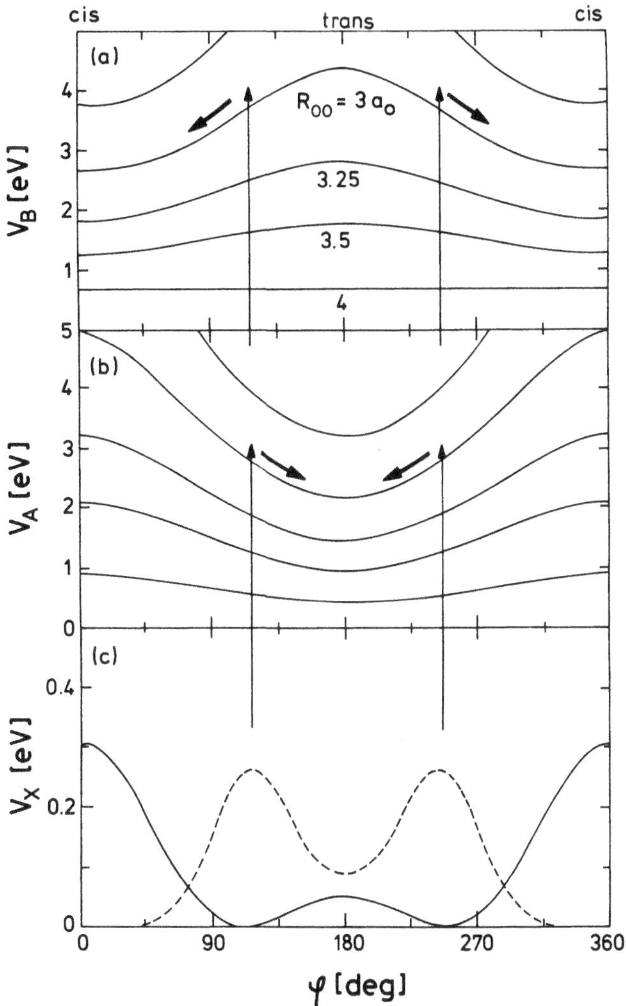

Fig. 10.8. (a) and (b) The dependence of the PESs for the two lowest excited electronic states of H_2O_2, $\tilde{A}$ and $\tilde{B}$, on the torsional angle φ for fixed R_{OO} bond distances. The two bending angles α_1 and α_2 are frozen at 94.8° (Staemmler, private communication). The arrows indicate the direction of rotation in the two excited states after absorption of the photon. (c) The φ dependence of the potential in the electronic ground state according to Hunt, Leacock, Peters, and Hecht (1965). The dashed curve shows the zero-point torsional wavefunction.

agreement illustrates the confluence of quantum chemistry, dynamical theory, and experiment. The influence of the rotational temperature of the parent molecule will be discussed in Section 10.5.

The photodissociation of H_2O_2 is a generic example of modest rotational excitation. The maximum of the distribution corresponds to only

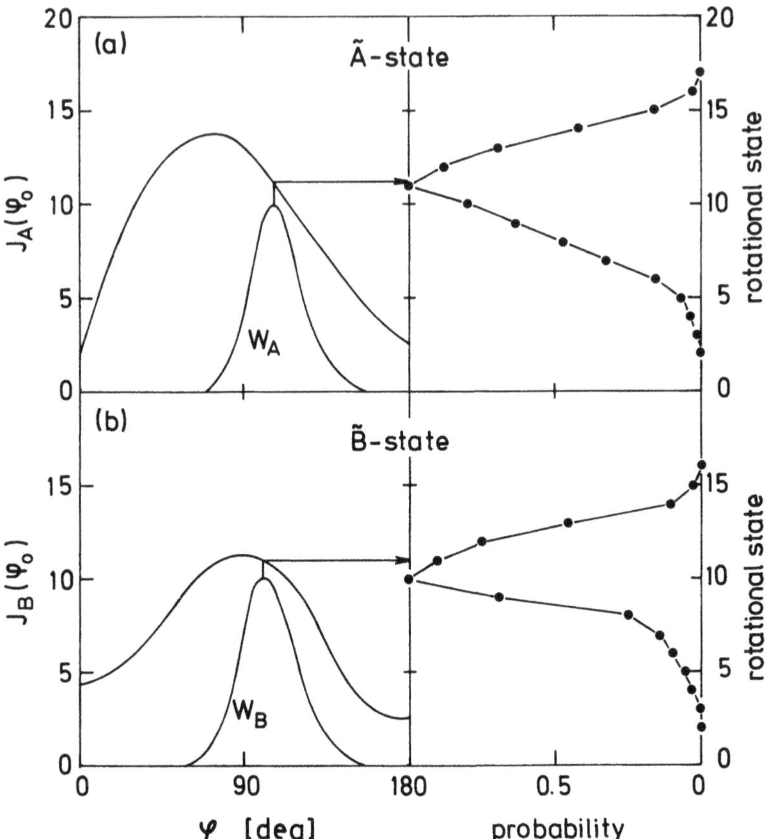

Fig. 10.9. Left-hand side: Rotational excitation functions J_A and J_B for the dissociation of H_2O_2 through the lowest two excited states, $\tilde{A}$ and $\tilde{B}$, as functions of the initial torsional angle φ_0. W_A and W_B represent the corresponding weighting functions. For definitions see Section 6.3. Right-hand side: The resulting final rotational state distributions of the OH products. Reproduced from Schinke and Staemmler (1988).

15% of the approximately 4.25 eV available for distribution. Like in the case of ClNO (Figure 10.6) the torque continuously excites the two rotors from their starting value ($j_0 \approx 0$) up to their maxima ($j \approx 12$) which are reached in not more than 20 fs (Kühl and Schinke 1989). The next example illustrates extremely strong rotational excitation.

10.2.2 *Photodissociation of* $H_2O(\tilde{B})$

A photon in the wavelength region around 125 nm promotes H_2O from the ground-state equilibrium to the steep slope of the $\tilde{B}^1A_1$ PES (see Figures 1.12 and 8.9) where it immediately experiences an extremely strong

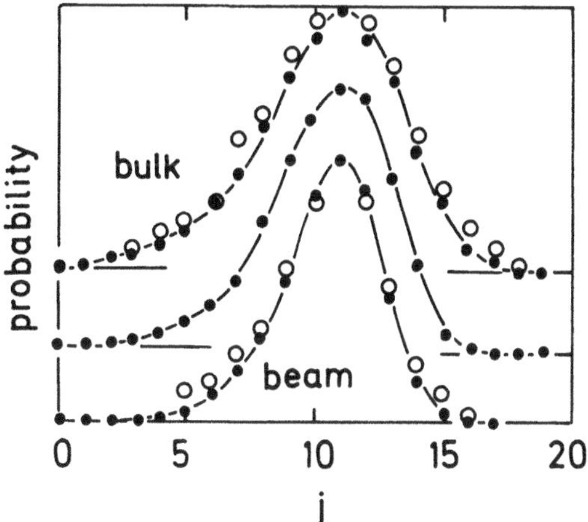

Fig. 10.10. Final rotational state distribution of the OH products following the photodissociation of H_2O_2 at 193 nm for three initial temperatures, $T_{H_2O_2} = 0$ (beam, lower curve), 150 K (middle curve), and 300 K (bulk, upper curve). The experimental data (open circles) for dissociation in the beam and in the bulk are taken from Grunewald, Gericke, and Comes (1988) and Jacobs, Wahl, Weller, and Wolfrum (1987), respectively. The theoretical results (filled circles) for the dissociation in the $\tilde{A}$ and in the $\tilde{B}$ states are averaged according to a ratio of 3 : 1. Adapted from Schinke (1988c).

torque which straightens the HOH molecule. The result is very efficient rotational excitation of the OH fragment. The light H atom is ejected like a discus, leaving the OH radical spinning very fast about its center-of-mass. Most of the available energy goes into rotation rather than translation as opposed to the dissociation of $H_2O(\tilde{A})$ or H_2O_2. Figure 10.11 depicts quantum mechanical and classical rotational state distributions for several total energies above the $H + OH(^2\Sigma)$ threshold. The majority of OH molecules are created in highly excited states close to the limit imposed by energy conservation. The classical trajectory calculations agree, on the average, well with the quantal distributions. The measured distributions (Carrington 1964; Simons, Smith, and Dixon 1984; Hodgson et al. 1985; Briggs et al. 1989) show the same qualitative behavior; they peak even closer to the highest accessible state (see Figure 1.8). Incidentally we note that the $OH(^2\Pi)$ products following excitation in the $\tilde{B}$ continuum are also generated with extremely large rotational energy (Krautwald, Schnieder, Welge, and Ashfold 1986). As $H_2O(\tilde{B})$ swings through the conical intersection with the $\tilde{X}$ state it very likely makes a transition to the lower electronic state. However, the large degree of rotational excitation of OH gained in the first part of the fragmentation is retained in this nonadiabatic transition and even increased in the second part of the breakup (Weide and Schinke 1987).

Monitoring the rotational energy content E_{rot} as a function of time reveals an interesting *multiple collision* effect: in the first part of the dissociation E_{rot} rises quickly to a value which significantly exceeds the asymptotically allowed rotational energy. This is possible because the system gains more than 3 eV of potential energy inside the deep well which can be additionally distributed among translation and rotation. In the second step of the fragmentation, the torque reverses its sign and therefore slows down the rotation. The main portion of E_{rot} converts back

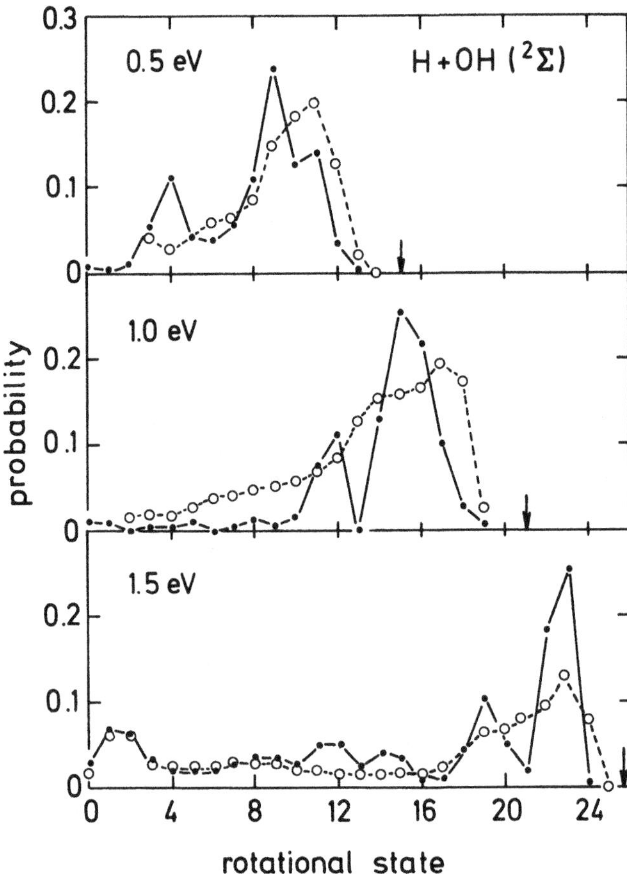

Fig. 10.11. Quantum mechanical (filled circles) and classical (open circles) rotational state distributions of $OH(^2\Sigma)$ following the dissociation of H_2O in the second continuum. The energies are measured relative to the $H + OH(^2\Sigma)$ threshold. The arrows mark the highest accessible rotational state. Reproduced from Weide and Schinke (1987).

into potential energy and just enough translational energy to enable the H-OH bond to break. $OH(^2\Sigma)$ is temporarily much more strongly excited than at the end of the bond fission. The photodissociation via the $\tilde{B}$ state represents a case of extremely strong internal energy redistribution (Kühl and Schinke 1989) and elucidates very clearly the role of exit channel dynamics, especially in comparison to dissociation in the first band.

Frequently one finds the argument that an atom as light as hydrogen *cannot* significantly excite the fragment molecule from which it recoils (see also Section 10.4). The dissociation of $H_2O(\tilde{B})$ convincingly proves that such simple arguments can seriously mislead: it is the torque $-\partial V/\partial\gamma$, i.e., the anisotropy of the dissociative PES, rather than kinematical factors, which ultimately controls the degree of rotational excitation.

10.3 Rotational distributions following the decay of long-lived states

The examples in the preceding two sections belong to the category of direct dissociation. The final rotational state distributions represent either a *direct* reflection of the ground-state wavefunction (FC mapping) or a reflection of the initial state mediated by the anisotropy of the upper-state PES (dynamical mapping). The situation is much more involved if the fragmentation proceeds via a long-lived intermediate complex. The internal vibration of the complex erases the memory on the initial level in the ground electronic state and before the molecule dissociates the energy may be randomly redistributed among all participating modes.

We distinguish two limiting cases: dissociation through a narrow transition state and dissociation through a wide transition state where we define the region that separates the reactants and the products (i.e., the "point of no return") as the transition state. The first case may be qualitatively considered as a direct process with the ultimate dissociation starting *at* the transition state. The second case may be treated by statistical methods without including dynamical constraints. We will discuss both limits separately and illustrate them with typical examples.

10.3.1 Mapping of the transition-state wavefunction

Let us consider, as a simple prototype, the photodissociation of the nitrites X-NO with X=CH_3O, HO, Cl, or F, for example. The dissociation of CH_3ONO has been discussed in detail in Sections 7.3 and 7.4. The excited-state PES has a barrier along the X-NO dissociation coordinate R, as sketched in Figure 10.12(a), which traps the excited complex for an appreciable time before it breaks apart. Figure 10.12(b) shows schematically a two-dimensional representation featuring the three essential re-

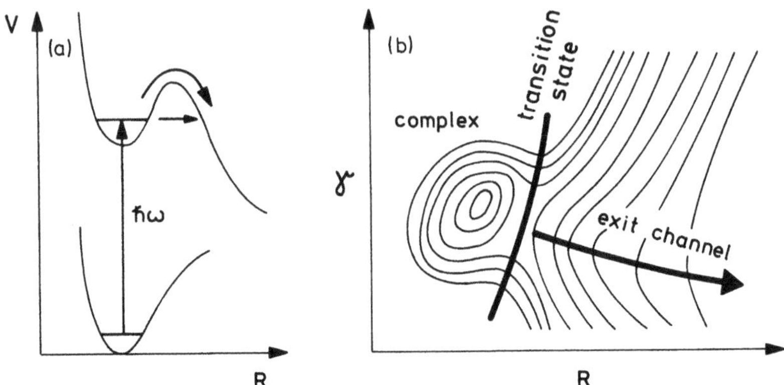

Fig. 10.12. (a) Schematic one-dimensional illustration of dissociation through a transition state. (b) The corresponding two-dimensional version with γ being the orientation angle of the product with respect to the scattering vector $\mathbf{R}$. See Figure 10.1(b) for the definition of Jacobi coordinates.

gions: the complex region inside the barrier, the transition state, i.e., the saddle point, and the exit channel beyond the barrier. As we described in Chapter 7, the photon selectively excites quasi-bound (resonance) states localized within the region of the shallow potential well. Due to the coupling to the exit channel the resonance states decay and dissociate through the bottleneck of the transition state. Beyond the transition state, the fragments recoil and quickly leave the interaction region. A sufficiently narrow bottleneck justifies considering the ultimate dissociation step as a direct process which starts on top of the barrier. Under such circumstances the wavefunction at the transition state rather than the wavefunction of the parent molecule in the electronic ground state defines the "initial" conditions for the evolution of a quantum mechanical wavepacket or the motion of classical trajectories in the exit channel. The final state distributions of the products are then determined solely in the exit channel whereas the motion before the barrier has no or only little impact.

In the following we will illustrate this qualitative picture by a two-dimensional model for the photodissociation of FNO(S_1) which is characteristic for other systems as well (Ogai et al. 1992). The coordinates which we include in the model are the Jacobi coordinates R, the distance between F and the center-of-mass of NO, and γ, the orientation angle of $\mathbf{R}$ with respect to the N-O bond; the latter is fixed. The two-dimensional PES (see the top of Figure 10.13), which is constructed from three-dimensional *ab initio* calculations, has the general shape illustrated in Figure 10.12(b), i.e., a shallow trough at short distances, where the excited complex is trapped for several internal periods, and a repulsive exit channel which describes the fast recoil of F and NO. As discussed in Section 7.6.1, the resulting absorption spectrum, shown in Figure 10.14, is composed of a progression of resonances corresponding to excitation of the four lowest bending states with quantum numbers $k^* = 0, \ldots, 3$. After the photon has promoted the molecule from the ground to the excited electronic state the complex first performs several oscillations along the angular coordinate before it eventually breaks apart.

The central question of this section is:

- Do the final rotational state distributions of NO depend on the particular bending resonance excited in the complex and, if yes, how does the distribution "reflect" the degree of bending excitation?

Bending excitation of the parent molecule transforms into rotation of the fragment molecule and therefore it is plausible to expect a relationship between the initial excitation and the final rotational state distribution.

The right-hand side of Figure 10.15 depicts the rotational state distributions of NO following the decay of the four lowest resonance states of FNO(S_1). Excitation of the lowest bending state yields a perfect Gaussian-type distribution with a maximum at a relatively high quantum state. This distribution closely resembles the kind of distributions observed in direct photodissociation. Excitation of the excited bending states, on the other hand, yields multimodal distributions with a clear relationship between the number of minima in the distribution on one hand and the bending quantum number k^* on the other. At the same time the distributions become significantly broader with increasing k^*. Figure 10.15 clearly manifests state-selective decay: the rotational distribution of the product depends uniquely on the resonance state in the excited complex. This is comparable to the state selectivity observed in Figure 9.12 for vibrational excitation.

The final rotational state distributions of NO can be qualitatively interpreted as a reflection of the stationary wavefunction at the transition state mediated by the dynamics in the exit channel. Figure 10.13 depicts the total stationary wavefunctions $\Psi_{tot}(R, \gamma; E)$ corresponding to the energies of the four lowest resonances. The internal excitation inside the shallow well is clearly visible. Because of the strong coupling between R and γ the F-NO stretching and bending motions are significantly mixed. Nevertheless, for convenience of the discussion we will term this type of excitation bending excitation. Remember that these wavefunctions represent the Fourier transforms of the evolving time-dependent wavepacket as we discussed in 4.1. The transition state, or better the transition line, may be defined as the saddle between the shallow potential well and the asymptotic channel. It is indicated by the straight line in Figure 10.13.

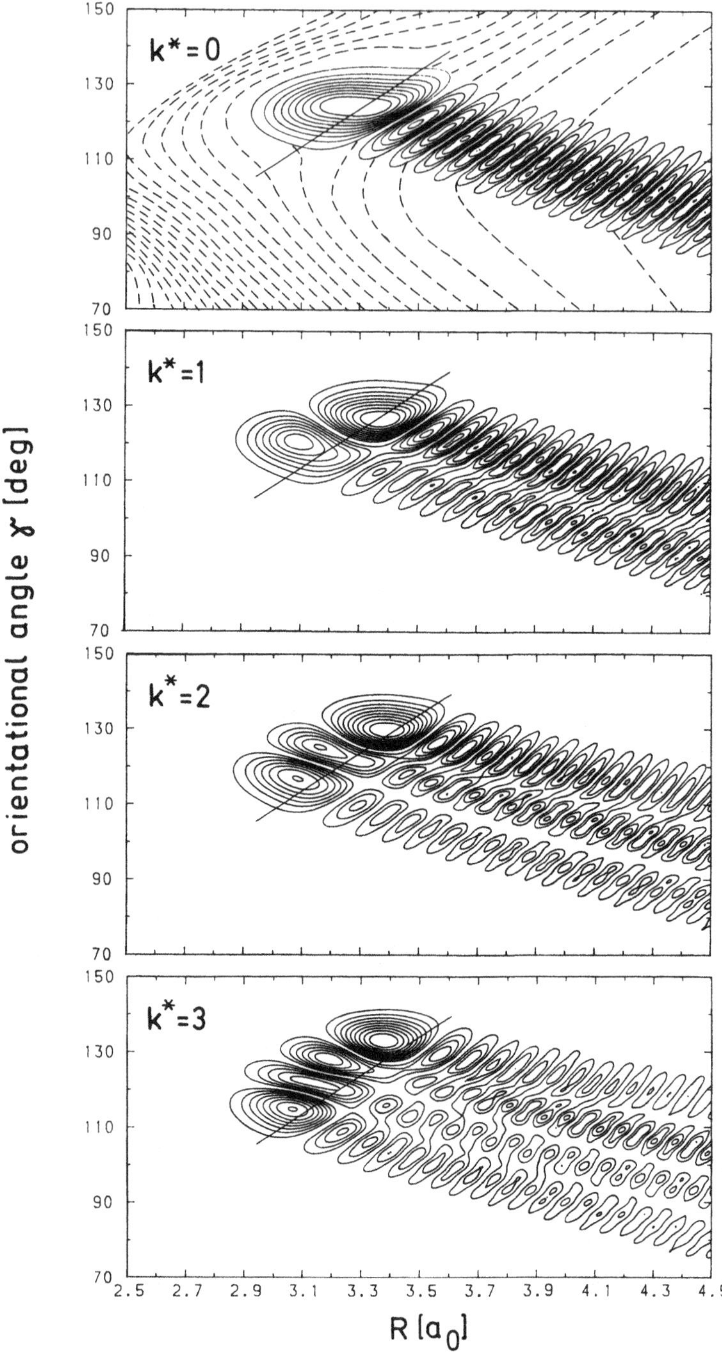

Fig. 10.13. Contour plots of the stationary total wavefunctions $\Psi_{tot}(R, \gamma)$ for the dissociation of FNO in the S_1 electronic state. R is the distance between F (*cont.*)

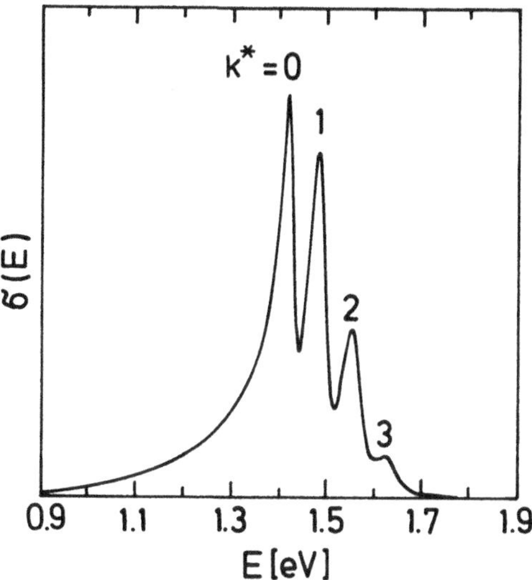

Fig. 10.14. Absorption spectrum for the two-dimensional model of the photodissociation of FNO via the S_1 state. E is the energy in the excited electronic state relative to F+NO(r_e). The resonances are assigned to excitation of the bending degree of freedom with quantum numbers $k^* = 0, \ldots, 3$ (see Figure 10.13). Adapted from Ogai et al. (1992).

Cuts of the stationary wavefunctions along the transition line, shown on the left-hand side of Figure 10.15 and denoted by $\Psi_{ts}(\gamma_0)$, clearly manifest that:

- The final rotational state distributions *qualitatively* reflect the shape of the resonance wavefunctions along the transition line, i.e., along a line which is roughly perpendicular to the minimum energy path.

Within the classical approach one proceeds in a way analogous to direct photodissociation:

1) Construct the rotational excitation function $J(\gamma_0)$ by starting classical trajectories *on the transition line* with initial angle γ_0.

Fig. 10.13. (*cont.*) and the center-of-mass of NO and γ is the orientation angle of NO with respect to the scattering vector $\mathbf{R}$. The energies correspond to the four resonance peaks ($k^* = 0, \ldots, 3$) in the absorption spectrum depicted in Figure 10.14. The straight lines indicate the transition line and the dashed contours in the upper part represent the two-dimensional PES. Reproduced from Ogai et al. (1992).

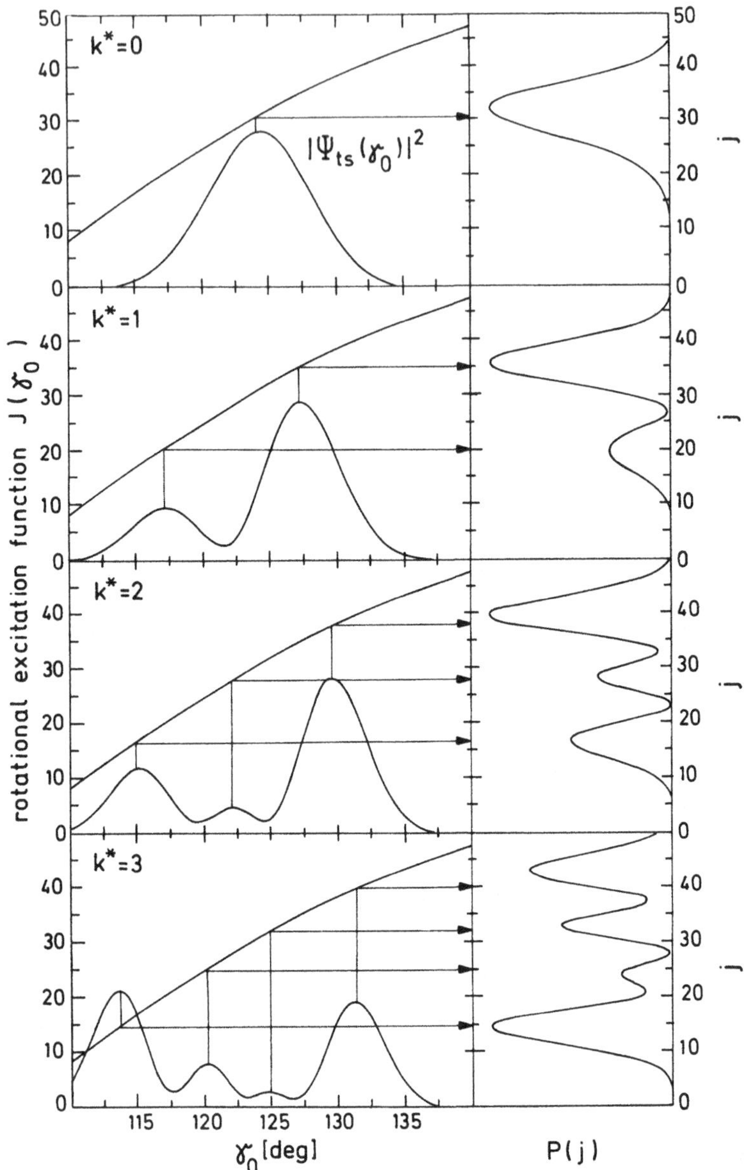

Fig. 10.15. Left-hand side: Rotational excitation function $J(\gamma_0)$ in the dissociation of FNO as a function of the initial bending angle γ_0. The classical trajectories used to construct $J(\gamma_0)$ are started at the transition line indicated by the straight line in Figure 10.13. Also shown are cuts along the transition line of the (stationary) resonance wavefunctions depicted in Figure 10.13. The structures reflect the nodal structures of the wavefunctions in the excited bending states. Right-hand side: The corresponding rotational state distributions following excitation in the various bending peaks obtained in the two-dimensional model. The arrows illustrate the rotational reflection principle. Reproduced from Ogai et al. (1992).

2) The square of the quantum mechanical wavefunction along the transition line, $|\Psi_{ts}(\gamma_0)|^2$, provides the appropriate weighting for each trajectory.
3) If the excitation function is monotonic it establishes a unique relation $\gamma_0(j)$ between the "initial" angle and the final angular momentum state.

The excitation function together with the transition-state wavefunctions for FNO are shown on the left-hand side of Figure 10.15. This figure underlines very clearly the success of the simple picture in interpreting the quantum mechanically calculated final state distributions. The reflection principle is obvious in the light of Section 6.3 and needs no further explanation. Each maximum and each minimum in the distribution has its counterpart in the transition state wavefunction.

In total analogy to the discussion in Section 6.3 we conclude that the rotational state distribution reflects — at least qualitatively — the quantum mechanical coordinate distribution at the transition state, before the wavepacket leaks out into the exit channel (Schinke, Untch, Suter, and Huber 1991; Vegiri, Untch, and Schinke 1992). A purely classical calculation with trajectories starting in the FC region rather than at the transition state inherently cannot reproduce the structures of the product state distributions which ultimately reflect the nodal behavior of the quantum mechanical wavefunctions. It should be clear from the above analysis that a statistical approach is inappropriate if the complex dissociates through a relatively narrow transition state with strong exit channel coupling, irrespective of how long the complex survives. By the same token, the final distribution is to a large extent independent of the structure of the parent molecule and its internal excitation in the *ground* electronic state (Schinke, Untch, Suter, and Huber 1991). As a consequence of the trapping within the intermediate well the memory on the initial state is erased by the time the molecule reaches the transition line.

The reflection-type structures found in the two-dimensional model calculations have been observed in reality. The true spectrum of FNO in the first absorption band consists of a major progression, corresponding to excitation of the NO stretching mode with the bending mode in its ground state $k^* = 0$, and a second less pronounced progression, built on the first excited bending state $k^* = 1$ (Suter, Huber, von Dirke, Untch, and Schinke 1992). Due to spectral overlap of the vibrational bands the higher members of the bending progression are not resolved. Excitation in the $k^* = 0$ bands leads to unimodal rotational state distributions while excitation in the $k^* = 1$ resonances yields bimodal distributions of the same kind as shown in Figure 10.15. Three-dimensional calculations reproduce the measured distributions satisfactorily except for slight details concerning the variation of the $k^* = 1$ distributions with the NO stretching quantum number n^* (Ogai et al. 1992).

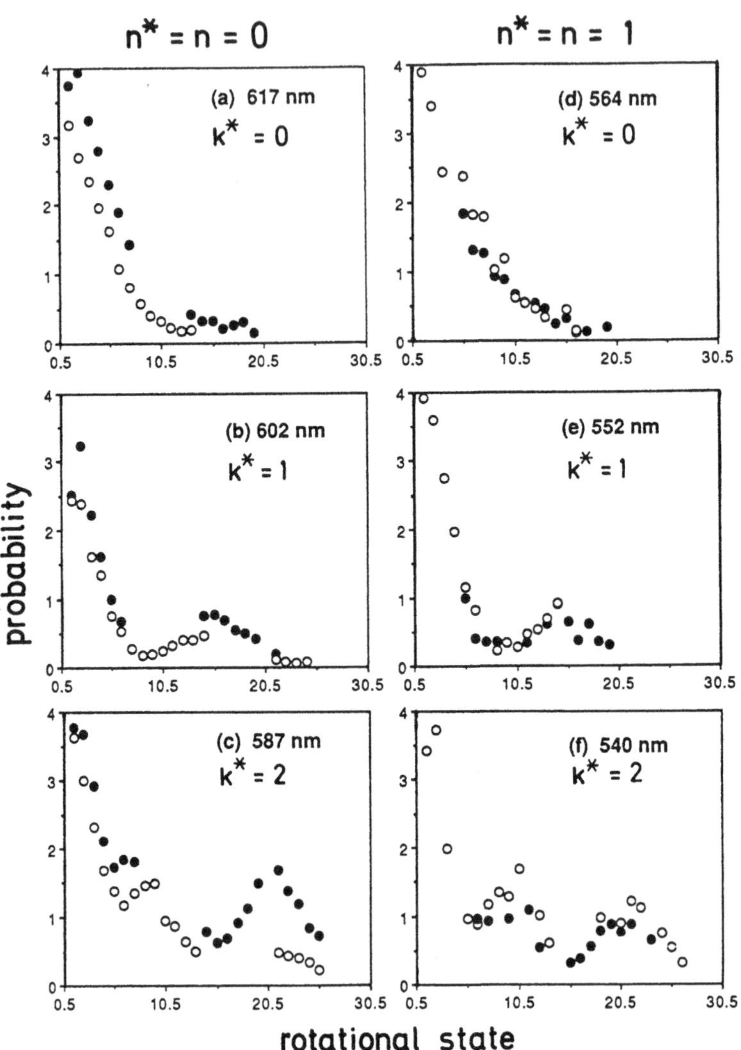

Fig. 10.16. Final rotational state distributions of NO following the dissociation of ClNO through the T_1 state. The quantum numbers n^* and k^* specify the initial vibrational and bending excitation of the ClNO(T_1) complex. The undulations for the excited bending states reflect the nodal structures of the dissociation wavefunction at the transition state. The open and the filled circles indicate different P and Q branches. The corresponding absorption spectrum is depicted in Figure 7.14. Adapted from Qian, Ogai, Iwata, and Reisler (1990).

The photodissociation of ClNO via the T_1 state is another example which intriguingly demonstrates the influence of initial bending excitation in the complex on the final rotational state distribution of the product. Figure 7.14 depicts the corresponding absorption spectrum together with selected partial photodissociation cross sections. The three narrow peaks for each vibrational band are associated with excitation of the three lowest bending states in the ClNO(T_1) complex. Figure 10.16 depicts the measured distributions following excitation in the three bending resonances. In accordance with Figure 10.15 the lowest bending state leads to a Gaussian-type distribution while the decay of the excited states yields bi- and trimodal distributions, respectively. Recent *ab initio* calculations by Sölter et al. (1992) reproduce the general trends and show unambiguously that the undulatory structures reflect the bending structure of the stationary wavefunctions at the transition state in the same sense as illustrated in Figure 10.15 for FNO. In contrast to FNO, however, the degree of rotational excitation is comparatively weak and therefore all three distributions peak near $j = 0$.

The photodissociation of formaldehyde,

$$H_2CO(S_0) + \hbar\omega \longrightarrow H_2CO(S_1) \longrightarrow H_2CO^*(S_0) \longrightarrow H_2(j_1) + CO(j_2),$$

exemplifies in a very clear way the decay through a narrow transition state. The photon first excites the S_1 electronic state which rapidly undergoes internal conversion (nonadiabatic transition) to a highly excited vibrational level in the electronic ground state S_0. The excited complex subsequently dissociates through a very narrow transition state yielding the fragments H_2 and CO (Moore and Weisshaar 1983). The final rotational state distributions of both products, H_2 and CO, are determined by the forces in the steep exit channel (Bamford et al. 1985; Debarre et al. 1985; Schinke 1985, 1986b; Chang, Minichino, and Miller 1992). The floppy motion *before* the transition state is not pertinent. Since the total energy is in the range of the barrier energy it is plausible to assume that the quantum mechanical distributions of the two orientation angles are Gaussian-like and therefore it comes as no surprise that the measured rotational state distributions have a bell shape, too.

Let us summarize this section. In the dissociation of long-lived resonance states through a relatively narrow bottleneck the final rotational state distributions of the products reflect the quantum mechanical wavefunctions at the transition state rather than the wavefunction of the parent molecule. If the fragmentation originates from the lowest bending state in the complex, which is the normal case, the final distribution is unimodal. The photodissociations of CH_3ONO (Brühlmann and Huber 1988) and HONO (Dixon and Rieley 1989) through their S_1 states are illustrative examples. However, if the dissociation starts from excited bending states, we expect structured rotational state distributions.

10.3.2 Statistical limit

If the transition state is very broad and if there is no appreciable inelastic interaction in the exit channel, one commonly assumes that the final product distributions follow essentially statistical laws [see, for example, Hase (1976), Pechukas (1976), Quack and Troe (1981), Wardlaw and Marcus (1988), Gilbert and Smith 1990, and Troe (1992)]. Energy and angular momentum conservation impose the only constraints whereas dynamical effects are assumed to be negligible. The most widely used statistical theories are the *Rice-Rampsperger-Kassel-Marcus theory* (RRKM; Forst 1973:ch.2; Smith 1980:ch.4; Hirst 1990:ch.6), the *phase space theory* (PST; Pechukas and Light 1965; Pechukas, Light, and Rankin 1966) and the *statistical adiabatic channel model* (SACM; Quack and Troe 1974, 1975; Troe 1988; Hirst 1990:ch.6). The *separate statistical ensemble method* (SSE; Nadler, Noble, Reisler, and Wittig 1985; Wittig et al. 1985) built on the PST restricts the free energy flow between some of the involved modes, vibration and rotation, for example. The so-called *prior distribution* takes into account only energy conservation while no constraint on the angular momentum states is imposed (Levine and Bernstein 1987:ch.5).

The photodissociation of nitroso compounds XNO, with X = H, CN, and CF_3, for example, demonstrates the implications for the final state distributions of the photofragments (Reisler, Noble, and Wittig 1987; Reisler and Wittig 1992). Figure 10.17 illustrates the general excitation and fragmentation scheme. The photon excites the parent molecule from the S_0 to the S_1 state. Since the corresponding PES has a high and broad barrier, dissociation is extremely slow in the S_1 state. Internal conversion, i.e., a radiationless transition from state S_1 to state S_0, is much faster and thus generates a highly excited vibrational level in the electronic ground state where ultimately the bond cleavage takes place.

The essential assumption of all statistical theories is that during the long lifetime the total available energy is uniformly distributed among

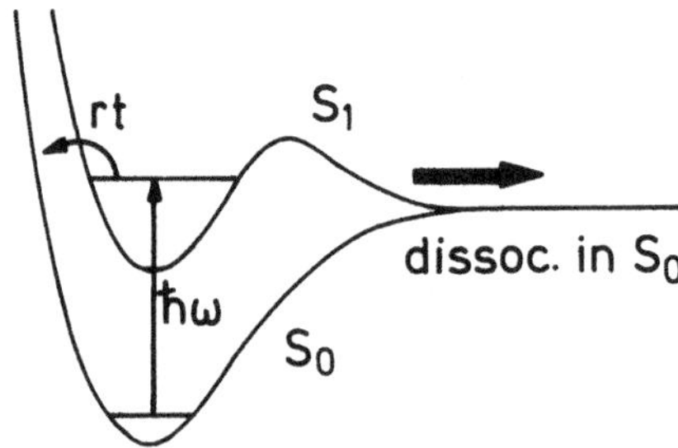

Fig. 10.17. Schematic illustration of the unimolecular dissociation of nitroso compounds. "rt" stands for *radiationless transition*.

all degrees of freedom before the complex breaks apart. As a result, all accessible quantum states in the complex, as long as they obey energy and angular momentum conservation, are uniformly populated. If inelastic forces in the exit channel are negligibly weak, the state distributions of the photofragments retain their statistical behavior. Calculating final state distributions in the statistical limit thus basically amounts to counting of states rather than solving the quantum mechanical or classical equations of motion.

The most thoroughly studied molecule in this respect is NCNO which has the advantage that the state distributions of both products, CN and NO, can be routinely determined by laser-induced fluorescence (Qian et al. 1985; Buelow et al. 1986). Figure 10.18 depicts the rotational distributions of $CN(X^2\Sigma^+)$ and $NO(X^2\Pi)$ for three photolysis wavelengths. All distributions are rather symmetric with the maximum shifting gradually to higher states as the available energy increases. The agreement with the statistical predictions is indeed remarkable. No information about the multi-dimensional PES (which is not known) enters the statistical calculations! The photodissociations of CF_3NO (Bower, Jones, and Houston 1983; Dyet, McCoustra, and Pfab 1988), C_2N_2 (Wannenmacher, Lin, and Jackson 1990), and H_2O_2 (Rizzo, Hayden, and Crim 1984; Brouwer et al. 1987; Crim 1987; Luo and Rizzo 1990, 1991), to name only a few, provide further examples of statistical distributions following dissociation in the electronic ground state.

It is worthwhile, however, pointing out that the existence of a long-lived intermediate state and the absence of a barrier in the exit channel do not necessarily imply statistical product state distributions. The fragment distributions in the dissociation of weakly bound van der Waals molecules are usually neither thermal nor statistical, despite the extremely long lifetime of the complex. We will come back to this in Chapter 12.

10.4 The impulsive model

As a consequence of the notorious lack of information on potential energy surfaces in excited electronic states, especially for larger polyatomic molecules, the so-called *impulsive model* is frequently employed to interpret final rotational state distributions (Holdy, Klotz, and Wilson 1970; Busch and Wilson 1972a; Tuck 1977). Within the impulsive model one assumes that the bond between the fragment atom A and its nearest neighbor B breaks *instantaneously* with C merely playing the role of a spectator (Figure 10.19). The sudden repulsion between A and B generates a torque which induces rotation of BC about its center-of-mass. After the bond rupture the molecule is considered to rotate freely and secondary collisions of C with A, for example, are ignored.

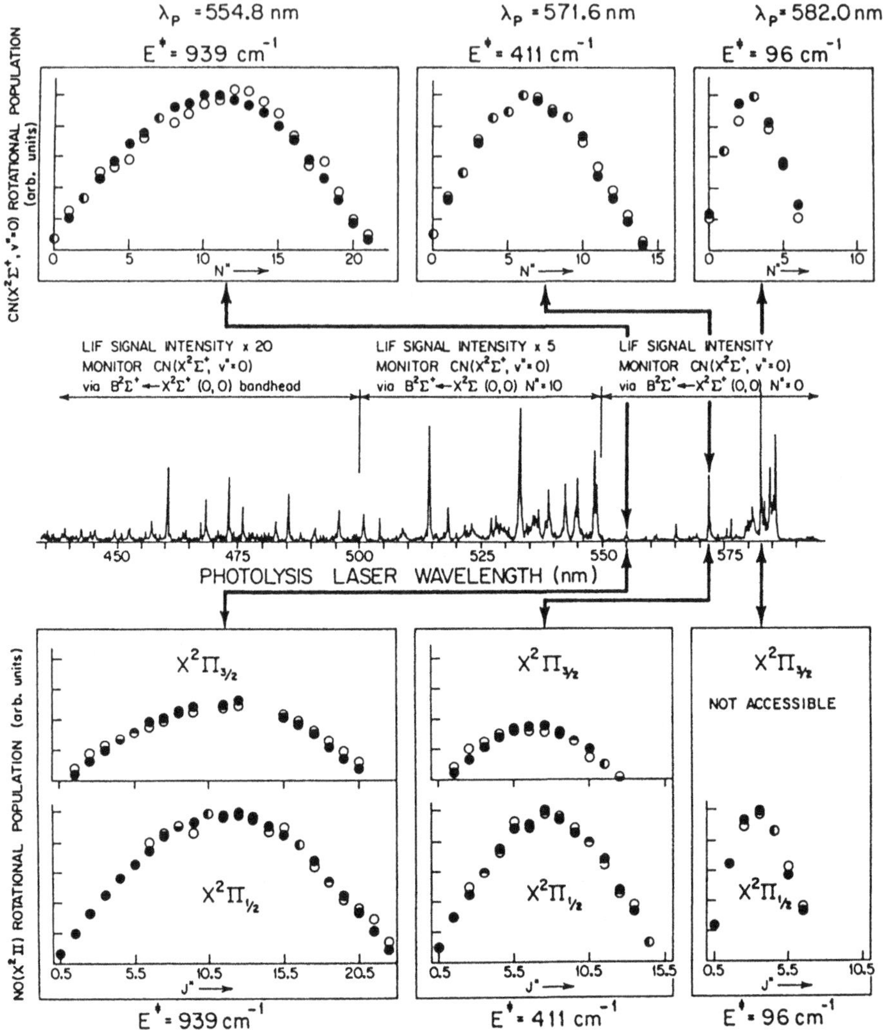

Fig. 10.18. Rotational state distributions of $CN(X^2\Sigma^+)$ (upper part) and $NO(X^2\Pi_{1/2})$ respectively $NO(X^2\Pi_{3/2})$ (lower part) following the photodissociation of NCNO at three wavelengths. The middle part depicts the absorption spectrum. Comparison of experimental data (open circles) with the predictions of the statistical theory (filled circles). Reproduced from Qian et al. (1985).

In the pure impulsive model, the total excess energy E_{excess} partitions into translational energy of both products, as well as vibrational and rotational energy of the diatomic fragment. In a modified version, Busch and Wilson (1972a) assumed that the B-C bond is infinitely stiff such that vibrational energy transfer is prohibited. Employing conservation of

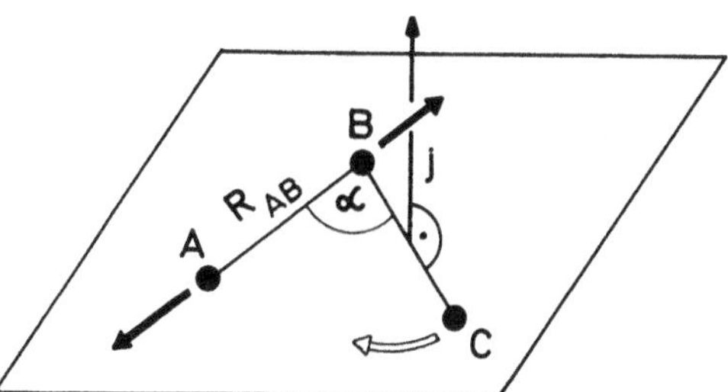

Fig. 10.19. Schematic illustration of the impulsive model for the dissociation of a triatomic molecule, ABC → A + BC(j). The heavy arrows indicate the repulsive force between atom A and its nearest neighbor, B, which generates rotation of BC about its center-of-mass.

energy and total angular momentum yields (Levene and Valentini 1987)

$$E_{rot} = E_{excess} \frac{m_C m_A \sin^2 \alpha_0}{(m_C + m_B)(m_A + m_B) - m_C m_A \cos^2 \alpha_0}, \tag{10.11}$$

for the rotational energy of the fragment, where m_A, m_B, and m_C are the atomic masses and α_0 is the initial ABC bond angle at the time when the A-B bond breaks. Using the classical relation $E_{rot} = B_{rot} j^2$ readily yields a simple correlation between the final angular momentum and the initial angle, which actually corresponds to the rotational excitation function defined in Section 5.1 and utilized in Section 6.3 to examine rotational state distributions. If α_0 is chosen to be the equilibrium angle in the electronic ground state, Equation (10.11) defines the most probable final angular momentum state of the fragment. Weighting each angle according to the distribution of bending angles of the parent molecule gives a distribution of final rotational states (Levene and Valentini 1987), in essentially the same way as outlined in Section 6.3 [rotational reflection principle, see Equation (6.25)].

Because of its simplicity the impulsive model is very appealing and frequently employed to model measured rotational state distributions (Dugan and Anthony 1987; Levene and Valentini 1987; Butenhoff, Carleton, and Moore 1990). In most applications, however, it is necessary to incorporate at least one fit parameter or some dynamical constraints in order to obtain agreement with experimental results, for example, the "equilibrium angle" in the excited electronic state or the point at which the repulsive force vector intersects the BC-axis. The impulsive model is *not* an *a priori* theory.

The main drawback of the impulsive model is the neglect of the α dependence of the interaction potential (Schinke 1989a). When the fragments separate the diatomic molecule starts to rotate and at the same time α decreases as indicated in Figure 10.19. Each PES shows some angular dependence which generates a torque and this additional torque can either slow down or accelerate the rotation of the diatom. As we pointed out at the beginning of this chapter, the PES may be expressed in terms of Jacobi coordinates R and γ or alternatively, in terms of the bond distance R_{AB} and the bond angle α. However, whichever coordinates we prefer to represent the potential, the torque $F_\gamma = -\partial V/\partial\gamma$ alone controls the rate of change of the molecular angular momentum [see Equation (5.4f)]. If the PES is expressed in terms of $(R_{\mathrm{AB}}, \alpha)$, F_γ is given by

$$F_\gamma = -\frac{\partial V}{\partial R_{\mathrm{AB}}}\frac{\partial R_{\mathrm{AB}}}{\partial\gamma} - \frac{\partial V}{\partial\alpha}\frac{\partial\alpha}{\partial\gamma} = F_\gamma^{(1)} + F_\gamma^{(2)}, \tag{10.12}$$

where the first term, $F_\gamma^{(1)}$, manifests the strong repulsion between A and B and the second term, $F_\gamma^{(2)}$, reflects the α dependence of the potential. The impulsive model incorporates only the first part of the torque and completely ignores the second one. A realistic description of rotational excitation in photodissociation, however, must take into account the complete torque.

Depending on the sign, $F_\gamma^{(2)}$ can either reduce or increase the angular momentum gained as a result of the action of $F_\gamma^{(1)}$. The photodissociation of ClNO via the S_1 state is an example for the first alternative. Figure 10.2(a) shows the corresponding PES in terms of bond coordinates. The equilibrium angle in the ground state is $\alpha_0 \approx 115°$ and therefore the Cl-N repulsion generates a large angular momentum. However, as the bond breaks NO revolves towards smaller angles, up the repulsive wall, which hinders its rotation. This has the effect that the net rotational excitation is significantly smaller than predicted by the impulsive model. In Figure 6.7 we compare the corresponding excitation functions and the resulting final state distributions. While the impulsive model yields the maximum at about $j \approx 40$, the calculation using the full PES as well as the experiment yields the maximum of the final state probability at $j \approx 30$. Of course, by changing α_0 one could easily fit the measured distribution.

The photodissociation of ClCN represents an example for the second alternative. ClCN is linear in the electronic ground state, $\alpha_0 \approx 0°$, and therefore the repulsion between Cl and C generates only relatively little rotational excitation (see Figure 10.20). The measured as well as the theoretical distributions, the latter being calculated using an *ab initio* PES, however, peak at very high rotational states, in striking disagreement with the prediction of the impulsive model. In this case, the high degree of rotational excitation originates mainly from $F_\gamma^{(2)}$ rather than $F_\gamma^{(1)}$. One must bear in mind that ClCN should be an ideal candidate for

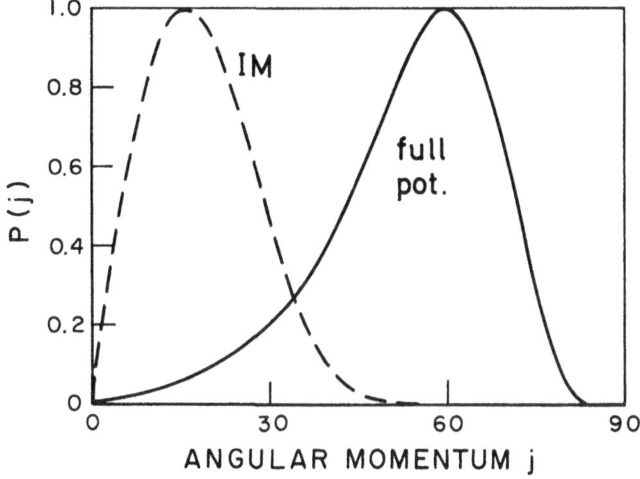

Fig. 10.20. Rotational state distribution of CN following the photodissociation of ClCN at 191.5 nm. Comparison between exact close-coupling calculations using the full *ab initio* PES of Waite and Dunlap (1986) (solid curve) and the impulsive model (IM, dashed curve). Adapted from Schinke (1990).

the impulsive model because the dissociation is direct and fast and the interaction between Cl and N should be negligibly small! In conclusion:

- The impulsive model is only applicable if the dissociative PES depends so weakly on the bond angle that the torque $-\partial V/\partial \alpha$ can be safely neglected.
- This presumption, however, is unrealistic; all potentials which we have analyzed on the basis of reliable *ab initio* calculations actually have a significant bond-angle dependence which must not be ignored (Schinke 1989a).

10.5 Thermal broadening of rotational state distributions

As discussed in the introduction to this chapter the final rotation of the photofragment reflects three sources: overall rotation of the parent molecule, which is preserved in the electronic excitation process, internal bending or torsional motion in the electronic ground state, and final state interaction. The latter two sources have been analyzed in Sections 10.1–10.4. The influence of overall rotation will be discussed in this section. Its relative contribution increases gradually with the magnitude of the total angular momentum **J** and therefore it increases with the temperature of the molecular sample. As we will demonstrate below, the conversion of overall rotation of the parent molecule in the electronic ground state into rotation of the fragment merely leads to a less interesting broadening effect which does not reveal much information about the actual dissociation dynamics.

10.5.1 *The photodissociation of* H_2O_2

The photodissociation of H_2O_2 instructively demonstrates the influence of thermal averaging on final angular momentum distributions (Schinke 1988c). For simplicity of the presentation we assume that the two OH radicals rotate like cartwheels in planes *perpendicular* to the O—O axis as illustrated in Figure 10.21. The corresponding angular momentum vectors $\mathbf{j}_1$ and $\mathbf{j}_2$ point both along the O—O axis. Let us further assume that before the fragmentation (index i) the two hydrogen atoms spin about the axis in the same direction and with the same angular momentum, i.e., $\mathbf{j}_{i,1} = \mathbf{j}_{i,2} = \mathbf{J}/2$, where $\mathbf{J} = \mathbf{j}_1 + \mathbf{j}_2$ represents the total angular momentum about the O—O axis. In doing so we neglect internal torsional motion in the parent molecule before the photoexcitation. In order to keep the discussion as simple as possible, rotations about other axes are ignored.

When the O-O bond breaks, the angular dependence of the upper-state PES generates a torque $-\partial V/\partial\varphi$ as discussed in Section 10.2.1. In analogy with (5.4f), the corresponding Hamilton equations for the evolution of $\mathbf{j}_1$ and $\mathbf{j}_2$ are

$$\frac{d\mathbf{j}_1}{dt} = -\frac{\partial V}{\partial\varphi} \quad , \quad \frac{d\mathbf{j}_2}{dt} = +\frac{\partial V}{\partial\varphi}, \tag{10.13}$$

where $\varphi = \varphi_1 - \varphi_2$ is the torsional angle with φ_1 and φ_2 being the azimuthal angles of the two rotamers. The integration of (10.13) yields for the final (index f) angular momenta

$$\mathbf{j}_{f,1} = \mathbf{j}_{i,1} + \jmath = \mathbf{J}/2 + \jmath \quad , \quad \mathbf{j}_{f,2} = \mathbf{j}_{i,2} - \jmath = \mathbf{J}/2 - \jmath, \tag{10.14}$$

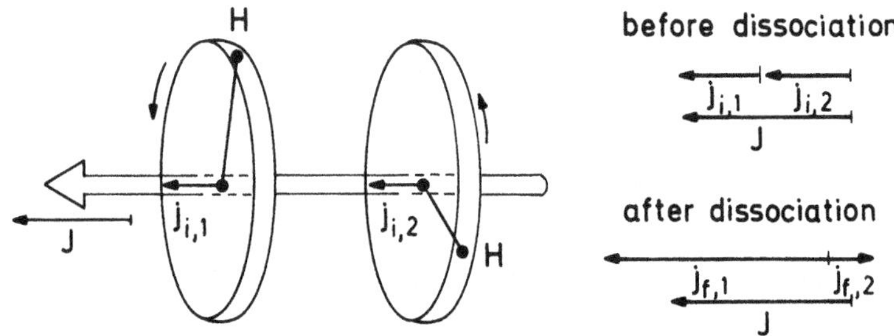

Fig. 10.21. Schematic illustration of the angular momentum vectors in the photodissociation of H_2O_2. The left-hand side depicts the situation before the dissociation. The right-hand side shows the appropriate vector diagrams before and after the dissociation.

where $\bar{\jmath} = -\int_0^\infty dt\ \partial V/\partial\varphi$. The total angular momentum is preserved, i.e., $\mathbf{J} = \mathbf{j}_{i,1} + \mathbf{j}_{i,2} = \mathbf{j}_{f,1} + \mathbf{j}_{f,2}$. The torque caused by the angle dependence of the dissociative PES *accelerates* the initial rotation of one rotor but it *decelerates* the motion of the other one. Consequently, the final rotational distribution of one of the OH fragments is shifted to larger j values, compared to the distribution for $\mathbf{J} = 0$, while the distribution of the other rotor is shifted to smaller quantum numbers. The shift amounts to approximately $\pm\mathbf{J}/2$.

In the macroscopic picture we have to average over $\mathbf{J}$, respectively $\mathbf{j}_{i,1}$ and $\mathbf{j}_{i,2}$, according to the Boltzmann distribution

$$P_T(j_{i,1}, j_{i,2}) = \mathrm{e}^{-B_{rot}\, j_{i,1}^2/k_B T}\ \mathrm{e}^{-B_{rot}\, j_{i,2}^2/k_B T}, \tag{10.15}$$

with $j_{i,1} = j_{i,2}$; B_{rot}, T, and k_B are the rotational constant of OH, the temperature of H_2O_2, and Boltzmann's constant, respectively, and the sign of $\mathbf{J}$ is uniformly distributed. The Boltzmann averaging can be conveniently incorporated into the Monte Carlo procedure for selecting the initial conditions of the classical trajectories in the upper electronic states. It leads to the temperature dependence of the final state distribution shown in Figure 10.10. As T increases, the distribution becomes continuously broader, but interestingly the peak does not move at all. The latter finding reflects that, irrespective of the temperature, $\mathbf{J} = 0$ always has the largest weight and that the two rotamers are equivalent.

10.5.2 Other examples

A rigorous modelling of thermal broadening is — in practice — quite cumbersome and tedious. Let us consider a general asymmetric top molecule such as H_2O, for example. Each total angular momentum state, specified by the quantum number J, splits into $(2J+1)$ *nondegenerate* substates with energies $E_{rot}^{J,K}$ $(K = 1, \ldots, 2J+1)$. Every one of these $(2J+1)$ rotational states corresponds to a different type of rotational motion and is described by a distinct rotational wavefunction $\Psi_{rot}^{J,K}$ (see Section 11.3). Under thermal conditions, the Boltzmann distribution

$$P_T(J, K) \propto (2J+1)\ \mathrm{e}^{-E_{rot}^{J,K}/k_B T} \tag{10.16}$$

governs their population with the factor $(2J+1)$ accounting for the $(2J+1)$ *degenerate* magnetic sublevels for each state (J, K). The photodissociation of each of the $(2J+1)$ nondegenerate rotational states represents an independent event leading, in principle, to a *unique* final rotational state distribution. The final distribution reflects the initial rotational wavefunction $\Psi_{rot}^{J,K}$, the bending wavefunction, and the exit channel dynamics. More of this follows in the next chapter.

If we perform the dissociation in the bulk with temperature T we have to average the final state distribution over all initially populated rotational states of the parent molecule with relative weights (10.16). For H_2O at room temperature, for example, rotational states up to $J \approx 10$ or so are significantly populated which amounts to about 100(!) different nondegenerate states. A rigorous quantum mechanical treatment would require the calculation of numerous bound-state wavefunctions for the parent molecule, the calculation of the corresponding continuum wavefunctions in the upper electronic state, the evaluation of the bound-continuum overlap integrals, and finally the averaging over all initial states. This represents a tremendous amount of work of high numerical complexity. The information gain about the dissociation mechanism, however, is at best modest.

The influence of thermal broadening depends very much on the particular system, especially on the strength of rotational excitation in the exit channel induced by the anisotropy of the dissociative PES. Let us consider the weak and the strong coupling cases separately. If the torque in the exit channel is very weak, the final rotational state distribution of the product essentially reflects the bending motion of the parent molecule, as elucidated in Section 10.1, and its overall rotation.[†] The consequence is a relatively strong temperature dependence: as T increases the photon excites higher and higher rotational states and the increased overall rotation shows up in the rotational distribution of the product.

The photodissociation of H_2O via the first continuum illustrates the intensification of OH rotation when the temperature of the H_2O sample rises (Figure 10.22). The distribution obtained at room temperature is noticeably broader and shifted to slightly higher quantum numbers relative to the distribution measured in the beam. While the beam distribution reflects essentially the zero-point bending motion of $H_2O(\tilde{X})$, the 300 K distribution manifests bending vibration *and* overall rotation. Plotting the two distributions in a Boltzmann representation yields OH "temperatures" of about 350 K and 900 K for $T_{H_2O} \approx 50$ K (beam) and 300 K, respectively. In terms of T_{OH}, the difference between the beam and the bulk appears much more dramatic than it actually is! Levene and Valentini (1987) investigated the effect of thermal broadening in the

---

[†] Morse, Freed, and Band (1979), Morse and Freed (1981, 1983), and Beswick and Gelbart (1980) investigated for several cases the influence of parent rotation on final rotational state distributions within the elastic limit. The neglect of translational-rotational coupling significantly simplifies the study of thermal effects because only Franck-Condon factors of the general type (10.3a), including, however, overall rotation, are required.

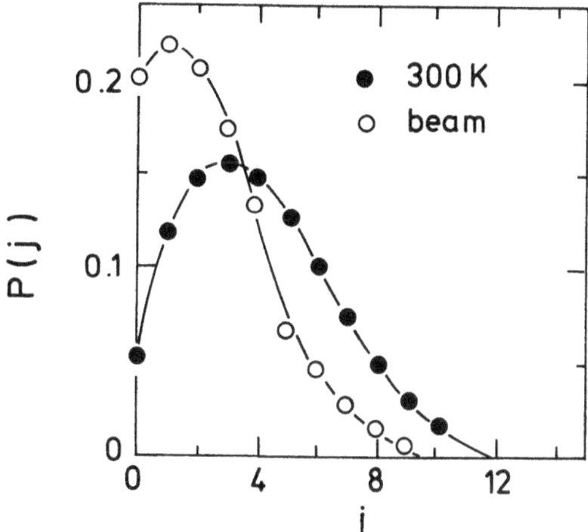

Fig. 10.22. Rotational state distributions of OH following the dissociation of H_2O through the $\tilde{A}$ state in a molecular beam and at room temperature. The distributions are averaged over the two spin states. The data are taken from Andresen et al. (1984).

photolysis of H_2O by describing the parent rotation classically and the fragmentation in the excited state within the impulsive model.†

If, on the other hand, strong rotational excitation in the exit channel determines the final rotation of the fragment, the motion of the parent molecule in the electronic ground state fails to keep its importance with the result that thermal broadening is insignificant: irrespective of how fast the parent molecule rotates before the photoexcitation, the torque induced by the anisotropy of the dissociative PES is the dominant factor and solely determines the final rotation of the product as discussed in Sections 6.3 and 10.2. The photodissociation of CH_3ONO via the S_1 state provides a clear example (Figure 10.23). The rotational distributions of NO obtained in a molecular beam and in the 300 K bulk differ only marginally; the latter distribution is broader by about one or two rotational quanta, but the general form is identical. Quite similar results have been observed for several other molecules. Incidentally we note that Figure 10.23 clearly justifies our previous restriction to study the dissociation of rotationless molecules, $J = 0$, in order to assess the influence of final state interaction.

Let us summarize this section:

† The impulsive model is applicable in the case of $H_2O\tilde{A}$ because the dependence of the PES on the HOH bending angle is indeed very weak throughout the bond fission.

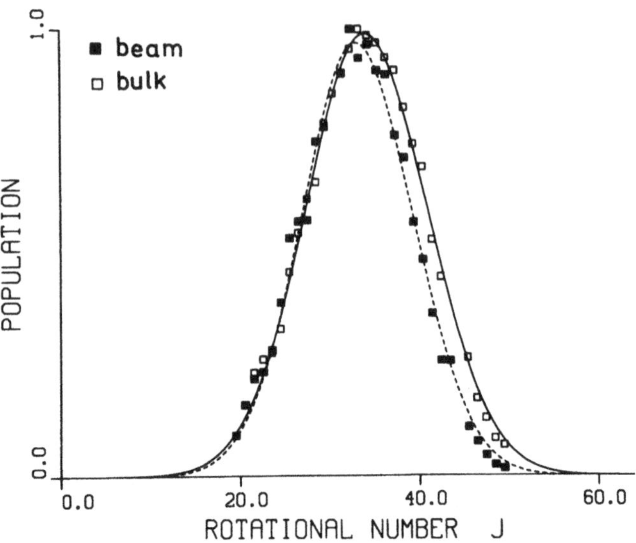

Fig. 10.23. Rotational state distributions of NO following the photodissociation of CH_3ONO through the S_1 state in the molecular beam and in the 300 K bulk. Adapted from Brühlmann and Huber (private communication).

- A very small torque in the exit channel implies a comparatively large effect of overall rotation of the parent molecule on final rotational state distributions.
- Strong final state interaction, on the other hand, leads to a relatively small impact of initial rotation on the final rotation of the fragment.

Cases for which both parent rotation as well as translational-rotational coupling have a more balanced weight are certainly more difficult to analyze.

11
Rotational excitation II

Rotational excitation of photofragments is a wide field with many subtleties. In the foregoing chapters we have considered exclusively the scalar properties of rotational excitation, i.e., the distributions of final rotational states of the products and the forces that control them. For this purpose, it was sufficient to study the case that the total angular momentum of the entire molecular system is zero, $\mathbf{J} = 0$. This restriction drastically facilitated the theoretical formulation and allowed us to concentrate on the main effects without being intimidated by complicated angular momentum coupling. In Section 11.1 we will extend the theory of rotational excitation to general total angular momentum states $\mathbf{J} \neq 0$. Our aim is the investigation of final rotational product states following the photodissociation of *single* rotational states of the parent molecule (Section 11.3). Before doing so, however, we discuss in Section 11.2 the distribution of the various electronic fine-structure states (Λ-doublet states) if the fragment possesses a nonzero electronic angular momentum which couples with the angular momentum of the nuclear motion. Important examples are OH and NO.

The vector of the electromagnetic field defines a well specified direction in the laboratory frame relative to which all other vectors relevant in photodissociation can be measured. This includes the transition dipole moment, $\boldsymbol{\mu}$, the recoil velocity of the fragments, $\mathbf{v}$, and the angular momentum vector of the products, $\mathbf{j}$. Vector correlations in photodissociation contain a wealth of information about the symmetry of the excited electronic state as well as the dynamics of the fragmentation. Section 11.4 gives a short introduction. Finally, we elucidate in Section 11.5 the correlation between the rotational excitation of the products if the parent molecule breaks up into two diatomic fragments.

11.1 General theory of rotational excitation for $\mathbf{J} \neq 0$

In this section we extend the theory of photodissociation and rotational excitation outlined in Section 3.2 for $\mathbf{J} = 0$ to general angular momentum states $\mathbf{J} \neq 0$ of a triatomic system ABC. We will closely follow the detailed presentation of Balint-Kurti and Shapiro (1981) [see also Hutson (1991), Glass-Maujean and Beswick (1989), Beswick (1991), and Roncero et al. (1990)]. The discussion in this section is not meant to be a substitute for reading the original literature; we merely want to outline the general methodology and underline the complexity of the theory.

11.1.1 Hamiltonian, expansion functions, and coupled equations

In the center-of-mass system, the triatomic molecule is described by the two vectors $\mathbf{R}$ and $\mathbf{r}$ as illustrated in Figure 11.1. $\mathbf{r}$ represents the internuclear vector of the fragment BC and $\mathbf{R}$ points from the center-of-mass of BC to the recoiling atom A. In accordance with the previous chapters we assume that ABC breaks up into BC and A, i.e., that $\mathbf{R}$ is the dissociation vector. The *space-fixed coordinate system* $\{xyz\}$ is chosen such that the z-axis is parallel to the electric field vector $\mathbf{E}_0$. The scattering vector $\mathbf{R}$ is described by its length R and the two polar angles $\hat{\mathbf{R}} \equiv (\theta_R, \varphi_R)$ with respect to the space-fixed coordinate system. Instead of describing the internal vector $\mathbf{r}$ in terms of the corresponding angles θ_r and φ_r, defined in the space-fixed system, it is more convenient to use the polar angles $\hat{\mathbf{r}} \equiv (\gamma, \psi)$ defined with respect to the *body-fixed coordinate system* $\{x'y'z'\}$ in which the z'-axis is *always* parallel to the internuclear vector $\mathbf{R}$. During the dissociation, the body-fixed coordinate system rotates with the vector $\mathbf{R}$. The angle γ is defined by $\cos\gamma = \mathbf{R} \cdot \mathbf{r}/Rr$ and ψ is the azimuthal angle of $\mathbf{r}$ in the (x', y')-plane. The advantage of the body-fixed coordinate system rests upon the fact that the interaction potential between A and BC depends, in addition to R and r, only on the internal orientation angle γ but not on the other angles ψ, θ_R, and φ_R.[†]

The total Hamiltonian including all nuclear degrees of freedom is given in the center-of-mass system by

$$\hat{H}(\mathbf{R}, \mathbf{r}) = -\frac{\hbar^2}{2m}\nabla_{\mathbf{R}}^2 - \frac{\hbar^2}{2\mu}\nabla_{\mathbf{r}}^2 + V(\mathbf{R}, \mathbf{r}), \tag{11.1}$$

where m and μ are the reduced masses defined in (2.38). The kinetic energy operators can be rewritten as (Cohen-Tannoudji, Diu, and Laloë

[†] The body-fixed coordinate system has been amply used in the past to study rotational excitation in full collisions. See, for example, Jacob and Wick (1959), Pack (1974), McGuire and Kouri (1974), Child (1974:ch.6), Rabitz (1976), Lester (1976), Launay (1976), and Gianturco (1979:ch.3).

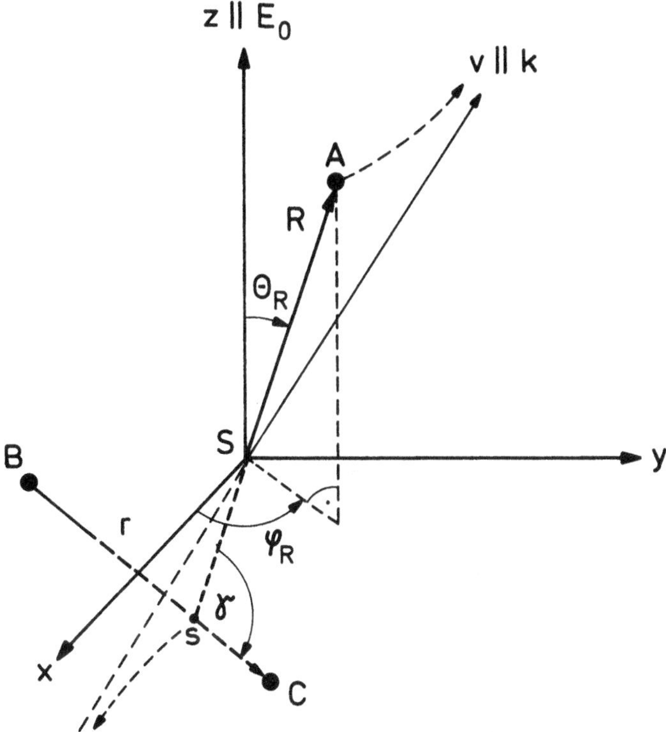

Fig. 11.1. Center-of-mass or Jacobi coordinates **R** and **r** used to describe the fragmentation of a triatomic molecule ABC into A and BC. S and s are the centers-of-mass of ABC and BC, respectively. **v** is the relative velocity of the recoiling fragments in the center-of-mass system. The space-fixed z-axis is parallel to the vector $\mathbf{E}_0$ of the electric field, while the body-fixed z'-axis is parallel to the scattering vector **R** at all times. The azimuthal angle ψ, which is not indicated in the figure, describes rotation in the plane perpendicular to **R**.

1977:ch.7)

$$-\frac{\hbar^2}{2m}\nabla^2_{\mathbf{R}} = -\frac{\hbar^2}{2m}\frac{1}{R}\frac{\partial^2}{\partial R^2}R + \frac{1}{2mR^2}\,\hat{\mathbf{l}}^2 \tag{11.2}$$

and

$$-\frac{\hbar^2}{2\mu}\nabla^2_{\mathbf{r}} = -\frac{\hbar^2}{2\mu}\frac{1}{r}\frac{\partial^2}{\partial r^2}r + \frac{1}{2\mu r^2}\,\hat{\mathbf{j}}^2, \tag{11.3}$$

where $\hat{\mathbf{l}}$ is the angular momentum operator of the rotation of A with respect to BC and $\hat{\mathbf{j}}$ is the angular momentum operator describing the rotation of BC. In terms of the differential operators with respect to the angle coordinates γ and ψ, $\hat{\mathbf{j}}^2$ is given by

$$\hat{\mathbf{j}}^2 = -\hbar^2 \left(\frac{\partial^2}{\partial \gamma^2} + \frac{1}{\tan\gamma} \frac{\partial}{\partial \gamma} + \frac{1}{\sin^2\gamma} \frac{\partial^2}{\partial \psi^2} \right);$$

a similar equation holds for $\hat{\mathbf{l}}^2$. Writing the total wavefunction Ψ as $\tilde{\Psi}/Rr$ yields the following Hamiltonian for $\tilde{\Psi}$,

$$\hat{H}(\mathbf{R}, \mathbf{r}) = -\frac{\hbar^2}{2m} \frac{\partial^2}{\partial R^2} + \frac{1}{2mR^2} \hat{\mathbf{l}}^2 - \frac{\hbar^2}{2\mu} \frac{\partial^2}{\partial r^2} + \frac{1}{2\mu r^2} \hat{\mathbf{j}}^2 + V(R, r, \gamma). \quad (11.4)$$

For simplicity of the presentation we will hereafter employ the rigid-rotor approximation in which r is fixed at the equilibrium bond distance r_e of BC.† Note that the potential energy surface (PES) V depends only on the two distances R and r and on the orientation angle γ.

The wavefunction depends on R and on the four angles θ_R, φ_R, γ, and ψ. In order to solve the Schrödinger equation, we expand $\tilde{\Psi}$ in terms of *parity adapted angular wavefunctions* of the form (Launay 1976)

$$\Theta_{j\Omega}^{JMp}(\hat{\mathbf{R}}, \hat{\mathbf{r}}) = t_\Omega \left(\frac{2J+1}{4\pi} \right)^{1/2}$$
$$\times \left\{ D_{\Omega M}^J(\varphi_R, \theta_R, 0) Y_{j\Omega}(\gamma, \psi) + p D_{-\Omega M}^J(\varphi_R, \theta_R, 0) Y_{j-\Omega}(\gamma, \psi) \right\}, \quad (11.5)$$

where $t_\Omega = 1/2$ for $\Omega = 0$ and $2^{-1/2}$ for $\Omega > 0$. The $D_{\Omega M}^J(\varphi_R, \theta_R, 0)$ are the elements of the rotation matrices in the convention of Edmonds (1974) and the $Y_{j\Omega}(\gamma, \psi)$ are spherical harmonics. The former describe the rotation of A+BC in the space-fixed system while the latter describe the rotation of BC in the body-fixed coordinate system, relative to $\mathbf{R}$. Note that:

- The angular expansion functions $\Theta_{j\Omega}^{JMp}$ are simultaneously eigenfunctions of $\hat{\mathbf{J}}^2$ and $\mathbf{J}_z$, with eigenvalues $J(J+1)$ and M, as well as eigenfunctions of $\hat{\mathbf{j}}^2$, with eigenvalues $j(j+1)$.

$\mathbf{J} = \mathbf{j} + \mathbf{l}$ is the total angular momentum of the system and M is its projection on the space-fixed z-axis. Furthermore:

- The angular expansion functions have the *parity* $(-1)^J p$ under the inversion operation; the parity parameter p has the values $p = \pm 1$.

The so-called *helicity quantum number* Ω ranges from 0 to Min(J, j) for $p = +1$ and from 1 to Min(J, j) for $p = -1$. The angular expansion functions (11.5) are orthogonal and complete.

† Including the vibrational motion associated with r as well is straightforward: the expansion in Equation (11.6) must be augmented by a summation over vibrational wavefunctions $\varphi_n(r)$. The derivation of the extended coupled equations is left as a problem for the reader.

The reduced wavefunction $\tilde{\Psi}$ is now expanded according to

$$\tilde{\Psi}(\mathbf{R}, \mathbf{r}) = \sum_{JMp} \sum_{j\Omega} \chi_{j\Omega}(R; Jp)\, \Theta_{j\Omega}^{JMp}(\theta_R, \varphi_R, \gamma, \psi). \tag{11.6}$$

Inserting (11.6) into the time-independent Schrödinger equation and utilizing the orthonormality of the expansion functions leads to the following set of coupled equations for the radial functions $\chi_{j\Omega}(R; JMp)$,[†]

$$\begin{aligned} \left[\frac{d^2}{dR^2} + k_j^2 - \frac{J(J+1) + j(j+1) - 2\Omega^2}{R^2}\right] \chi_{j\Omega}(R; Jp) \\ = \frac{2m}{\hbar^2} \sum_{j'} V_{jj'}^{(\Omega)}(R)\, \chi_{j'\Omega}(R; Jp) \\ + C_{\Omega\Omega+1}^{Jp}\, \chi_{j\Omega+1}(R; Jp) + C_{\Omega\Omega-1}^{Jp}\, \chi_{j\Omega-1}(R; Jp). \end{aligned} \tag{11.7}$$

The $C_{\Omega\Omega'}^{Jp}$ are generalized centrifugal potentials which are proportional to R^{-2}; the proportionality constants depend on J, p, j, and Ω and they are explicitly given by Balint-Kurti and Shapiro (1981). The wavenumbers k_j are defined in (3.21) and the potential coupling elements are given by

$$V_{jj'}^{(\Omega)}(R) = \int d\hat{\mathbf{R}} \int d\hat{\mathbf{r}} \left[\Theta_{j\Omega}^{JMp}(\hat{\mathbf{R}}, \hat{\mathbf{r}})\right]^* V(R, \gamma)\, \Theta_{j'\Omega}^{JMp}(\hat{\mathbf{R}}, \hat{\mathbf{r}}). \tag{11.8}$$

If we expand the potential in terms of Legendre polynomials $P_\lambda(\cos\gamma)$ according to (3.23), the four-fold angular integral can be evaluated analytically to yield

$$\begin{aligned} V_{jj'}^{(\Omega)}(R) = (-1)^\Omega \left[(2j+1)(2j'+1)\right]^{1/2} \\ \times \sum_\lambda V_\lambda(R) \begin{pmatrix} j & \lambda & j' \\ 0 & 0 & 0 \end{pmatrix} \begin{pmatrix} j & \lambda & j' \\ -\Omega & 0 & \Omega \end{pmatrix}, \end{aligned} \tag{11.9}$$

where the V_λ are the R-dependent expansion coefficients.

There are several points which must be noted:

1) The coupled equations do not depend on the magnetic quantum number M; therefore, the radial functions $\chi_{j\Omega}(R; Jp)$ need not to be labeled by M.
2) The coupled equations split up into separate blocks each belonging to a specific total angular momentum quantum number J; blocks belonging to different values of J do not couple.

[†] In deriving (11.7) one exploits that $l_{z'} = 0$ in the body-fixed system and uses the identity [see, for example, Lester (1976)]

$$\hat{\mathbf{l}}^2 = (\hat{\mathbf{J}} - \hat{\mathbf{j}})^2 = \hat{\mathbf{J}}^2 + \hat{\mathbf{j}}^2 - 2\hat{\mathbf{J}}_{z'}^2 - \hat{\mathbf{J}}_+\hat{\mathbf{j}}_- - \hat{\mathbf{J}}_-\hat{\mathbf{j}}_+,$$

where $\mathbf{J}_\pm$ and $\mathbf{j}_\pm$ are the raising and lowering operators (Cohen-Tannoudji, Diu, and Laloë 1977:ch.6).

3) Furthermore, each J block separates into two blocks belonging to different parities $(-1)^J p$ where $p = \pm 1$; wavefunctions with the same value of J but different parity parameters p do not couple.
4) The potential matrix Equation (11.9) couples states with different molecular rotational quantum numbers j but it is diagonal in the helicity quantum number Ω.
5) In contrast, the *Coriolis coupling* [i.e., the last two entries on the right-hand side of Equation (11.7)] couples states with different helicity quantum numbers Ω but it is diagonal in the rotational quantum number j.

The first point reflects the fact that the dynamics is independent of the orientation in space. Points 2 and 3 manifest that both the total angular momentum and the parity are conserved. The Coriolis coupling arises from the continuous rotation of the body-fixed system with the scattering vector. Finally we stress that for $J = \Omega = 0$ Equation (11.7) goes over into (3.20).

11.1.2 Rotational states of asymmetric top molecules

The theory outlined above can be used to calculate the exact bound-state energies and wavefunctions for any triatomic molecule and for any value J of the total angular momentum quantum number. We can solve the set of coupled equations (11.7) subject to the boundary conditions $\chi_{j\Omega}(R; Jp) \to 0$ in the limits $R \to 0$ and $R \to \infty$ (Shapiro and Balint-Kurti 1979). Alternatively we may expand the radial wavefunctions in a suitable set of one-dimensional oscillator wavefunctions $\varphi_m(R)$,

$$\chi_{j\Omega}(R; Jp) = \sum_m G_{j\Omega}^{(m)}(Jp)\, \varphi_m(R), \tag{11.10}$$

and diagonalize the resulting Hamilton matrix (Tennyson and Henderson 1989; Mladenović and Bačić 1990; Leforestier 1991). The latter approach is usually more convenient because it yields all eigenvalues for a given set of quantum numbers J and p *at once.*

Let us initially assume that the Coriolis coupling is negligibly small, i.e., that $C_{\Omega\Omega'}^{Jp} \approx 0$ for all helicities. Then, the coupled equations split up into $J+1$ *uncoupled* blocks if $p = +1$ respectively J blocks if $p = -1$. Each block of equations depends, however, *parametrically* on the helicity quantum number Ω. This leads for any total angular momentum state to $J+1$ respectively J *nondegenerate* eigenvalues depending on whether $p = +1$ or -1. Switching the Coriolis coupling on will modify the eigenvalues without, however, changing their number in each of the two parity blocks. As a consequence:

- A general asymmetric top molecule has $2J+1$ nondegenerate eigenstates for any given total angular momentum quantum number J; $J+1$ states belong to the $(-1)^J$ parity manifold and J states have parity $(-1)^{J+1}$.
- Each state is $(2J+1)$-fold degenerate according to the projection quantum number M.

Figure 11.2 illustrates the scheme of rotational energies for H_2O in the electronic ground state and $J = 4$. In Section 11.3 we will elucidate the photodissociation of single rotational states and the resulting final rotational state distributions of the OH fragment. For this purpose it is important to stress that:

- Each nondegenerate eigenstate has a distinct wavefunction which is uniquely characterized by the totality of expansion coefficients $G_{j\Omega}^{(m)}(Jp)$ in Equation (11.10).

11.1.3 Selection rules and detailed state-to-state cross sections

Next we outline the calculation of fully resolved differential photodissociation cross sections for any arbitrary initial rotational state of the parent

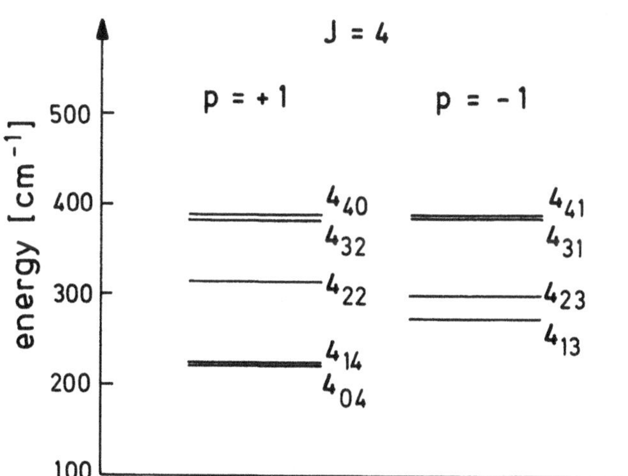

Fig. 11.2. Energy-level diagram of $H_2O(\tilde{X})$ in the lowest vibrational state for total angular momentum quantum number $J = 4$; $E = 0$ corresponds to the lowest rotational level 0_{00}. The nomenclature $J_{K_-K_+}$ with $K_+ = J, J-1, \ldots, 0$ and $K_- = 0, 1, \ldots, J$ follows the standard spectroscopic convention (Levine 1975:ch.5; Zare 1988:ch.6).

molecule:

$$\frac{d\sigma}{d\hat{\mathbf{k}}}(\omega, \hat{\mathbf{k}}, j, m_j | E_i, J_i, M_i, p_i).$$

$\hat{\mathbf{k}} \equiv (\theta_k, \varphi_k)$ is a unit vector in the direction of the recoil velocity; the angles are measured with respect to the space-fixed system whose z-axis is parallel to the electric field vector (Figure 11.1). The quantum number j represents the rotational state of the fragment and m_j is the projection quantum number of $\mathbf{j}$ with respect to the space-fixed z-axis. The quantum numbers J_i and M_i, the parity parameter p_i, and the energy E_i *uniquely* specify the initial rotational state of ABC. For simplicity of presentation, the vibrational quantum numbers of the parent as well as the fragment molecule are ignored in the subsequent discussion.

The total continuum wavefunction defined in the space-fixed (sf) coordinate system must satisfy the boundary condition

$$\Psi_{\mathrm{sf}}(\mathbf{R}, \hat{\mathbf{r}}; \hat{\mathbf{k}} E j m_j) \propto \left(\frac{m}{\hbar k_j}\right)^{1/2} \Big[\mathrm{e}^{i\mathbf{k}_j \cdot \mathbf{R}}\, Y_{jm_j}(\hat{\mathbf{r}}) + \frac{1}{R} \sum_{j'm_j'} f_{jm_jj'm_j'}(\hat{\mathbf{k}}, \hat{\mathbf{R}})\, Y_{j'm_j'}(\hat{\mathbf{r}})\, \mathrm{e}^{-ik_{j'}R} \Big] \tag{11.11}$$

in the limit $R \to \infty$. $\mathbf{k}_j$ is a vector with direction $\hat{\mathbf{k}}$ and length k_j defined in (3.21). The first term represents an outgoing plane wave in the direction $\hat{\mathbf{k}}$ with the rotor being in state (j, m_j) while the second term represents a superposition of incoming waves in *all* rotational states.

In order to facilitate the evaluation of the overlap between the bound and the continuum wavefunctions it is advisable to represent the continuum wavefunction, like the wavefunction of the parent molecule, in the body-fixed (bf) coordinate system although the boundary conditions are most simply expressed in the space-fixed system. The transformation from the space- to the body-fixed system is rather lengthy and has been worked out in detail by Balint-Kurti and Shapiro (1981). The result is

$$\Psi_{\mathrm{bf}}(\mathbf{R}, \hat{\mathbf{r}}; \hat{\mathbf{k}} E j m_j) \propto \left(\frac{m}{\hbar k_j}\right)^{1/2} \sum_{JMp} \sum_{lm_l} Y_{lm_l}^*(\hat{\mathbf{k}})\, (jlJM | jm_j l m_l) \times \frac{1}{R} \sum_{j'\Omega'} (-1)^{j'+\Omega'} \sum_{\Omega} (Jjl0 | J\Omega j - \Omega)\, \Theta_{j'\Omega'}^{JMp}(\hat{\mathbf{R}}, \hat{\mathbf{r}}) \chi_{j'\Omega'}(R; EJpj\Omega), \tag{11.12}$$

where $(\cdot\cdot\cdot\cdot|\cdot\cdot\cdot\cdot)$ is a Clebsch-Gordan coefficient in the notation of Edmonds (1974). The radial expansion functions $\chi_{j'\Omega'}(R)$ fulfill coupled equations similar to (11.7).[†] The corresponding boundary conditions in the limit $R \to \infty$ are given by Balint-Kurti and Shapiro (1981).

The dipole operator $\hat{\mathbf{d}}$ is a vector defined in the body-fixed frame of the molecule. Consequently, the transition dipole moment $\boldsymbol{\mu}$ defined in (2.35) is a vector field with three components each depending — like the potential — on R, r, and γ. For a *parallel transition* the transition dipole lies in the plane defined by the three atoms and for a *perpendicular transition* it is perpendicular to this plane. Following Balint-Kurti and Shapiro, the projection of $\boldsymbol{\mu}$, which is normally calculated in the body-fixed coordinate system, on the space-fixed z-axis, which is assumed to be parallel to the polarization of the electric field, can be written as

$$\mu^{(e)}(R, \hat{\mathbf{R}}, \hat{\mathbf{r}}) = \sum_L \sum_{K=0,1} B_{LK}(R)\, \Theta_{LK}^{101}(\hat{\mathbf{R}}, \hat{\mathbf{r}}). \tag{11.13}$$

The coefficients B_{LK} are related to the components of the transition dipole moment in the molecule-fixed system and the Θ_{LK}^{101} are the angular expansion functions defined in (11.5). The dipole moment transforms like a tensor of rank 1 which explains why it is expanded in terms of the angular functions for an angular momentum $J = 1$. Since its projection on the space-fixed z-axis is independent of the azimuthal angle φ_R, only functions with $M = 0$ are allowed. Furthermore, the dipole moment has the parity -1 so that the parameter p is restricted to $+1$ [remember that the parity is given by $(-1)^J p$].

The calculation of photodissociation cross sections requires the overlap of the continuum wavefunctions with the bound-state wavefunction multiplied by the transition dipole function. Employing for both wavefunctions the expansion in terms of the $\Theta_{j\Omega}^{JMp}$ and utilizing (11.13) leads to radial integrals of the form

$$\int_0^\infty dR\, \chi_{j'\Omega'}^*(R; EJpj\Omega)\, B_{LK}(R)\, \chi_{j''\Omega''}(E_i J_i p_i),$$

which must be evaluated numerically, and to angular integrals of the form

$$\left\langle \Theta_{j'\Omega'}^{JMp} \middle| \Theta_{LK}^{101} \middle| \Theta_{j''\Omega''}^{J_i M_i p_i} \right\rangle.$$

The latter can be done analytically by employing the standard expressions for the integrals of three rotation matrix elements $D_{\Omega M}^J$ respectively three spherical harmonics $Y_{j\Omega}$ (Edmonds 1974:ch.4). Without explicitly quoting the result for the angular integral we note the following selection rules:

[†] Note, however, that $(-1)^{j'+\Omega'}\Theta_{j'\Omega'}^{JMp}$ rather than $\Theta_{j'\Omega'}^{JMp}$ are the angular expansion functions in (11.12) which leads to slight but important modifications of the coupled equations.

- $p(-1)^J = -p_i(-1)^{J_i}$, i.e., the parity is changed in the excitation process.
- $M = M_i$, i.e., the projection of the total angular momentum on the space-fixed axis is conserved.
- $J = J_i$ or $J_i \pm 1$, i.e., the total angular momentum changes at most by one quantum.
- $\Omega' = \Omega''$ or $\Omega'' \pm 1$, i.e., the helicity quantum number of the rotor changes at most by one quantum.

The first selection rule is a consequence of the fact that the transition dipole moment has negative parity. The second reflects that the quantization axis is parallel to the polarization of the electric field. The third follows from the fact that the transition dipole moment is a tensor of rank 1 (corresponding to an "angular momentum" with quantum number 1).

It is not difficult to surmise that the final expression for the fully resolved differential photodissociation cross section is extremely complicated (Balint-Kurti and Shapiro 1981). It contains vast quantities of sums and angular momentum coupling elements. Note that the cross section depends explicitly on the magnetic quantum numbers M_i and m_j. Somewhat simpler cross section expressions can be derived by *averaging over the initial projection quantum number* M_i and *summing over the final projection quantum number* of the rotor, m_j. As shown by Balint-Kurti and Shapiro, the angle-resolved cross section then takes on the general form

$$\frac{d\sigma}{d\theta_k}(\omega, \theta_k, j) = \frac{\sigma(\omega, j)}{4\pi} [1 + \beta(\omega, j) P_2(\cos\theta_k)], \tag{11.14}$$

where $\beta(\omega, j)$ is the anisotropy parameter . Note that the latter depends on the final state of the product (see also Section 11.4.1). $\sigma(\omega, j)$ is the integral cross section obtained by integration over θ_k. All quantities in (11.14) naturally depend on the initial state of the parent molecule.

11.2 Population of Λ-doublet and spin-orbit states

Up to now we have considered exclusively the nuclear angular momentum of the product molecule. In many cases, however, the fragment is generated in an electronic state which possesses a nonzero electronic angular momentum as well as an electronic spin. The coupling of the nuclear and the electronic angular momenta and the spin gives rise to splitting of each rotational level into several fine-structure components. The hydroxy radical OH is a prototype (Atkins 1983:ch.12; Lefebvre-Brion and Field 1986:ch.2; Graybeal 1988:ch.15). The population of the possible electronic sublevels reveals a lot of information about the electronic states involved in the dissociation and the geometry of the fragmentation process.

11.2.1 The eigenstates of $OH(^2\Pi)$

The configuration of the outer electrons of OH in the electronic ground state, $X^2\Pi$, is

$$\ldots (2s\sigma)^2\ (2p\sigma)^2\ (2p\pi)^3.$$

This implies that three electrons are in the two $p\pi$ orbitals with lobes perpendicular to the internuclear O–H axis which serves as the z-axis of the body-fixed frame of the diatom (not to be confused with the body-fixed system of the triatomic molecule). Following Andresen et al. (1984) we assume that two of the $p\pi$ electrons are paired and one is unpaired, the latter determining the open-shell character of the OH radical. For a more refined analysis see Alexander and Dagdigian (1984).

The unpaired $p\pi$ electron has an orbital angular momentum $\mathbf{L}$ with projections $\Lambda = \pm 1$ on the body-fixed z-axis which — in a classical sense — describe rotation of the electron in opposite directions about the internuclear axis. Furthermore, it possesses a spin $\mathbf{S}$ with components $\Sigma = \pm 1/2$. Coupling of Λ and Σ leads to two $(2|\Sigma| + 1 = 2)$ manifolds with $|\Lambda + \Sigma| = 1/2$ and $3/2$ which are represented by $^2\Pi_{1/2}$ and $^2\Pi_{3/2}$, respectively. The splitting is in the range of several hundredths cm^{-1}.

If OH is not rotating, both states are doubly degenerate according to the two projections $\Lambda = \pm 1$. The degeneracy, however, is lifted by the coupling of the electronic angular momentum with the nuclear rotation with the splitting being of the order of several hundredths of a cm^{-1} only. This energy separation is, of course, negligibly small compared to the total energy release in fragmentation processes, which is typically of the order of 10,000 cm^{-1}, and therefore it is irrelevant for the filling of the two Λ-doublet states. However, the symmetry of the corresponding electronic wavefunctions determined by the orientation of the orbital of the unpaired $p\pi$ electron is important. For one Λ-doublet, which we will denote by $\Lambda(A')$, the lobe is in the plane of rotation while for the other one, denoted by $\Lambda(A'')$, the lobe is perpendicular to the plane of rotation as illustrated schematically in Figure 11.3. In the first case, the electronic wavefunction is symmetric with respect to reflection at the plane of rotation and in the second case it is anti-symmetric.

These considerations are strictly applicable, however, only if the symmetry plane is well defined, which for $OH(^2\Pi)$ as an example of Hund's coupling case (b) is valid only for large rotational quantum numbers j. In a more refined treatment (Andresen and Rothe 1985; Bigio and Grant 1987b) one expands the true Λ-doublet states in terms of the idealized wavefunctions depicted in Figure 11.3 and obtains for the electronic densities

$$\rho(A') \propto (2 - c_j)\sin^2\varphi + c_j \cos^2\varphi \qquad (11.15a)$$

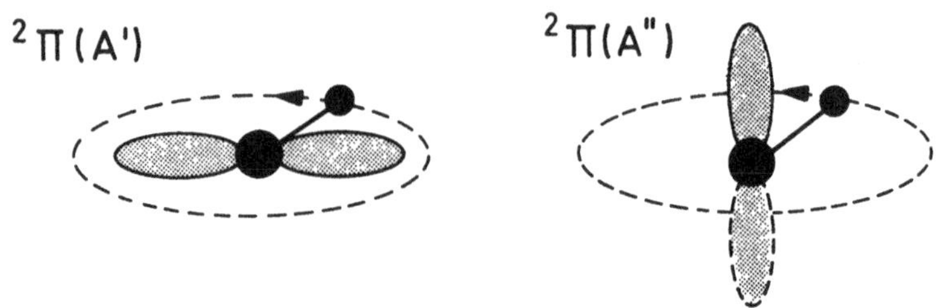

Fig. 11.3. Schematic illustration of the wavefunction of the unpaired $p\pi$ electron in the symmetric, $^2\Pi(A')$, and in the anti-symmetric, $^2\Pi(A'')$, Λ-doublet states in the limit of large angular momentum quantum numbers j.

$$\rho(A'') \propto c_j \sin^2\varphi + (2 - c_j)\cos^2\varphi \tag{11.15b}$$

with

$$c_j = 1 + \left[1 + \frac{(A/B-2)^2}{(2j-1)(2j+3)}\right]^{-1/2} . \tag{11.16}$$

Here, A and B are the fine-structure and the rotational constants of OH and φ is the azimuthal angle out of the rotation plane with $\varphi =$ 0 on the internuclear vector. $\cos^2\varphi$ and $\sin^2\varphi$ represent, respectively, lobes in the rotation plane and perpendicular to it. The c_j approach the value 2 in the limit of large j which has the consequence that $\rho(A')$ is preferentially oriented in the rotation plane while $\rho(A'')$ is preferentially oriented perpendicular to the plane of rotation. Equation (11.15) gives the relative weights of the in-plane and the out-of-plane contributions of the real Λ-doublet states. They play a vital role in the photolysis of water through the $\tilde{A}^1B_1$ state (see Section 11.2.2).

Figure 11.4 depicts the energy-level diagram of OH($^2\Pi$), excluding the additional splitting due to the nuclear spin. The $^2\Pi_{3/2}$ manifold is lower in energy than the $^2\Pi_{1/2}$ manifold. Here, j is the total angular momentum of OH including nuclear rotation, the electronic angular momentum, and the electronic spin; as a consequence of $\Sigma = \pm 1/2$ it is half-integer. $N = 1, 2, \ldots$ represents the angular momentum which results from coupling only the nuclear and the electronic angular momenta, $N = j + 1/2$ for $^2\Pi_{1/2}$ and $N = j - 1/2$ for $^2\Pi_{3/2}$.

11.2.2 *Preferential Λ-doublet population in the photodissociation of* $H_2O(\tilde{A}^1B_1)$

In the photodissociation of H_2O in the first absorption band, performed in a supersonic beam, the anti-symmetric Λ-doublet state $^2\Pi(A'')$ of the OH fragment is preferentially populated and the degree of preference increases markedly with the rotational quantum number. Figure 11.5 depicts the

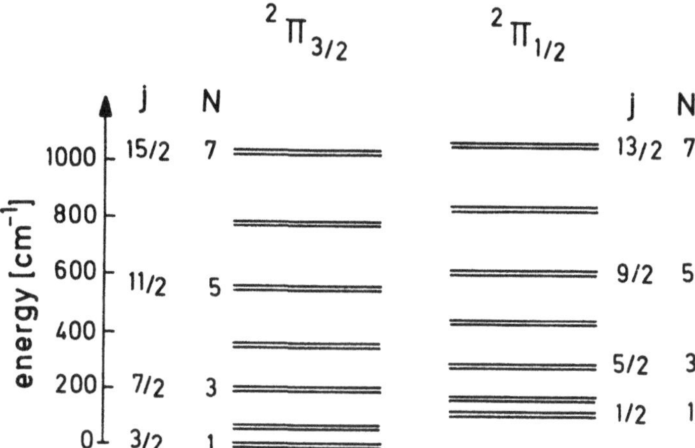

Fig. 11.4. Energy levels of OH($^2\Pi$). The splitting of the two Λ-doublets is not drawn to scale!

ratio $^2\Pi_{3/2}(A'') \,/\, ^2\Pi_{3/2}(A')$ as a function of $N = j - 1/2$. It rises from about 2 for $N = 1$ to about 20 for $N = 9$, i.e., for high rotational states OH is almost exclusively generated in the anti-symmetric Λ-doublet state. The population difference can easily be understood, qualitatively as well as quantitatively, in terms of the geometrical arrangement of the lobe of the unpaired $p\pi$ electron.

The two most weakly bound electrons in $H_2O(\tilde{X})$ occupy a $1b_1$ orbital which is essentially a pure $p\pi$ lobe perpendicular to the plane defined by the three nuclei. On excitation to the $\tilde{A}^1B_1$ state one of them is excited to the strongly antibonding $4a_1^*$ orbital leaving one unpaired $p\pi$ electron with its lobe perpendicular to the nuclear plane. Let us assume that H_2O does not initially rotate ($\mathbf{J} = 0$) thus that the dissociation plane remains fixed in space for all times. As a consequence of the fragmentation, which proceeds perfectly in-plane, the OH($^2\Pi$) fragment is created with an unpaired electron reminiscent of the $\Lambda(A'')$ state. Therefore:

- The preference for the $\Lambda(A'')$ state in the dissociation of H_2O in the first absorption band simply reflects conservation of the wavefunction of the unpaired electron as illustrated in the inset of Figure 11.5.

A more quantitative analysis must take into account that the electron densities of the two Λ-doublets, $\rho(A')$ and $\rho(A'')$, are not strictly perpendicular to the OH rotation plane or in that plane. According to (11.15), the contributions of the perfectly perpendicular lobe to $\rho(A')$ and $\rho(A'')$ are $(2 - c_j)$ and c_j, respectively, which yields a Λ-doublet ratio of

$$^2\Pi(A'') \,/\, ^2\Pi(A') = c_j/(2 - c_j). \tag{11.17}$$

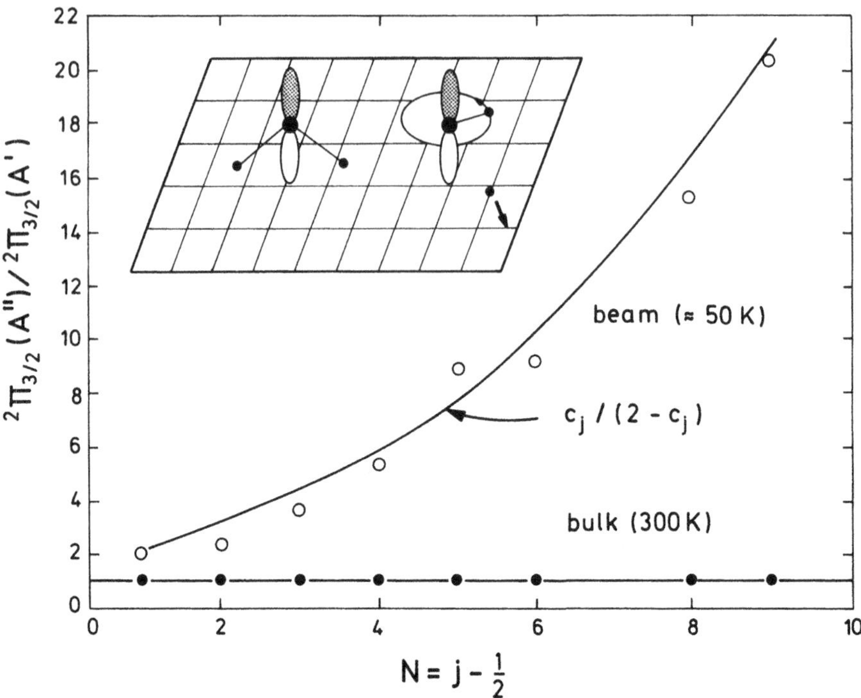

Fig. 11.5. Population ratio of the two Λ-doublet states of $OH(^2\Pi_{3/2})$ generated in the photodissociation of H_2O via the $\tilde{A}$ state in a nozzle beam ($T \approx 50$ K) and in the bulk ($T = 300$ K). $N = j - 1/2$ for the $^2\Pi_{3/2}$ spin-orbit manifold. The inset illustrates schematically the (approximate) conservation of the antisymmetric $p\pi$ lobe as one of the OH bonds breaks. Adapted from Andresen and al. (1984).

It reproduces the ratio measured in the nozzle beam *quantitatively* as manifested in Figure 11.5. In the limit of large j, the coefficient c_j goes to 2 and the Λ-doublet ratio goes to infinity. The higher j the better defined is the rotation plane of OH and the higher is the alignment of the $p\pi$ lobe of the unpaired electron perpendicular to this plane. Formally, the population of the two Λ-doublets can be interpreted as a Franck-Condon-type overlap between the wavefunction of the unpaired electron in the excited state of H_2O and the two distinct wavefunctions of the unpaired electron in OH.

Initial rotation of H_2O before the absorption of the photon gradually destroys the preferential population of $\Lambda(A'')$ with the consequence that the $^2\Pi(A'')\,/\,^2\Pi(A')$ ratio slowly converges towards the statistical limit of 1 as the temperature of H_2O increases. It is exactly 1 for the photodissociation in the room-temperature bulk (Figure 11.5). More about

Λ doublets in the photodissociation of H_2O can be found in the review articles of Andresen and Schinke (1987) and Andresen (1988).

Although the picture developed above describes the beam data extremely well we must point out that it fails to account for all the details observed if a *single* rotational state is dissociated. While the simple FC-type model predicts a preferential population of the $^2\Pi(A'')$ state for all j, in the dissociation of single initial quantum states of $H_2O(\tilde{X})$, on the other hand, one finds an oscillatory $^2\Pi(A'')\,/\,^2\Pi(A')$ ratio (see Section 11.3).

The selective population of one Λ-doublet state has been observed for many systems: NH_3 (Alberti and Douglas 1978; Quinton and Simons 1981), HONO (Vasudev, Zare, and Dixon 1984; Dixon and Rieley 1989), $(CH_3)_2NNO$ (Dubs, Brühlmann, and Huber 1986; Lavi and Rosenwaks 1988), and $(CH_3)_3CONO$ (Schwartz-Lavi, Bar, and Rosenwaks 1986), for example. In the case of CH_3ONO, Lahmani, Lardeux, and Solgadi (1986) and Brühlmann, Dubs and Huber (1987) measured for the NO product a clear preference for the Λ-doublet whose $p\pi$ lobe is anti-symmetric with respect to the plane of rotation. This result is interpreted in the sense that the dissociation proceeds in a mainly planar configuration because significant out-of-plane motion is believed to lower the degree of preferential population of the $^2\Pi(A'')$ Λ-doublet in the same way as overall rotation of the parent molecule. In contrast to the many experimental results on Λ-doublet ratios, there are only very few theoretical attempts to predict the population of the two Λ-doublet states (Alexander, Werner, and Dagdigian 1988; Alexander, Dagdigian, and Werner 1991; see also Section 11.3).

In summary, the observation of a strong propensity for one of the two Λ-doublet states indicates that the dissociation is relatively fast and planar.

11.2.3 Statistical and nonstatistical population of spin-orbit manifolds

While in the photodissociation of H_2O through the $\tilde{A}^1B_1$ state the two possible Λ-doublets of $OH(^2\Pi)$ are populated in a highly nonstatistical way, the two spin-orbit states, $OH(^2\Pi_{1/2})$ and $OH(^2\Pi_{3/2})$, are perfectly statistically populated. Unlike for the Λ-doublets there is *a priori* no geometrical reason to expect a difference in the spin-orbit states other than that given by the $2j+1$ statistical weighting factor. Since $j = N + 1/2$ for $^2\Pi_{3/2}$ and $j = N - 1/2$ for $^2\Pi_{1/2}$, the statistical weighting factor is $(N+1)/N$. Therefore, the population ratio $^2\Pi_{3/2}\,/\,^2\Pi_{1/2}$ multiplied by $N/(N+1)$ must be 1 for a statistical distribution as it is indeed measured in the bulk, in the beam, as well as in the dissociation of single rotational states of H_2O (Andresen and Schinke 1987). The reason for the statistical distribution of the spin states is presumably the absence of any forces in the exit channel acting on the spin of the unpaired $p\pi$ electron.

A preferential population of one spin-orbit state has been found for OH in the photodissociation of HONO, where the higher state is more populated than predicted by statistics (Vasudev, Zare, and Dixon 1984; Shan, Vorsa, Wategaonkar, and Vasudev 1989). In NO($^2\Pi$) generated via photolysis of CH_3ONO (Lahmani, Lardeux, and Solgadi 1986; Brühlmann, Dubs, and Huber 1987), $(CH_3)_2NNO$ (Dubs, Brühlmann, and Huber 1986; Lavi and Rosenwaks 1988), or $(CH_3)_3CONO$ (Schwartz-Lavi, Bar, and Rosenwaks 1986) the lower state is slightly more populated than the upper state. The reason for this preference is not yet clear.

Probably the most spectacular difference in the spin-orbit distributions has been observed in CN($^2\Sigma$) originating from the photodissociation of ICN.[†] Figure 11.6 depicts the quantity

$$f(N) = \frac{n(F_1) - n(F_2)}{n(F_1) + n(F_2)}, \tag{11.18}$$

where $n(F)$ is the population of the two spin states labelled by F_1 and F_2, respectively. A value $f(N) \approx 0$ would indicate a statistical distribution. However, in the present case $f(N)$ oscillates as a function of N and in addition it shows a pronounced wavelength dependence.

Joswig, O'Halloran, Zare, and Child (1986) interpreted this intriguing difference in the spin-orbit populations in terms of a semiclassical model. The keystone of this model is the assumption that CN($^2\Sigma$) can be reached via two different electronic states whose potential energy surfaces cross near the Franck-Condon region. The phase difference between trajectories or quantum mechanical wavepackets following different routes but leading to the same fragment causes the oscillatory structure of $f(N)$. The same scenario would also explain the distinct variation with the photon energy. This model is not implausible because it is known experimentally (Nadler, Mahgerefteh, Reisler, and Wittig 1985) that the photodissociation of ICN indeed involves at least two, possibly even three electronic states (Black, Waldeck, and Zare 1990; Yabushita and Morokuma 1990).

Another example of a dramatic difference in the population of the two spin-orbit states is the dissociation of HN_3 in the electronic ground state induced by overtone pumping (Foy, Casassa, Stephenson, and King 1988;

[†] The photodissociation of ICN has been extensively studied in several laboratories. See Black, Waldeck, and Zare (1990) for a comprehensive discussion and an extensive list of references. The complexity of this photodissociation system arises from the fact that both fragments can be formed in several electronic states.

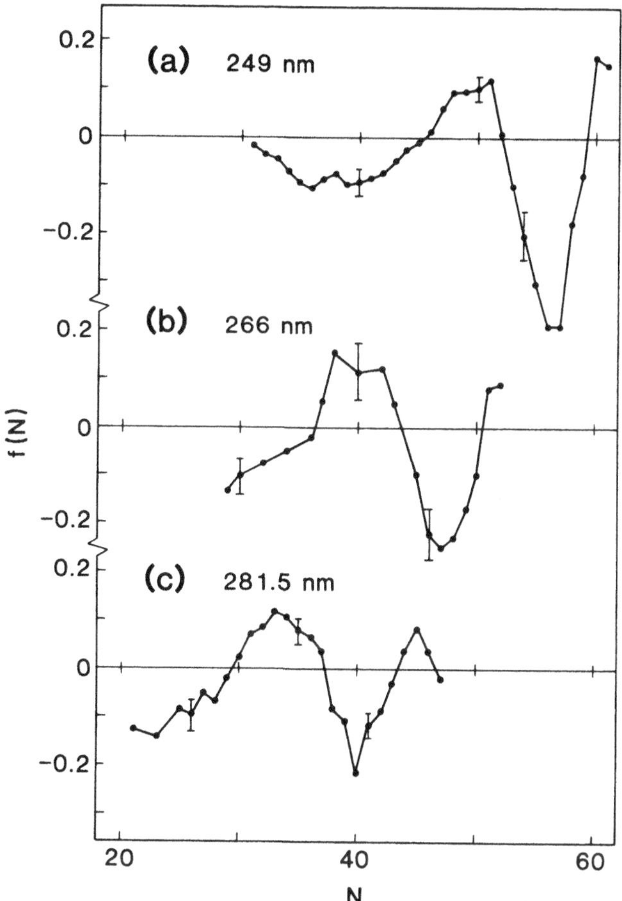

Fig. 11.6. The spin-orbit difference function $f(N)$ *vs.* N for the photolysis of ICN at (a) 249 nm, (b) 266 nm, and (c) 281.5 nm. Reproduced from Joswig, O'Halloran, Zare,and Child (1986).

Stephenson, Casassa, and King 1988). An intriguing explanation of this spin selectivity based on the planarity of the transition state and the symmetry of the electronic wavefunction at the transition state has been provided by Alexander, Werner, and Dagdigian (1988) and Alexander, Werner, Hemmer, and Knowles (1990).

11.3 Dissociation of single rotational states

As we emphasized at the beginning of Chapter 10, the rotational excitation of the fragment originates from the overall rotation of the parent molecule, the bending motion inside the parent molecule, and from the torque during the breakup in the excited electronic state. If the final state interaction is weak, the first two sources are the determining factors and consequently the final rotational state distribution depends uniquely on the particular bending or rotational state of the parent molecule in the electronic ground state. H_2O is an ideal system to demonstrate this quantum-state dependence.

In Section 10.1 we showed that the rotational distribution of $OH(^2\Pi)$ following the dissociation through excitation of the $\tilde{A}^1B_1$ state is well described within the Franck-Condon limit. The coupling between the rotational and the dissociation coordinates can be ignored (see Figure 10.4) and the final rotational state distribution is simply the modulus square of the overlap of the free OH wavefunction with the bending wavefunction of the parent molecule [see Equation (10.4)]. Basically the same theory can also be applied if overall rotation of $H_2O(\tilde{X})$ as well as the electronic fine-structure states of $OH(^2\Pi)$ is included. The resulting expression, however, becomes exceedingly more complicated!

The necessary angular momentum coupling theory has been worked out by Balint-Kurti (1986). It includes:

1) The orbital angular momentum of the recoiling H atom with respect to OH.
2) The nuclear angular momentum of OH.
3) The orbital angular momentum of the unpaired electron of $OH(^2\Pi)$.
4) The electronic spin of OH.
5) The electronic spin of the H atom.

The final expression for the population of OH in a particular rotational state j (which, as a consequence of the electronic spin, is half-integer in this case) and in one of the four possible electronic fine-structure states, $^2\Pi_{1/2}(A', A'')$ and $^2\Pi_{3/2}(A', A'')$, which we designate by the index $l =$ 1–4, is given by

$$P_{jl}^{(i)} \sim \sum_{Jk\Omega} (2J+1)(2k+1)\, t_\Omega^2 \,\Big| \sum_{j''\Omega''} \chi_{j''\Omega''}^{(i)}(R_e; E_i J_i p_i)\, t_{\Omega''} \tag{11.19}$$
$$\Big\{(-1)^\Omega [J-\Omega|J_i\Omega''] + p_i(-1)^{J+J_i}[J\Omega|J_i\Omega'']\Big\} F_l(jk\Omega|j''\Omega'')\Big|^2,$$

where the normalization constants t_Ω were defined below Equation (11.5). Furthermore, we defined

$$[J\Omega|J_i\Omega''] = \begin{pmatrix} J & 1 & J_i \\ \Omega & 1 & \Omega'' \end{pmatrix} + \begin{pmatrix} J & 1 & J_i \\ \Omega & -1 & \Omega'' \end{pmatrix} \tag{11.20}$$

and the four functions $F_l(jk\Omega|j''\Omega'')$ are given by Schinke, Engel, Andresen, Häusler, and Balint-Kurti (1985). The $\chi_{j''\Omega''}^{(i)}(R_e; E_i J_i p_i)$ are the radial expansion functions for the particular initial rotational state of the parent molecule (labeled by the superscript i), evaluated at the ground-state equilibrium. They are the solutions of the coupled equations (11.7) and are different for each total angular momentum state designated by E_i, J_i, and p_i. Equation (11.19) is cited here only in order to demonstrate the complexity of the final FC expression.

Fully resolved rotational state distributions of $OH(^2\Pi)$ following from the photodissociation of single rotational states have been measured by Andresen et al. (1985) and Häusler, Andresen, and Schinke (1987). Figure 11.7 illustrates the essential steps of this pioneering experiment. First, an IR laser with frequency ω_1 excites H_2O from the vibrational ground state $|00^+0\rangle$ into the $|10^-0\rangle$ vibrational manifold (local mode nomenclature; see Section 13.2). Since this excitation process is resonant, only a single, well defined rotational state represented by $|E_iJ_ip_i\rangle$ is filled in the excited vibrational manifold. Subsequently, a second laser with frequency ω_2 photolyzes H_2O by excitation into the $\tilde{A}^1B_1$ continuum. Choosing ω_2 small enough so that the states in the ground vibrational state *cannot* be dissociated at the same time, because there is not sufficient energy, ensures that the resultant OH products stem exclusively from the single rotational level in the excited vibrational manifold. They are subsequently probed with a third laser with frequency ω_3. This experiment is state-specific on the basis of six (!) "quantum numbers": three "quantum numbers" — E_i, J_i, and p_i — specify the initial rotational state of the

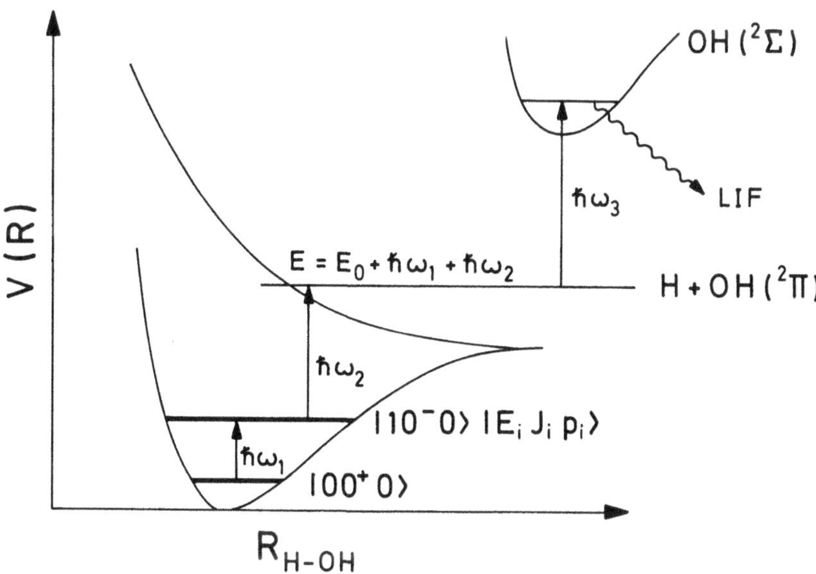

Fig. 11.7. Schematic illustration of the experiment of Andresen et al. (1985) to photolyze a single rotational state of H_2O.

parent molecule and three "quantum numbers" — j, $\Omega = 1/2$ or $3/2$, and A' or A'' — define the final state of the product.†

Figures 11.8 and 11.9 show rotational state distributions following the dissociation of the lowest rotational state $J_i = 0$, i.e., 0_{00}, and of rotational states with $J_i = 4$, i.e., $4_{K_-K_+}$; the corresponding energies are depicted in Figure 11.2. We note the following general observations:

- Each initial rotational state yields a distinct final rotational state distribution. This holds true even if the total angular momentum J is the same and merely the "projection quantum numbers" K_- and K_+, are different.
- The rotational state distributions following the dissociation of a single initial state oscillate as a function of j, whereas the distributions obtained in the molecular beam or in the 300 K bulk are smooth functions.
- The strict preference of the upper (anti-symmetric) Λ-doublet state observed in the beam experiment is lost, i.e., the $^2\Pi(A'')\,/\,^2\Pi(A')$ ratio may be larger than 1 for one rotational state but smaller than 1 for the next one. It is valid only on the average.

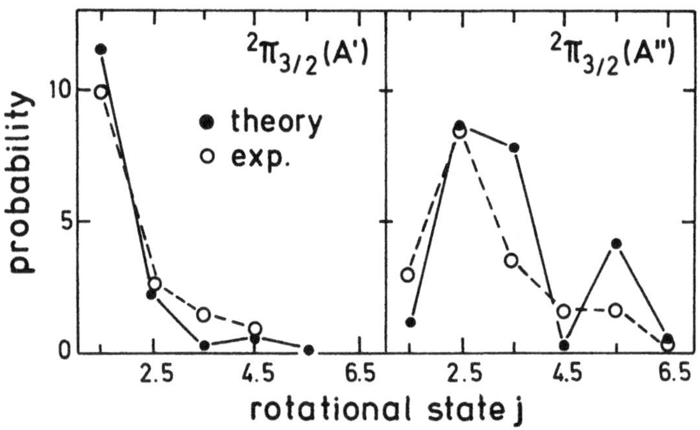

Fig. 11.8. Rotational state-distributions of $OH(^2\Pi_{3/2})$ originating from the 193 nm photolysis of H_2O in the 0_{00} rotational ground state within the $|10^-0\rangle$ vibrational manifold. The results for the A' and the A'' Λ-doublets are drawn on the same scale. Adapted from Häusler, Andresen, and Schinke (1987).

† Vander Wal, Scott, and Crim (1991) repeated this experiment by pre-exciting H_2O into vibrational levels corresponding to about four instead of one quanta of the OH stretching mode. They obtained, within the experimental uncertainties, the same rotational distributions which underscores that stretching vibration and overall rotation are to a large extent decoupled in H_2O.

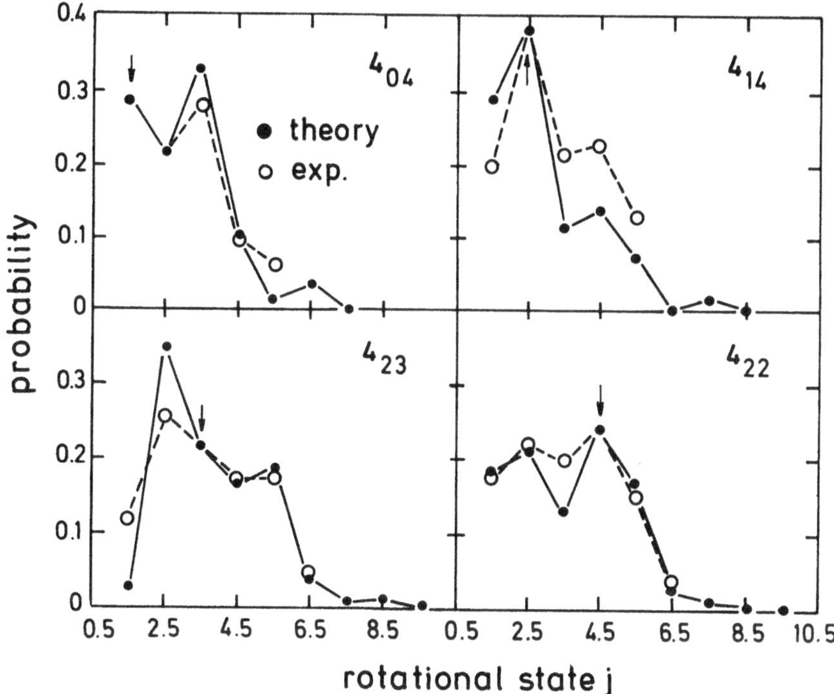

Fig. 11.9. Rotational state distributions of OH in the $^2\Pi_{3/2}(A')$ Λ-doublet for four initial rotational states of H_2O all having the same total angular momentum quantum number $J = 4$. The normalization between experiment and theory is made for each case separately and is indicated by the arrows. The corresponding energy level diagram is shown in Figure 11.2. Adapted from Häusler, Andresen, and Schinke (1987).

- All experimental observations are *quantitatively* reproduced by the extended Franck-Condon theory outlined above; this includes the rotational state distributions for all initial states as well as the population ratios of the two spin-orbit manifolds and the two Λ-doublet states.

The rotational state distributions shown in Figures 11.8 and 11.9 uniquely reflect the wavefunctions of the respective rotational states of the parent molecule, represented through the coefficients $\chi^{(i)}_{j''\Omega''}(R_e; E_i J_i p_i)$ in Equation (11.19). The energy of the individual levels, however, is not pertinent. The dependence on the initial rotational state manifests a kind of Franck-Condon mapping which is in essence similar to the analysis given in Section 10.1. In the previous discussion we included only the nuclear angular momentum of the fragment molecule. In order to compare with the fully state-resolved measurements, however, it is essential to include *all* angular momenta as well as *all* spins and to couple them correctly. Taking into account merely the nuclear angular momenta would yield rotational distributions which satisfactorily reproduce the overall shape of the distributions but not the finer details. Ignoring the electronic degrees of freedom leads to oscillations which are too pronounced (see Figure 10.3 in comparison to Figure 10.5).

Incidentally we note that the structures seen in Figures 11.8 and 11.9 are the remnants of the oscillations due to the $\sin^2(j\gamma_e)$ term in Equation (10.7). The inclusion of the electronic fine-structure states tends to damp them. Since the structures are out-of-phase for the two Λ-doublet states, averaging over $^2\Pi(A')$ and $^2\Pi(A'')$ smears them out and leads to relatively smooth rotational distributions.

With the extended FC theory at hand one can easily generate *temperature-dependent* probabilities $P_{jl}(T_{H_2O})$ by summing the *state-resolved* probabilities $P_{jl}(E_i, J_i, p_i)$ over all initial rotational states according to a Boltzmann distribution for a temperature T_{H_2O}. In this way one can study the influence of the initial temperature on the distribution of the final states (Häusler, Andresen, and Schinke 1987). The averaging has the following effects: First, raising the initial temperature from zero rapidly smears out the oscillatory structures of the state-resolved distributions with the result that $P_{jl}(T_{H_2O})$ can be better and better described by a Boltzmann distribution with a "temperature" T_{OH}. Remember, however, the discussion in Section 10.1.2 on the real origin of the Boltzmann distribution in the dissociation of H_2O! Since an increase of the overall rotation of H_2O leads to the population of higher and higher rotational states of OH, it does not come as a surprise to find that T_{OH} steadily increases with T_{H_2O}. The temperature of the fragment, however, is consistently much larger than the temperature of the parent molecule. At the same time, the preference for the anti-symmetric Λ-doublet state gradually diminishes with increasing initial temperature. The extended Franck-Condon theory reproduces the state-resolved measurements as well as the temperature dependence of the product state distributions.

Similar results have been obtained for the photodissociation of NO_2 (Bigio and Grant 1987a; Robra, Zacharias, and Welge 1990). In concluding this section we must emphasize that the unique dependence of the final state distribution on the initial rotational state of the parent molecule is possible only if significant rotational excitation in the exit channel is absent. Increasing the final state interaction would gradually destroy the sensitivity on the particular parent state and the final distribution would more and more reflect the dynamics in the excited state as discussed in Section 6.3 and Chapter 10.

11.4 Vector correlations

Up to now we have exclusively considered the *scalar* properties of the photodissociation products, namely the vibrational and rotational state distributions of diatomic fragments, i.e., the energy that goes into the various degrees of freedom. Although the complete analysis of final state distributions reveals a lot of information about the bond breaking and the forces in the exit channel, it does not completely specify the dissociation process. Photodissociation is by its very nature an anisotropic process — the polarization of the electric field $\mathbf{E}_0$ defines a unique direction relative to which all vectors describing both the parent molecule and the products can be measured. These are:

1) The transition dipole moment of the parent molecule, $\boldsymbol{\mu}$.
2) The recoil velocity of the products, $\mathbf{v}$.
3) The rotational angular momenta of the fragments, $\mathbf{j}$.

Analysis of the correlations of $\boldsymbol{\mu}$, $\mathbf{v}$, and $\mathbf{j}$ with $\mathbf{E}_0$ and among each other is necessary for a full understanding of photodissociation dynamics, especially for polyatomic molecules with more than three atoms. Vector correlations play an increasingly important role in experimental investigations. This section covers only the most elementary aspects and for deeper insight the reader is referred to the numerous reviews on this fascinating topic (Zare 1972; Simons 1977; Houston 1987, 1989; Simons 1987; Hall and Houston 1989).

11.4.1 $\mathbf{E}_0$–$\boldsymbol{\mu}$–$\mathbf{v}$ *correlation*

Let us consider the fragmentation of a diatomic molecule with transition dipole moment $\boldsymbol{\mu}$ which can be either in the direction of the internuclear

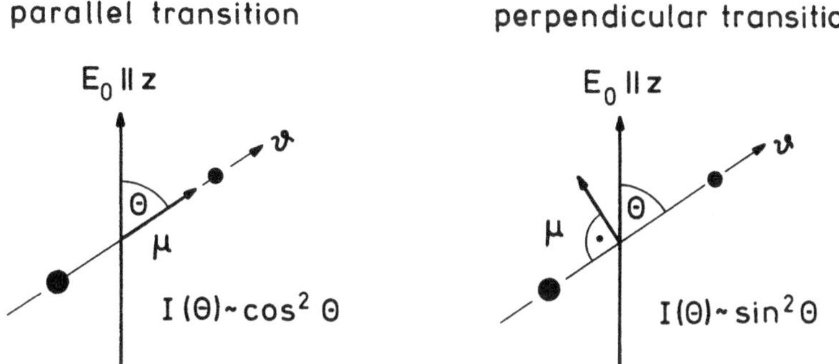

Fig. 11.10. Schematic illustration of the alignment of a diatomic molecule owing to a parallel and a perpendicular transition. The polarization of the electric field vector, $\mathbf{E}_0$, is parallel to the laboratory z-axis.

bond (*parallel transition*) or perpendicular to it (*perpendicular transition*) as sketched in Figure 11.10. The polarization of the electric field defines the z-axis in the laboratory frame. In the electric dipole approximation (see Section 2.1.2) the probability for absorbing the photon is proportional to $\cos^2\theta$ where θ is the angle between $\boldsymbol{\mu}$ and $\mathbf{E}_0$. The photon preferentially excites those molecules with their dipole moments parallel to the polarization vector. This leads, immediately after the absorption, to an alignment of the excited molecule in the laboratory system.

If the bond breaking occurs on a time scale which is short compared to overall rotation of the excited complex, the alignment of the molecule implies an alignment of the recoil velocity $\mathbf{v}$ which is usually directed along the internuclear vector. Thus, if the excited molecule dissociates immediately after the absorption of the photon, the angular distribution of the photofragments, $I(\theta)$, is proportional to $\cos^2\theta$ for a parallel transition and proportional to $\sin^2\theta$ for a perpendicular transition. More generally, we may write the angular distribution as

$$I(\theta) \propto \frac{1}{4\pi}[1 + \beta P_2(\cos\theta)], \tag{11.21}$$

where $P_2(x) = 3x^2/2 - 1/2$ is the second-order Legendre polynomial. The *anisotropy parameter* β ranges between -1 for a perpendicular transition and $+2$ for a parallel transition. Thus, measuring the angular distribution of the fragments can provide information about the type of electronic transition and hence the electronic symmetry of the excited state (Zare and Hershbach 1963; Bersohn and Lin 1969; Zare 1972; Busch and Wilson 1972b; Zare 1988:ch.3). Equation (11.21) also describes the angular distribution in the dissociation of a triatomic molecule provided one averages over the initial magnetic quantum number M_i and sums over the final magnetic quantum number m_j of the fragment molecule which is generated in rotational state j (Balint-Kurti and Shapiro 1981). In that case, the anisotropy parameter β depends on j.

β takes on the limiting values of -1 and $+2$ only if the dissociation time is negligibly small compared to the rotational period of the parent molecule. Rotation of the excited complex before the bond completely ruptures diminishes the alignment of $\mathbf{v}$ with respect to $\mathbf{E}_0$ with the consequence that β differs from these limits. The degree of incomplete alignment can be used to estimate the lifetime of the excited complex (Busch and Wilson 1972b; Yang and Bersohn 1974). However, since the relation between the lifetime on one hand and the anisotropy parameter on the other hand is rather indirect, one should regard such estimates with caution.

11.4.2 $\mathbf{E}_0$-$\boldsymbol{\mu}$-$\mathbf{j}$ correlation

A second correlation exists between the transition dipole moment $\boldsymbol{\mu}$, which is preferentially aligned parallel to $\mathbf{E}_0$, and the angular momentum vector of the photofragment, $\mathbf{j}$. This correlation concerns the direction of the final angular momentum vector $\mathbf{j}$ with respect to the space-fixed axis $\mathbf{E}_0$. In quantum mechanics the projection of $\mathbf{j}$ on the axis defined by $\mathbf{E}_0$ is quantized with quantum numbers $m_j = -j, -j+1, \ldots, j-1, j$ and the distribution $P(m_j)$ is a measure of the average angle between $\mathbf{j}$ on one hand and $\mathbf{E}_0$ on the other.

Let us for simplicity discuss a triatomic molecule, for example H_2O, with $\boldsymbol{\mu}$ perpendicular to the plane defined by the three atoms. In that case, the photon will mainly excite molecules that are perpendicularly aligned to the $\mathbf{E}_0$ vector, i.e., that lie in a plane perpendicular to $\mathbf{E}_0$. If the dissociation time is small compared to the rotational period of the parent molecule, the rotational vector of OH will be preferentially directed parallel to the laboratory z-axis because the recoil of H and OH proceeds in-plane. This would lead to a distribution in the projection quantum number m_j which is strongly peaked near $m_j \approx \pm j$. For a parallel transition, on the other hand, we would expect the opposite situation, i.e, $\mathbf{j}$ would be aligned perpendicularly to the z-axis and $P(m_j)$ would peak near $m_j = 0$.

In general, one distinguishes between *orientation* and *alignment* as illustrated in Figure 11.11. The fragment diatom is said to be oriented if the distribution of magnetic states is not symmetric with respect to $m_j = 0$; the totality of the fragments has a resulting angular momentum direction. If $P(m_j)$ is symmetric with respect to $m_j = 0$, the molecule is aligned and the net angular momentum averaged over all products is

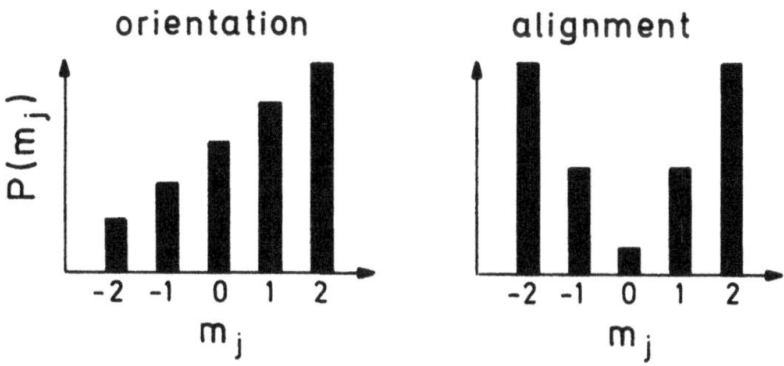

Fig. 11.11. Schematic illustration of orientation and alignment of the fragment rotational angular momentum vector in terms of the distribution $P(m_j)$ where m_j is the projection or magnetic quantum number, i.e., the eigenvalue of the z-component of $\mathbf{j}$.

zero. If the direction of **j** is isotropically distributed, it follows that $P(m_j)$ is constant.

If the fragment molecule is created in an electronically excited state, the alignment of **j** leads to polarized emission. Conversely, if it is produced in the electronic ground state, the fragment will preferentially absorb light of a particular polarization. An elegant and useful way to describe the directional distribution of the angular momentum vector follows from the multipole moment expansion of **j** in terms of spherical tensor operators (Greene and Zare 1982; Zare 1988:ch.5). Alignment of the photofragments has been observed for a number of molecules including H_2O, HONO, ICN, and CH_3ONO , for example. Hasselbrink, Waldeck, and Zare (1988) and Black, Hasselbrink, Waldeck, and Zare (1990) have seen orientation of the CN fragment in the photodissociation of ICN with circularly polarized radiation. A theoretical treatment has been recently provided by Beswick, Glass-Maujean, and Roncero (1992). As in the case of the $\mathbf{E}_0$–**v** correlation, the $\mathbf{E}_0$–**j** correlation can be highly profitable in assigning the symmetry of the electronic states involved. Since orientation and alignment are defined with respect to the space-fixed system, the same effects which diminish the alignment of **v** with respect to $\mathbf{E}_0$, namely overall rotation of the parent molecule, also lead to a loss of alignment of **j**.

11.4.3 **v**–**j** *correlation*

Since **v** and **j** are both correlated with $\boldsymbol{\mu}$ respectively $\mathbf{E}_0$ it does not come as a surprise that they are correlated with each other as well. However, there is one fundamental difference between the correlations discussed above and the **v**–**j** correlation:

- While the $\mathbf{E}_0$–**v** and the $\mathbf{E}_0$–**j** correlations relate to the laboratory frame, the correlation between the recoil velocity **v** and the fragment angular momentum **j** is made in the molecular frame and it is thus completely independent of the laboratory system.

This has very important consequences! The alignment of **v** and **j** with respect to the space-fixed z-axis is diminished by overall rotation before the excited complex dissociates. A long lifetime destroys the alignment. The correlation of **j** with **v**, on the other hand, is not established until the bond breaks and the two fragments recoil such that rotation of the parent molecule prior to dissociation is irrelevant. It represents a new observable which can provide additional information about the bond rupture and the exit channel dynamics other than the final state distributions.

Let us illustrate two examples of **v**–**j** correlation, the dissociation of a triatomic and of a tetratomic molecule, H_2O and H_2O_2 , for example. In the first case, the correlation between recoil velocity and final fragment rotation is trivial provided overall rotation is small, whereas in the second example it is essential for a true understanding of the dissociation. The photodissociation of a triatomic molecule is illustrated in Figure 3.1. For total angular momentum $\mathbf{J} \approx 0$ the collision plane remains basically fixed in space throughout the fragmentation and $\mathbf{l} \approx -\mathbf{j}$, where $\mathbf{l}$ is the orbital angular momentum of A with respect to BC. Since $\mathbf{l}$ is perpendicular to the scattering vector $\mathbf{R}$ it follows immediately that $\mathbf{j}$ also stands perpendicularly on the plane defined by the three atoms. Furthermore, the recoil velocity lies in this plane such that $\mathbf{j} \perp \mathbf{v}$, i.e., the angular momentum of BC and the recoil velocity $\mathbf{v}$ are always perpendicular to each other. This holds true, however, only for low overall rotation of the parent molecule.

Figure 10.7 illustrates H_2O_2 in the electronic ground state. Apart from overall rotation, there are, in principle, two possible sources of rotational excitation of the two OH fragments after promotion into the two relevant excited electronic states: bending motion associated with the two HOO angles, α_1 and α_2, and torsional motion associated with the angle φ. The recoil velocity is directed along the breaking O-O bond. The mere measurement of scalar properties, i.e., the rotational state distributions of the two OH fragments, *cannot* discriminate between these two possibilities! Determining the $\mathbf{v}$–$\mathbf{j}$ vector correlation, however, brings to light the relative contributions of excitation through the bending and the torsional degrees of freedom. Bending motion would lead to rotation of each rotamer in the corresponding HOO-plane with $\mathbf{j}$ being mainly oriented *perpendicularly* to $\mathbf{v}$. Torsional motion, on the other hand, would induce rotation in a plane perpendicular to the recoil velocity such that the $\mathbf{j}$s are preferentially oriented *parallel* to $\mathbf{v}$.

The measurement yields average angles between $\mathbf{j}$ and $\mathbf{v}$ of 26° and 35° in the 193 nm photodissociation in the bulk and in the molecular beam, respectively (Grunewald, Gericke, and Comes 1987, 1988). Thus, torsional motion is undisputedly the main source of rotation of the OH fragments. Trajectory calculations including both the bending angles as well as the torsional angle and using a calculated PES give an average angle of 23° (Schinke and Staemmler 1988). Figure 10.8 depicts the φ dependence of the two upper-state potential energy surfaces; it readily makes clear why torsional motion is the dominant factor. Other polyatomic examples for which the analysis of the $\mathbf{v}$–$\mathbf{j}$ vector correlation has been proven to be extremely helpful are HN_3 (Gericke, Theinl, and Comes 1990), CHOCHO (Burak et al. 1987) and $(CH_3)_2NNO$ (Dubs, Brühlmann, and Huber 1986).

11.5 Correlation between product rotations

In the conclusion of our discussion of rotational excitation we will consider an effect which has been noted only in the last few years. Let us imagine the breakup of a tetratomic molecule into two diatomic products, for example $HN_3 \rightarrow HN + N_2$. Usually one measures *independently* the rotational state distribution of one rotor, $\tilde{P}_1(j_1)$, and the distribution of the other rotor, $\tilde{P}_2(j_2)$. In the case of HN_3 one obtains relatively little rotational excitation for HN but very strong excitation for the sibling fragment N_2 (Gericke, Theinl, and Comes 1989; Chu, Marcus, and Dagdigian 1990). Other examples for which the state distributions of both products are obtained in this way are NCNO (Qian et al. 1985), H_2CO (Bamford et al. 1985; Debarre et al. 1985), and HNCO (Spiglanin, Perry, and Chandler 1987; Spiglanin and Chandler 1987).

One should notice, however, that $\tilde{P}_1(j_1)$ and $\tilde{P}_2(j_2)$ are — in principle — highly averaged quantities. The most detailed information is provided by the two-dimensional transition probability $P(j_1, j_2)$:

- $P(j_1, j_2)$ is the probability of finding the first rotor in state j_1 and *simultaneously* the second rotor in state j_2. It contains the full correlation between the two recoiling rotamers.

The conventionally measured probabilities are defined by

$$\tilde{P}_1(j_1) = \sum_{j_2} P(j_1, j_2) \tag{11.22}$$

and likewise for $\tilde{P}_2(j_2)$. They represent, for each rotational state of one rotor, an average over *all* rotational levels of the sibling fragment.

Let us distinguish two extremes. The fragments are completely uncorrelated if the probability for rotor 1 is independent of j_2 and *vice versa*, i.e., if the detailed probability separates according to

$$P(j_1, j_2) \approx P_1(j_1) P_2(j_2). \tag{11.23}$$

An example of (almost) uncorrelated product rotations is the photodissociation of H_2CO in which the rotational distribution of CO is largely independent of j_{H_2} and likewise the distribution of H_2 is independent of j_{CO}. The reason for the lack of correlation rests upon the fact that two distinct mechanisms determine the rotational excitations of CO and H_2, respectively. While the distribution of H_2 is mainly determined by the Franck-Condon principle (no final state interaction), CO experiences a relatively strong torque which leads to a highly inverted distribution (Schinke 1986b; Chang, Minichino, and Miller 1992). At the other extreme, the two fragments are strictly correlated if

$$P(j_1, j_2) \approx \delta_{j_1 j_2} P_1(j_1), \tag{11.24}$$

i.e., if the fragments are produced in identical states. In both cases it suffices merely to measure the averaged probabilities $\tilde{P}_1$ and $\tilde{P}_2$. In general, however, one cannot *a priori* expect (11.23) or (11.24) to hold.

The photodissociation of H_2O_2 is an attractive example to study the correlation between two photofragments. The general dissociation dynamics has been discussed in Section 10.2.1. Classical trajectory calculations provide no problems in generating the fully resolved distribution matrix $P(j_1, j_2)$; one merely needs to run more trajectories than required for the averaged probabilities $\tilde{P}_1$ and $\tilde{P}_2$ (Schinke 1988c). Owing to the symmetry of the system, $P(j_1, j_2)$ is symmetric in j_1 and j_2 and the averaged probabilities are identical for the two fragments. The left-hand side of Figure 11.12 depicts for zero temperature of the parent molecule the distribution of rotor 2 for several final states j_1 of rotor 1. These distributions are narrower than the averaged distribution shown in Figure 10.10 (beam) and with increasing j_1 the peak moves continuously to larger values of j_2. Except for the highest rotational states, the two-dimensional probability matrix $P(j_1, j_2)$ is maximal on or near the diagonal $j_1 = j_2$.

The results for $T = 0$ K in Figure 11.12 manifest strong correlation between the rotational excitation of the two OH radicals. This accords well with the fact that torsional motion about the O-O axis is the main source of rotational excitation as emphasized in previous sections. According to Equation (10.13) both rotors experience the same torque $|\partial V/\partial\varphi|$ with φ being the torsional angle. If bending motion associated with the HOO bending angles were absent, the correlation would be complete and the two cartwheels would rotate about the common axis with the same angular velocity but in opposite directions, i.e., $\mathbf{j}_1 = -\mathbf{j}_2$ (see Figure 10.21). Both radicals would be generated in the same rotational state. Bending motion partly diminishes the strong pair-pair correlation so that the distributions have finite widths.

Overall rotation about the O-O axis destroys the symmetry of the two OH products; the rotation of both rotamers in the same direction leads to acceleration of one rotor but deceleration of the other when the O-O bond breaks [see Equation (10.14)] with the result that the correlation is partly lost. This effect obviously becomes more prominent with increasing temperature because rotational states of the parent molecule belonging to higher and higher total angular momenta are dissociated. As a consequence, the distributions for dissociation at room temperature shown on the right-hand side of Figure 11.12 are significantly broadened compared to the beam results and the maximum is nearly independent of j_1.

Improvements in the frequency resolution of the laser used to probe OH by laser-induced fluorescence has made it feasible to study the effect of pair-pair correlation also experimentally (Gericke 1988; Gericke, Grunewald, Klee, and Comes 1988; Gericke, Gläser, Maul, and Comes 1990; Dixon, Nightingale, Western, and Yang 1988). The basic idea is

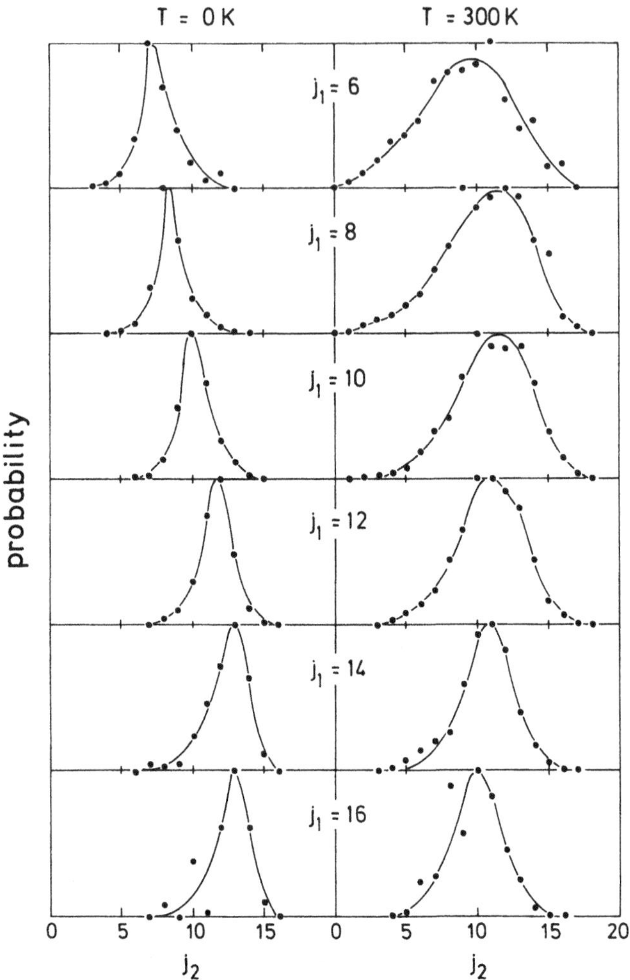

Fig. 11.12. Rotational pair correlation in the photodissociation of H_2O_2 at 193 nm. Cuts through the two-dimensional probability matrix $P(j_1, j_2)$ for several values of j_1 are shown. $P(j_1, j_2)$ is the probability that OH fragment 1 is produced in state j_1 if simultaneously fragment 2 is created in rotational state j_2. All distributions are normalized to the same values at their maxima. About 10,000 classical trajectories have been calculated for each of the two electronic states involved in the photodissociation at 193 nm. See Section 10.2 for a more detailed discussion of the dissociation dynamics. The initial temperature of H_2O_2 is 0 K (beam) and 300 K (bulk). Reproduced from Schinke (1988c).

based on energy conservation. For any pair of rotational states j_1 and j_2 the translational recoil energy takes on the value

$$E_{trans} = E_{excess} - B_{rot}\, j_1(j_1 + 1) - B_{rot}\, j_2(j_2 + 1), \tag{11.25}$$

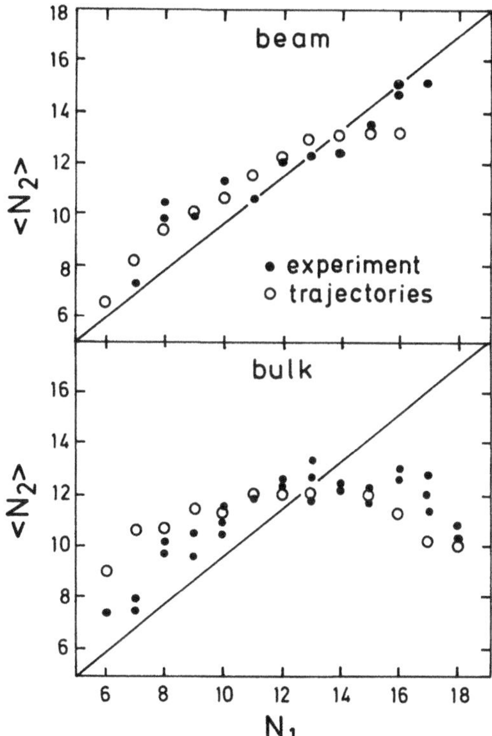

Fig. 11.13. The mean rotational state $< N_2 >$ of rotor 2 if the sibling rotor 1 is produced in rotational state N_1; comparison between experimental results and classical trajectory calculations. Here, N is the total angular momentum of $OH(^2\Pi)$ excluding the spin. With increasing temperature of the H_2O_2 parent molecule the correlation gradually diminishes. The straight line indicates complete correlation. Adapted from Maul, Gläser, and Gericke (1989).

where E_{excess} is the total energy available for distribution and B_{rot} is the rotational constant of OH. If we probe the rotational state of rotor 1 the possible states of the sibling fragment lead, as a consequence of the Doppler effect, to a broadening of the laser-induced fluorescence line from which — in principle — the distribution of rotational states of rotor 2 can be extracted by deconvolution.

The experimental results for the dissociation at 193 nm confirm the general predictions of the trajectory calculations: high correlation for dissociation in the beam and significantly reduced correlation if H_2O_2 is dissociated at room temperature. Because of insufficient frequency resolution, the comparison with theory is made only for the average rotational state $< N_2 >$ as a function of N_1 rather than the fully resolved distribution matrix $P(N_1, N_2)$. The results shown in Figure 11.13 reveal satisfactory agreement between experiment and calculation and illustrate clearly the loss of correlation with increasing temperature. The same experimental technique has been also applied to analyze the pair-pair correlation in the photodissociation of HN_3 (Gericke, Theinl, and Comes 1990; Gericke, Haas, Lock, Theinl, and Comes 1991).

12
Dissociation of van der Waals molecules

Van der Waals molecules are complexes held together by the weak long-range forces between closed-shell atoms and/or molecules, i.e., electrostatic forces, dispersive forces, hydrogen bonding, and charge transfer interactions (Van der Avoird, Wormer, Mulder, and Berns 1980; Buckingham, Fowler, and Hutson 1988). Typical examples are $X \cdots I_2$, $X \cdots ICl$, $X \cdots Cl_2$, $X \cdots HF$, $HF \cdots HF$ with X being He, Ne, Ar, etc. The dots indicate the weak physical bonding in contrast to a strong chemical bonding. The main characteristics of van der Waals molecules are:

1) The small dissociation energy ranging from a few cm^{-1} to about 1000 cm^{-1}.
2) The relatively large bond length, R_e, of typically 4 Å.
3) The retention of the properties of the individual entities within the van der Waals complex.
4) Relatively weak coupling between the van der Waals mode and the internal coordinates of the molecular entity.

Owing to the small dissociation energy, van der Waals molecules exist mainly at very low temperatures as they prevail in the interstellar medium or in supersonic jets.

Van der Waals molecules are rather floppy complexes which can be best described by the usual Jacobi coordinates which we used throughout all previous chapters (see the inset of Figure 12.1): the van der Waals bond distance R, the intramolecular separation of the chemically bound molecule r, and the orientation angle γ. Figure 12.1 depicts the R dependence of a typical potential energy surface (PES) $V(R, r, \gamma)$ for fixed values of r and γ. The long-range part of the potential is attractive and governed by the forces mentioned above. The leading term of a multipole expansion is typically proportional to R^{-6}. The short-range branch

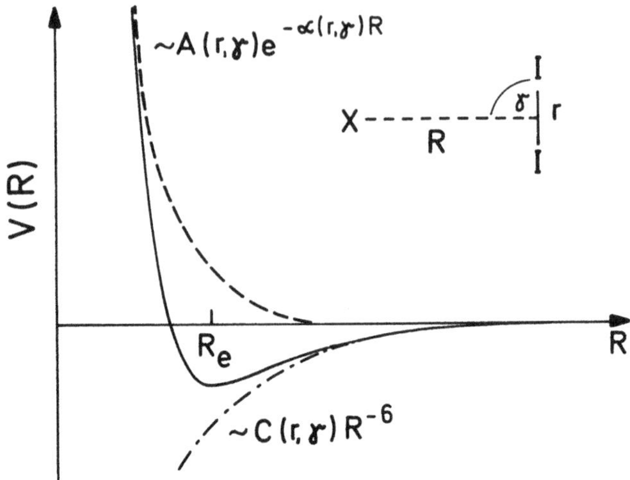

Fig. 12.1. Schematic illustration of a van der Waals potential as a function of the intermolecular bond distance R measured from the atom X to the center-of-mass of the diatom, I_2 in the present case. The orientation angle γ and the intramolecular I_2 bond distance are fixed. The potential parameters governing the long-range attractive and the short-range repulsive branches, namely A, α, and C, depend, in principle, on r and γ.

reflects mainly the repulsion between the electron clouds of the two fragments and can be well approximated by an exponential function. The superposition of both parts leads to the generic form of the potential shown in Figure 12.1 with a shallow minimum at a relatively large equilibrium distance R_e. Because of the small well depth, the potential is rather anharmonic in R and γ, thus normal modes are inappropriate to describe the vibration of such floppy molecules.

Potential energy surfaces of van der Waals molecules have — in comparison to the PESs of excited states of chemically bound molecules like H_2O, H_2S, or CH_3ONO — a relatively simple general appearance. There are no barriers due to avoided crossings and no saddle points etc. Moreover, the coupling between R, on one hand, and r and γ, on the other hand, is usually weak so that a representation of the form

$$V(R,r,\gamma) = \sum_{\lambda,k} V_{\lambda k}(R)\, P_\lambda(\cos\gamma)\, (r-r_e)^k + v_{\mathrm{BC}}(r) \tag{12.1}$$

is suitable; $P_\lambda(\cos\gamma)$ is the Legendre polynomial of order λ and $v_{\mathrm{BC}}(r)$ is the vibrational potential of the free diatom with equilibrium separation r_e. Since the intramolecular potential of the diatom changes only little when the two entities approach each other, usually only few expansion terms in r are required in (12.1). Furthermore, the *radial strength functions* $V_{\lambda k}(R)$ can be presented by simple functions (exponentials or inverse powers in R, for example) with the long-range behavior being determined by perturbation theory (Buckingham 1967; Levine and Bernstein 1987:ch.3). By virtue of fitting detailed experimental data such as excitation spectra (Le Roy and Carley 1980; Hutson 1990; Cohen and Saykally 1991) and elastic and inelastic scattering cross sections at thermal energies (Buck 1975, 1982; Faubel et al. 1982) it is, in principle, possible to determine the open parameters of such a potential model. Fitting an excited-state PES for ClNO, for example, is undoubtedly more difficult if possible at all, despite the availability of many detailed experimental data.

Because of the small dissociation energies, the absorption of an infrared (IR) photon corresponding to one vibrational or even only a single rotational quantum of the diatom suffices to break the van der Waals bond. A typical example is the breakup of rare gas–hydrogen compounds,

$$\mathrm{He}\cdots\mathrm{H}_2(n=0)+\hbar\omega_{\mathrm{IR}}\rightarrow\mathrm{He}\cdots\mathrm{H}_2(n=1)\rightarrow\mathrm{He}+\mathrm{H}_2(n=0,j).$$

Alternatively, one may pump an excited electronic state of the diatomic moiety with subsequent dissociation within the excited state. Representative examples are the rare gas–halogen complexes like

$$\mathrm{He}\cdots\mathrm{Cl}_2(n=0)+\hbar\omega_{\mathrm{UV}}\rightarrow\mathrm{He}\cdots\mathrm{Cl}_2^*(n')\rightarrow\mathrm{He}+\mathrm{Cl}_2^*(n'',j''),$$

where the * indicates electronic excitation of Cl_2. Energy conservation requires that $n'' < n'$.

As in the dissociation of chemically bound molecules with UV light, the main observables, from which one may infer both the structure of the complex and the fragmentation mechanism, are the absorption spectrum recorded under high resolution and the final state distribution of the product molecule. From the formal point of view, the breakup of weakly bound molecules is completely equivalent to the dissociation processes described in the foregoing chapters. The quite diverse time scales, however, demand different models of interpretation.

The last decade has witnessed an explosion of very detailed both experimental and theoretical studies of van der Waals fragmentation. The degree of nicety with which the breakup of individual quantum states can be investigated and the confluence with essentially exact quantum mechanical studies is fascinating (Levy 1980, 1981; Le Roy and Carley 1980; Beswick and Jortner 1981; Smith et al. 1982; Janda 1985; Legon and Millen 1986; Weber 1987; Miller 1988; Nesbitt 1988a,b; Halberstadt and Janda 1990; Bernstein 1990; Beswick and Halberstadt 1993). In the course of this monograph we have room to outline only the most elementary aspects. We will particularly elucidate the similarities and dissimilarities with UV photodissociation discussed in the preceding chapters. The many facets of van der Waals molecules and their theoretical description easily fill a whole monograph on their own (Beswick and Halberstadt 1993). In Sections 12.1 and 12.2 we will briefly outline the two basic mechanisms of dissociation, *vibrational* and *rotational predissociation* of weakly bound species. Finally, in Section 12.3 we will discuss the state distributions of the photofragments.

12.1 Vibrational predissociation

In this section we will explain the essential mechanism of *vibrational predissociation* by virtue of a linear atom–diatom complex such as $Ar \cdots H_2$. Figure 12.1 illustrates the corresponding Jacobi coordinates.† In particular, we consider the excitation from the vibrational ground state of H_2 to the first excited state as illustrated in Figure 12.2. The close-coupling approach in the diabatic representation, summarized in Section 3.1, provides a convenient basis for the description of this elementary process. For simplicity of presentation we assume that the coupling between the van der Waals coordinate R and the vibrational coordinate r is so weak that it suffices to include only the two lowest vibrational states, $n = 0$ and $n = 1$, in expansion (3.4) for the total wavefunction,

$$\Psi(R, r) = \chi_0(R)\, \varphi_0(r) + \chi_1(R)\, \varphi_1(r), \tag{12.2}$$

which leads to the two coupled equations for the radial functions $\chi_n(R)$ [see Equation (3.5)]

$$\begin{aligned} \left(\frac{d^2}{dR^2} + k_0^2\right) \chi_0(R) &= \frac{2m}{\hbar^2} \left[V_{00}(R)\, \chi_0(R) + V_{01}(R)\, \chi_1(R)\right] \\ \left(\frac{d^2}{dR^2} + k_1^2\right) \chi_1(R) &= \frac{2m}{\hbar^2} \left[V_{10}(R)\, \chi_0(R) + V_{11}(R)\, \chi_1(R)\right]. \end{aligned} \tag{12.3}$$

The wavenumbers k_n are defined in Equation (2.52) and the potential coupling matrix elements are given in (3.6) with V_I being the interaction potential, i.e., Equation (12.1) without the asymptotic oscillator potential $v_{BC}(r)$. Figure 12.2 shows the diagonal matrix elements $V_{00}(R)$ and $V_{11}(R)$. As a result of the weak dependence of V_I on r, they are at all distances R vertically displaced by approximately the energy difference between the two vibrational states, i.e., they run roughly parallel.

For energies below the first threshold, $Ar + H_2(n = 0)$, both vibrational channels are asymptotically closed and the coupled equations must be

† Van der Waals molecules which are composed of an atom and a homonuclear molecule such as $Ar \cdots H_2$ have a T-shape rather than a collinear geometry. But that is irrelevant for the present discussion.

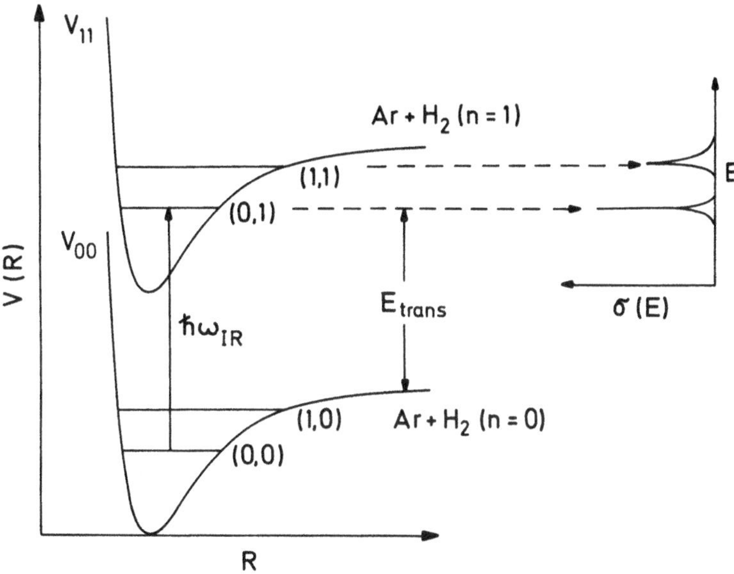

Fig. 12.2. Schematic illustration of the absorption of an IR photon by the van der Waals complex $Ar \cdots H_2$ and the subsequent dissociation into $Ar{+}H_2(n = 0)$. $V_{00}(R)$ and $V_{11}(R)$ are the diagonal elements of the potential coupling matrix defined in Equation (3.6) which serve to define the zero-order radial wavefunctions employed in the Golden Rule expression for the dissociation rate. The assignment of the bound levels is (m, n), where m and n denote the number of quanta of excitation in the dissociation mode R and the vibrational mode of H_2, respectively.

solved subject to the boundary conditions $\chi_0 \to 0$ and $\chi_1 \to 0$ in the limit $R \to \infty$. Between the $n = 0$ and the $n = 1$ thresholds, χ_0 is a continuum wavefunction and therefore must fulfil a boundary condition similar to (3.10) while χ_1 behaves like a bound-state wavefunction. Finally, above the $n = 1$ threshold both channels are energetically open.

The rigorous calculation of the absorption spectrum now proceeds in exactly the same way as outlined in Section 3.1.2 [see Equation (3.14)] with the exception that the transition dipole function is replaced by the dipole function [i.e., the diagonal element of the dipole matrix defined in Equation (2.35)] because the absorption takes place in the same electronic state. Below the first threshold, the spectrum is discrete because both vibrational states are true bound states. Above the $n = 0$ threshold, however, the system becomes open and can dissociate yielding Ar and $H_2(n = 0)$. The spectrum is consequently a continuous spectrum with sharp resonances near the quasi-bound states temporarily trapped by the

V_{11} potential curve. The inclusion of higher vibrationally excited states is straightforward.†

The breakup of a van der Waals molecule is a special example of indirect dissociation discussed in detail in Chapter 7. The photon first excites the vibrational mode of the molecular entity. Energy redistribution within the complex slowly transforms vibrational energy into translational energy associated with the van der Waals bond which eventually breaks. The autocorrelation function shows distinct recurrences with a period which corresponds to the period of the intramolecular vibration. The coupling element $V_{01}(R)$ and ultimately the r dependence of the interaction potential determine the rate of energy exchange and therefore the lifetime of the excited complex. Because the coupling between R and r is very weak for van der Waals molecules the lifetime of the resonance states is very long, usually much longer than for the examples discussed in Chapter 7.

Resonances of the type illustrated in Figure 12.2 are called *Feshbach resonances* (Child 1974:ch.4; Fano and Rao 1986:ch.8; see also Figure 12.5). The quasi-bound states trapped by the $V_{11}(R)$ potential can only decay via coupling to the lower vibrational state because asymptotically the $n = 1$ channel is closed and therefore cannot be populated. This is different from the dissociation of $CH_3ONO(S_1)$, for example, [see Figure 7.10(a)] where the resonances can either decay via tunneling or alternatively by nonadiabatic coupling to the lower states.

Predissociation of van der Waals molecules is ideally suited for the application of the general expressions for the decay of resonance states derived in Section 7.2, especially Equation (7.12) for the dissociation rate. The reason is that the coupling is so small that we can rigorously define accurate zero-order states

$$\Psi_n^{(0)}(R, r; E) = \chi_n^{(0)}(R; E)\, \varphi_n(r). \tag{12.4}$$

with $n = 0$ or Λ where the two radial wavefunctions $\chi_0^{(0)}$ and $\chi_1^{(0)}$ solve the first respectively the second equation in (12.3) with the coupling elements $V_{10} = V_{01}$ set to zero. $\chi_0^{(0)}$ is a one-dimensional continuum wavefunction with outgoing boundary conditions and $\chi_1^{(0)}$ is a one-dimensional

---

† If one is only interested in the positions and the widths of the resonances and not in the intensities of the spectrum one may perform a conventional scattering calculation yielding the S-matrix for transitions from initial vibrational state n to all open vibrational states n' [see Equation (2.59)]. For an isolated and narrow resonance, $\mathbf{S}(E)$ can be fitted to the Breit-Wigner form given in (7.27) which provides the resonance energy E_R and the dissociation rate Γ directly (Ashton, Child, and Hutson 1983). Resonances in scattering cross sections and in photodissociation cross sections have exactly the same origin; they occur at the same positions and have identical widths.

bound-state wavefunction as illustrated in Figure 12.3. Note that both wavefunctions belong to the same total energy.

For the sake of simplicity we assume that the interaction potential has the form $V_I(R,r) = V_0(R) + V_1(R)\,(r - r_e)$. Only the second term can induce vibrational-translational coupling and therefore initiate the decay of the quasi-bound state. With $\hat{W} \equiv V_1(R)\,(r - r_e)$ inserted in Equation (7.12) the dissociation rate becomes

$$\begin{aligned}\Gamma &= \hbar^{-2} |\langle \Psi_1^{(0)} \mid V_1(R)\,(r - r_e) \mid \Psi_0^{(0)} \rangle|^2 \\ &= \hbar^{-2} |\langle \chi_1^{(0)} \mid V_1(R) \mid \chi_0^{(0)} \rangle|^2 \, |\langle \varphi_1 \mid (r - r_e) \mid \varphi_0 \rangle|^2 .\end{aligned} \tag{12.5}$$

It factorizes into an *intermolecular* (the first term) and an *intramolecular* (the second term) factor. As predicted in Section 7.2 in general terms, the dissociation rate depends quadratically on the coupling potential $V_1(R)$.

The intermolecular factor involves the overlap of the bound-state wavefunction, the oscillatory continuum wavefunction, and the coupling potential $V_1(R)$. All three ingredients are illustrated in Figure 12.3. The coupling is usually small and mainly restricted to the repulsive branch of the potential. The smallness of the coupling potential and the cancellation due to the positive and the negative parts of the continuum wavefunction cause Γ to be exceedingly small under realistic conditions with the result that the resonance state originally excited by the IR photon can easily have lifetimes as large as microseconds. The actual value of the dissociation rate depends crucially on the parameters of the system, e.g., the reduced mass m and the translational energy E_{trans} which becomes available after the breakup, as well as subtleties of the PES. The larger m and/or E_{trans}, i.e., the smaller the de Broglie wavelength, the

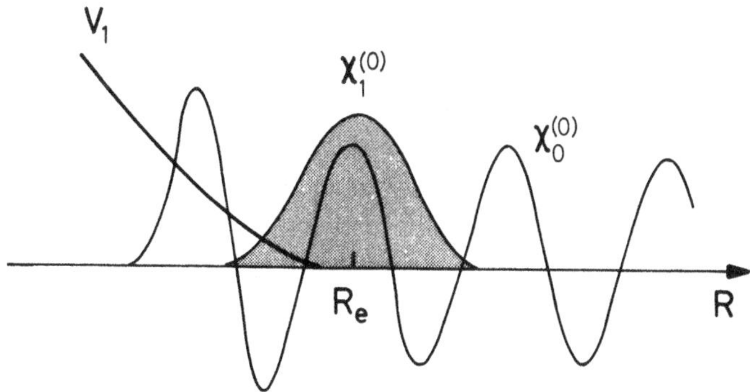

Fig. 12.3. Schematic illustration of the vibrational-translational coupling term $V_1(R)$, the bound-state wavefunction $\chi_1^{(0)}(R)$, and the continuum wavefunction $\chi_0^{(0)}(R)$. The overlap of these three functions determines the dissociation rate Γ according to Equation (12.5).

more rapidly oscillates the scattering wavefunction and consequently the smaller is the dissociation rate.

Assuming a Morse oscillator in R in order to obtain approximate expressions for the one-dimensional wavefunctions, Beswick and Jortner (1977, 1981) and Ewing (1979, 1981, 1982, 1987) derived a very handy formula for the dissociation rate which is known as the *energy gap* or *momentum gap law* ,

$$\Gamma \propto n \exp\left[-\frac{\pi P_{trans}}{\alpha\hbar}\right] = n \exp\left[-\frac{\pi(2mE_{trans})^{1/2}}{\alpha\hbar}\right]; \tag{12.6}$$

α is the range parameter of the Morse oscillator and n is the vibrational quantum number of the decaying state ($n = 1$ in the present case).† For the subsequent discussion it is important to emphasize that:

- The decay rate decreases (approximately) exponentially with the momentum corresponding to the van der Waals mode. The larger the reduced mass and the larger the excess energy the longer survives the compound state.

Whether Equation (12.6) is applicable for accurate predictions is not so important and certainly depends very much on the particular system. Its asset is its interpretative power. In particular, it elucidates the strong dependence of Γ on the final translational momentum P_{trans} which we assessed above in qualitative terms. Equation (12.6) clearly explains why the lifetimes of excited van der Waals molecules can differ by orders of magnitudes.

Due to the anharmonicity of the intramolecular potential $v_{BC}(r)$ the energy spacing between adjacent vibrational levels and therefore E_{trans} is not constant but *decreases* with n. This leads, in turn, to a drastic *increase* of the decay rate and therefore to a *decrease* of the lifetime with increasing n, in addition to the linear dependence induced by the intramolecular factor. Figure 12.4 shows measured dissociation rates for the fragmentation of $HeCl_2$, plotted in a way which emphasizes the linear dependence on n and the exponential dependence on the square root of the energy spacing $\epsilon_n - \epsilon_{n-1}$ between adjacent vibrational levels. The straight-line behavior convincingly proves the general applicability of Equation (12.6).

Because of the sensitive dependence of Γ on the final translational energy, excitation of a low-energy mode of one of the fragments, which

† The multiplication with n stems from the intramolecular term in Equation (12.5); see Section 12.3.1 and especially Equation (12.10).

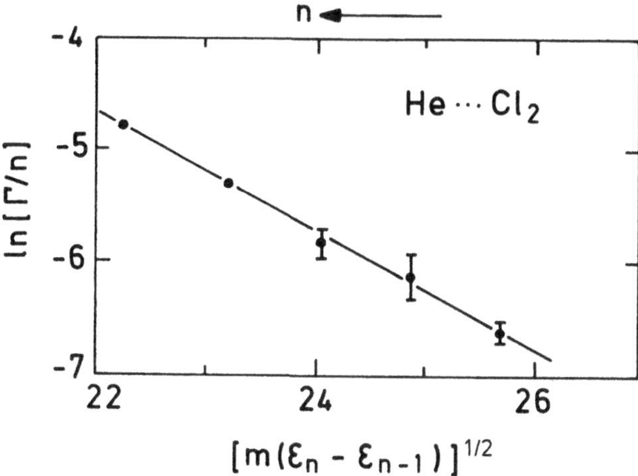

Fig. 12.4. $\ln(\Gamma/n)$ plotted versus $[m(\epsilon_n - \epsilon_{n-1})]^{1/2}$ for the vibrational predissociation of $HeCl_2$. m is the reduced mass of the van der Waals molecule and $\epsilon_n - \epsilon_{n-1}$ is the energy spacing between adjacent levels of the Morse oscillator. Note that n increases from the right- to the left-hand side! Adapted from Cline, Evard, Thommen, and Janda (1986).

necessarily decreases E_{trans}, leads to a significant enhancement of the predissociation rate. If the complex consists of an atom and a diatomic molecule, rotation is the only other channel which can consume energy. Therefore, any realistic treatment of an atom-diatom system must include in addition to the vibrational mode, which directly mediates the resonance and its decay, the rotation of the fragment molecule. In that case we must extend the expansion of the total wavefunction, Equation (12.2), by also including the rotor states Y_{j0} if $J = 0$ or the generalized angular basis functions $\Theta_{j\Omega}^{JMp}$ if $J \neq 0$. Deriving the corresponding coupled equations including vibration and rotation of the fragment should not cause any problem after reading the previous chapters (Le Roy 1984; Hutson 1991).

12.2 Rotational predissociation

In the same way as an excited van der Waals compound can dissociate through energy transfer from the vibrational to the translational mode, it can also break apart through internal energy redistribution between rotation and translation. To be specific, let us discuss a particular system, He $\cdots$ HF, on which highly resolved spectroscopic investigations as well as detailed calculations have been performed (Lovejoy and Nesbitt 1990). The confluence of both has led to an accurate two-dimensional PES $V(R, \gamma)$ where γ is the Jacobi angle defined in Figures 3.1 and 12.1.

The fascination of this and similar systems [NeHF (Clary, Lovejoy, ONeil, and Nesbitt 1988; Nesbitt et al. 1989; ONeil et al. 1989), NeDF (Lovejoy and Nesbitt 1991), and ArHF (Nesbitt, Child, and Clary 1989)] arises from the possibility of studying molecular breakup on the level of individual quantum states (Nesbitt 1988a,b).

In the experiment the HF molecule within the complex is excited by an IR photon from the vibrational-rotational ground state ($n = 0, j = 0$) to particular rotational states within the vibrationally excited manifold, ($n = 1, j$) (see Figure 12.5). Since vibrational predissociation is extremely weak in hydrogen containing systems (because the energy gap between adjacent states is relatively large) the He $\cdots$ HF($n = 1, j$) complex can be considered to be stable if the total energy is below the HF($n = 1, j = 0$) asymptote. However, if the He + HF($n = 1, j = 0$) channel becomes energetically accessible, the system will ultimately dissociate with a rate depending on the anisotropy of the PES *and* on the particular overall rotational state. It is this intimate quantum state dependence of the dissociation rate respectively the lifetime which we will focus our attention on in this section.

12.2.1 Characterization of rotational eigenstates

In order to understand the subsequent discussion it is helpful first to elucidate the eigenvalue spectrum of the triatomic compound [see also Nesbitt (1988a,b)]. The body-fixed theory for total angular momenta $J \neq 0$ outlined in Section 11.1, especially the set of coupled equations (11.7), provides a convenient starting point. As shown there for any arbitrary asymmetric top, each angular momentum state with quantum number J has $2J+1$ nondegenerate sublevels, $J+1$ for the parity quantum number $p = 1$ and J states for $p = -1$. Remember that the total parity of the nuclear wavefunction is given by $p(-1)^J$.

In contrast to strongly bound molecules like H_2O, for example, the rotational quantum number j and the helicity quantum number Ω are almost "good" quantum numbers (i.e., the corresponding observables are almost conserved) for weakly bound van der Waals molecules and therefore they can be used to label the various eigenstates. Ω is the projection quantum number of the molecular angular momentum $\mathbf{j}$ onto the body-fixed z-axis $\mathbf{R}$. According to (11.6), the total wavefunction is expanded in terms of the parity-adapted angular basis functions (11.5) with the radial functions $\chi_{j\Omega}(R; Jp)$ solving the resulting set of coupled equations (11.7).

Zeroth-order energies and wavefunctions may be obtained by neglecting the coupling between different j- and different Ω-states and solving the

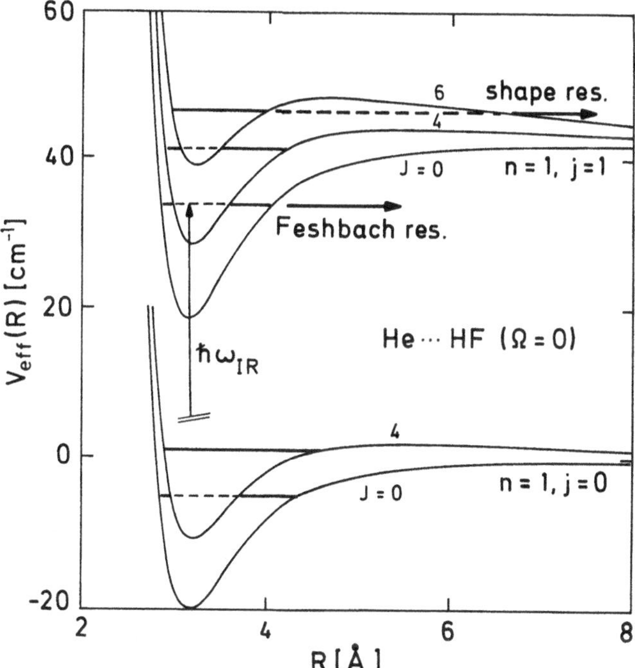

Fig. 12.5. Zeroth-order potentials $V_{eff}(R; j, \Omega, J)$ defined in (12.7) for $\Omega = 0$ and several total angular momentum quantum numbers J. The excited rotational states can decay either by tunneling (shape resonances) or by rotational predissociation (Feshbach resonances) as indicated by the horizontal arrows. The excitation through the IR photon originates from the vibrational ground state $n = 0$ which is not shown in the figure.

one-dimensional Schrödinger equation with one-dimensional potentials

$$V_{eff}(R; j, \Omega, J) = B_{rot}\, j(j+1) \\ + \hbar^2 \frac{J(J+1) + j(j+1) - 2\Omega^2}{2mR^2} + V_{jj}^{(\Omega)}(R), \quad (12.7)$$

where B_{rot} is the rotational constant of HF and the $V_{jj}^{(\Omega)}$ are the diagonal elements of the potential coupling matrix defined in (11.9). The effective potentials $V_{eff}(R; j, \Omega, J)$ are simply the diagonal elements of the coupled equations (11.7). They depend *parametrically* on J, j, as well as the helicity quantum number Ω. Figure 12.5 depicts examples for $j = 0$ and 1, $\Omega = 0$, and some selected values of the total angular momentum quantum number J. Because of the large van der Waals bond distances, the rotational constant for overall rotation of the complex is significantly smaller than the rotational constant of the diatom. This causes the energy spacing between the various J levels to be much smaller than the spacing between the rotational levels of HF. Potentials similar to those in Figure 12.5 also exist for $\Omega = 1$. Remember, however, that the helicity is a projection quantum number and as such it must be smaller or equal to the minimum of J and j.

Each of the potentials shown in Figure 12.5 supports at least one bound or quasi-bound state which can be labeled by quantum numbers (j, Ω, J). These zeroth-order states correspond to almost free rotation of HF within the van der Waals complex with quantum numbers $j = 0, 1, 2, \ldots$ and $\Omega = 0, 1, 2, \ldots, \min(j, J)$. In analogy with the nomenclature for electronic states, they are termed Σ and Π for $\Omega = 0$ and 1, respectively. For $j = 1$ and $\Omega = 0$ the diatom rotates in the plane defined by the three atoms. In contrast, for $j = 1$ and $\Omega = 1$ it rotates in a plane perpendicular to the intramolecular vector $\mathbf{R}$. As J increases, the centrifugal potential $\hbar^2[J(J+1) + j(j+1) - 2\Omega^2]/2mR^2$ increases as well and eventually $V_{eff}(R; j, \Omega, J)$ becomes purely repulsive and the sequence of bound or quasi-bound states breaks off.

If the coupling between the different j-states, caused by the off-diagonal elements of the potential matrix (11.9), and the Coriolis coupling between different helicity states Ω, induced by the third line in (11.7), are incorporated the energies change slightly without, however, altering the general pattern. In addition, all states with helicities $\Omega \geq 1$ split into two sublevels, one having positive and the other one having negative parity. The splitting is caused by the Coriolis coupling and for HeHF and $J = 1$ it is of the order of only 3 cm^{-1}.

12.2.2 Decay mechanisms

Measuring the vibrational-rotational energies by absorption spectroscopy allows one accurately to fit the interaction potential which for weakly bound systems may be represented as

$$V(R, \gamma) = V_0(R) + V_1(R)\, P_1(\cos\gamma) + V_2(R)\, P_2(\cos\gamma) + \ldots \qquad (12.8)$$

(Nesbitt 1988a,b; Hutson 1990). For the purpose of this monograph we will subsequently feature the decay of the quasi-bound levels which contains additional information on the PES, especially its anisotropy.

In the experiment of Lovejoy and Nesbitt (1990) the IR photon excites the complex into the $(n = 1, j = 1)$ manifold. There are two pathways for fragmentation within the $n = 1$ state:

1) Rotational predissociation by a transition from $j = 1$ to $j = 0$.
2) Tunneling through the centrifugal barrier provided the $j = 1$ channel is asymptotically open.

In the first case, energy is transferred from the rotational to the translational mode which enables the system to rupture the weak van der Waals bond. Rotational predissociation is completely equivalent to vibrational predissociation.[†] The second type of fragmentation can occur only for higher total angular momentum states when the effective potential has a well and a barrier. In contrast to Feshbach resonances, this type of resonance is known as *orbiting* or *shape resonance* in full collisions (Child 1974:ch.4; Pauly 1979). Note that transitions between different total angular momentum or different parity states are, of course, not allowed!

The dissociation rate Γ for each quasi-bound level can be extracted from the homogeneous widths of the absorption lines. It depends on j, J, Ω as well as the parity of the considered state. Figure 12.6 depicts experimental and theoretical linewidths for rotational state $j = 1$ and

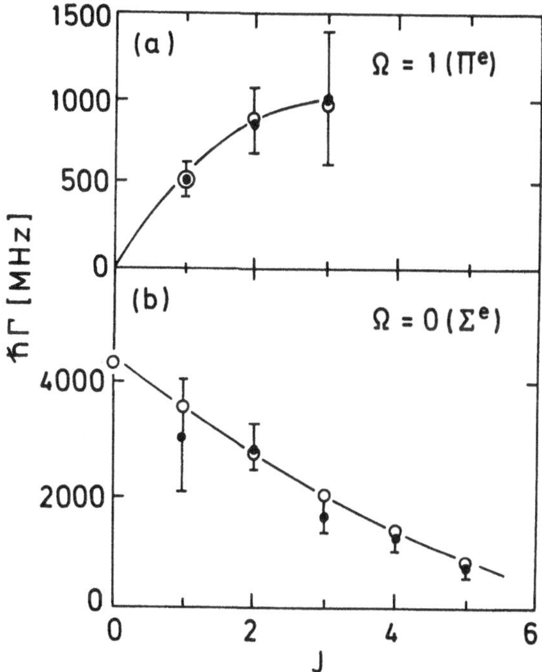

Fig. 12.6. Measured (full circles) and calculated (open circles) linewidths $\hbar\Gamma$ for He$\cdots$HF($j = 1$) as a function of the total angular momentum J for helicity states $\Omega = 0$ and 1 and even parity. The corresponding spectroscopic notation is Σ^e and Π^e, respectively. 1000 MHz corresponds to 0.033 cm^{-1}. Adapted from Lovejoy and Nesbitt (1990).

† However, because the energy spacing between rotational states is typically one order of magnitude smaller than between vibrational states, the rotational predissociation rates are generally much larger than the rates for vibrational predissociation (energy gap law). An illuminating example is the dissociation of ArH_2 (Kidd and Balint-Kurti 1985, 1986).

both helicity states; for $\Omega = 1$ only states with even parity are considered here. The theoretical data are obtained in a fully converged close-coupling calculation, Equation (11.7), for fixed J and parity parameter p. The two-dimensional PES is a slight modification of the *ab initio* PES of Rodwell, Sin Fain Lam, and Watts (1981). In order to extract the resonance widths from the close-coupling calculations the full-collision S-matrix determined on a narrow grid of total energies is fitted to the Breit-Wigner form (7.27).

Let us start the discussion with $\Omega = 0$. These resonances can *directly* decay to the $j = 0$ rotational state which automatically has the same helicity $\Omega = 0$. The coupling is mainly provided by the $V_{10}^{(\Omega=0)}(R)$ potential matrix element [see Equation (11.9)]. Because of the triangular condition $|j - j'| \leq \lambda \leq j + j'$ following from the Wigner $3j$-symbol, $V_{10}^{(\Omega=0)}(R)$ is proportional to the potential term with $\lambda = 1$ in expansion (12.8). In other words, the linewidths are most sensitive to $V_1(R)$. If we calculate the transition rate according to the perturbation theory limit similar to (12.5), we have to evaluate the overlap between a bound-state wavefunction in rotational channel $j = 1$ and a continuum wavefunction in rotational channel $j = 0$ multiplied with the coupling term $V_1(R)\, P_1(\cos\gamma)$. Because of the increase of the centrifugal potential with J, the corresponding radial wavefunctions shift slightly but gradually to larger distances (*centrifugal stretching*) which has the effect that the overlap with the coupling potential $V_1(R)$ continuously diminishes. This explains the decrease of $\hbar\Gamma$ with increasing J for $\Omega = 0$ in Figure 12.6(b).

The linewidth for the $\Omega = 1$ states shows quite the opposite dependence on J. It first rises with J and reaches a maximum around $J = 3$. In addition, it is significantly smaller than the linewidths for the $\Omega = 0$ resonances. Unlike the latter, the $\Omega = 1$ quasi-bound states *cannot directly* decay to the $j = 0$ asymptotic channel which has the helicity quantum number $\Omega = 0$. Transitions between different helicity states can be promoted only through Coriolis coupling, i.e., the last two terms on the right-hand side of (11.7). Coriolis coupling, on the other hand, is diagonal in the rotational quantum number j. Therefore, the $j = 1, \Omega = 1$ resonances can only decay via a two-step process of the form

$$\begin{aligned} \mathrm{He}\cdots\mathrm{HF}(j = 1, \Omega = 1) &\stackrel{\text{Coriolis}}{\longrightarrow} \mathrm{He}\cdots\mathrm{HF}(j = 1, \Omega = 0) \\ &\stackrel{\text{potential}}{\longrightarrow} \mathrm{He} + \mathrm{HF}(j = 0), \end{aligned}$$

where the first step is induced by Coriolis coupling and the final decay is caused by potential coupling. The functions $C_{\Omega\Omega\pm1}^{Jp}(R)$ in (11.7), which provide the coupling between different helicity states, are approximately proportional to J giving rise to the increase of $\hbar\Gamma$ with J. On the other hand, the rate for the final decay step decreases with J as discussed above.

The interplay between the two individual rates explains the overall J dependence observed in Figure 12.6(a).

In conclusion, the quantum-state dependence of the dissociation rate reveals a lot of detailed information on the decay of van der Waals complexes and allows to accurately ascertain parts of the multi-dimensional PES.

12.3 Product state distributions

The last question that we will address concerns the final vibrational and rotational state distributions of the diatomic fragments. Although the excited resonance states can live up to nanoseconds or even microseconds, the final distributions do not follow simple statistical laws which were briefly spoken about in Section 10.3.2. On the contrary, they manifest either prominent propensity rules or dynamical features similar to those discussed in the context of direct dissociation.

12.3.1 Final vibrational state distributions

Let us assume that vibrational state n of the diatom is originally excited in the complex. Then, all channels $n' < n$ can, in principle, be populated after the fragmentation with partial decay rates $\Gamma_{n'}^{(n)}$. The $\Gamma_{n'}^{(n)}$ represent the rates for the breakup of the complex *and* for filling vibrational channel n' of the diatomic fragment. The total rate is the sum of all partial rates and the final vibrational state distribution is proportional to $\Gamma_{n'}^{(n)}$ [see Chapter 7 and especially Equations (7.15) and (7.23)]. In perturbation theory, outlined in Section 12.1, the partial rates are given by

$$\Gamma_{n'}^{(n)} = \hbar^{-2} |\langle \chi_n^{(0)} \mid V_1(R) \mid \chi_{n'}^{(0)} \rangle|^2 \, |\langle \varphi_n \mid (r - r_e) \mid \varphi_{n'} \rangle|^2, \tag{12.9}$$

where we assumed that the interaction potential is linear in $(r - r_e)$. Equation (12.9) is the analogue to (12.5) for an arbitrary initial vibrational state with quantum number n.

If we assume the diatom to be harmonic, the matrix element of the displacement operator in (12.9) can be shown to yield (Cohen-Tannoudji, Diu, and Laloë 1977:ch.V)

$$\langle \varphi_n \mid (r - r_e) \mid \varphi_{n'} \rangle = \left(\frac{\hbar}{2\mu\omega} \right)^{1/2} \left[(n'+1)^{1/2} \, \delta_{n,n'+1} + (n')^{1/2} \, \delta_{n,n'-1} \right], \tag{12.10}$$

where ω and μ are the frequency and the reduced mass of the oscillator, respectively. Access of the next higher state, $n' = n + 1$, is forbidden by energy conservation and thus it follows that:

- Under the assumptions of a harmonic oscillator and an interaction potential, which is linear in $(r - r_e)$, only the next lowest vibrational channel $n = n^* - 1$ can be populated. The propensity rule $n^* \rightarrow n^* - 1$ is a strict selection rule.

In accordance with our previous notation we indicate the quasi-bound state by a star.

In reality, the oscillator is not harmonic nor is the interaction potential linear which has the consequence that other channels with $\Delta n = n^* - n > 1$ are also filled in the fragmentation. However, discrepancies from the ideal situation can be assumed to be small so that the channel $n^* - 1$ clearly dominates. Furthermore, the intermolecular factor in Equation (12.9) depends crucially on the final vibrational quantum number as discussed in Section 12.1. The larger Δn the more energy is released as translational energy and the smaller is, according to the energy gap law, the overlap of the bound and the continuum wavefunctions. The drastic decrease of the bound-continuum overlap factor with decreasing n' additionally favors the population of the next lowest state $n' = n^* - 1$. Note the similarity with the UV photodissociation of CH_3ONO in the S_1 absorption band where we also observed a $n^* \rightarrow n^* - 1$ propensity (see Figure 9.12).

12.3.2 Final rotational state distributions

Rotational state distributions following the decay of long-lived van der Waals complexes reveal more interesting (but also more complicated) dynamical features than vibrational distributions. The energy gap or the momentum gap model predicts that the highest accessible rotational states have the largest probability since the corresponding energy in the van der Waals mode is, by energy conservation, minimal so that the overlap with the bound-state wavefunction of the resonance state is largest. On the other hand, the population of high rotational states requires significantly anisotropic potential energy surfaces to support large Δj transitions whereas the anisotropy in van der Waals molecules is usually rather weak.

In the following we will discuss two examples which have been studied in considerable detail both experimentally and theoretically: $X \cdots Cl_2$ and $X \cdots ICl$ with X = He, Ne, and Ar (Evard et al. 1988; Cline et al. 1988, 1989; Skene, Drobits, and Lester 1986; Drobits and Lester 1988a,b; Reid, Janda, and Halberstadt 1988; Waterland, Lester, and Halberstadt 1990; Waterland, Skene, and Lester 1988; Halberstadt, Roncero, and Beswick 1989; Gray and Wozny 1989, 1991; Gray 1991). For a review see Janda and Bieler (1990). In both cases, the halogen molecule is excited from the electronic ground state to a particular vibrational state n^* in the electronic B state which subsequently breaks apart via vibrational predissociation. Rotational state distributions of the halogen fragments have been measured by laser detection methods. By changing the initial vibrational state n^* one can vary the released energy and thus study the influence of the total available energy on the degree of rotational excitation. Similarly, one can determine rotational state distributions in various vibrational levels $n^* - 1$, $n^* - 2$, etc.

$HeCl_2$ has a T-shape, i.e., the equilibrium angle is $\gamma_e = 90°$, and therefore it does not come as a surprise that Cl_2 is produced in relatively low rotational states. Figure 12.7 shows measured rotational state distributions following the decay of the $n^* = 12$ level with $\Delta n = n^* - n = 1$, 2,

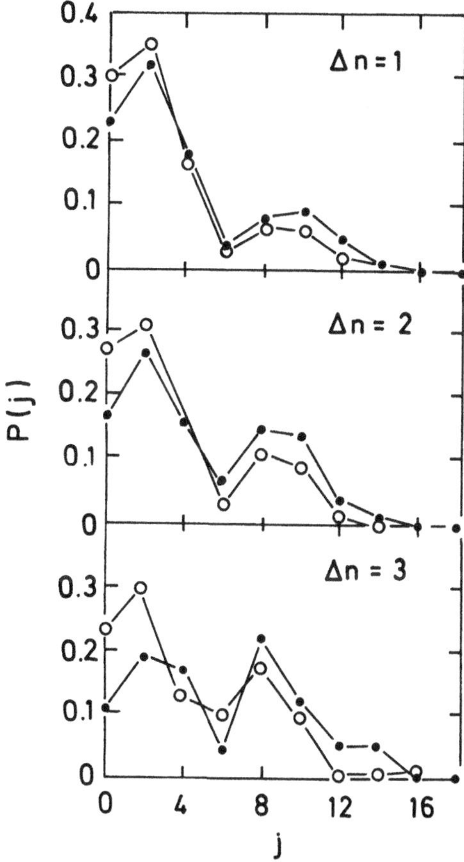

Fig. 12.7. Rotational state distributions of Cl_2 following the dissociation of the $n^* = 12$ vibrational level of $He \cdots Cl_2$. The final vibrational levels of Cl_2 are $n = 11$, 10, and 9, respectively. Comparison of measured (filled circles) and calculated (open circles) distributions. Adapted from Cline et al. (1988).

and 3, respectively. They peak at $j = 1$, have a minimum at $j = 6$, and a second, less intense maximum at $j = 8$–10. Also shown are the results of exact close-coupling calculations which employ a three-dimensional PES fitted to reproduce all the available data, namely the energy levels of the resonance states, the lifetimes, as well as the rotational state distributions of the Cl_2 fragment. The agreement of the calculated distributions with the measured results is excellent, not only for those transitions shown in Figure 12.7 but also for many others. The results for Ne $\cdots$ Cl_2 are similar except that the range of rotational states excited in the fragmentation is noticeably larger. This is very likely due to the significantly increased reduced mass m for Ne.

Most striking in Figure 12.7 is the bimodality of the distribution, despite the fact that the dissociation initiates from the lowest bending state. Remember that in the UV photodissociations of ClNO(T_1) and FNO(S_1) bi- and even trimodal distributions have been observed, however, only if the initial state of the complex is an excited bending state (see Figures 10.15 and 10.16). Also remarkable is the relative insensitivity of the final rotational state distribution on the final vibrational state and hence on the total available energy which varies from 120 cm^{-1} for $\Delta n = 1$ to 410 cm^{-1} for $\Delta n = 3$. This is in striking contrast to direct dissociation where the range of excited rotational states usually depends roughly on the square root of E_{excess} [see Equations (6.31) and (10.11), for example].

The general behavior of the rotational state distributions of ICl following the dissociation of He $\cdots$ ICl and Ne $\cdots$ ICl is quite similar to Cl_2 with the exception that the degree of rotational excitation is noticeably larger. For example, the main maximum occurs at significantly higher states than for Cl_2. The substantial torque, necessary to excite such high rotational states, probably reflects the large offset of the center-of-mass from the midpoint between I and Cl. Figure 12.8 depicts an example for the dissociation of Ne $\cdots$ ICl (Skene, Drobits, and Lester 1986) in comparison with the results of an exact close-coupling calculation which employs a fitted PES (Roncero et al. 1990). Note the remarkable similarity with the distributions following from direct dissociation (see, for example, Figures 10.9 and 10.10 for the dissociation of H_2O_2). As in the case of He $\cdots$ Cl_2, the distributions resulting from the fragmentation of He $\cdots$ ICl reveal a bimodal shape.

Both the structured distributions for He as partner atom and the highly inverted distributions for Ne strongly suggest that a direct scattering mechanism as discussed in Chapters 6 and 10 rather than statistical laws mentioned in Section 10.3.2 determines the final rotational state distributions. This is, at first glance, astonishing because the van der Waals complex lives for more than a thousand internal vibrational periods and

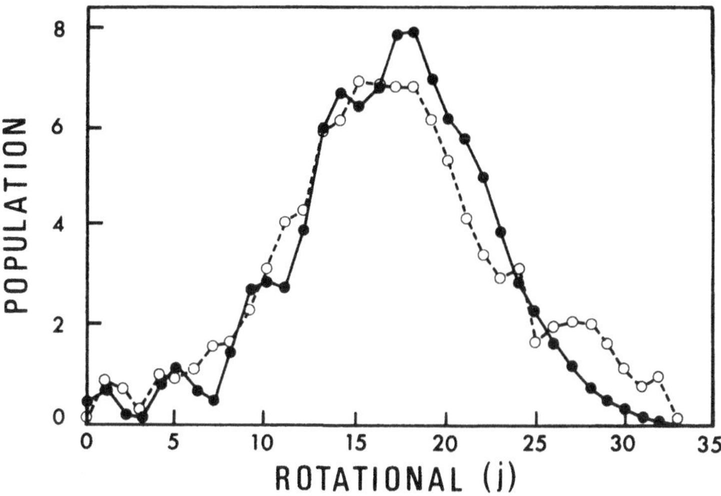

Fig. 12.8. Measured (open circles; Skene, Drobits, and Lester 1986) and calculated (filled circles; Roncero et al. 1990) rotational state distributions following the predissociation of the $n^* = 2$ vibrational state of Ne $\cdots$ ICl. Reproduced from Roncero et al. (1990).

thus one is tempted to assume that the energy is randomized before the final breakup.

All relevant data, i.e., the resonance energies, the linewidths, and the final state distributions, can be obtained in exact time-independent close-coupling calculations as outlined in Section 11.1.1 with the addition that the vibrational degree of freedom of the fragment molecule is also taken into account (Hutson 1991; Roncero et al. 1990). Deeper insight is gained, however, if one calculates the partial dissociation rates $\Gamma_{nj}^{(n^*)}$ in perturbation theory with the translational-vibrational coupling considered as the perturbation $\hat{W}$ (see Section 7.2 and especially Figure 7.5). $\Gamma_{nj}^{(n^*)}$ is the rate with which the resonance state n^* decays and the various rotational states j within the final vibrational state n of Cl_2 are filled. The necessary prerequisite for the perturbation theory to be valid is a very weak coupling between R and r. The coupling between translation and rotation, on the other hand, is treated exactly.

If we again assume that the interaction potential is linear in $(r - r_e)$, i.e.,

$$V_I(R, r, \gamma) = V_0(R, \gamma) + V_1(R, \gamma)(r - r_e), \tag{12.11}$$

the partial decay rates become

$$\Gamma_{nj}^{(n^*)} = \hbar^{-2} \left| \langle \Psi_{n^*}^{(0)}(R, \gamma) \mid V_1(R, \gamma) \mid \Psi_n^{(0)}(R, \gamma; E, j) \rangle \right|^2 \times \left| \langle \varphi_{n^*} \mid (r - r_e) \mid \varphi_n \rangle \right|^2. \tag{12.12}$$

The wavefunctions $\Psi_n^{(0)}(R, \gamma; E, j)$ and $\Psi_{n^*}^{(0)}(R, \gamma)$ are zeroth-order wavefunctions. $\Psi_n^{(0)}(E, j)$ is the solution of the two-dimensional Schrödinger equation in R and γ with Hamiltonian (3.16) and outgoing boundary conditions in rotational channel j. The appropriate potential is the diagonal potential matrix element $V_{nn}(R, \gamma)$ in vibrational channel n as defined in (3.6). In the same manner, $\Psi_{n^*}^{(0)}$ is a bound-state wavefunction and the corresponding two-dimensional PES is $V_{n^*n^*}(R, \gamma)$. For simplicity we assumed that $J = 0$. The intramolecular factor $|\langle \varphi_{n^*} \mid (r - r_e) \mid \varphi_n \rangle|^2$ is independent of j and therefore unimportant for the analysis of the final rotational state distributions. It merely determines the overall dissociation rate.

The intermolecular term has the same general form as the absorption cross section in the case of direct photodissociation, namely the overlap of a set of continuum wavefunctions with outgoing free waves in channel j, a bound-state wavefunction, and a coupling term. For absorption cross sections, the coupling between the two electronic states is given by the transition dipole moment function $\mu^{(e)}(R, r, \gamma)$ whereas in the present case the coupling between the different vibrational states n^* and n is provided by $V_1(R, \gamma) = \partial V_I(R, r, \gamma)/\partial r$ evaluated at the equilibrium separation $r = r_e$.

In view of this similarity it seems appealing to analyze the rotational state distribution following the vibrational predissociation of van der Waals molecules in the same way as outlined in Section 6.3 for direct dissociation, essentially using classical mechanics and the rotational reflection principle. There is, however, one very important difference: while the coordinate dependence of the transition dipole function can usually be ignored, the coupling potential $V_1(R, \gamma)$ depends strongly on both R and γ. The angle dependence is especially crucial for the understanding of final state distributions (Waterland, Lester, and Halberstadt 1990; Gray and Wozny 1991). It is generally largest in the collinear geometry ($\gamma = 0$) and smallest for the perpendicular configuration ($\gamma = 90°$). This concurs with the experience that vibrational excitation or quenching in neutral atom-molecule collisions is usually strongest in the collinear approach. In addition, the sign of $V_1(R, \gamma)$ may change as a function of γ.

In the context of classical mechanics one calculates trajectories in the four-dimensional phase-space including R and γ and the corresponding momenta P and j (Halberstadt, Beswick, and Schinke 1991; Gray and Wozny 1991). The trajectories are started with the appropriate total energy at the classical turning point $R_t(\gamma_0)$ and followed until the complex is dissociated. Since the vibrational degree of freedom is excluded from

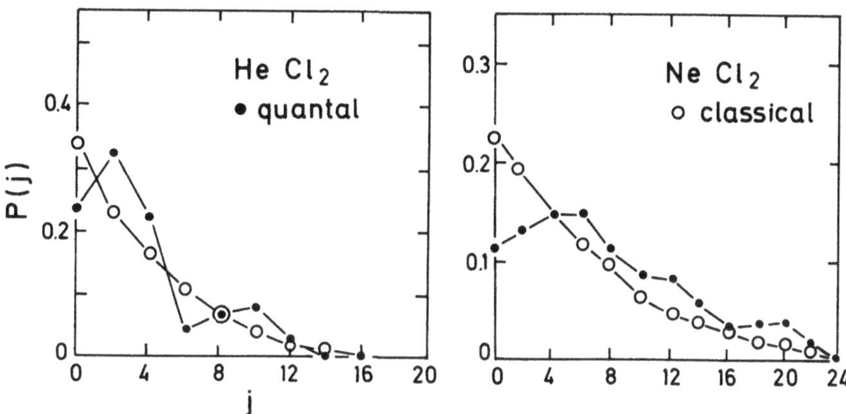

Fig. 12.9. Comparison of final rotational state distributions of Cl_2 as obtained from classical Monte Carlo calculations and exact quantum mechanical close-coupling calculations for $HeCl_2$ and $NeCl_2$. Adapted from Halberstadt, Beswick, and Schinke (1991).

this zeroth-order analysis, the fragmentation is fast and direct. Each trajectory is then weighted by

$$W(\gamma_0) = \sin\gamma_0 \left|V_1(R_t, \gamma_0)\, \Psi_{n^*}^{(0)}(R_t, \gamma_0)\right|^2 \tag{12.13}$$

where $\Psi_{n^*}^{(0)}(R, \gamma)$ is the resonance wavefunction and V_1 is the potential term which couples the vibrational motion of the diatom and the van der Waals mode.

Figure 12.9 depicts a comparison between classical trajectory results and exact close-coupling calculations for $He \cdots Cl_2$ and $Ne \cdots Cl_2$, respectively. In both cases, the classical procedure reproduces the overall behavior of the final state distributions satisfactorily. Subtle details such as the weak undulations particularly for He are not reproduced, however. As shown by Gray and Wozny (1991), who treated the dissociation of van der Waals molecules in the time-dependent framework, the bimodality for $He \cdots Cl_2$ is the result of a quantum mechanical interference between two branches of the evolving wavepacket and therefore cannot be obtained in purely classical calculations.

In summary, the overall behavior of final state distributions in the decay of van der Waals complexes may be well described by direct dissociation mechanisms and classical mechanics, despite the exceedingly large lifetime.

13
Photodissociation of vibrationally excited states

So far we have consistently assumed — with only very few exceptions — that the photodissociation starts from the lowest vibrational state of the parent molecule. The corresponding bound-state wavefunction is typically a narrow multi-dimensional Gaussian-like function centered at the equilibrium configuration in the electronic ground state. This wavefunction defines the starting zone for the motion of the time-dependent wavepacket or the swarm of classical trajectories on the excited-state potential energy surface (PES). If the dissociation proceeds in a direct way, the forces near the Franck-Condon region determine to a large extent the fate of the wavepacket and ultimately the energy and state dependence of the dissociation cross sections. Since the initial wavefunction for a normal, chemically bound molecule such as H_2O has a typical width of the order of 0.1–0.2 Å, the evolving wavepacket explores only a relatively small portion of the dissociative PES (see Figures 3.2, 9.8, and 9.9 for example).

By starting the photodissociation from an excited vibrational level one can access a significantly wider range of the upper-state PES (see Figure 13.1) and to some extent manipulate and steer the reaction path. One example has already been discussed in Section 10.1: dissociation of excited bending states of H_2O through the $\tilde{A}$ state probes a much wider angular region of the corresponding PES than dissociation of the lowest bending state. The influence of the increased anisotropy for smaller HOH angles clearly shows up in the final rotational state distribution of the OH product (see Figure 10.5).

Theoretically, the calculation of photodissociation cross sections for excited vibrational states proceeds in exactly the same way as for the dissociation of the lowest level. The basic quantities are the photodissociation amplitudes (2.68) with the initial wavefunction $\Psi_i(E_i)$ being an oscillatory rather than a bell-shaped function like in the dissociation

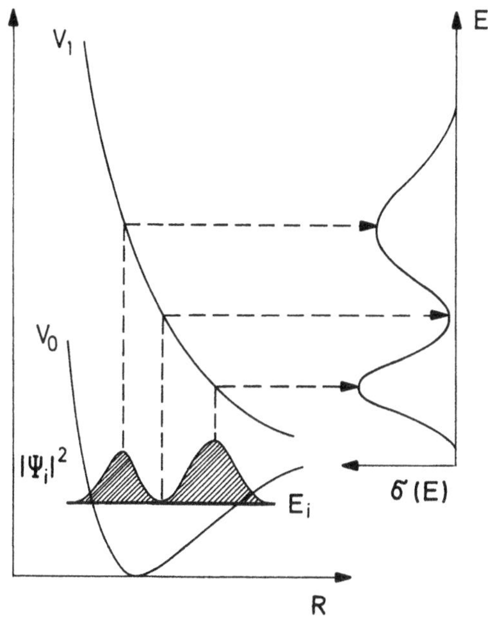

Fig. 13.1. Schematic illustration of the origin of reflection structures in the absorption spectrum if the parent molecule is initially excited.

of the vibrational ground state. In the time-independent formulation we can simultaneously determine cross sections for many initial states without much further work, because the calculations of the bound and of the continuum wavefunctions are independent of each other. Figure 11.7 illustrates schematically how the photodissociation of a single excited state of the parent molecule can be accomplished experimentally by employing two lasers, one resonantly to excite a selected vibrational level in the electronic ground state and a second one to photolyze this and only this state (Häusler, Andresen, and Schinke 1987; Luo and Rizzo 1990, 1991; Vander Wal, Scott, and Crim 1991).

In this chapter we will focus on the following questions:

1) How does the absorption spectrum reflect the nodal structure of the wavefunction of the parent molecule in the electronic ground state?
2) How does the initial excitation influence the final product state distributions and the chemical branching provided two different dissociation channels can be accessed?

We first briefly discuss in Section 13.1 the one-dimensional case and a simple two-dimensional model. The photodissociation of highly excited vibrational states of H_2O and HOD, for which both experimental and theoretical results are available, will be reviewed in more detail in Sections 13.2 and 13.3.

13.1 Reflection structures

In this section we will elucidate — on a *qualitative* basis — the implications of excitation in the electronic ground state on the energy dependence of the absorption spectrum.

13.1.1 One-dimensional case

Let us consider a one-dimensional model system with coordinate R as illustrated in Figure 13.1. $V_0(R)$ and $V_1(R)$ are the potentials in the ground and in the excited electronic states and $\Psi_i(R)$ represents the initial wavefunction. In analogy to Equation (6.3), the classical absorption cross section may be approximately written as

$$\sigma(E) \approx |\Psi_i(R_t)|^2 \left|\frac{dV_1}{dR}\right|^{-1}_{R=R_t(E)}, \tag{13.1}$$

where the classical turning point $R_t(E)$ is defined through

$$V_1(R_t) = E. \tag{13.2}$$

According to the reflection principle, discussed in detail in Section 6.1, the absorption cross section is simply a reflection of the coordinate distribution, $|\Psi_i(R)|^2$, onto the energy axis mediated directly by the upper-state potential $V_1(R)$. Equation (13.2) establishes the unique relation between R and E. Within this simple classical picture each maximum and each minimum of $|\Psi_i(R)|^2$ is uniquely mapped onto the energy axis as illustrated in Figure 13.1.

This ultrasimple classical theory is, of course, too crude for practical applications, especially for highly excited states of the parent molecule. Its usefulness gradually diminishes as the degree of vibrational excitation increases, i.e., as the initial wavefunction becomes more and more oscillatory. If both wavefunctions oscillate rapidly, they can be approximated by semiclassical WKB wavefunctions and the radial overlap integral of the bound and the continuum wavefunctions can subsequently be evaluated by the method of *steepest descent.* This leads to analytical expressions for the spectrum (Child 1980, 1991:ch.5; Tellinghuisen 1985, 1987). In particular, relation (13.2), which relates the coordinate R to the energy E, is replaced by

$$E_i - V_0(R^*) + V_1(R^*) = E, \tag{13.3}$$

where E_i is the energy of the initial level. The expression on the left-hand side is called the *Mulliken difference potential.*[†] Equation (13.3)

[†] Equation (13.3) follows from the condition that the difference of the two WKB phases, $\eta_0(R) = \int^R dR' \, [2m(E_i - V_0(R'))]^{1/2}$ and likewise for the phase of the

goes over into (13.2) if the excitation of the parent molecule is not too high, because then $E_i - V_0(R) \approx 0$ for all bond distances. If the difference potential is a monotonic function of R, (13.3) has only one solution, denoted by $R^*(E)$, and the analysis of reflection structures proceeds *qualitatively* as described in the simple classical limit. If the relation between R and E is not unique, however, the interpretation becomes much more involved because in that case two (or even more) stationary phase points R^* contribute for the same energy to the overlap integral of the bound and the continuum wavefunction. As a consequence, additional quantum mechanical interference patterns arise which make the energy dependence of the spectrum more puzzling [see Tellinghuisen (1985) for references and examples].

13.1.2 Two-dimensional case

If two (or more) degrees of freedom are involved, it is important which mode is excited whether the spectrum shows reflection structures or not. Let us consider the linear triatomic molecule, ABC $\rightarrow$ A+BC, with Jacobi coordinates R and r as illustrated in Figure 2.1. Figure 13.2 depicts an elastic PES of the form (6.35) with coupling parameter $\epsilon = 0$.

The wavefunction of the parent molecule in the electronic ground state is assumed to be a product of two harmonic oscillator wavefunctions with m and n quanta of excitation along R and r, respectively. In Figure 13.2(b) only the vibrational mode of BC is excited, $n = 3$, while the dissociation mode is in its lowest state, $m = 0$. The corresponding spectrum is smooth without any reflection structures. Conversely, the wavefunction in Figure 13.2(a) shows excitation in the dissociation mode, $m = 3$, while the vibrational mode of BC is unexcited. The resulting spectrum displays very clear reflection structures in the same way as in the one-dimensional case. Thus, we conclude that, in general:

- Multimodal reflection structures exist only (or most prominently) if the bound-state wavefunction has one or several nodes along the dissociation path or, expressed in different words, if the parent molecule is excited in the direction of the dissociation path.

Realistic examples are shown in Figure 13.3.

The spectrum in Figure 13.2(a) extends over a much wider energy range than the spectrum in (b). In particular, because the ground-state

---

continuum wavefunction, is stationary in R, i.e.,

$$d[\eta_0(R) - \eta_1(R)]/dR = 0.$$

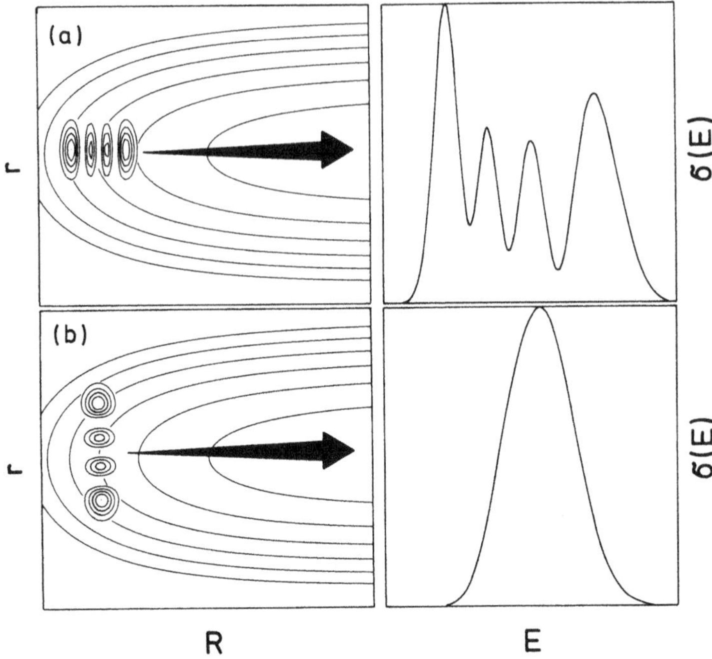

Fig. 13.2. Illustration of the origin of reflection structures for polyatomic molecules. The potential energy surface is of the form (6.35) with $\epsilon = 0$. The wavefunction of the parent molecule is simply the product of two harmonic oscillator wavefunctions. The heavy arrows illustrate the dissociation path.

wavefunction extends further out into the exit channel, the spectrum starts at considerably lower energies. The breadth and the structure of the R-dependent part of the wavefunction in the initial state determine the width and the general appearance of the spectrum. A simple adiabatic analysis on the basis of (3.42) readily explains the obvious differences between cases (a) and (b) in Figure 13.2.

The shape of reflection structures becomes more difficult to imagine, however, if the two modes are considerably coupled in the lower and/or in the upper state. In such cases, the lines of nodal structures of the two wavefunctions are not rectangular to each other so that the positive and the negative portions of the wavefunctions overlap in a less regular manner. In general, strong coupling between the relevant coordinates either in the ground or in the excited state tends to smear out the reflection structures. The same holds true if the degree of excitation of the parent molecule increases.

An approximate yet accurate analysis of general multi-dimensional overlap integrals is undoubtedly very complicated. A purely classical approach as outlined in Chapter 5 becomes rather uncertain because the

Wigner distribution function for excited states has negative parts. On the other hand, the extension of the semiclassical theory, established for one-dimensional systems, is not straightforward especially if the two degrees of freedom are appreciably coupled.[†]

13.2 Photodissociation of vibrationally excited H_2O

Photodissociation of excited vibrational states has been examined mainly for H_2O_2 (Ticich et al. 1987; Likar et al. 1988; Brouard, Martinez, O'Mahony, and Simons 1988, 1989, 1990; Brouard, Martinez, and O'Mahony 1990), H_2O (Andresen et al. 1985; Häusler, Andresen, and Schinke 1987; Vander Wal and Crim 1989; Weide, Hennig, and Schinke 1989; Vander Wal, Scott, and Crim 1991) and for HOD (Engel and Schinke 1988; Zhang and Imre 1988b; Imre and Zhang 1989; Zhang, Imre, and Frederick 1989; Vander Wal, Scott, and Crim 1990; Bar et al. 1990, 1991; Vander Wal et al. 1991). In all these examples one or several OH vibrational quanta are used to pump excited states within the parent molecule. For the two isotope variants of water detailed experimental results have been compared with the predictions of *ab initio* calculations; in the following we elucidate the highlights of this confluence of experiment and theory.

13.2.1 Calculation and characterization of the bound states of H_2O

H_2O in its electronic ground state is best described by a *local mode* expansion (Child and Halonen 1984; Child 1985; Halonen 1989). For the purpose of this chapter it suffices to consider a simple two-dimensional model in which the bending angle is frozen at its equilibrium value 104° and the oxygen atom is assumed to be infinitely heavy. For an exact three-dimensional treatment see Bačić, Watt, and Light (1988), for example. The approximate two-dimensional Hamiltonian reads

$$\hat{H}(R_1, R_2) = -\frac{\hbar^2}{2m_H}\frac{\partial^2}{\partial R_1^2} - \frac{\hbar^2}{2m_H}\frac{\partial^2}{\partial R_2^2} + V(R_1, R_2), \qquad (13.4)$$

where m_H is the mass of hydrogen and R_1 and R_2 are the two O-H bond distances.

In order to calculate the bound-state energies one expands the two-dimensional wavefunction in terms of products of suitably chosen, one-dimensional basis functions $\varphi_n(R_i)$, $i = 1, 2$, which describe the vibration of the two OH entities within $H_2O(\tilde{X})$. Because the Hamiltonian is symmetric with respect to the interchange of the two bond distances, the

[†] A semiclassical analysis of two-dimensional bound-free matrix elements has been provided by Child and Shapiro (1983) assuming, however, that the two potential energy surfaces are separable.

eigenfunctions must be either symmetric or anti-symmetric. In order to account for the symmetry one constructs *gerade* and *ungerade* basis functions according to

$$\Upsilon_{mn}^{\pm}(R_1, R_2) = C[\varphi_m(R_1)\,\varphi_n(R_2) \pm \varphi_m(R_2)\,\varphi_n(R_1)] \qquad (13.5)$$

with $C = 2^{-1/2}$ if $m \neq n$ and 2^{-1} if $m = n$. These symmetry-adapted basis functions fulfil the relations

$$\begin{aligned} \Upsilon_{mn}^{+}(R_1, R_2) &= +\Upsilon_{mn}^{+}(R_2, R_1) \\ \Upsilon_{mn}^{-}(R_1, R_2) &= -\Upsilon_{mn}^{-}(R_2, R_1) \end{aligned} \qquad (13.6)$$

and are orthogonal and normalized to one. Next, we expand the total wavefunction in terms of the $\Upsilon_{mn}^{\pm}$,

$$\Psi^{\pm}(R_1, R_2) = \sum_{mn} a_{mn}^{\pm}\,\Upsilon_{mn}^{\pm}(R_1, R_2), \qquad (13.7)$$

and calculate the energies and the coefficients $a_{mn}^{\pm}$ by diagonalization of the Hamiltonian matrix with elements $\langle \Upsilon_{mn}^{\pm} \mid \hat{H} \mid \Upsilon_{m'n'}^{\pm} \rangle$ as described in Section 2.4.2. Since $\hat{H}$ is symmetric the Hamiltonian matrix splits into two separate blocks, one for the gerade and one for the ungerade symmetry, i.e., the gerade and the ungerade states are not coupled by the symmetric Hamiltonian.

Figure 13.3 depicts the lowest four eigenfunctions of the ungerade symmetry (multiplied with the $\tilde{X} \rightarrow \tilde{A}$ transition dipole function). They are anti-symmetric with respect to the interchange of R_1 and R_2 and therefore they have a node on the symmetry line $R_1 = R_2$. Some examples for gerade states will be shown in Figure 14.4. The assignment $|mn^{\pm}\rangle$ reflects the leading term in expansion (13.7). For example, $|21^{-}\rangle$ means that the function Υ_{21}^{-} dominates the expansion while the coefficients for the other basis functions are considerably smaller. The corresponding wavefunction is approximately given by

$$\Psi_{21}^{-}(R_1, R_2) \approx \varphi_2(R_1)\,\varphi_1(R_2) - \varphi_2(R_2)\,\varphi_1(R_1).$$

The energy of this state corresponds to approximately 2+1 OH stretching quanta.

13.2.2 Absorption spectra

The right-hand side of Figure 13.3 depicts the corresponding absorption spectra as functions of the energy in the excited electronic state.† As

† Since the absorption cross sections are mainly determined by the energetics and the dynamics in the excited electronic state, it is advantageous to plot all cross sections on the same energy scale irrespective of the initial level in the electronic ground state. In this way the structures in the spectra can be better compared and related to characteristic features of the excited-state PES.

a result of the larger breadth of the ground-state wavefunction with increasing grade of vibrational excitation they become generally broader and at the same time their overall amplitude decreases. In order to understand the general shape of the spectra it is mandatory also to take into account the coordinate dependence of the transition dipole function μ_{AX} which continuously decreases as one of the OH bonds increases. It is relatively unimportant for the photodissociation of the lowest vibrational state because the corresponding wavefunction extends only over a narrow region of the coordinate space. This does not hold, however, for excited vibrational states. Since μ_{AX} rises towards smaller bond distances, the spectra generally increase with growing energy [Figures 13.3 (b) and (c)].

To understand the development or the absence of reflection structures one must imagine — in two dimensions — how the continuum wavefunction for a particular energy E overlaps the various ground-state wavefunctions and how the overlap changes with E. This is not an easy task! Figure 9.9 shows two examples of continuum wavefunctions for H_2O. Alternatively, one must imagine how the time-dependent wavepacket, starting from an excited vibrational state, evolves on the upper-state PES and what kind of structures the autocorrelation function develops as the wavepacket slides down the potential slope.

In accordance with the simple example shown in Figure 13.2, Figure 13.3 demonstrates that only the nodes along the reaction path, illustrated by the arrows, cause structures in the spectrum. For example, the wavefunction for the $|30^-\rangle$ state has two nodes along the reaction route which clearly show up in the spectrum as well. Nodes along a line perpendicular to the reaction path, on the other hand, do not lead to reflection structures as Figures 13.3 (a) and (d) evidently manifest [see also Sheppard and Walker (1983), Shapiro (1986), and Guo, Lao, Schatz, and Hammerich (1991)]. Without showing further examples we note that, in general, the interpretation of the spectra becomes gradually less obvious as the degree of vibrational excitation increases.

Using the two-laser excitation scheme illustrated in Figure 11.7 Vander Wal, Scott, and Crim (1991) measured part of the absorption spectrum for excitation of the $|40^-\rangle$ state. Figure 1.9 shows the comparison with the result of a two-dimensional calculation. The corresponding bound-state wavefunction, depicted in Figure 13.4, qualitatively resembles the $|30^-\rangle$ wavefunction with one additional node along the reaction path. Accordingly, the absorption spectrum has three instead of two pronounced

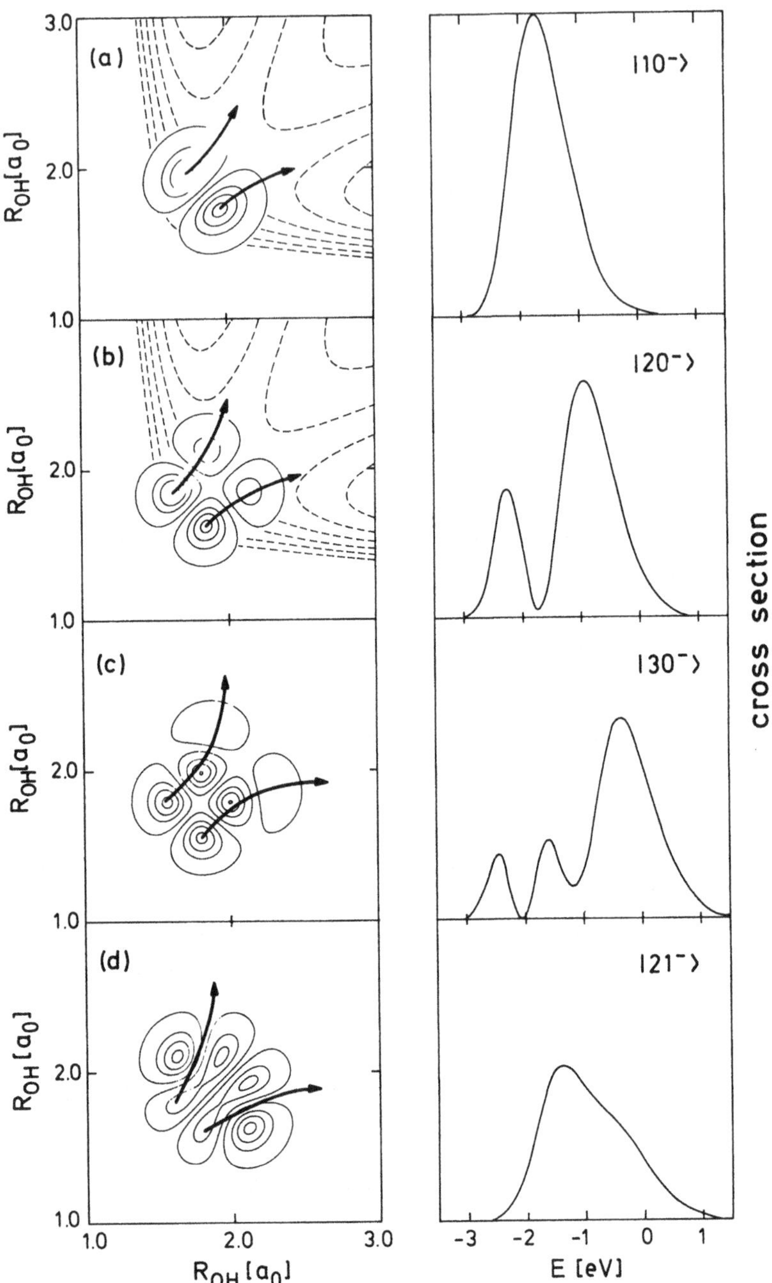

Fig. 13.3. Left-hand side: Contour plots of the modulus square of the lowest four anti-symmetric eigenfunctions of H_2O in the electronic ground state, multiplied with the $\tilde{X} \to \tilde{A}$ transition dipole function. They have a node on the symmetric stretch line. The assignment $|mn^-\rangle$ is based on the local mode expansion (13.5) (*cont.*)

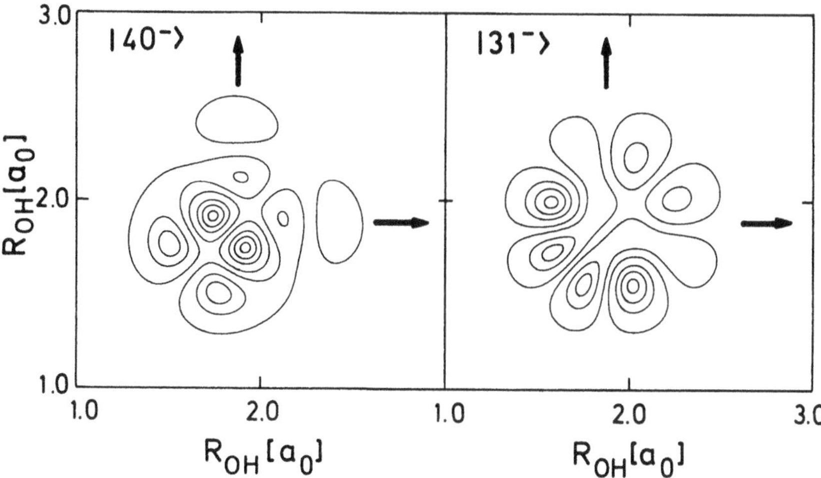

Fig. 13.4. Contour plots of the wavefunctions for the $|40^-\rangle$ and the $|31^-\rangle$ states of $H_2O(\tilde{X})$ multiplied by the $\tilde{X} \rightarrow \tilde{A}$ transition dipole function; $|\mu_{AX}\Psi^-_{mn}|^2$ is shown. The heavy arrows indicate the route of the evolving wavepacket in the exit channel of the excited-state PES. Although these two states correspond to roughly the same total energy, the resulting absorption spectra are drastically different (Weide, Hennig, and Schinke 1989).

minima. The experimental data in Figure 1.9 cover only the first one at the onset of the spectrum.[†]

13.2.3 Final vibrational state distributions

The partial photodissociation cross sections for producing OH in different vibrational states n also depend — like the absorption spectrum — sensitively on the initial vibrational level of the parent molecule. Let us

---

Fig. 13.3. (*cont.*) and (13.7). The wavefunctions are calculated using the empirical fit PES of Sorbie and Murrell (1975). The arrows indicate qualitatively the route of the evolving time-dependent wavepacket in the excited state. The dashed contours in (a) and (b) represent the $\tilde{A}$-state PES. Right-hand side: The corresponding absorption spectra as functions of the energy in the excited electronic state. $E = 0$ corresponds to three ground-state atoms, H + O + H. Within this normalization the lowest eigenvalue of $H_2O(\tilde{X})$ is -9.500 eV.

[†] Reflection-type structures can also be investigated by means of emission spectra for transitions from a bound excited electronic state to a lower lying repulsive state, provided the molecule is initially vibrationally excited (Zhang, Abramson, and Imre 1991).

consider the $|40^-\rangle$ and the $|31^-\rangle$ states which both correspond to four quanta of OH stretching vibration (approximately 1.75 eV). Their energies are only 0.060 eV apart, but are differently partitioned between the two degrees of freedom. Figure 13.5 shows the corresponding partial absorption cross sections for the three lowest vibrational states of OH as functions of the energy. While the dissociation of $|40^-\rangle$ preferentially yields OH($n = 0$) products for essentially the entire energy region, the dissociation of $|31^-\rangle$ generates, at least in the first part of the spectrum, mainly OH($n = 1$).

The individual partial cross sections are even more structured than the total cross sections and a simple explanation of the energy dependences is probably impossible, except at the low-energy tail of the spectrum. For total energies *below* the barrier of the $\tilde{A}$-state PES ($E = -2.644$ eV in this normalization), the dissociative wavefunction is mainly confined to the two H + OH channels with little amplitude in the intermediate region. It therefore overlaps only that part of the ground-state wavefunction which extends well into the two exit channels. There, however, the $|40^-\rangle$ and the $|31^-\rangle$ states behave completely differently (see Figure 13.4),

$$\lim_{R_1 \to \infty} \Psi_{40}^-(R_1, R_2) \propto \varphi_4(R_1)\, \varphi_0(R_2)$$
$$\lim_{R_1 \to \infty} \Psi_{31}^-(R_1, R_2) \propto \varphi_3(R_1)\, \varphi_1(R_2).$$

Here, R_1 is the dissociation coordinate and R_2 is the OH vibrational coordinate.

The wavefunction for the $|40^-\rangle$ state, considered in the direction perpendicular to the dissociation path, has the character of an OH($n = 0$) vibrational wavefunction and according to the elastic limit discussed in Section 9.1 the overlap with the continuum wavefunction yields mainly OH($n = 0$) (Franck-Condon mapping). The $|31^-\rangle$ wavefunction, on the other hand, behaves more like an OH($n = 1$) wavefunction and therefore preferentially produces OH($n = 1$). The measurements of Vander Wal, Scott, and Crim (1991) for two photolysis wavelengths, 239.5 nm and 218.5 nm, fully confirmed these theoretical predictions. Figure 13.5 manifests, in a simple way, "steering" of a chemical reaction by choosing the "right" initial state and the appropriate wavelength. More of this follows in the next section.

13.3 State-selective bond breaking

The control of the wavefunction of the parent molecule over the outcome of the fragmentation is especially illuminating in the photodissociation of HOD which can break up into two *different* product channels: H + OD(n)

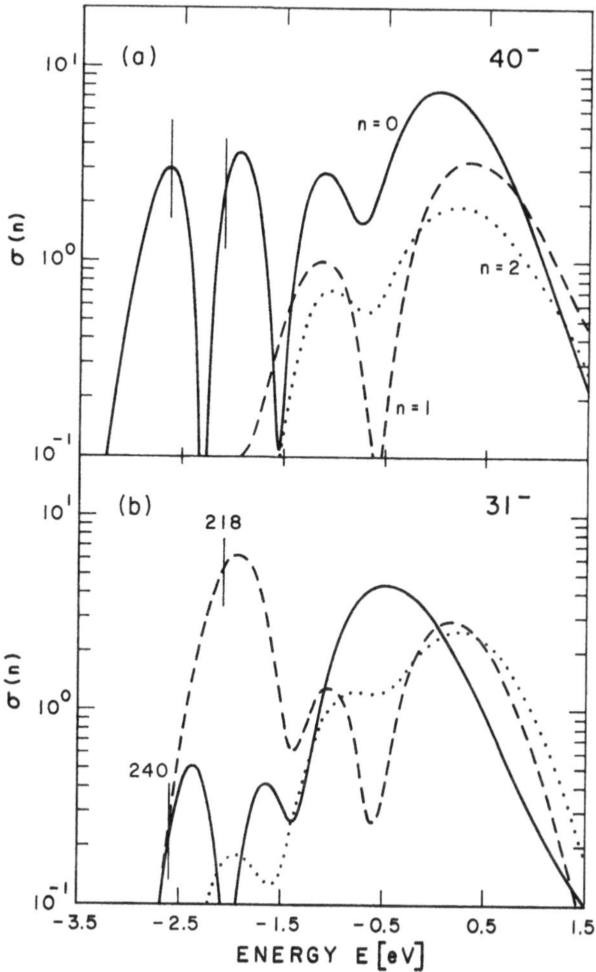

Fig. 13.5. Partial photodissociation cross sections $\sigma(E, n)$ following the photodissociation of the $|40^-\rangle$ and the $|31^-\rangle$ vibrational states of $H_2O(\tilde{X})$ as functions of the energy in the $\tilde{A}$ state. The quantum number n specifies the vibrational state of the OH product: $n = 0$ (solid line), $n = 1$ (dashed line), and $n = 2$ (dotted line). $E = 0$ corresponds to three ground-state atoms. The vertical lines mark the total energies in the excited state corresponding to the two photolysis wavelengths $\lambda_2 = 239.5$ and 218.5 nm in the experiment of Vander Wal, Scott, and Crim (1991). Reproduced from Weide, Hennig, and Schinke (1989).

and HO(m) + D. Let us start the discussion with a simple question: how large is the probability for generating the products HO and D if we considerably pre-excite the H-O bond within the parent molecule? Intuitively, one is tempted to presume that excitation of the O-H bond drastically increases the possibility for creating the alternative products

H and OD. If we put, for example, four vibrational quanta into the HO entity and leave OD unexcited one obviously expects the H-O bond to break first and therefore the preferential production of OD.

This naive picture is only partly correct. The probability for producing OD and HO depends strongly on the photolysis wavelength as well as the degree of internal excitation. Figure 13.6 depicts the branching ratio $\sigma_{H+OD}/\sigma_{D+OH}$ as a function of the total energy E in the $\tilde{A}$ state for three initial vibrational states $|nm\rangle = |00\rangle$, $|02\rangle$, and $|04\rangle$ of HOD($\tilde{X}$). The quantum numbers n and m indicate excitation of the O-D and the

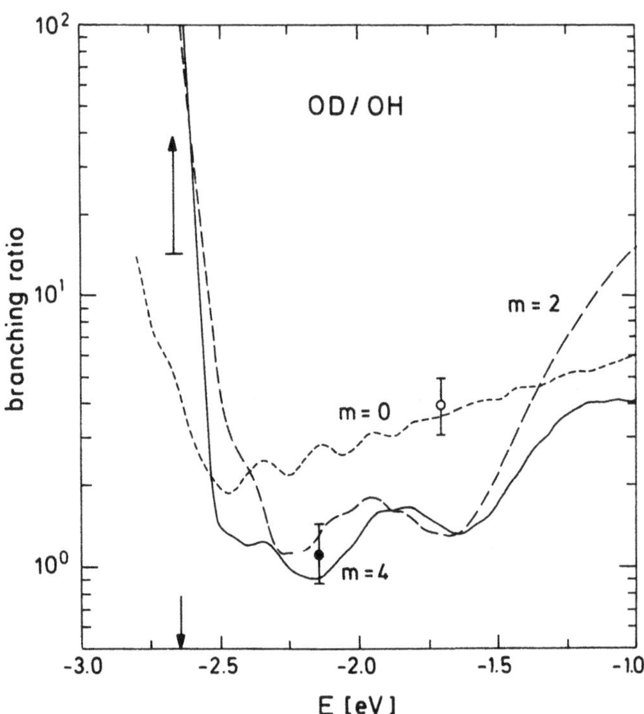

Fig. 13.6. Calculated branching ratios $\sigma_{H+OH}/\sigma_{D+OH}$ following the photodissociation of the $|0m\rangle$ ($m = 0, 2$ and 4) vibrational states of HOD through the first continuum. The first quantum number gives the excitation of the O-D bond ($n = 0$ in the present case) and the second one, m, indicates excitation of the O-H bond. The open circle is the experimental result of Shafer, Satyapal, and Bersohn (1989) for the photolysis of $|00\rangle$ at 157 nm. The filled circle is the result for initial state $|04\rangle$ and photolysis wavelength $\lambda_2 = 218.5$ nm and the (upward) arrow indicates the lower limit for state $|04\rangle$ and $\lambda_2 = 239.5$ nm. See Figure 11.7 for an illustration of the experimental set-up. The arrow on the energy axis marks the energy of the barrier of the $\tilde{A}$-state PES. $E = 0$ corresponds to three atoms in their ground state. Reproduced from Vander Wal et al. (1991).

H-O bond, respectively; the bending quantum number is irrelevant for the present discussion and is zero in all cases. Because of the lack of symmetry, the superscripts $\pm$ are obsolete for HOD. Let us focus our attention on the $m = 4$ case. As surmised above, $\sigma_{H+OD}/\sigma_{D+OH}$ is very large at low energies. However, it rapidly drops to values between one and two in the intermediate energy regime before it rises again to relatively large values. Around -2.2 eV, OH and OD are produced with about equal probability *although* the O-H bond contains four OH quanta corresponding to roughly 1.7 eV of internal vibrational energy (Vander Wal, Scott, and Crim 1990; Vander Wal et al. 1991).

The overall shape of the corresponding bound-state wavefunction and its position relative to the upper-state PES explains qualitatively the overall energy dependence. Figure 13.7 illustrates the wavefunction of the $|04\rangle$ state of HOD multiplied by the $\tilde{X} \to \tilde{A}$ transition dipole function. The wavefunction by itself would be largest at the outermost maximum around 2.5 Å. However, since μ_{AX} gradually diminishes with increasing H-O bond separation, the product $|\mu_{AX}\Psi_{04}|^2$ on the average decreases from the inner to the outer region. The lack of symmetry and the absence of Fermi-type resonance effects, since the HO and the OD vibrational frequencies are quite dissimilar, cause the excited HOD wavefunctions to have a much simpler behavior than the excited H_2O wavefunctions.

In the following we will denote the continuum wavefunctions in the upper state with outgoing flux in either the H+OD or the D+OH channel by Ψ_{H+OD} and Ψ_{D+OH}, respectively. At very low energies, significantly below the barrier of the upper-state PES, Ψ_{H+OD} is restricted to the H + OD channel while Ψ_{D+OH} is solely confined to the D + OH channel. Tunneling through the barrier is marginal and can be neglected. The partial cross sections for producing OD or OH are determined by the overlap of these partial dissociation wavefunctions with the ground-state wavefunction and the branching ratio is proportional to

$$\frac{\sigma_{H+OD}}{\sigma_{D+OH}} \propto \frac{|\langle\Psi_{H+OD} \mid \mu_{AX} \mid \Psi_{04}\rangle|^2}{|\langle\Psi_{D+OH} \mid \mu_{AX} \mid \Psi_{04}\rangle|^2}.$$

Because of the appreciable excitation along the H-O bond in the parent molecule, the bound-state wavefunction Ψ_{04} extends much further into the H+OD than into the D+OH channel. The consequence is that it overlaps Ψ_{H+OD} at relatively low energies where the overlap with the other wavefunction, Ψ_{D+OH}, is essentially zero. The result is a relatively small but finite cross section for the production of OD while the cross section for OH is practically zero.

The classical picture of photodissociation outlined in Chapter 5 provides an alternative explanation of the preference for the H+OD channel

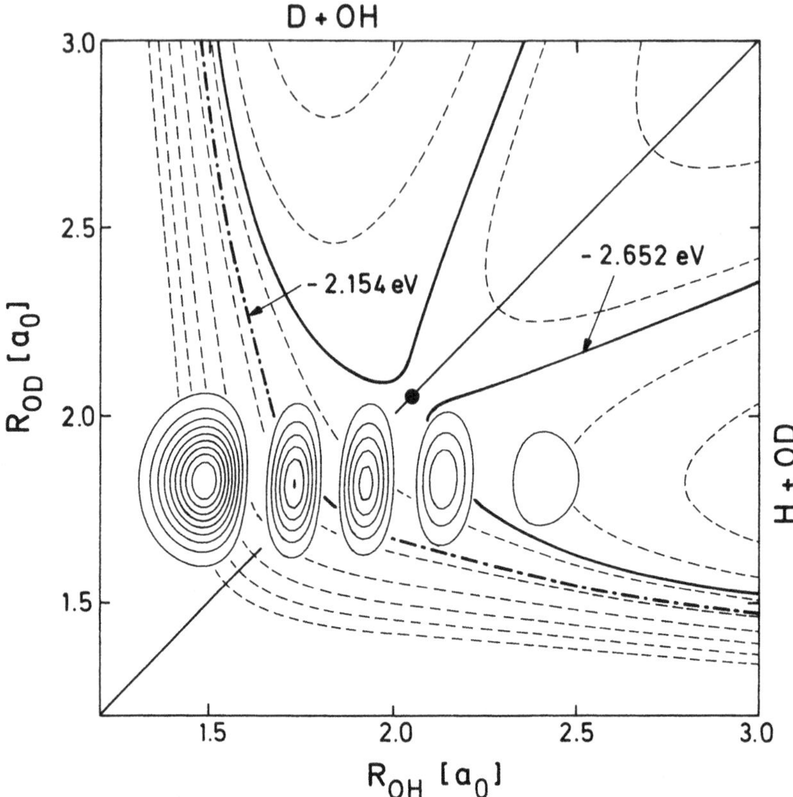

Fig. 13.7. Contour plot of the $\tilde{A}$-state PES for a bending angle $\alpha_e = 104°$. Energy normalization is such that $E = 0$ corresponds to H+O+H. Superimposed are contours of $|\mu_{AX}\Psi_{04}|^2$ where μ_{AX} is the $\tilde{X} \rightarrow \tilde{A}$ transition dipole function and Ψ_{04} is the bound-state wavefunction of HOD with four quanta of excitation in the O-H bond. The filled circle indicates the barrier and the two especially marked contours represent the energies for the two photolysis wavelengths $\lambda_2 =$ 239.5 and 218.5 nm used in the experiment. Adapted from Vander Wal et al. (1991).

at low energies. In order to calculate the partial cross sections for a particular energy $E = E_i + \hbar\omega$ one starts classical trajectories at the classical turning point $R_t(E)$ and follows their evolution into either of the two product channels. If the corresponding momenta are initially set to zero, each trajectory is weighted by $|\mu_{AX}\Psi_{04}|^2$. At energies below the saddle point the trajectories begin either in the H+OD or in the D+OH channel and thus yield directly the corresponding products without crossing the saddle-point region. For example, the two thick contours in Figure 13.7 indicate the lines of turning points for the energy $E = -2.652$ eV (corresponding to a photolysis wavelength of $\lambda_2 = 239.5$ nm). Because this energy is smaller than the barrier energy of -2.644 eV, there are two contour lines. Trajectories which start on the right-hand side of the dividing line $R_{OH} = R_{OD}$ lead immediately to H+OD and trajectories starting on the left-hand side yield D+OH. Since the wavefunction of the $|04\rangle$ state stretches much further along the O-H than along the O-D bond, trajectories starting in the H+OH channel have a finite weight while the trajectories launched in the other channel have negligibly small weight at this low energy. Therefore, in the Monte Carlo sampling almost all trajectories originate in the H+OD channel which readily explains the extremely large branching ratio at low energies.

If the total energy increases and eventually exceeds the barrier, the two dissociation wavefunctions begin to spread over *both* channels. This causes the overlap of $\mu_{AX}\Psi_{04}$ with Ψ_{D+OH} to increase while at the same time the overlap with Ψ_{H+OD} gradually diminishes because the positive and the negative portions of the bound-state and the continuum-state wavefunctions partly cancel each other. Therefore, the branching ratio decreases and approaches roughly unity at around -2.2 eV. In terms of classical mechanics, trajectories start in both channels if the energy is larger than the barrier energy and lead to both H+OD as well as D+OH. The second especially marked contour in Figure 13.7, for example, indicates the line of turning points for $E = -2.652$ eV (corresponding to $\lambda_2 = 239.5$ nm).

At still higher energies, the trajectories start mainly on the D+OH side of the dividing line, in the region of the main maximum of $|\mu_{AX}\Psi_{04}|^2$ in Figure 13.7. In this region, however, the accelerating force points primarily into the H + OD channel. Thus, on their way towards fragmentation many trajectories cross over into the other channel with the result that H and OD are again the main products and the branching ratio rises with energy.

The chemical branching ratio depends, in the same way as the final state distributions of the products, on subtle details of the overlap of two oscillatory quantum mechanical wavefunctions. Because of the oscillations the net result of the overlap integral is generally difficult to predict *a priori*, except for limiting cases. The very large propensity for producing H + OD rather than D + OH at low energies does not at all come as a surprise; it follows readily from the form of the $|04\rangle$ wavefunction (multiplied by the transition dipole function) and the global shape of the upper-state PES. The nearly equal probability for producing OH and OD at intermediate energies is astonishing. The measurements of Vander Wal et al. (1991) for two photolysis wavelengths confirm the theoretical predictions (Figure 13.6). Note, that the experimental value of $\sigma_{H+OD}/\sigma_{D+OH} \approx 15$ for $\lambda_2 = 239.5$ nm represents only a lower limit!

The theoretical calculations together with the few experimental data shown in Figure 13.6 manifest state-specific rupture of molecular bonds and the possibility of "steering" molecular decay through the variation of the initial vibrational state and the wavelength of the photolysis laser. This idea was first pursued by Zhang and Imre (1988b) [see Manz and Parmenter (1989) for other references concerning mode- and state-specific fragmentation]. More important from the chemical point of view would be the dissociation of molecules with two or even more different dissociation channels, CH_3OH for example. Is it possible to control the production of H and OH by selecting the initial vibrational state?

Let us summarize this chapter. The absorption spectrum as well as the partial cross sections in direct photodissociation depend sensitively on the precursor state of the parent molecule from which the promotion into the excited electronic state takes place. It is important to underline that:

- The wavefunction in the electronic ground state and its overlap with the continuum wavefunctions in the upper state rather than the corresponding vibrational energy are the determining factors which control the outcome of the dissociation.

Even if two vibrational states are degenerate they can yield completely different cross sections. The dissociation of excited vibrational states samples a considerably wider region of the upper-state PES than dissociation of the ground vibrational state. However, because the two quantum mechanical wavefunctions both have an oscillatory behavior, the interpretation of the various cross sections is not always obvious. The photodissociation of excited vibrational states is closely related to the emission spectroscopy of the dissociating molecule which is the topic of the following chapter.

14

Emission spectroscopy of dissociating molecules

If the potential in the excited electronic state is purely repulsive without a barrier hindering direct dissociation, the complex survives for only about 10^{-14} seconds before it breaks apart. For comparison, the radiative lifetime, i.e., the average time before the excited complex decays through emission of a photon, is of the order of 10^{-8} seconds and therefore dissociation is an efficient destruction mechanism for fluorescence. Nevertheless, there is a tiny probability for the molecule to emit a photon while it breaks apart and thereby to return to the lower electronic state. This process is known as *radiative recombination* or *Raman scattering* and it is illustrated schematically in Figure 14.1. The emission spectrum shown on the right-hand side gives the probability for filling the various vibrational bound states in the lower electronic state.

Although the photon yield is exceedingly small, it can be detected with modern techniques (Imre, Kinsey, Sinha, and Krenos 1984). Figure 14.2 depicts an intriguing example, the emission spectrum of CH_3I excited with a 266 nm photon. It consists essentially of a very long progression in the C-I mode (ν_3), i.e., the mode that leads to dissociation. Emission spectra contain two basic types of information:

1) The energies of the vibrational eigenstates in the electronic ground state, which are related to the frequencies of the emitted photons by $E_i + \hbar\omega = E_f + \hbar\omega'$.
2) The intensities of the emission lines which reflect the dissociation dynamics in the excited electronic state and the vibrational dynamics in the ground electronic state.

The example of CH_3I and similar results for O_3 (Imre, Kinsey, Field, and Katayama 1982; Imre, Kinsey, Sinha, and Krenos 1984) clearly demonstrate the possibility of accessing vibrational levels of polyatomic

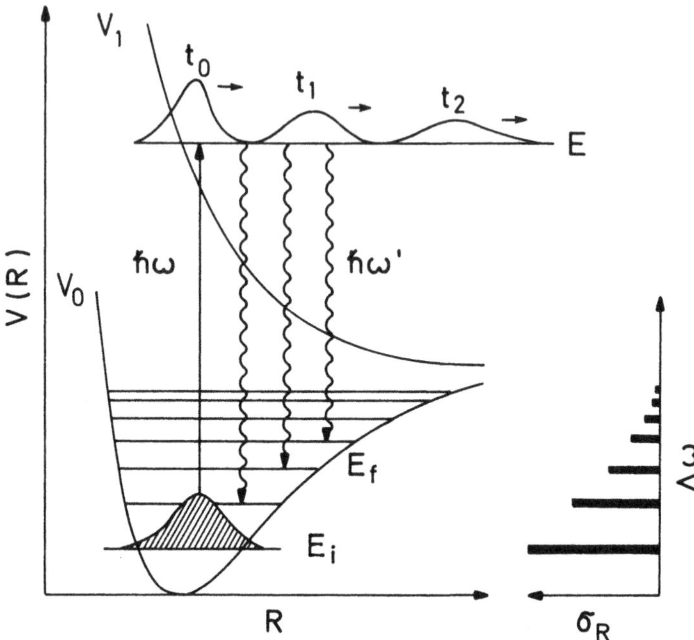

Fig. 14.1. Schematic illustration of radiative recombination of a dissociating molecule. The lower part depicts the electronic ground state and the corresponding vibrational energy levels. The upper part shows the excited electronic state and the evolving time-dependent wavepacket for various instants $t_0 = 0$, t_1, etc. ω is the frequency of the photolysis laser which excites the molecule and ω' is the frequency of the emitted photon. The right-hand side shows the emission or Raman spectrum.

molecules up to very high overtones, near the dissociation limit. This allows the construction of multi-dimensional PESs for the electronic ground state, which are valid up to large displacements from equilibrium (Johnson, Kinsey, and Shapiro 1988). In the context of this monograph, however, we are more interested in the intensity pattern of the emission spectrum and its relation to the dissociative motion in the upper electronic state.

Like absorption spectra and final state distributions, emission or Raman spectra can be calculated either in the time-independent or the alternative time-dependent framework (Heller 1981a,b; Williams and Imre 1988a). The two methods will be outlined in Section 14.1. At the present time a joint experimental and fully *ab initio* theoretical investigation has been performed only for H_2O excited into the first continuum. Section 14.2 summarizes the main results which are characteristic for other molecules as well. Finally, we briefly discuss in Section 14.3 some recent

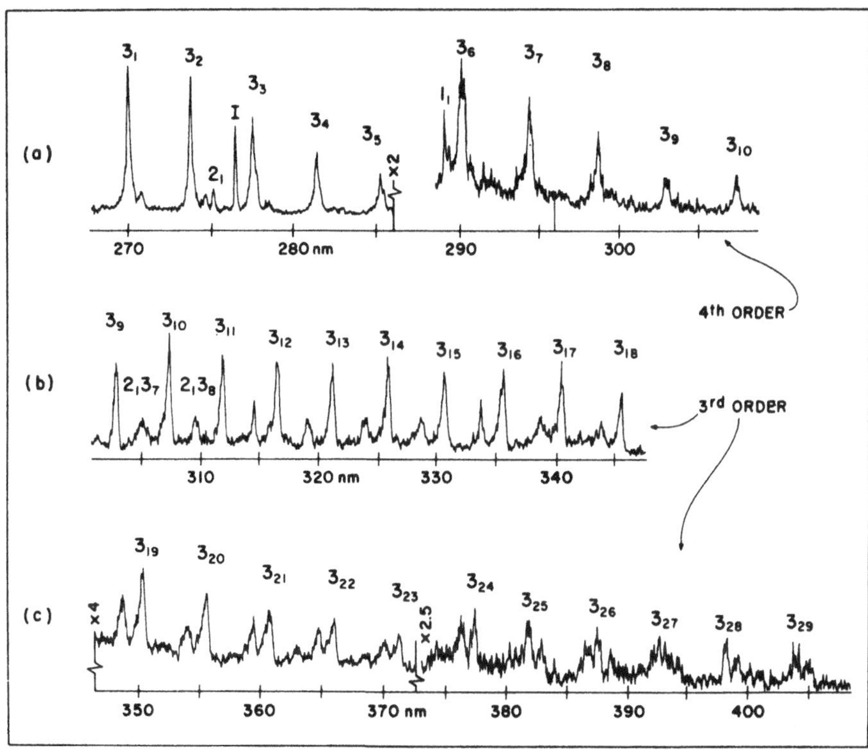

Fig. 14.2. Raman spectrum for CH_3I obtained by excitation at 266 nm. 3_n stands for the nth eigenstate of the ν_3-mode, for example. ν_3 and ν_2 are the C-I and the CH_3-umbrella mode, respectively. Reproduced from Imre, Kinsey, Sinha, and Krenos (1984).

results for H_2S which are strikingly different and which, in contrast to H_2O, point to the participation of two electronic states.

14.1 Theoretical approaches

The process illustrated in Figure 14.1 is a special example of *Raman scattering* (Weissbluth 1978:ch.24): the molecule absorbs one photon with frequency ω and emits a second photon with frequency ω'. The initial (index i) and the final (index f) vibrational states both belong to the electronic ground state. However, the transition from vibrational state i to vibrational state f is mediated by excitation of the upper electronic state. If the initial and the final states are identical, we speak of *Rayleigh scattering* with $\omega = \omega'$. Otherwise, $\omega' < \omega$ provided that the excitation starts from the lowest vibrational level in the electronic ground state. In order to illustrate the calculation of the emission spectrum we consider again the two-dimensional collinear model for a triatomic molecule ABC with dissociation channel A + BC(n) where n denotes the final vibrational state of the BC fragment (see Section 2.5). The appropriate coordinates are R and r defined in Figure 2.1.

14.1.1 The time-independent view

Raman scattering is a two-photon process and must be described by second-order perturbation theory. The cross section for a transition from state $|\Psi_{0i}(E_i)\rangle$ with energy E_i to state $|\Psi_{0f}(E_f)\rangle$ with energy E_f (in the following the indices 0 and 1 will label the lower and the upper electronic state, respectively) is given by the *Kramers-Heisenberg-Dirac* formula (Kramers and Heisenberg 1925; Dirac 1927; for a sufficiently detailed derivation see, for example, Weissbluth 1978:ch.24)

$$\sigma_R(\omega, i \to f) \propto \omega\, \omega'^3 \left| t_{res} + t_{nonres} \right|^2, \tag{14.1}$$

where the resonant term is defined by

$$\begin{aligned} t_{res} = \lim_{\Gamma \to 0} \int dE \sum_n (E_i + \hbar w - E + i\Gamma)^{-1} \\ \times \langle \Psi_{0f}(E_f) \mid \mu_{01} \mid \Psi_1(E,n)\rangle \, \langle \Psi_1(E,n) \mid \mu_{10} \mid \Psi_{0i}(E_i)\rangle. \end{aligned} \tag{14.2}$$

The nonresonant term t_{nonres} is identical to the resonant one but with the denominator $E_i - \hbar\omega - E + i\Gamma$; it is naturally much smaller than the resonant term and therefore it will be omitted in the subsequent discussion. In order to simplify the presentation, all unimportant prefactors are disregarded in (14.1).

The $\Psi_1(E,n)$ are continuum wavefunctions in the upper electronic state as defined in Section 2.5; the quantum number n specifies the particular solution of the time-independent Schrödinger equation with outgoing free wave in vibrational channel n.† The integration over the continuous variable E and the summation over the discrete variable n are necessary in order to account for the excitation of all states in the upper electronic manifold. In the case that the upper electronic state is bound the integration over E must be replaced by summation over all possible bound vibrational states and the summation over the various product channels n becomes redundant. The term $i\Gamma$ in the denominator is introduced for numerical convenience. We must strongly emphasize at this point that reducing the integration over all continuum states to merely the resonant state with energy $E = E_i + \hbar\omega$ is inadequate. Although it has certainly the largest contribution to the integral, the nonresonant states must not

† The states in the upper electronic state are often called *virtual states*. This terminology is rather misleading because it falsely suggests that these states do not really exist. That is, however, not the case!

be omitted! Equation (14.2) has the general form of a quantum mechanical amplitude in second-order perturbation theory (Cohen-Tannoudji, Diu, and Laloë 1977:ch.XI).

The matrix elements $\langle \Psi_1(E,n) \mid \mu_{10} \mid \Psi_{0f}(E_f)\rangle$ are just the partial photodissociation amplitudes (2.68) which are required in the calculation of dissociation cross sections for vibrationally excited parent molecules. The actual calculation proceeds in the following way:

1) Calculation of all bound-state wavefunctions $\Psi_{0f}(E_f)$ in the electronic ground state.
2) Calculation of continuum wavefunctions $\Psi_1(E,n)$ in the excited electronic state for all possible final states n and their overlap with all bound-state wavefunctions for a sufficiently narrow grid of energies.
3) Evaluation of the integral over E.

If the computer program is well organized, the actual calculations consume only marginally more time than the determination of the absorption spectrum (Rousseau and Williams 1976; Atabek, Lefebvre, and Jacon 1980a,b; Hennig, Engel, Schinke, and Staemmler 1988). However, if more than two degrees of freedom are involved, the time-independent approach becomes impractical unless severe approximations are incorporated. The main bottleneck is the need for calculating stationary wavefunctions for many energies and all final channels, although at the end one integrates over E and sums over n, i.e., one generates much more information than is actually required for the calculation of emission spectra. In addition to these more technical aspects we find the interpretation of emission spectra in terms of the Kramers-Heisenberg-Dirac expression rather difficult and obscure (as is generally true for second-order formulas). The following time-dependent approach is better suited to elucidate the relation between the emission spectrum on one hand and the dynamics in the dissociative state on the other.

14.1.2 The time-dependent view

The time-dependent formulation of Raman scattering has been introduced by Lee and Heller (1979), Heller, Sundberg, and Tannor (1982), Tannor and Heller (1982), and Myers, Mathies, Tannor, and Heller (1982). Its derivation is strikingly simple. We start from the Kramers-Heisenberg-Dirac formula (14.1) and (14.2) without the nonresonant term and transform it into an integral over time by using the identity

$$(E_i + \hbar\omega - E + i\Gamma)^{-1} = -\frac{i}{\hbar}\int_0^\infty dt\; e^{i(E_i+\hbar\omega-E+i\Gamma)t/\hbar} \tag{14.3}$$

with Γ chosen to be positive. Inserting (14.3) into (14.2) yields

$$t_{res} = -\frac{i}{\hbar} \lim_{\Gamma \to 0} \int dE \sum_n \int_0^\infty dt \, \mathrm{e}^{i(E_i + \hbar\omega + i\Gamma)t/\hbar} \, \mathrm{e}^{-iEt/\hbar} \times \langle \Psi_{0f}(E_f) \mid \mu_{01} \mid \Psi_1(E,n) \rangle \, \langle \Psi_1(E,n) \mid \mu_{10} \mid \Psi_{0i}(E_i) \rangle \quad (14.4)$$

which can be rewritten as

$$t_{res} = -\frac{i}{\hbar} \lim_{\Gamma \to 0} \int_0^\infty dt \, \mathrm{e}^{i(E_i + \hbar\omega + i\Gamma)t/\hbar} \times \langle \Psi_{0f}(E_f) \mid \mu_{01} \int dE \sum_n |\Psi_1(E,n)\rangle\langle\Psi_1(E,n) \mid \mu_{10} \mid \Psi_{0i}(E_i)\rangle \times \mathrm{e}^{-iEt/\hbar}. \quad (14.5)$$

In view of Equations (4.3)–(4.5) we immediately recognize that

$$\Phi_i(t) \propto \int dE \sum_n |\Psi_1(E,n)\rangle\langle\Psi_1(E,n) \mid \mu_{10} \mid \Psi_{0i}(E_i)\rangle \, \mathrm{e}^{-iEt/\hbar} \quad (14.6)$$

is a time-dependent wavepacket evolving in the excited state with the initial condition $\Phi_i(0) \equiv \mu_{10}\Psi_{0i}(E_i)$. Finally the Raman cross section becomes

$$\sigma_R(\omega, i \to f) \propto \omega\omega'^3 \left| \int_0^\infty dt \, \mathrm{e}^{i(E_i + \hbar\omega)t/\hbar} \, \langle \Phi_f(0) \mid \Phi_i(t) \rangle \right|^2, \quad (14.7)$$

where, for consistency, we defined

$$\Phi_f(0) \equiv \mu_{01}\Psi_{0f}(E_f). \quad (14.8)$$

The matrix element

$$C_{if}(t) \equiv \langle \Phi_f(0) \mid \Phi_i(t) \rangle \quad (14.9)$$

is the so-called *cross-correlation function.* Realistic examples are depicted in Figure 14.5.

In the time-dependent wavepacket formulation the Raman process is visualized in the following way (Heller 1981a,b; see also Figure 14.1). The excitation laser promotes the initial state Ψ_{0i}, multiplied by the corresponding transition dipole moment function, *instantaneously* to the upper electronic state, where it immediately starts to move under the influence of the Hamiltonian $\hat{H}_1$. During its journey out into the exit channel it accumulates overlap with *all* excited vibrational wavefunctions in the lower electronic state; the time-dependences and the relative magnitudes of the cross-correlation functions reflect the motion in the upper state as well as the shape of the vibrational wavefunctions in the ground electronic state. In analogy to Section 4.1 we can state:

- The cross-correlation functions C_{if} are the links between the motion of the wavepacket in the upper state, on one hand, and the Raman spectrum on the other hand. Its behavior in time controls the fluorescence intensities into the vibrational states of the electronic ground state.

Because the excited vibrational wavefunctions usually have a complicated nodal pattern (examples are given in Figure 14.4) it is rather difficult to draw specific conclusions about the time dependence of the $C_{if}(t)$.

The calculations proceed in a similar way as for the calculation of absorption cross sections with only slight modifications:

1) Evolution of the wavepacket in the excited state with initial condition $\mu_{10}\Psi_{0i}$.
2) Evaluation of the cross-correlation functions with all vibrational states in the lower electronic state.
3) Half-Fourier transformation of the cross-correlation functions.

The great asset of the time-dependent picture rests on the fact that no reference is made to the (many) stationary continuum states in the excited electronic state. The propagation of a single wavepacket, which contains the entire history of the dynamics in the upper state, thus yields the absorption cross section, all Raman cross sections, and the final state distributions of the photofragments.

14.2 Emission spectroscopy of dissociating $\mathbf{H_2O}(\tilde{A})$

Emission spectra of dissociating small polyatomics have been measured and theoretically analyzed for several examples in recent years:[†]

1) CH_3I (Imre, Kinsey, Sinha, and Krenos 1984; Hale, Galica, Glogover, and Kinsey 1986; Sundberg et al. 1986; Shapiro 1986).
2) O_3 (Imre, Kinsey, Field, and Katayama 1982; Atabek, Bourgeois, and Jacon 1986; Johnson and Kinsey 1987; Chasman, Tannor, and Imre 1988).
3) NH_3 (Gregory and Lipski 1976; Ziegler and Hudson 1984; Rosmus et al. 1987).
4) H_2O (Sension, Brudzynski, and Hudson 1988; Hennig, Engel, Schinke, and Staemmler 1988; Zhang and Imre 1989; Sension et al. 1990).
5) H_2S (Kleinermanns, Linnebach, and Suntz 1987; Person, Lao, Eckholm, and Butler 1989; Brudzynski, Sension, and Hudson 1990).
6) CH_2I_2 (Zhang and Imre 1988a; Zhang, Heller, Huber, and Imre 1988).
7) ClNO (Bell, Pardon, and Frey 1989).

† See Hartke (1991) for a more complete list of references.

H_2O excited into the first continuum is particularly interesting because both high-resolution experiments and parameter-free *ab initio* theory are feasible. It reveals detailed insight into the interrelation between fast fragmentation on a repulsive potential and the emission spectrum.

The experimental Raman spectrum does not show any excitation of motion in the bending mode. This does not come as a surprise because, as we discussed in detail in Section 10.1, the $\tilde{A}$-state PES is nearly independent of the HOH bending angle around the ground-state equilibrium of 104°. Thus, upon promotion to the upper state the center of the wavepacket remains near 104°, its angular shape remains almost unchanged, and therefore the overlap with excited bending wavefunctions in the electronic ground state is negligibly small *for all times*. The lack of bending excitation in the Raman spectrum has the same origin as the FC-type rotational state distribution of the $OH(^2\Pi)$ product. Therefore, in the theoretical analysis one can fix the bending angle at the ground-state equilibrium and include only the two OH stretching modes.†

The calculations of the energies and the wavefunctions of excited vibrational states of H_2O have been described in Section 13.2.1. While we considered in Section 13.2 the ungerade states, i.e., those which are anti-symmetric with respect to exchange of the hydrogen atoms, the emission spectrum involves only the gerade states. Since the transition dipole function is symmetric, the Raman process preserves the symmetry. Thus, starting from the ground vibrational state $|00^+\rangle$, the evolving wavepacket remains symmetric for all times and therefore only transitions to symmetric states $|mn^+\rangle$ are allowed.

Figure 14.3 shows an energy-level diagram of the gerade states of H_2O; the corresponding wavefunctions, multiplied by the $\tilde{X} \rightarrow \tilde{A}$ transition dipole function, are depicted in Figure 14.4. The energies are clustered according to the total number $N = m + n$ of OH vibrational quanta. The splitting, e.g., between $|11^+\rangle$ and $|20^+\rangle$, manifests the coupling between the two O-H bonds. The wavefunctions show an interesting nodal pattern. For the subsequent discussion of the high resolution emission spectrum, one should note the quite different behavior of the $|N0^+\rangle$ and the $|(N -$

† Incidentally we note that the excitation into the second absorption band, however, yields Raman spectra with excitation of the bending mode which is in full accord with the overall picture of the dissociative motion in the $\tilde{B}$ state as illustrated in Figure 8.9 (Sension, Brudzynski, and Hudson 1988). Immediately after the wavepacket begins to move in the upper state, the HOH bending angle opens under the influence of the strong torque $-\partial V/\partial\alpha$. The emission to excited bending states in the electronic ground state and the observation of highly excited rotational states of OH both confirm this picture.

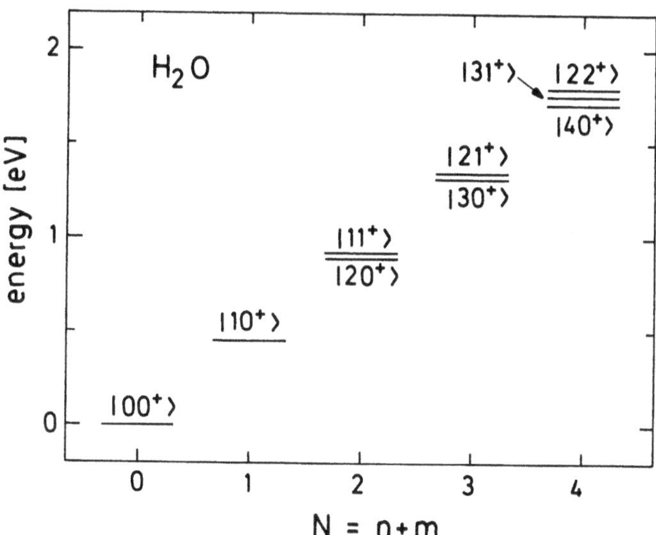

Fig. 14.3. Energy-level diagram of the *gerade* vibrational states of H_2O in the electronic ground state for fixed bending angle $\alpha = 104°$. The local mode assignment is explained in Section 13.2. $N = m + n$ denotes the total number of OH stretching quanta.

$1)1^+\rangle$ wavefunctions. As a side comment we mention that for $N \geq 2$ a description in terms of normal modes is no longer obvious.

The upper-right panel of Figure 14.4 shows the initial wavepacket prepared by the (infinitely short) laser pulse in the excited electronic state. In the very first moments, the wavepacket moves basically in the direction of the saddle of the $\tilde{A}$-state potential. Simultaneously, it spreads, bifurcates, and each part starts to slide down into either of the two exit channels. While the wavepacket evolves in the upper state, it accumulates overlap with the vibrationally excited wavefunctions in the lower state which is manifested by the time dependence of the cross-correlation function shown in Figure 14.5. The overlap of the wavepacket with itself ($i = f$), i.e., the autocorrelation function, is unity at the beginning and rapidly drops to zero as we discussed in the preceding chapters. If the transition dipole function were coordinate-independent, all cross-correlation functions with $i \neq f$ would be exactly zero at $t = 0$ because of orthogonality. The finite values thus manifest the influence of the coordinate dependence of μ_{AX}. The cross-correlation functions first rise, reach a maximum, and descend to zero. With increasing excitation the nodal pattern of the vibrational wavefunctions becomes more complicated and since positive and negative regions partly cancel each other the maximum values of the $C_{if}(t)$ decrease with N. This trend is additionally sustained by the fact

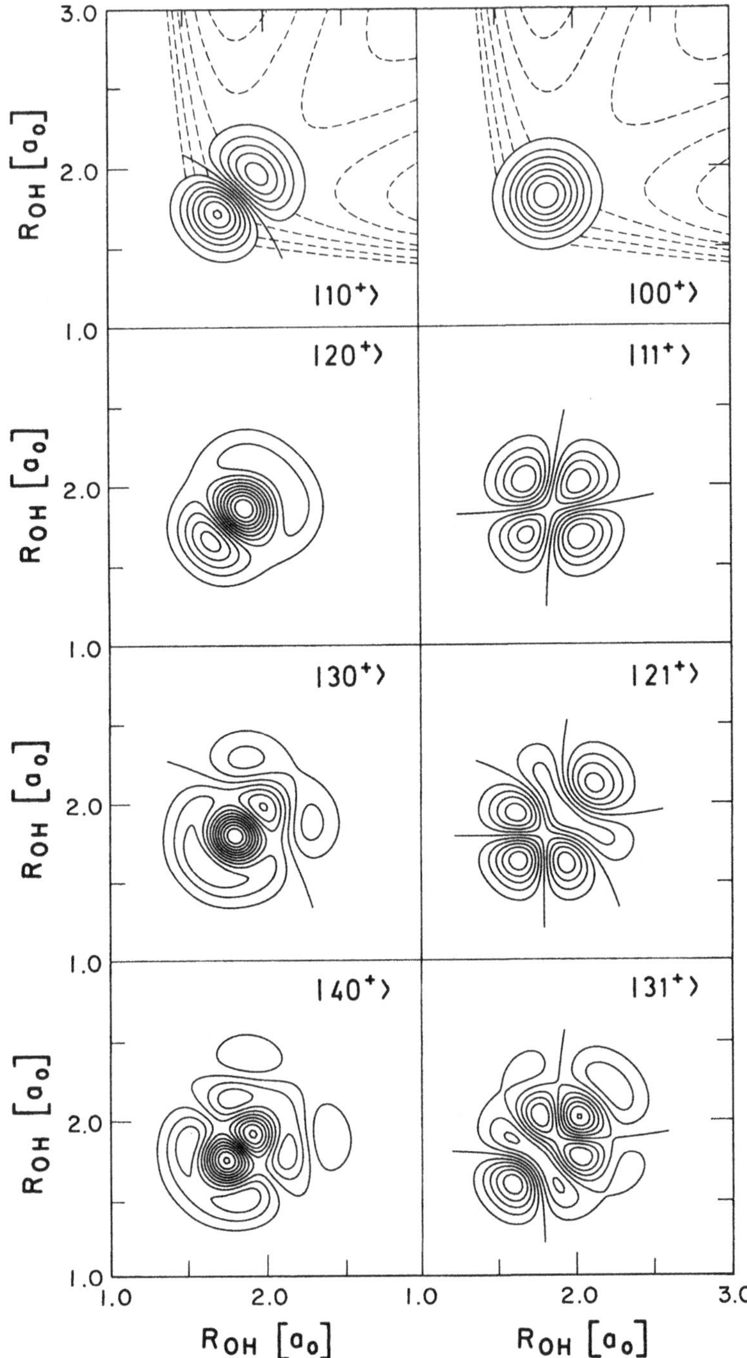

Fig. 14.4. Vibrational eigenfunctions, multiplied by the $\tilde{X} \rightarrow \tilde{A}$ transition dipole function, for the gerade states of $H_2O(\tilde{X})$. All wavefunctions are symmetric with (*cont.*)

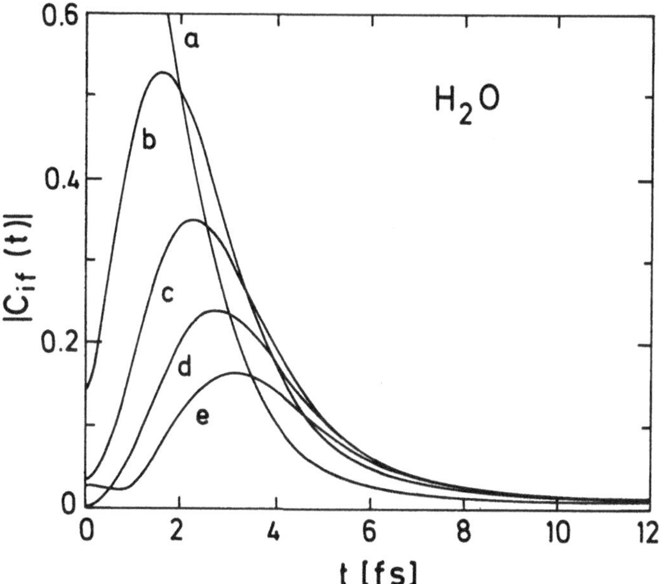

Fig. 14.5. Cross-correlation functions $|C_{if}(t)|$ for H_2O excited in the first continuum: (a) $|00^+\rangle$, (b) $|10^+\rangle$, (c) $|20^+\rangle$, (d) $|30^+\rangle$, and (e) $|40^+\rangle$.

that the overall amplitude of the wavefunctions *decreases* with the degree of excitation.

In view of Figure 14.1 we expect that the cross correlation functions become maximal when the evolving wavepacket overlaps the maximum of the excited vibrational wavefunction at the outer turning point. This implies that, on the average, the maxima of the $C_{if}(t)$ occur at increasingly later times with increasing excitation. Although the effect is small, the examples in Figure 14.5 confirm this prediction. Note, however, that in the case of H_2O, which exemplifies very fast dissociation, the overlap is essentially finished after only 10 fs.

Figure 14.6 shows the measured low resolution emission spectrum for two excitation wavelengths together with the results of two calculations. Here, low resolution means that the intensities for different final states with the same total number N of OH stretching quanta, e.g., $|20^+\rangle$ and $|11^+\rangle$, are summed. The smooth fall-off indicates the increase of the

---

Fig. 14.4. (*cont.*) respect to exchange of the two hydrogen atoms. The dashed contours in the top panel represent the PES in the excited electronic state. In contrast to Figure 13.3 we show here the wavefunction rather than its modulus square!

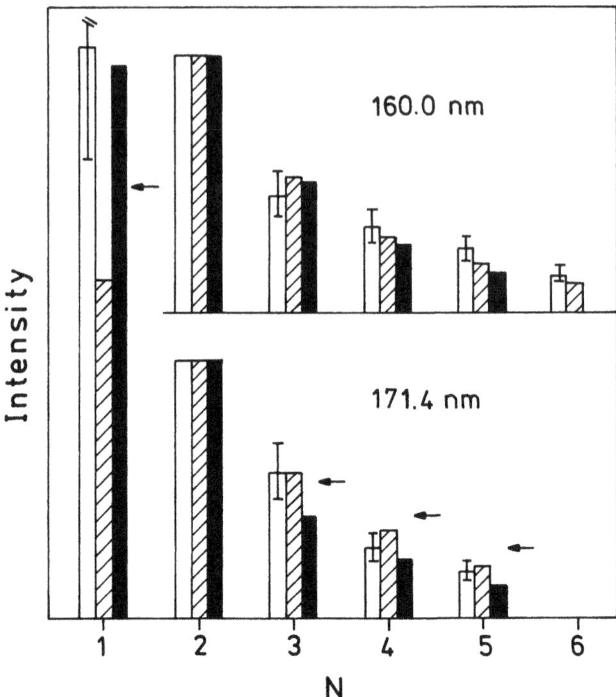

Fig. 14.6. Low resolution Raman spectrum for H_2O excited in the first continuum for two excitation wavelengths. N is the total number of OH stretching quanta. The open and the shaded bars represent the measurements and the calculations of Sension et al. (1990) and the black bars are the calculations of Hennig et al. (1988). The arrows mark the intensities calculated with constant transition dipole function. The experimental and the two theoretical spectra are normalized to the same intensity for $N = 2$. Adapted from Sension et al. (1990).

oscillatory behavior of the vibrational wavefunctions with N and therefore the decrease of the overlap with the evolving wavepacket. It also crucially depends on the coordinate dependence of the transition dipole. In the present case, μ_{AX} diminishes as the O-H bond elongates which has the net consequence that the spectrum declines more rapidly than the spectrum calculated with a constant transition dipole moment.

The measured high resolution spectrum, shown in Figure 14.7 for D_2O, reveals an interesting fine-structure. Each cluster of lines belonging to the same $N = m + n$ is split into basically two lines, $|N0^+\rangle$ and $|(N-1)1^+\rangle$. The main progression steadily decreases with N while the second one first rises and eventually falls off again with a maximum at $N = 5$. The calculations reproduce this peculiar behavior (Hennig et al. 1988; Sension et al 1990). Why are the cross sections for $|11^+\rangle$ and $|21^+\rangle$ so much smaller than for their counterparts $|20^+\rangle$ and $|30^+\rangle$, respectively? The answer

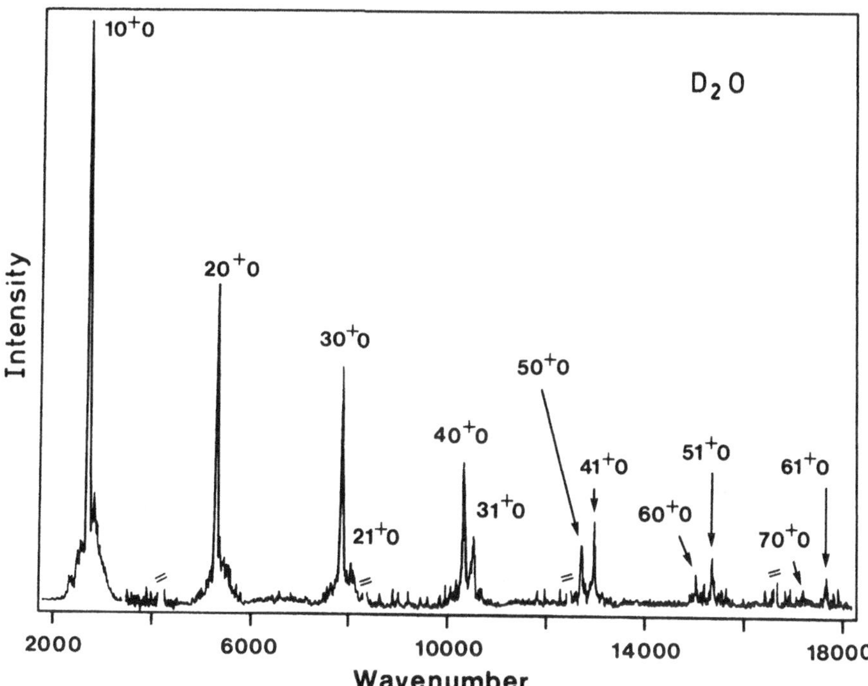

Fig. 14.7. High resolution Raman spectrum for D_2O excited with a 171 nm photon. Local mode assignment is used; the first two quantum numbers indicate the OH stretching modes, the plus sign indicates the symmetry, and the third number represents the bending quantum number. The latter is zero in all cases and therefore not mentioned in the text. The energies are measured with respect to the $|00^+0\rangle$ vibrational ground state. Reproduced from Sension et al. (1990).

lies in the structural behavior of the corresponding wavefunctions and the dissociation path of the evolving wavepacket. The $|(N-1)1^+\rangle$ wavefunctions behave for larger O-H bond distances roughly like an excited OH($n = 1$) wavefunction with a node in the direction perpendicular to the dissociation path (see Figure 14.4 and also Figure 13.4 for wavefunctions with ungerade symmetry). As the evolving wavepacket slides into the exit channel it therefore overlaps a positive and a negative portion with the result that the cross-correlation function is comparatively small. The $|N0^+\rangle$ states, on the other hand, behave in the exit channel more like an OH($n = 0$) wavefunction with only one maximum in the direction perpendicular to the dissociation mode. The cross-correlation functions are consequently larger than for the $|(N-1)1^+\rangle$ states. With increasing N, however, the differences become less prominent. The finer structures of the emission spectrum are the fingerprints of the vibrationally excited wavefunctions in the ground electronic state and how they overlap with the evolving wavepacket in the upper electronic state. H_2O shows the same effect, however, slightly less pronounced.

14.3 Raman spectra for H_2S

The qualitative behavior of the emission spectrum for H_2O excited in the first absorption band does not significantly depend on the excitation wavelength. Such a behavior usually indicates that only one excited electronic state is involved.† This contrasts sharply the emission spectroscopy of H_2S excited in the 195 nm band (Brudzynski, Sension, and Hudson 1990). Figure 14.8 depicts Raman spectra for a sequence of excitation wavelengths covering the main portion of the absorption spectrum. Unlike for H_2O, the $|20^+0\rangle\,/\,|11^+0\rangle$ and the $|30^+0\rangle\,/\,|21^+0\rangle$ ratios depend drastically on the wavelength. In addition, excitation of the first bending overtone, in combination with one or two stretching quanta, is also prominent, whereas for H_2O absolutely no action is detected in the bending mode. The rise and the subsequent decline of the bending peaks as the photolysis wavelength is scanned across the spectrum has the same qualitative wavelength dependence as the intensity ratio of the fine-structure transitions.

Since the vibrationally excited states of $H_2O(\tilde{X})$ and $H_2S(\tilde{X})$ have qualitatively the same overall behavior, the strong dissimilarities in the emission spectra must be caused by strong differences in the motion in the upper state. The eye-catching energy dependence in the case of H_2S has been interpreted by Brudzynski et al. as the consequence of two overlapping electronic states. *Ab initio* calculations indeed confirm this hypothesis (Weide, Staemmler, and Schinke 1990; Heumann, Düren, and Schinke 1991; Theodorakopoulos and Petsalakis 1991). The fragmentation involves two electronic states, one being bound and the other one being dissociative (in the diabatic picture). Their energy separation in and near the symmetric stretch configuration is comparatively small which points to strong nonadiabatic coupling. Remember also that the differences in the vibrational state distributions of the OH and the SH products (see Figures 9.10 and 9.11) indicate distinct differences in the fragmentation dynamics. The mixing of two electronic states, i.e., the breakdown

† A strong wavelength dependence is, however, expected if the excited electronic state has an appreciably long lifetime so that the absorption spectrum shows pronounced resonances. The photodissociations of $CH_3ONO(S_1)$ or NH_3, for example, naturally yield different emission spectra for different excitation wavelengths, despite the fact that only one electronic state participates. The Raman spectrum obviously depends on the particular vibrational band excited in the upper state (see Figures 7.12 and 7.17). If the lifetime of the excited complex is sufficiently long, the Raman spectrum is equivalent to an "ordinary" emission spectrum originating from a (quasi-)bound upper state.

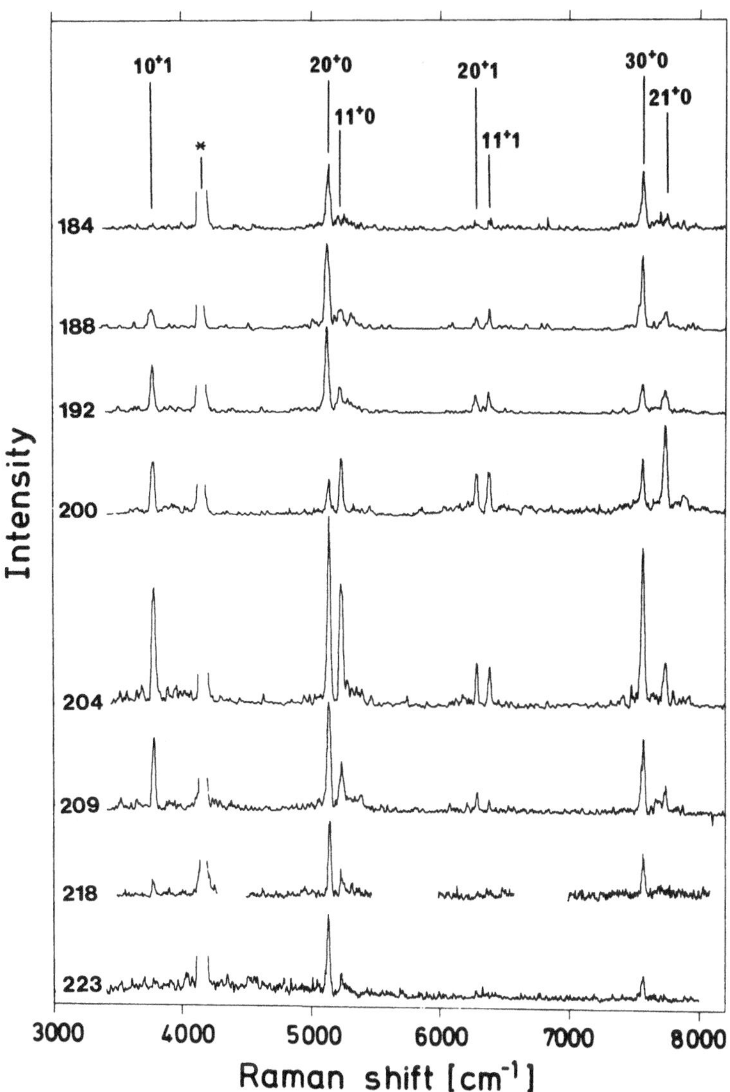

Fig. 14.8. Raman spectra for H_2S measured for several excitation wavelengths. The maximum of the absorption spectrum is around 195 nm. Local mode assignment is used as described in Figure 14.7 for H_2O. Note the striking dependence on the excitation wavelength which is given at the left-hand side of each spectrum! Reproduced from Brudzynski, Sension, and Hudson (1990).

of the Born-Oppenheimer approximation, exceedingly complicates any rigorous treatment of the nuclear motion. Nonadiabatic effects in photodissociation is the topic of Chapter 15, where we will discuss H_2S as one particular example.

Let us summarize this chapter. The emission spectrum of a dissociating molecule represents, in addition to the absorption spectrum and the final state distributions, an alternative means for the study of molecular motion in excited electronic states. Provided the upper-state PES is repulsive and the fragmentation time is short, the absorption spectrum is mainly sensitive to the very early times of the evolution of the wavepacket and therefore it reflects primarily the shape of the multi-dimensional PES close to the Franck-Condon region. As the wavepacket starts to move out of its initial position it explores a larger region of the PES. The Raman spectrum is the fingerprint of the motion in the excited state over a somewhat longer period and therefore it is sensitive to larger portions of the PES. Finally, the state distributions of the photofragments reflect the entire history of the dissociation process. In this (certainly somehow too simplistic) picture emission spectroscopy is the link between the short- and the long-time dynamics.

Raman spectra are naturally closely related to the photodissociation cross sections for vibrationally excited parent molecules. The latter contain, without any doubt, more details about the potential energy surfaces in the lower as well as the upper states, but on the other hand, they are more difficult to measure. Compared to the experiments described in Chapter 13, which requires three lasers, Raman spectra are rather cheap to obtain.

15
Nonadiabatic transitions in dissociating molecules

Dissociation via a single excited electronic state is the exception rather than the rule. The remarkable success with which all experimental results for the dissociation of H_2O, for example, have been reproduced by rigorous calculations without any adjustable parameter rests mainly upon the fact that only one electronic state is involved (Engel et al. 1992). Many other photodissociation processes, however, proceed via two or even more electronic states with the possibility of transitions from one state to another. Figure 15.1 illustrates a common situation: the photon excites the molecule from the electronic ground state (index 0) to a dipole-allowed upper state (index 1) which further out in the exit channel interacts with a second electronic state (index 2). The latter may be dipole-forbidden and therefore not directly accessible by the photon. The corresponding *diabatic* potentials V_1 and V_2 cross at some internuclear distance R_c. In the proximity of this point the coupling between the two electronic states, which was ignored throughout all of the preceding chapters, can be large with the consequence that a transition from state 1 to state 2 and/or *vice versa* becomes possible (*radiationless transition, electronic quenching*). Electronic transitions manifest the break-down of the Born-Oppenheimer approximation, i.e., the motion of the electrons and the heavy particles can no longer be adiabatically separated.

Let us imagine a wavepacket starting in the Franck-Condon region on potential V_1. When it reaches the crossing region it splits under the influence of the coupling into two parts. Beyond R_c we will find two wavepackets, one evolving on V_1 which leads to products A^* and B and the other one moving on potential V_2 which yields both fragments A and B in their electronic ground states. The electronic branching ratio A^*/A depends intimately on the coupling strength between states 1 and 2.

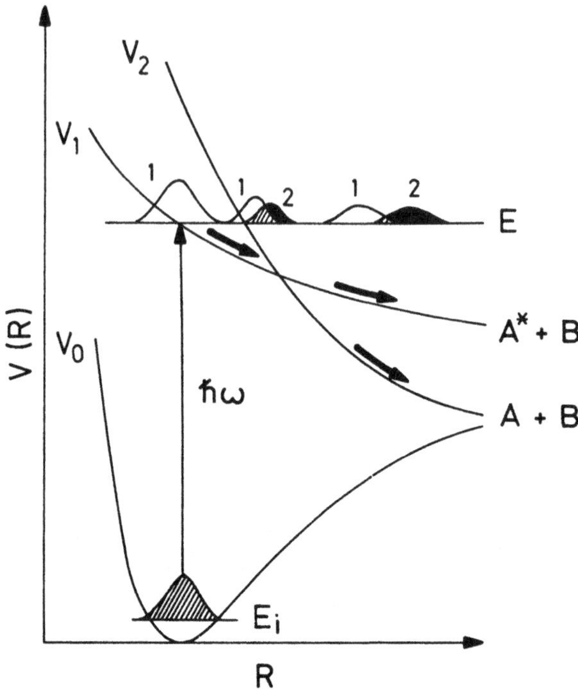

Fig. 15.1. Schematic illustration of an electronic transition in the neighborhood of the crossing of two *diabatic* potential curves V_1 and V_2. The upper part shows the splitting of the wavepacket which was originally generated by the photon in electronic state 1. The part labeled by index 1 (white) evolves on V_1 and the part labeled by index 2 (black) evolves on V_2.

Transitions between electronic states are *formally* equivalent to transitions between different vibrational or rotational states which were amply discussed in Chapters 9–11. Computationally, however, they are much more difficult to handle because they arise from the coupling between electronic and nuclear motions. The rigorous description of electronic transitions in polyatomic molecules is probably the most difficult task in the whole field of molecular dynamics (Siebrand 1976; Tully 1976; Child 1979; Rebentrost 1981; Baer 1983; Köppel, Domcke, and Cederbaum 1984; Whetten, Ezra, and Grant 1985; Desouter-Lecomte et al. 1985; Baer 1985b; Lefebvre-Brion and Field 1986; Sidis 1989a,b; Coalson 1989). The reasons will become apparent below. The two basic approaches, the *adiabatic* and the *diabatic representations*, will be outlined in Sections 15.1 and 15.2, respectively. Two examples, the photodissociation of CH_3I and of H_2S, will be discussed in Section 15.3.

15.1 The adiabatic representation

In order to illustrate electronic transitions we discuss the simple two-dimensional model of a linear triatomic molecule ABC as depicted in Figure 2.1. R and r are the appropriate Jacobi coordinates to describe the nuclear motion and the vector $\mathbf{q}$ comprises all electronic coordinates. The total molecular Hamiltonian $\hat{H}_{mol}$, including all nuclear and electronic degrees of freedom, is given by Equation (2.28) with $\hat{H}_{el}$ and $\hat{T}_{nu}$ being the electronic Hamiltonian and the kinetic energy of the nuclei, respectively.

According to Equation (2.29), in the *adiabatic representation* (index a) one expands the total molecular wavefunction $F(R, r, \mathbf{q})$ in terms of the Born-Oppenheimer states $\Xi_k^{(a)}(\mathbf{q}; R, r)$ which solve the electronic Schrödinger equation (2.30) for fixed nuclear configuration (R, r). In this representation, the electronic Hamiltonian is diagonal,

$$\langle \Xi_k^{(a)} \mid \hat{H}_{el} \mid \Xi_{k'}^{(a)} \rangle = V_k^{(a)} \, \delta_{kk'}, \tag{15.1}$$

whereas $\hat{T}_{nu}$ is not diagonal, i.e., it couples different electronic states [see Equation (15.4) below]. The $V_k^{(a)}$ are the adiabatic potentials defined in (2.30). The electronic wavefunctions $\Xi_k^{(a)}$ depend — like the potentials — *parametrically* on all nuclear degrees of freedom (R, r). The stationary wavefunctions (including all electronic and all nuclear degrees of freedom) which represent the initial and the final molecular states for the situation sketched in Figure 15.1, are written in full detail as

$$F_i(R, r, \mathbf{q}; E_i) = \Psi_0(R, r; E_i) \, \Xi_0^{(a)}(\mathbf{q}; R, r) \tag{15.2a}$$

$$F_f(R, r, \mathbf{q}; E, e, n) = \sum_{k=1,2} \Psi_k(R, r; E, e, n) \, \Xi_k^{(a)}(\mathbf{q}; R, r). \tag{15.2b}$$

Ψ_0 is the bound nuclear wavefunction in the electronic ground state corresponding to energy E_i and Ψ_1 and Ψ_2 describe the nuclear motion in the two excited electronic states with outgoing unit flux in electronic channel e and vibrational channel n. The energy in the excited states is given by $E = E_i + \hbar\omega$. Ψ_1 and Ψ_2 are continuum wavefunctions of the kind constructed in Section 2.5. F_f is a coherent superposition of the wavefunctions for electronic states 1 and 2. For simplicity of presentation nonadiabatic coupling is assumed to exist only between the two excited states but not between the ground and the excited states.

The nuclear wavefunctions for states 1 and 2 must solve the coupled equations (2.31) with

$$\hat{T}_{nu} = -\frac{\hbar^2}{2m} \frac{\partial^2}{\partial R^2} - \frac{\hbar^2}{2\mu} \frac{\partial^2}{\partial r^2} \tag{15.3}$$

in the case of the collinear system. Inserting (15.3) into (2.31) yields

$$\begin{aligned}&\Big[-\frac{\hbar^2}{2m}\frac{\partial^2}{\partial R^2}-\frac{\hbar^2}{2\mu}\frac{\partial^2}{\partial r^2}+V_k^{(a)}-E\Big]\Psi_k(E,e,n)\\&+\sum_{k'=1,2}\Big[Q_{kk'}^{(R)}\frac{\partial}{\partial R}+Q_{kk'}^{(r)}\frac{\partial}{\partial r}+U_{kk'}^{(R)}+U_{kk'}^{(r)}\Big]\Psi_{k'}(E,e,n)=0\end{aligned}\quad(15.4)$$

with $k, k' = 1$ and 2. The nonadiabatic elements defined by

$$\begin{aligned}Q_{kk'}^{(R)}(R,r)&=-\frac{\hbar^2}{m}\left\langle\Xi_k^{(a)}\,\middle|\,\frac{\partial}{\partial R}\,\middle|\,\Xi_{k'}^{(a)}\right\rangle\\Q_{kk'}^{(r)}(R,r)&=-\frac{\hbar^2}{\mu}\left\langle\Xi_k^{(a)}\,\middle|\,\frac{\partial}{\partial r}\,\middle|\,\Xi_{k'}^{(a)}\right\rangle\\U_{kk'}^{(R)}(R,r)&=-\frac{\hbar^2}{2m}\left\langle\Xi_k^{(a)}\,\middle|\,\frac{\partial^2}{\partial R^2}\,\middle|\,\Xi_{k'}^{(a)}\right\rangle\\U_{kk'}^{(r)}(R,r)&=-\frac{\hbar^2}{2\mu}\left\langle\Xi_k^{(a)}\,\middle|\,\frac{\partial^2}{\partial r^2}\,\middle|\,\Xi_{k'}^{(a)}\right\rangle\end{aligned}\quad(15.5)$$

provide the coupling between the two electronic states 1 and 2. They reflect how strongly the electronic wavefunctions depend on the nuclear coordinates.

The set of coupled equations must be solved subject to boundary conditions similar to (2.59) with unit outgoing flux in only one particular electronic channel, which we designate by e, and one particular vibrational state n of the diatomic fragment, e.g., $e = 1, n = 5$ for $A + BC(n = 5)$ and $e = 2, n = 6$ for $A^* + BC(n = 6)$. In the actual calculation one would subsequently expand the nuclear wavefunctions $\Psi_k(R, r; E, e, n)$ in a set of vibrational basis functions ($n = 0, \ldots, n_{max}$) as described in Section 3.1 which leads to a total of $2(n_{max} + 1)$ coupled equations. It is not difficult to surmise how complicated the coupled equations will become if the rotational degree of freedom is also included.

Following the general rules given in Chapter 2, the partial cross sections for absorbing a photon with frequency ω and at the same time producing the fragments in electronic state e and vibrational state n are given by

$$\sigma(\omega,e,n)\propto\omega|\langle F_f(R,r,\mathbf{q};E,e,n)\mid\mathbf{e}\cdot\hat{\mathbf{d}}\mid F_i(R,r,\mathbf{q};E_i)\rangle|^2,\quad(15.6a)$$

where $\hat{\mathbf{d}}$ is the dipole operator and $\mathbf{e}$ is a unit vector in the direction of the electric field. Inserting (15.2a) and (15.2b) into (15.6a) yields

$$\sigma(\omega,e,n)\propto\omega|\langle\Psi_1(E,e,n)\mid\mu_{10}^{(e)}\mid\Psi_0\rangle+\langle\Psi_2(E,e,n)\mid\mu_{20}^{(e)}\mid\Psi_0\rangle|^2,\quad(15.6b)$$

where $\mu_{10}^{(e)}$ and $\mu_{20}^{(e)}$ are the components of the corresponding transition dipole functions in the direction of the electric field vector. The latter are obtained after integration over the electronic degrees of freedom [see

Equation (2.35)]. Equation (15.6b) is formally equivalent to (2.66) with the exception that in the present case the outgoing channel also includes, in addition to the vibrational state, the particular electronic state. It is important to realize that because of the nonadiabatic coupling both excited electronic states and both electronic product channels are populated, even if one transition dipole moment is exactly zero for all nuclear geometries. Furthermore, the superposition of two complex-valued amplitudes in the case that both transition moments are non-zero can lead to interesting interference patterns.

Solving the coupled vibronic equations is the minor problem. In principle, they can be treated in the same way as Equation (3.5) or (3.20) for pure vibrational or pure rotational excitation, respectively. The calculation of the kinetic coupling elements represents the major problem! They require first- and second-order derivatives of the electronic wavefunctions with respect to all nuclear degrees of freedom. The $\Xi^{(a)}(\mathbf{q}; R, r)$ are many-body wavefunctions and usually depend on several electronic coordinates. It is therefore not difficult to surmise that the evaluation of the nonadiabatic coupling elements demands an incredible amount of computational work which exceeds by far the calculation of potential energy surfaces. Numerical differentiation by finite differences as well as analytical gradient methods is used in practical applications (Buenker, Peric, Peyerimhoff, and Marian 1981; Petrongolo, Buenker, and Peyerimhoff 1982; van Dishoek, van Hemert, Allison, and Dalgarno 1984; Lengsfield, Saxe, and Yarkony 1984; Lengsfield and Yarkony 1986; Saxe and Yarkony 1987). The lack of nonadiabatic coupling elements represents *the* major bottleneck for realistic investigations of electronic transitions in polyatomic molecules.

The kinetic coupling originates from the coordinate dependence of the adiabatic electronic wavefunctions in the neighborhood of an avoided crossing. Let us consider the situation depicted in Figure 15.2(a). The two adiabatic potentials $V_1^{(a)}$ and $V_2^{(a)}$ (dashed curves) are not allowed to cross if the corresponding wavefunctions have the same electronic symmetry and therefore they show an avoided crossing at a bond distance R_c. Although the energy separation at R_c may be very small, the potential for adiabatic state 1 is consistently the lower eigenvalue of the electronic Schrödinger equation while the potential for state 2 is for all distances the upper eigenvalue. In such a case it is typical that the adiabatic wavefunctions $\Xi_1^{(a)}$ and $\Xi_2^{(a)}$ interchange their overall character in the vicinity of the avoided crossing. For example, before R_c the lower adiabatic state may have dominantly $^1B_1$ character while the upper adiabatic state may be better represented by an electronic wavefunction with $^1A_2$ symmetry;

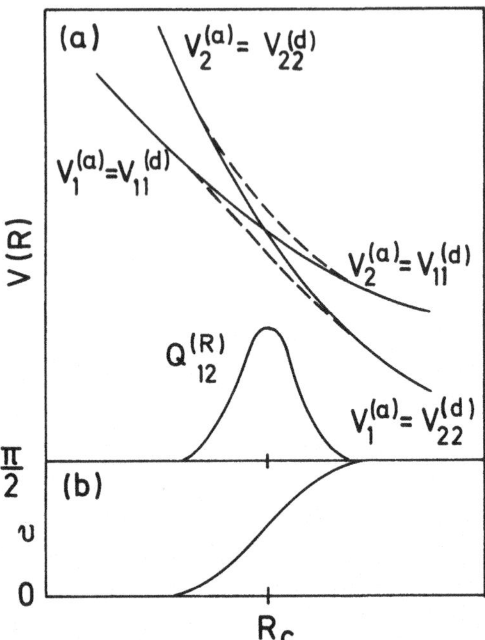

Fig. 15.2. (a) Adiabatic potentials, $V_1^{(a)}$ and $V_2^{(a)}$ (dashed curves), and the diagonal elements of the diabatic potential matrix, $V_{11}^{(d)}$ and $V_{22}^{(d)}$ (solid curves), near an avoided crossing. $Q_{12}^{(R)}$ schematically represents the coordinate dependence of the nonadiabatic coupling elements. (b) Variation of the mixing angle υ defined in Equation (15.7) in the region of the avoided crossing.

beyond the avoided crossing, however, the assignment is reversed (see Section 15.3.2 for H_2S). The ordering of the adiabatic states is determined by their energies and not by the behavior of the electronic wavefunctions like in the diabatic representation! As a consequence, the more or less abrupt change of the electronic wavefunctions over a relatively short distance leads to a strong nonadiabatic coupling element centered around R_c as illustrated schematically in Figure 15.2. Nonadiabatic transitions therefore occur most prominently near an avoided crossing.

15.2 The diabatic representation

The sudden change of the adiabatic wavefunctions near an avoided crossing and the resultant large nonadiabatic coupling elements, which may be confined to an exceedingly narrow region in coordinate space, make the adiabatic representation rather inconvenient for numerical applications. An electronic basis which changes smoothly across the region of the avoided crossing is therefore more desirable. Such so-called *diabatic states* (index d) can be constructed in several ways [for a general discussion see Köppel, Domcke, and Cederbaum (1984), for example]. The general idea is to compose diabatic wavefunctions $\Xi_k^{(d)}$ such that the coupling matrices $\mathbf{Q}$ and $\mathbf{U}$ defined in (15.5) are so small in this new representation that they can be neglected. The $\Xi_k^{(d)}$ are obviously no longer eigenstates of the electronic Hamiltonian $\hat{H}_{el}$. As opposed to the adiabatic picture, the matrix representation of the nuclear kinetic energy operator, $\hat{T}_{nu}$, is (approximately) diagonal in the diabatic representation, whereas the matrix of the electronic Hamiltonian is nondiagonal:

- The coupling between the adiabatic states is provided by the off-diagonal elements of the matrix representation of $\hat{T}_{nu}$ (kinetic coupling) while the coupling between the diabatic states arises from the off-diagonal elements of the matrix representation of $\hat{H}_{el}$ (potential coupling).

Formally, we can write the unitary transformation from the adiabatic to the diabatic basis as

$$\begin{pmatrix} \Xi_1^{(d)} \\ \Xi_2^{(d)} \end{pmatrix} = \begin{pmatrix} \cos\upsilon & -\sin\upsilon \\ \sin\upsilon & \cos\upsilon \end{pmatrix} \begin{pmatrix} \Xi_1^{(a)} \\ \Xi_2^{(a)} \end{pmatrix}, \tag{15.7}$$

with $\upsilon(R,r)$ being the coordinate-dependent *mixing angle.* It changes from 0 to $\pi/2$ across the avoided crossing as illustrated in Figure 15.2(b). The mixing angle reflects the relative contributions of the two adiabatic states for the construction of the diabatic ones and *vice versa.* The special form of (15.7) guarantees that the diabatic basis is orthogonal and normalized. With the help of (15.7) we can represent the diabatic potentials in terms of the adiabatic ones,

$$\begin{aligned} V_{11}^{(d)} &\equiv \langle \Xi_1^{(d)} \mid \hat{H}_{el} \mid \Xi_1^{(d)} \rangle = \cos^2\upsilon\ V_1^{(a)} + \sin^2\upsilon\ V_2^{(a)} \\ V_{22}^{(d)} &\equiv \langle \Xi_2^{(d)} \mid \hat{H}_{el} \mid \Xi_2^{(d)} \rangle = \sin^2\upsilon\ V_1^{(a)} + \cos^2\upsilon\ V_2^{(a)}, \end{aligned} \tag{15.8}$$

and likewise the off-diagonal elements become

$$V_{12}^{(d)} = V_{21}^{(d)} \equiv \langle \Xi_1^{(d)} \mid \hat{H}_{el} \mid \Xi_2^{(d)} \rangle = \cos\upsilon\ \sin\upsilon\ [V_1^{(a)} - V_2^{(a)}]. \tag{15.9}$$

Similar relations hold for the transition dipole functions with the electronic ground state if we make the transformation from the adiabatic to the diabatic representation. Figure 15.2(a) illustrates the interrelation between the adiabatic potentials and the diagonal elements of the potential matrix in the diabatic representation. Unlike the adiabatic potentials, $V_{11}^{(d)}$ and $V_{22}^{(d)}$, which are not eigenvalues of the electronic Schrödinger equation, are allowed to cross. If the mixing angle is zero or $\pi/2$, the diabatic and the adiabatic potentials coincide and the coupling between the electronic states is exactly zero irrespective of the energy separation of the two adiabatic states.

While the adiabatic representation is unique and well defined, the diabatic representation is not (Mead and Truhlar 1982). Explicit equations to derive the mixing angle v from the solutions of the electronic Schrödinger equation (2.30) have been proposed by several authors (Top and Baer 1977; Baer and Beswick 1979; Baer 1983; Köppel, Domcke, and Cederbaum 1984). These methods require, however, knowledge of the nonadiabatic coupling elements which, on the other hand, is extremely difficult to obtain from *ab initio* calculations. Other, more approximate methods to extract the mixing angle from large-scale *ab initio* calculations have been suggested, which avoid the direct calculation of the kinetic coupling elements (Marcías and Riera 1978; Werner and Meyer 1981; Durand and Malrieu 1987; Desouter-Lecomte, Dehareng, and Lorquet 1987; Pacher, Cederbaum, and Köppel 1988; Werner, Follmeg, and Alexander 1988; Werner, Follmeg, Alexander, and Lemoine 1989; Petrongolo, Hirsch, and Buenker 1990; Hirsch, Buenker, and Petrongolo 1990).

Because it is computationally more convenient, most researchers choose the diabatic rather than the adiabatic representation if they try to fit experimental data. In addition to the two diagonal elements of the potential matrix, $V_{11}^{(d)}$ and $V_{22}^{(d)}$, one merely needs a third potential (surface) $V_{12}^{(d)}$ which provides the coupling between the diabatic states (Shapiro 1986; Guo and Schatz 1990a,b; Dixon, Marston, and Balint-Kurti 1990).

So far we have invoked the time-independent formulation to describe electronic transitions. In the same manner as described in Section 4.1 we can also derive the time-dependent picture of electronic transitions, using either the adiabatic or the diabatic representation. In the following we feature the latter which is more convenient for numerical applications (Coalson 1985, 1987, 1989; Coalson and Kinsey 1986; Heather and Metiu 1989; Jiang, Heather, and Metiu 1989; Manthe and Köppel 1990a,b; Broeckhove et al. 1990; Schneider, Domcke, and Köppel 1990; Weide, Staemmler, and Schinke 1990; Manthe, Köppel, and Cederbaum 1991; Heumann, Weide, and Schinke 1992).

In analogy to (15.2) we define a time-dependent wavepacket by

$$\mathcal{F}_f(R, r, \mathbf{q}; t) = \sum_{k=1,2} \Phi_k(R, r; t)\, \Xi_k^{(d)}(\mathbf{q}; R, r), \tag{15.10}$$

where $\Phi_1(t)$ and $\Phi_2(t)$ are the corresponding nuclear wavepackets in diabatic states 1 and 2, respectively. Inserting (15.10) into the time-dependent Schrödinger equation with Hamiltonian $\hat{H}_{mol}$ given by (2.28) and exploiting the orthogonality of the (diabatic) electronic wavefunctions leads to coupled equations for the nuclear wavepackets,

$$\begin{aligned} i\hbar\frac{\partial}{\partial t}\Phi_1 &= \left[\hat{T}_{nu} + V_{11}^{(d)}\right]\Phi_1 + V_{12}^{(d)}\Phi_2 \\ i\hbar\frac{\partial}{\partial t}\Phi_2 &= \left[\hat{T}_{nu} + V_{22}^{(d)}\right]\Phi_2 + V_{21}^{(d)}\Phi_1, \end{aligned} \tag{15.11}$$

where the diabatic matrix elements are defined in (15.8) and (15.9). In deriving (15.11) we explicitly assumed that the off-diagonal matrix elements of $\hat{T}_{nu}$ are negligibly small. Equations (15.11) must be solved subject to the initial conditions

$$\Phi_k(R, r; t = 0) = \mu_{k0}^{(e)}\Psi_0(R, r), \tag{15.12}$$

where Ψ_0 is the nuclear wavefunction in the electronic ground state and $\mu_{k0}^{(e)}$ is the coordinate-dependent transition dipole function between the ground and the kth excited electronic state (in the diabatic representation). The autocorrelation function is given by

$$S(t) = \sum_{k=1,2} \langle \Phi_k(0) \mid \Phi_k(t) \rangle. \tag{15.13}$$

The photon instantaneously creates two nuclear wavepackets, one in each excited electronic state. Let us assume that, because of symmetry arguments, $\mu_{20}^{(e)}$ is exactly zero for all nuclear configurations so that $\Phi_2(0) = 0$ as illustrated in Figure 15.1. Φ_1 slides down the potential slope and as it reaches the region of the avoided crossing it couples to the second electronic state with the consequence that the probability for finding the system in state 1, $P_1(t) = \langle \Phi_1(t) \mid \Phi_1(t) \rangle$, diminishes while the probability for finding the system in the second state, $P_2(t) = \langle \Phi_2(t) \mid \Phi_2(t) \rangle$, increases. Beyond the crossing the coupling is again zero and $P_1(t)$ and $P_2(t)$ become both constant in time. The variation of the two probabilities reflects the breakdown of the Born-Oppenheimer approximation. The strength of the coupling element $V_{12}^{(d)}$ controls the transition probability and ultimately the branching ratio for producing the fragments in the two different electronic states. Determination of the product state distributions in the various electronic channels proceeds in exactly the same manner as described in Section 4.1.

The set of coupled equations (15.11) represents an example of time-dependent close-coupling as described in Section 4.2.3. It is formally equivalent to (4.25), for example, and can be solved by exactly the same numerical recipes. The dependence on the two stretching coordinates R and r is treated by discretizing the two nuclear wavepackets on a two-dimensional grid and the Fourier-expansion method is employed to evaluate the second-order derivatives in R and r. If we additionally include the rotational degree of freedom, we may expand each wavepacket in terms of the eigenfunctions of the angular momentum operator ($j = 0, \ldots, j_{max}$) which then yields a system of coupled two-dimensional partial differential equations with dimension $2(j_{max} + 1)$ (Manthe and Köppel 1991; Manthe, Köppel, and Cederbaum 1991; Guo 1991; Heumann, Weide, and Schinke 1992). The time-dependent picture is particularly fruitful because it clearly illustrates how the electronic coupling depletes one electronic state and fills the other one as the system evolves through an avoided crossing.

15.3 Examples

In this section we will elucidate the influence of nonadiabatic coupling on the dissociation dynamics for two systems which have been extensively studied both by experiment and by theory in the last decade: CH_3I and H_2S.

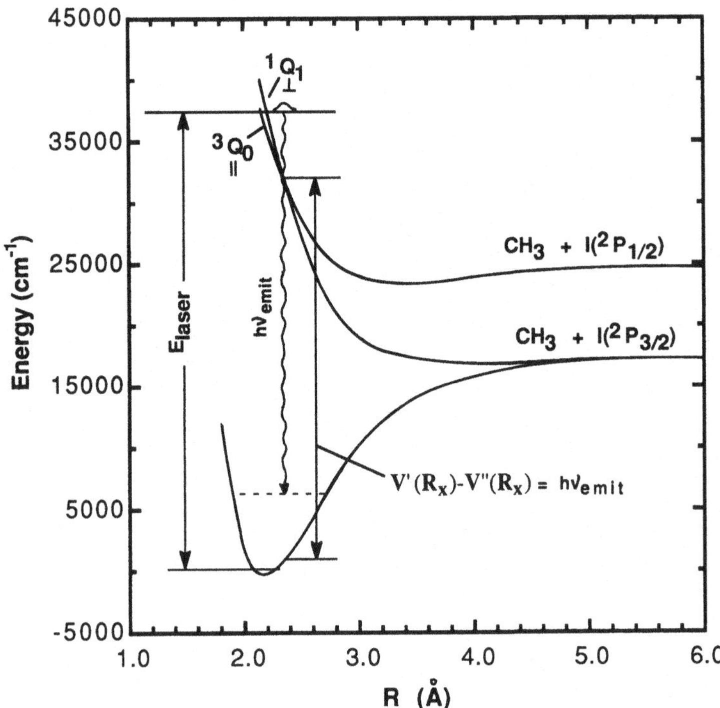

Fig. 15.3. One-dimensional representation of the electronic states involved in the photodissociation of CH_3I with R being the C-I bond distance. The potentials are based on the *ab initio* calculations of Tadjeddine, Flament, and Teichteil (1987). Reproduced from Lao, Person, Xayariboun, and Butler (1990).

15.3.1 The photodissociation of CH_3I

The importance of the photodissociation of CH_3I for the development of chemical lasers has already been emphasized in Section 1.2. Figure 9.6(a) schematically illustrates the geometry in the ground state and Figure 15.3 depicts the electronic states involved and the corresponding potentials.† Absorption is primarily due to excitation of the $^3Q_0$ state which correlates, in the diabatic sense, with CH_3 and the excited state of iodine, $I(^2P_{1/2})$. At a C-I bond distance of about 2.35–2.40 Å the $^3Q_0$ potential crosses the $^1Q_1$ potential which leads diabatically to the ground state of iodine, $I(^2P_{3/2})$. Without any coupling between the two electronic states the photodissociation of CH_3I would almost exclusively yield excited iodine. The experimental observation of an $I(^2P_{1/2})$ quantum yield of only 0.6–0.8 therefore points to an efficient coupling between the $^3Q_0$ and the $^1Q_1$ states somewhere on the way from the parent molecule to the fragments. The wavepacket starts on the $^3Q_0$ potential. When it reaches the crossing region it splits into two parts as a consequence of the coupling with the $^1Q_1$ state. The larger portion continues in the $^3Q_0$ state and yields products $CH_3 + I(^2P_{1/2})$ and the smaller one evolves on the $^1Q_1$ potential energy surface and leads to $CH_3 + I(^2P_{3/2})$.

Using a two-dimensional two-state model suggested by Shapiro (1986), Guo and Schatz (1990b) fitted the total absorption spectrum, the measured I^*/I branching ratio, and the final vibrational state distributions of the CH_3 radical to recent experimental data (see also Section 9.2.2). They found, for example, that the electronic branching ratio is an increasing function of the photon energy. For energies below the $I(^2P_{1/2})$ threshold it is necessarily zero. Shapiro as well as Guo and Schatz employed the diabatic picture. The probability for producing $I(^2P_{3/2})$ is more or less a direct measure of the off-diagonal element of the diabatic potential matrix, $V_{12}^{(d)}$. A similar study has been performed for the photodissociation of ICN (Guo and Schatz 1990a).

The evolution of the wavepacket through the curve crossing is nicely reflected in the polarization of the photons emitted to the electronic ground state during dissociation (Lao, Person, Xayariboun, and Butler, 1990). The transition dipole moments of the two excited states, $^3Q_0$ and $^1Q_1$, with the ground state are parallel and perpendicular to the C-I bond, respectively. The initial excitation is due to a parallel transition. The subsequent emission, however, involves both parallel and perpendicular transitions because the $^1Q_1$ state becomes populated during the breakup.

---

† Quite recently, Amatatsu, Morokuma, and Yabushita (1991) generated six-dimensional potential energy surfaces for the two excited states in *ab initio* calculations. The adiabatic potentials were transformed into the diabatic representation and subsequently used in classical trajectory calculations.

Analyzing the polarization of the emitted light, Lao et al. (1990) were able to determine the fraction $N_\perp$ of photons emitted from ${}^1Q_1$, which in the absence of electronic coupling would be zero because the dissociation starts solely in the ${}^3Q_0$ state. $N_\perp$, depicted in Figure 15.4, is, however, not zero but gradually increases with the number of quanta in the C-I stretching mode. Remember that the Raman spectrum of CH_3I consists basically of a long progression in the C-I stretching mode as illustrated in Figure 14.2.

The time-dependent picture of emission spectroscopy reveals an intriguing explanation of the general trend of $N_\perp$. The lower excited states of C-I stretching are populated relatively early, before the wavepacket in the excited state has moved considerably away from its starting position and before it has reached the curve crossing. The emitted photons therefore stem predominantly from the ${}^3Q_0$ state and $N_\perp$ is negligibly small. As the wavepacket slides down the potential slope it evolves through the curve crossing and the ${}^1Q_1$ state becomes populated as well. Since the higher vibrational states of C-I stretch are filled when the wavepacket has traveled an appreciable distance and thus traversed the crossing region, the corresponding photons are emitted from both electronic states. As a result, $N_\perp$ begins to deviate from zero and gradually increases with the

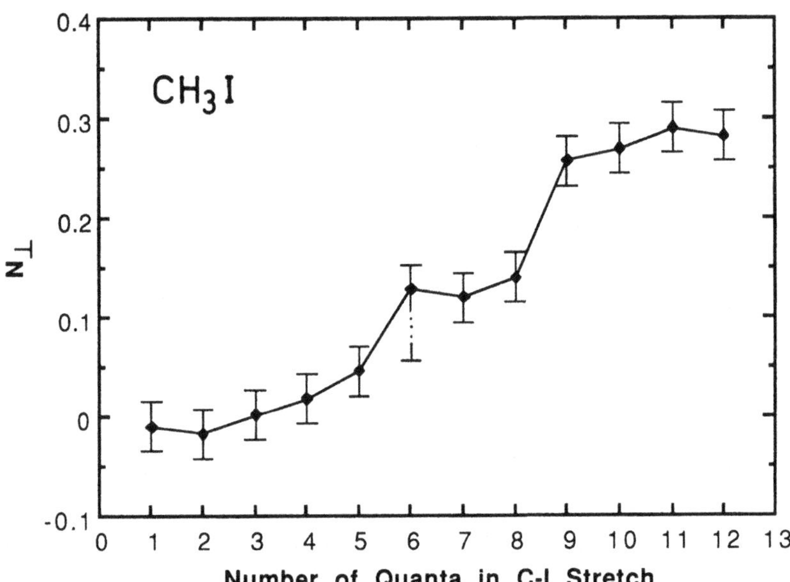

Fig. 15.4. The fraction of photons emitted by the dissociating CH_3I molecule via a perpendicular transition moment, plotted versus the C-I stretching level to which emission occurred. Reproduced from Lao, Person, Xayariboun, and Butler (1990).

number of quanta in the C-I stretching mode. Figure 15.4 manifests — in an indirect way — the splitting of the evolving wavepacket as it traverses the curve crossing region and the population of the two electronic states along the reaction path.

15.3.2 *The photodissociation of* H_2S

The photodissociation of H_2S in the 195 nm band has already been discussed in Chapters 9 and 14 in relation to vibrational excitation and Raman spectroscopy. At first glance, one might be tempted to think that it evolves similarly to the dissociation of H_2O in the first absorption band. That is not the case, however! While the fragmentation of H_2O proceeds via a single electronic state, with electronic symmetry $^1B_1$ in C_{2v}-configuration, the dissociation of H_2S involves two states, $^1B_1$ and $^1A_2$ (Weide, Staemmler, and Schinke 1990; Theodorakopoulos and Petsalakis 1991; Heumann, Düren, and Schinke 1991).

Figure 15.5 depicts for three HSH bending angles α the corresponding potentials as a function of the H-S bond distance in C_{2v}-symmetry, i.e., the two H-S separations are equally long. The equilibrium angle in the ground state is $\alpha_e = 92°$. In C_{2v}-symmetry, the two electronic states have different electronic symmetries and therefore the (adiabatic) potential curves are allowed to cross. For 92°, for example, they intersect twice, at H-S separations of about 2.45 $a_\circ$ and 3.7 $a_\circ$. The first crossing occurs in the Franck-Condon region which makes a rigorous theoretical analysis rather complicated. The energetic ordering of the two states depends crucially on the bending angle which implies a considerable angular dependence of the coupling between the two electronic states as we shall discuss below. For α smaller than α_e the $^1A_2$ potential is consistently below the $^1B_1$ potential while the opposite holds for angles larger than α_e.

If we vary the H-S bond lengths asymmetrically, both states belong to the same symmetry group, $^1A''$, which implies that the corresponding Born-Oppenheimer potentials are no longer allowed to cross. The *ab initio* calculations yield two *adiabatic* potential energy surfaces which we will call $1^1A''$ and $2^1A''$, respectively. They are shown in Figure 15.6 for $\alpha = 92°$. The lower one is dissociative with an overall shape similar to the analogous surface for H_2O (see Figure 9.9, for example). The upper one, on the other hand, is binding with a relatively small force constant along the symmetric stretch coordinate and a large force constant in the direction of the anti-symmetric stretch coordinate. The two-dimensional potential energy surfaces for other bending angles show a qualitatively similar behaviour. For fixed α, let us say 92°, the potential energy sur-

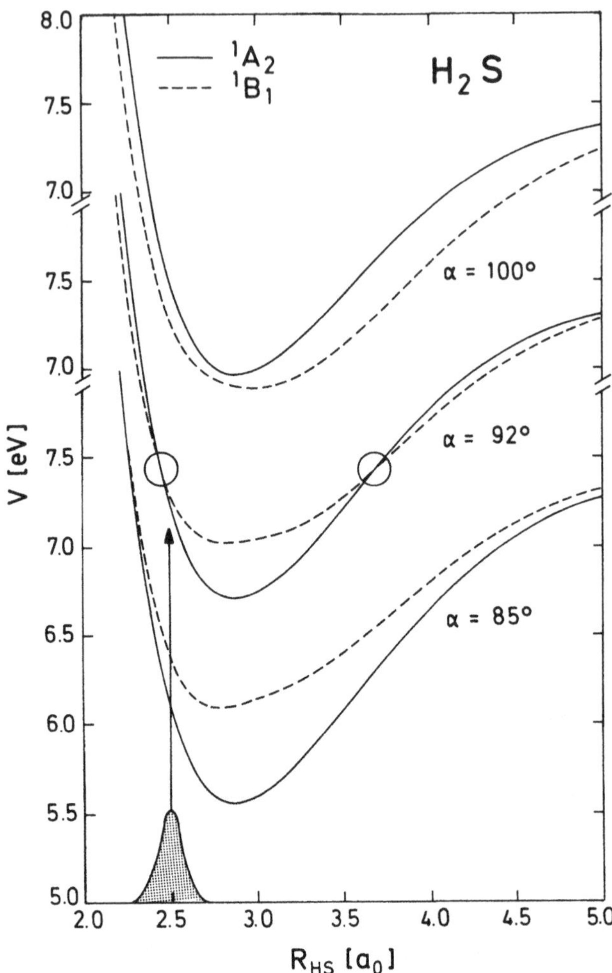

Fig. 15.5. Calculated potential energy curves for the $^1B_1$ and $^1A_2$ states of H_2S in C_{2v}-symmetry, i.e., the two H-S bond distances are varied symmetrically. The HSH bending angles are α=85°, 92°, and 100°. Note the different vertical axes for the three pairs of potential curves. Adapted from Heumann, Düren, and Schinke (1991).

faces have two *conical intersections* on the C_{2v}-symmetry line, i.e., the $1^1A''$ and the $2^1A''$ states are degenerate at two nuclear configurations.

Although the dissociative PES has an overall shape similar to the corresponding PES in H_2O, there is one important difference concerning the discussion of diffuse structures in Chapter 8. Due to the "interaction" with the other electronic state, the potential rim between the two dissociation channels is significantly narrower than for H_2O. Therefore, the motion in the direction of the symmetric stretch mode is so unstable —

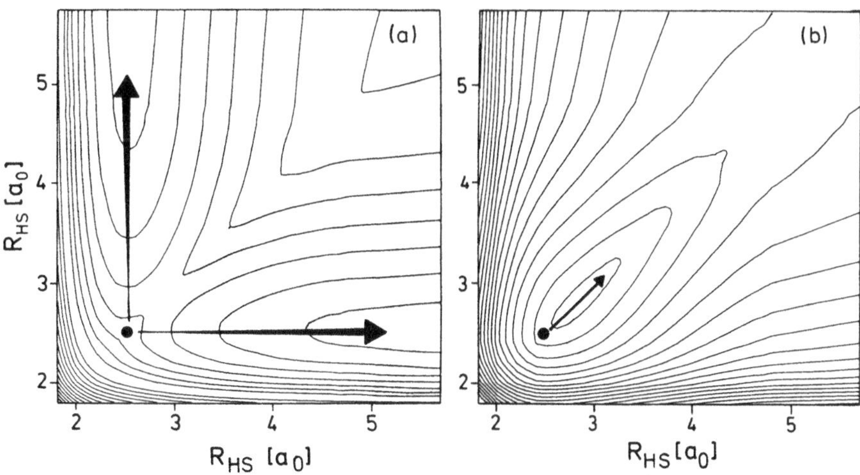

Fig. 15.6. Calculated potential energy surfaces for the two lowest electronic states of $^1A''$ symmetry of H_2S, (a) $1^1A''$ and (b) $2^1A''$; the $1^1A''$ PES is the lower one. The bending angle is fixed at 92°. The arrows schematically indicate the initial motions of the nuclear wavepackets started in the dissociative and in the binding states after the photon has promoted the system to the excited states. Adapted from Heumann, Düren, and Schinke (1991).

in the sense of Chapter 8 — that an appreciable recurrence cannot be developed and the spectrum, calculated taking only this PES into account, is completely smooth without showing any vibrational structures. The experimental spectrum, on the other hand, shows diffuse structures superimposed on a broad background which actually are more pertinent than for H_2O [Figure 15.7(a)]. What kind of internal molecular motion, if not symmetric stretch motion on top of the saddle region, causes these diffuse structures?

In contrast to H_2O, the $^1B_1$ state correlates — in the diabatic picture — with the binding state while the $^1A_2$ state becomes the dissociative state as we pull one H atom away. This correlation is made in view of the dominant molecular orbitals in the CI wavefunctions. If we use in the subsequent discussion the terminology $^1B_1$ and $^1A_2$ in order to distinguish the two states, we think in the diabatic rather than the adiabatic representation. In C_{2v}-symmetry, the $^1B_1$ state is dipole-allowed whereas the $^1A_2$ state is dipole-forbidden. In C_s-symmetry, however, both are dipole-allowed and thus can be, in principle, accessed by excitation with a photon.

Wavepacket calculations (Schinke, Weide, Heumann, and Engel 1991; Heumann, Weide, and Schinke 1992) including both electronic states and using entirely data from extensive *ab initio* calculations reveal the follow-

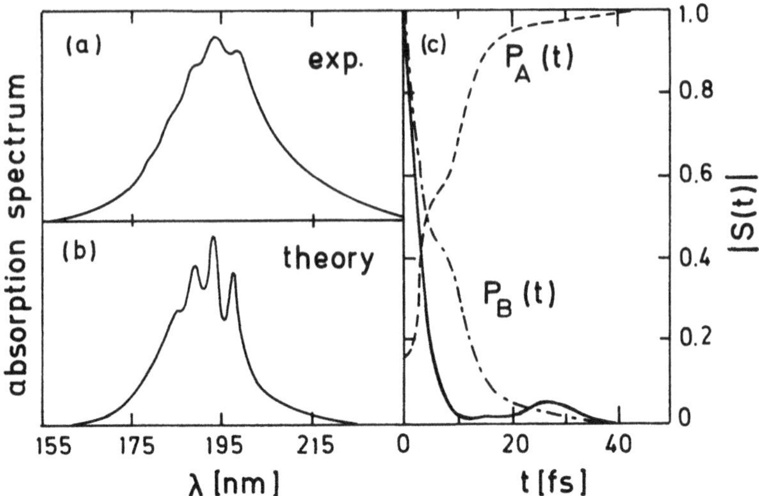

Fig. 15.7. (a) The measured absorption spectrum of H_2S in the first absorption band (Lee, Wang, and Suto 1987). (b) The theoretical absorption spectrum calculated by Heumann, Weide, and Schinke (1992). It is artificially shifted on the wavelength scale in order to account for a deficient dissociation energy in the *ab initio* calculation. The diffuse structures are due to symmetric stretch vibration in the (binding) ${}^1B_1$ state and correspond to quantum numbers $v_{ss}^* = 0, 1, \ldots$. (c) The autocorrelation function $|S(t)|$ and the probabilities $P_A(t)$ and $P_B(t)$ for finding the system in the states ${}^1A_2$ and ${}^1B_1$, respectively, as functions of time.

ing picture about the excitation and the subsequent dissociation [see also Figure 15.7(c)]:[†]

1) The photon excites predominantly the ${}^1B_1$ state, i.e., the motion in the upper electronic states begins in the *binding* state with $P_B(t = 0) = \langle \Phi_B(t) \mid \Phi_B(t) \rangle \approx 0.85$ and $P_A(t = 0) = \langle \Phi_A(t) \mid \Phi_A(t) \rangle \approx 0.15$ where Φ_A and Φ_B are the nuclear wavepackets in the two diabatic states. $P_A(t)$ and $P_B(t)$ are the time-dependent probabilities that the system is in states ${}^1B_1$ and ${}^1A_2$, respectively.
2) The wavepacket in the ${}^1B_1$ state, Φ_B, performs large-amplitude symmetric stretch motion leading to recurrences in the autocorrelation function; the recurrences in turn cause vibrational structures in the absorption spectrum.
3) Strong coupling with the dissociative ${}^1A_2$ state gradually diminishes Φ_B and at the same time builds up a wavepacket Φ_A in the dissociative state.

[†] For a contrary view of the photodissociation of H_2S see Dixon, Marston, and Balint-Kurti (1990).

4) Rapid dissociation in the $^1A_2$ state.

This scenario represents Herzberg's type I predissociation (electronic predissociation, electronic quenching; Herzberg 1967:ch.IV) as schematically illustrated in Figure 15.8. It is, of course, homologous to vibrational predissociation as exemplified in Chapter 7 for the dissociation of CH_3ONO, for example. Crucial for the development of resonance structures in the absorption spectrum is the survival time of the wavepacket Φ_B in the binding state which, on the other hand, is solely controlled by the diabatic coupling potential $V_{12}^{(d)}$ in Equation (15.11). The weaker the coupling the longer is the survival time of H_2S in the $^1B_1$ state and the more frequently Φ_B recurs to its origin. The absorption spectrum consequently exhibits comparatively narrow resonances. Conversely, a stronger coupling depletes the population in the $^1B_1$ state more rapidly, the autocorrelation function shows merely one recurrence [Figure 15.7(c)], and the spectrum exhibits rather broad resonances. In other words:

- The diffuseness of the vibrational structures in the absorption spectrum depends directly on the strength of the coupling between the two electronic states.

In the case of H_2S, the electronic coupling is substantial and the superimposed structures are very diffuse. The *ab initio* calculation seems to underestimate this coupling slightly, because the structures are too sharp in comparison with the measured spectrum. Furthermore, the onset of the spectrum at long wavelengths is not satisfactorily reproduced.

The wavepacket calculations reveal a remarkably strong dependence of the coupling strength and therefore of the diffuseness of the vibrational structures on the bending angle. Electronic quenching is much less efficient for angles smaller than 92° than for larger angles. Fixing the bending angle at 85°, for example, yields a spectrum with narrow resonances while $\alpha = 100°$ leads to an almost structureless spectrum. Thus, only if the angular dependence is fully taken into account, either in the rotational sudden approximation or in an exact three-dimensional treatment, is the diffuseness of the measured spectrum satisfactorily reproduced. This is astonishing at first glance because the motion of the wavepacket in the angle coordinate is rather trifling leading to a relatively "cold" rotational state distribution of SH, in full accord with the experimental data of Weiner, Levene, Valentini, and Baronavski (1989). Incidentally we note that the calculated final rotational and vibrational state distributions of SH agree rather well with the measured distributions (Heumann, Weide, and Schinke 1992). The calculations include two diabatic PESs, the mixing angle, and two dipole moment functions, all

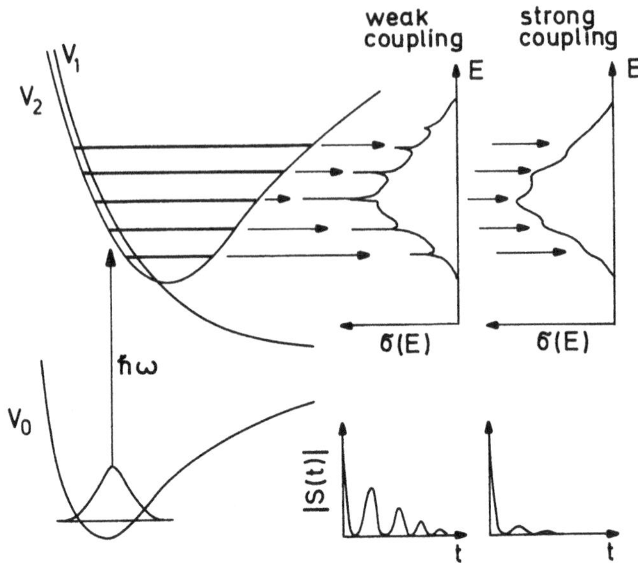

Fig. 15.8. Schematic one-dimensional illustration of electronic predissociation. The photon is assumed to excite simultaneously both excited states, leading to a structureless absorption spectrum for state 1 and a discrete spectrum for state 2, provided there is no coupling between these states. The resultant is a broad spectrum with sharp superimposed spikes. However, if state 2 is coupled to the dissociative state, the discrete absorption lines turn into resonances with lineshapes that depend on the strength of the coupling between the two excited electronic states. Two examples are schematically drawn on the right-hand side (weak and strong coupling). Due to interference between the non-resonant and the resonant contributions to the spectrum the resonance lineshapes can have a more complicated appearance than shown here (Lefebvre-Brion and Field 1986:ch.6). In the first case, the autocorrelation function $|S(t)|$ shows a long sequence of recurrences, while in the second case only a single recurrence with small amplitude is developed. The diffuseness of the resonances or vibrational structures is a direct measure of the electronic coupling strength.

depending on the two HS stretching coordinates and the HSH bending angle.

Although the diffuse structures for H_2O and for H_2S look rather alike, they reflect quite different dynamical situations. In both cases, they are caused by symmetric stretch motion. However, in the case of H_2O the wavepacket performs symmetric stretching motion *on the rim of the dissociative* PES between the two fragment channels, whereas in the case of H_2S the wavepacket oscillates *in the well of the binding* PES. In the first case, the instability of the trajectory on top of the saddle [see Figure 8.6(a)] damps the oscillatory motion while in the second case the damping is caused by coupling to a dissociative state. The net result is obviously the same. Incidentally we note that as for H_2O, the diffuse structures were originally (and as we know now, erroneously) assigned to excitation of the bending motion in the excited state (Thompson et al. 1966). The photodissociation of H_2S intriguingly documents that despite their vagueness diffuse structures can reveal interesting molecular dynamics. Electronic quenching of the type described for H_2S is very likely responsible for diffuse vibrational structures in absorption spectra of many polyatomic systems (see, for example, the studies of Braunstein and Pack (1992) and Banichevich, Peyerimhoff, Beswick, and Atabek (1992)).

Let us summarize this chapter. Nonadiabatic effects in photodissociation are very important for many molecular systems. The higher the excitation energies the more likely the necessity of considering more than one electronic states becomes. The propagation of wavepackets in two or even several electronic states is rather straightforward today in comparison with the construction of the relevant adiabatic or diabatic potentials and the coupling between them.

16
Real-time dynamics of photodissociation

In the preceding fifteen chapters of this monograph we have described how one can infer information about the dissociation process and ultimately about the multi-dimensional potential energy surface(s) (PES) in the excited electronic state(s) from the observables which one measures in "conventional" experiments, namely the absorption spectrum, the emission spectrum of the transient molecule, and the various final state distributions of the fragments. By "conventional" we mean those experiments in which the molecule is irradiated by a long, more or less monochromatic laser pulse. The last open question, which we will address here, concerns the true time dependence of the molecular system as it evolves from the Franck-Condon region, through the transition state, into the possible fragment channels and how this motion can be made transparent in the laboratory.

The lifetime of the complex in the excited electronic state is either in the range of 10^{-15}–10^{-13} seconds for direct dissociation respectively 10^{-12} seconds or longer for indirect fragmentation. In contrast, conventional experiments are performed with pulse lengths of the photolysis laser of the order of 10^{-9} seconds. Thus, no matter how refined such experiments are — full preparation of the initial state and complete resolution of the fragment states — they are inherently unable to resolve the *real* time dependence of the breakup process. They need accompanying theoretical studies (classical trajectories or quantum mechanical wavepackets) in order to disentangle the interaction among the various degrees of freedom and to disclose the evolution of the system.†

† If the dissociation is indirect, one can, in principle, deduce the lifetime of the compound from the widths of the resonances. In practice, however, such estimates are questionable because, as a consequence of insufficient resolution and thermal broadening, the measured widths very often do not reflect the true homogeneous line widths. Estimating the lifetime from the anisotropy parameter β of the an-

Direct probing of the temporal evolution of the dissociating system requires ultrashort laser pulses which have become available only in recent years. The necessary prerequisites for "time-resolved" experiments are two lasers with pulse lengths in the femtosecond range: the first laser generates a wavepacket, i.e., a coherent superposition of stationary states, in the excited electronic state and a second laser probes the evolving wavepacket after some delay time τ [see Figure 16.1(a) for a schematic illustration]. The probing can be accomplished by exciting the transient molecule either to an upper electronic state or into the ionization continuum. The laser-induced fluorescence signal or the ion signal as a function of the delay time mirrors the real time dependence of the evolving wavepacket [for reviews see Zewail and Bernstein (1988), Zewail (1988), Khundkar and Zewail (1990), Smith (1990), Gruebele and Zewail (1991), and Zewail (1991); see also the book of Bandrauk (1988)].

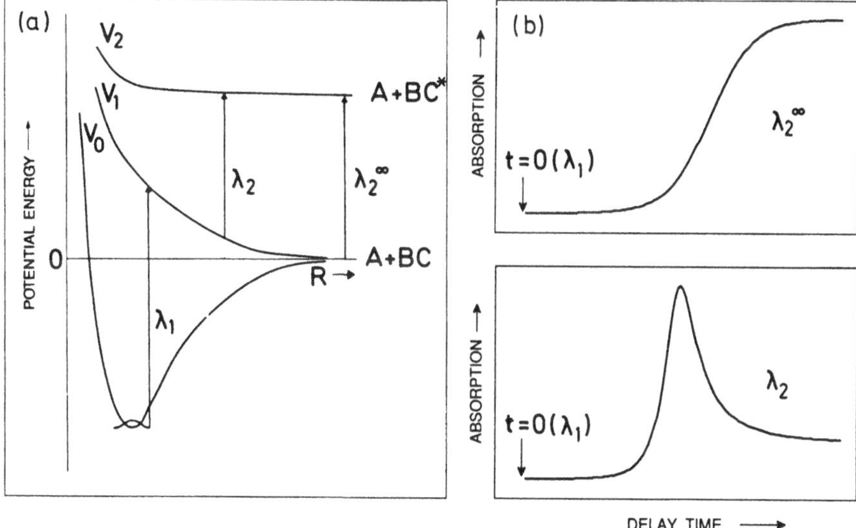

Fig. 16.1. (a) Schematic illustration of a femtosecond pump-probe experiment. The pump laser with wavelength λ_1 excites the molecule from the ground-state potential V_0 to an excited state-potential V_1. After a delay τ the probe laser with wavelength λ_2 excites the transient molecule to a second excited-state potential V_2. (b) Absorption signal of the transient molecule if the wavelength of the probe laser is tuned to the asymptotic wavelength λ_2^∞ (upper part) or to a wavelength shifted to the red of λ_2^∞ (lower part). Reproduced from Zewail (1988).

gular distribution of the products is also rather indirect and can lead to erroneous implications.

In order to model and to analyze time-resolved measurements with finite pulse lengths it is essential to construct the "true" wavepacket that the laser generates in the excited state and to explore how this wavepacket evolves in time. Chapter 16 closes our survey of molecular photodissociation. On the other hand, it brings us back to Chapter 2, namely to the question of how the photon beam excites the molecule. Section 16.1 discusses the excitation process under more general conditions than previously assumed. In Section 16.2 we will review two particular examples of real-time probing of molecular motion, one being representative of direct dissociation and the other one exemplifying indirect or delayed dissociation.

16.1 Coherent excitation

The time-dependent wavepacket constructed in Section 4.1 is *not* the wavepacket that a laser with finite duration creates in the excited electronic state. It represents the wavepacket created by a pulse with *infinitely narrow width in time.* In order to construct the real wavefunction of the molecular system we must go back to Section 2.1. For simplicity of presentation, let us consider a diatomic molecule with internuclear separation R. We assume that the excitation takes place from the electronic ground state (index 0) to a bound upper state (index 1). The extension to a dissociative state, several coupled excited states, or several degrees of freedom is formally straightforward.

16.1.1 Coupled equations

The total time-dependent molecular wavefunction, which fully defines the system at each instant, is represented by

$$\mathcal{F}(R, \mathbf{q}; t) = \mathcal{F}_0(R, \mathbf{q}; t) + \mathcal{F}_1(R, \mathbf{q}; t) \tag{16.1}$$

with

$$\begin{aligned} \mathcal{F}_0(t) &\equiv \Phi_0(R;t)\,\Xi_0 = a_0(t)\,\Psi_0(R)\,\mathrm{e}^{-iE_0t/\hbar}\,\Xi_0(\mathbf{q};R) \\ \mathcal{F}_1(t) &\equiv \Phi_1(R;t)\,\Xi_1 = \sum_l a_{1l}(t)\,\Psi_{1l}(R)\,\mathrm{e}^{-iE_lt/\hbar}\,\Xi_1(\mathbf{q};R). \end{aligned} \tag{16.2}$$

Ξ_0 and Ξ_1 are the electronic wavefunctions in states 0 and 1, respectively. Ψ_0 is the initial nuclear wavefunction in the electronic ground state and the Ψ_{1l} are the bound vibrational wavefunctions in electronic state 1. The corresponding energies are E_0 and E_l, respectively. The vector $\mathbf{q}$ comprises all electronic coordinates. $\Phi_0(t)$ and $\Phi_1(t)$ represent the corresponding nuclear wavepackets in state 0 and 1, respectively. Equations (16.1) and (16.2) are the analogues of (2.9). Within the Born-Oppenheimer approximation it is assumed that there is no nonadiabatic coupling between the two electronic manifolds.

In Section 2.1 we assumed that the electromagnetic pulse is infinitely long and that the amplitude of the field is constant in time [see Equation (2.13)]. In order to treat excitation with short laser pulses we must extend the theory to a more general electromagnetic field whose amplitude $E_0(t)$ varies with time. The total Hamiltonian describing the evolution of the molecular system is given by

$$\hat{H}_{tot}(t) = \hat{T}_{nu} + \hat{H}_{el} + \hat{\mathbf{d}} \cdot \mathbf{e}\, E_0(t) \cos \omega t, \tag{16.3}$$

where $\hat{T}_{nu}$ and $\hat{H}_{el}$ are the nuclear kinetic energy and the electronic energy, respectively. $\hat{\mathbf{d}}$ is the dipole operator and $\mathbf{e}$ is a unit vector in the direction of the field. Inserting (16.1) into the time-dependent Schrödinger equation with Hamiltonian $\hat{H}_{tot}(t)$ and exploiting the orthogonality of the electronic wavefunctions yields the following set of coupled equations for the nuclear wavepackets Φ_k $(k = 0, 2)$,

$$\begin{aligned} i\hbar \frac{\partial}{\partial t} \Phi_0(t) &= \hat{H}_0 \Phi_0(t) + E_0(t)\, \mu_{01}^{(e)} \cos \omega t\, \Phi_1(t) \\ i\hbar \frac{\partial}{\partial t} \Phi_1(t) &= E_0(t)\, \mu_{10}^{(e)} \cos \omega t\, \Phi_0(t) + \hat{H}_1 \Phi_1(t) \end{aligned} \tag{16.4}$$

with $\hat{H}_0 = \hat{T}_{nu} + V_0$ and $\hat{H}_1 = \hat{T}_{nu} + V_1$ being the nuclear Hamiltonians in states 0 and 1, respectively. $\mu_{01}^{(e)} = \mu_{10}^{(e)}$ is the component of the transition dipole function in the direction of the electric field as defined in (2.35). Equation (16.4) must be solved subject to the initial conditions

$$\Phi_0(t = 0) = \Psi_0 \qquad \text{and} \qquad \Phi_1(t = 0) = 0 \tag{16.5}$$

if the field is switched on at time $t = 0$ and if initially the system is completely in the lower state. Note the formal similarity of (16.4) with (15.11). While in (15.11) the off-diagonal elements of the diabatic potential matrix couple the two electronic states, in (16.4) it is the electromagnetic field that provides the interaction.†

If the field is zero, the wavepacket Φ_0 in the ground electronic state moves under the influence of $\hat{H}_0$, completely undisturbed by the other electronic states. As the electromagnetic field is switched on it continuously promotes some fraction of Φ_0 to the excited electronic state which immediately starts to move according to the upper-state nuclear Hamiltonian $\hat{H}_1$. This creates a so-called *wavetrain* whose extension in space depends on the length of the pulse. If the field strength is small, recoupling of Φ_1 to the ground state can be ignored.

† If the Born-Oppenheimer approximation is not valid, Φ_0 and Φ_1 are additionally coupled by nonadiabatic coupling elements on the right-hand sides of Equations (16.4).

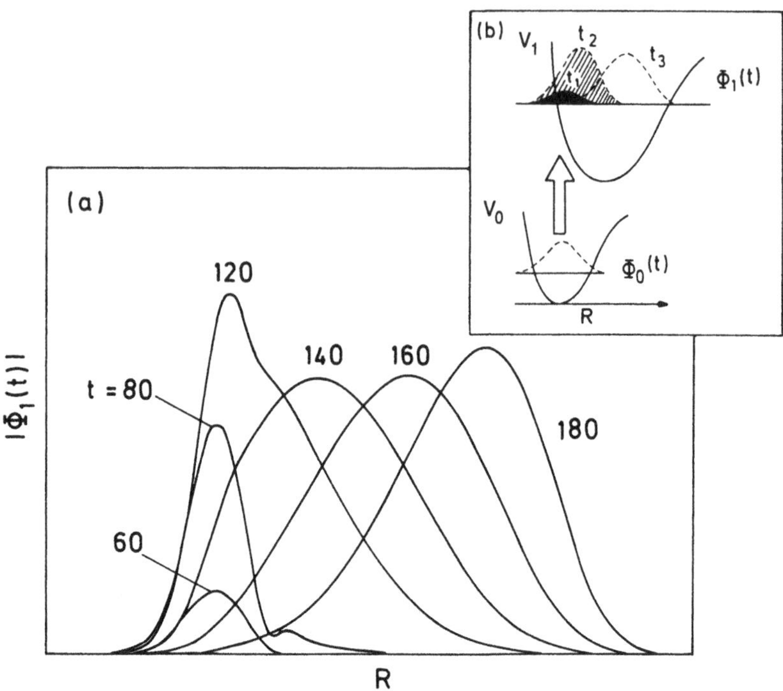

Fig. 16.2. (a) Evolution of the wavepacket in the excited electronic state created by a laser pulse centered at $t_0 = 90$ fs with a width of 50 fs. The times are given in femtoseconds. (b) Schematic illustration of the potentials in the lower and upper electronic states and of the excitation process. By courtesy of V. Engel.

Figure 16.2(b) schematically illustrates this scenario and part (a) depicts the time dependence of a calculated wavepacket in the excited state. The pulse has a Gaussian shape centered at $t_0 = 90$ fs with width $\Delta t = 50$ fs. The external field steadily pumps part of the ground-state wavepacket Φ_0 into the excited electronic state with the consequence that the maximum of Φ_1 at first increases with time. At the same time the wavepacket moves to larger distances and becomes significantly broader with increasing time because more and more probability amplitude arrives in the upper state. When $E_0(t)$ has decayed to zero, the promotion from the ground to the upper electronic state stops and $\Phi_1(t)$ travels in the upper-state potential without any coupling to Φ_0. At around 180 fs the wavepacket hits the outer wall of potential V_1 and starts to return to the Franck-Condon region from which it came. For the interpretation of pump-probe experiments it is important to note that:

- The wavepacket created by a laser pulse with finite duration is significantly broader than the wavepacket created by a δ-pulse in time which we discussed in the preceeding chapters.

16.1.2 Pulse duration and spectral width

Using expansion (16.2) for the wavepacket in terms of the stationary wavefunctions Ψ_{1l} we can derive a set of coupled equations for the expansion coefficients $a_{1l}(t)$ similar to (2.16). In the limit of first-order perturbation theory [see Equation (2.17)] the time dependence of each coefficient is then given by

$$i\hbar \frac{d}{dt} a_{1l} = E_0(t) \langle \Psi_{1l} \mid \mu_{10}^{(e)} \mid \Psi_0 \rangle \cos \omega t \, e^{i\omega_{l0} t} \qquad (16.6)$$

with transition frequencies $\omega_{l0} = (E_l - E_0)/\hbar$. Integration of (16.6) and insertion into (16.2) yields for the wavepacket in the upper state

$$\Phi_1(R;t) = -\frac{i}{\hbar} \sum_l \langle \Psi_{1l} \mid \mu_{10}^{(e)} \mid \Psi_0 \rangle \, \Psi_{1l}(R) \, e^{-iE_l t/\hbar} \, I(t;\omega,\omega_{l0}) \qquad (16.7)$$

with

$$I(t;\omega,\omega_{l0}) = \int_0^t dt' \, E_0(t') \cos \omega t' \, e^{i\omega_{l0} t'}. \qquad (16.8)$$

Equations (16.7) and (16.8) govern the spectral width of the absorption, i.e., how many stationary states $\Psi_{1l}(R)$ are *coherently* excited by the laser pulse. Let us assume, for simplicity, that $E_0(t')$ is a Gaussian in time centered around t_0,

$$E_0(t') = \tilde{E}_0 \, e^{-\alpha^2 (t'-t_0)^2}. \qquad (16.9)$$

Evaluation of (16.8) with the integration extended over the full length of the pulse yields

$$I(t=\infty;\omega,\omega_{l0}) \propto \tilde{E}_0 \, e^{-(\omega-\omega_{l0})^2/4\alpha^2}, \qquad (16.10)$$

i.e., a Gaussian in the frequency domain. The widths in the time and the energy domains are interrelated according to the familiar relation [Cohen-Tannoudji, Diu, and Laloë 1977:ch.III; see also Equation (6.12)]

$$\Delta E \, \Delta t = 8 \, \hbar \ln 2 \approx h. \qquad (16.11)$$

Figure 16.3 illustrates the interdependence of the pulse length, on one hand, and the spectral width, on the other hand:

- A light beam with finite duration *coherently* excites several stationary states in the upper electronic manifold — despite the fact that the frequency of the photon is fixed.
- The shorter the light pulse, the more stationary states are excited and *vice versa*.

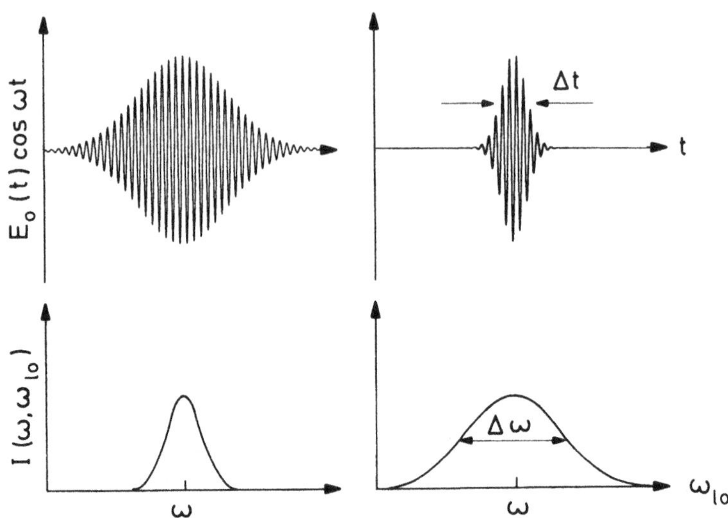

Fig. 16.3. Schematic illustration of the interrelation between the temporal width of the laser pulse and the spectral width in the frequency domain.

It is informative to consider the two limiting cases, an infinitely short and an infinitely long laser pulse. With $E_0(t')$ replaced by $\tilde{E}_0\,\delta(t')$ in (16.8) the wavepacket becomes

$$\Phi_1(R;t) \propto \sum_l \langle \Psi_{1l} \mid \mu_{10}^{(e)} \mid \Psi_0 \rangle\, \Psi_{1l}(R)\, \mathrm{e}^{-iE_l t/\hbar} \tag{16.12}$$

which is similar to Equation (4.3) with the coefficients given by Equation (4.5). As noted above:

- The wavepacket that we constructed in Section 4.1 corresponds to excitation with a δ-pulse in time.
- It contains *all* stationary states weighted by the overlap with the ground-state wavefunction multiplied by the transition dipole function.

On the other hand, replacing $E_0(t')$ by an infinitely long pulse with constant amplitude $\tilde{E}_0$ yields a delta-function in the frequency domain and therefore the sum in (16.7) collapses to a single term,

$$\Phi_1(R;t) \propto \langle \Psi_{1l} \mid \mu_{10}^{(e)} \mid \Psi_0 \rangle\, \Psi_{1l}(R)\, \mathrm{e}^{-iE_l t/\hbar}\, \delta(\omega - \omega_{l0}). \tag{16.13}$$

- A very long laser pulse excites only that particular stationary state in the upper manifold which is in resonance with the frequency of the electromagnetic field, i.e., for which $\omega = \omega_{l0}$; because only one state is excited, there is no motion.

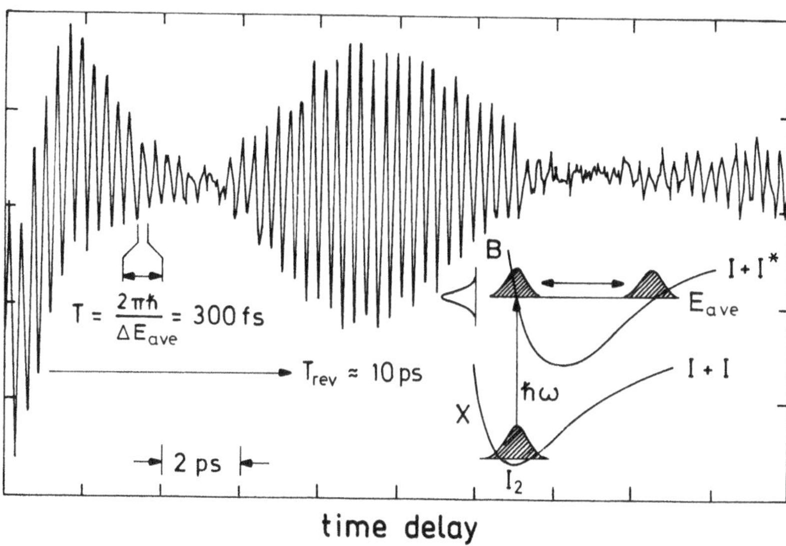

Fig. 16.4. Typical transient spectrum showing the wavepacket dynamics in the strongly bound B state of I_2. The period of 300 fs reflects the oscillatory motion between the two turning points. Adapted from Bowman, Dantus, and Zewail (1990).

Let us consider the motion of a coherently created wavepacket in a bound potential as illustrated in the inset of Figure 16.4 for I_2 . If the pump pulse is much narrower than the internal vibrational period, the wavepacket in the B state remains rather localized in space and oscillates back and forth between the two turning points. By absorption of a second photon, which promotes $I_2(B)$ to a higher state, one can monitor the motion of the wavepacket and a typical transient spectrum is depicted in the main part of Figure 16.4. The fast oscillations reflect the bound motion of the wavepacket with the period T being related to the average energy spacing ΔE_{ave} of the stationary states excited by the pump photon. Excitation by photons with different wavelengths leads to different periods because the energy spacing varies with E_{ave}. These data can be inverted to give the potential as has been shown for I_2 and ICl (Bowman, Dantus, and Zewail 1990; Gruebele et al. 1990; Janssen, Bowman, and Zewail 1990; Bernstein and Zewail 1990).

Anharmonic effects lead to a destruction of the initially localized wavepacket with the consequence that the transient signal is damped. However, there is the possibility for the wavepacket to regain its initial shape after long times resulting in so-called *revivals* [see, for example, Alber and Zoller (1991)]. The revival period T_{rev} contains additional information about the shape of the potential.

Using the identity

$$e^{-iE_l(t-t')/\hbar}\ \Psi_{1l} = e^{-i\hat{H}_1(t-t')/\hbar}\ \Psi_{1l} \tag{16.14}$$

and the completeness of the stationary wavefunctions one derives the following equation (valid in the first-order perturbation limit)

$$\Phi_1(R;t) = -\frac{i}{\hbar}\int_0^t dt'\ e^{-i\hat{H}_1(t-t')/\hbar}\mu_{10}^{(e)}\ E_0(t')\cos\omega t'\ \Phi_0(R;t') \tag{16.15}$$

for the wavepacket evolving in the upper state where

$$\Phi_0(R;t') \equiv \Psi_0(R)\ e^{-iE_0t'/\hbar} \tag{16.16}$$

is the wavepacket in the electronic ground state. Equation (16.15) is more convenient for the numerical implementation than (16.7). For further discussions about the wavepacket excited by a short-pulse laser see Engel et al. (1988), Williams and Imre (1988b,c), Tannor and Rice (1988), Rama Krishna and Coalson (1988), Pollard, Lee, and Mathies (1990), Fried and Mukamel (1990), Metiu and Engel (1990), Stock and Domcke (1990), Seel and Domcke (1991), Engel (1991b), Metiu (1991), and Krause, Shapiro, and Bersohn (1991). The excitation of wavepackets through the coherent excitation of stationary states is also a topic of high current interest in the field of atomic physics (see Alber and Zoller 1991 for a recent review). The general theory is, of course, equivalent to the applications in molecular physics.

16.2 Examples

Figure 16.1(a) illustrates the basic concept of an experiment probing the real-time dependence of a dissociating molecule. A short laser pulse with wavelength λ_1 excites the parent molecule into a repulsive state and thereby creates a wavepacket that starts to evolve on potential V_1. After the molecule has stretched for some time τ it is irradiated by a second short laser pulse with wavelength λ_2 which excites the transient molecule to another excited state with potential V_2. According to the general picture of light absorption developed in Section 2.1, the probe pulse can be significantly absorbed only if the photon energy is (roughly) in resonance with the energy difference $\Delta V(R) = V_2(R) - V_1(R)$ between the two excited states.

Varying the delay time τ yields an absorbtion signal of the form shown in the lower part of Figure 16.1(b). Shortly after the pump pulse is fired the wavepacket has not yet moved far away from the starting region and the energy difference ΔV is too small for the probe photon to be absorbed; therefore, the absorption spectrum is essentially zero for small values of τ. With increasing delay time the wavepacket traverses the appropriate region of the potential surface, where absorbtion is possible, just at the time when the second pulse is fired. The absorption signal consequently rises with τ and reaches a maximum. For even longer delay times, the wavepacket has already passed the region, where absorption can occur, when the second laser is fired and the absorption signal decreases again.

Absorption of the second pulse is only possible if the wavepacket crosses the appropriate region where $\Delta V \approx \hbar\omega_2$ when the second laser is switched on. This defines a "window" in the evolution of the transient molecule. Changing the wavelength of the probe laser opens another "window" on the potential energy surface. By tuning λ_2 to the wavelength absorbed by one of the fragments one measures the arrival time of the wavepacket in the asymptotic region. The absorption signal as a function of the delay time steadily rises and finally reaches a plateau as illustrated in the upper part of Figure 16.1(b).

Variation of the frequency as well as the delay of the probe laser allows snapshots of the wavepacket as it evolves on the potential energy surface. The resolution in the time domain is determined by the sharpness of the two lasers. The narrower the pulses the clearer is the temporal picture of the evolution of the molecule in the excited state. However, according to Equation (16.11) the less favorable is the resolution in the energy domain. In order to make the dynamics of the dissociating molecule transparent, a wavepacket created with a δ-pulse, which we pursued in the main part of the monograph or, even simpler, a few classical trajectories, is sufficient. However, if we want to mimic and deconvolute a transient absorption spectrum, that was recorded with pulse lengths of the order of 50–100 fs, the "real" wavepacket must be calculated either by solving (16.4) or by integration of (16.15).

The photodissociation of ICN is a direct process and the corresponding potential energy surface along the reaction coordinate qualitatively resembles V_1 in Figure 16.1(a). Figure 16.5 depicts absorption signals as functions of time delay for four wavelengths of the probe laser (Dantus, Rosker, and Zewail 1988). As λ_2 is gradually tuned to the red of the asymptotic CN wavelength, i.e., as the difference of potential energies $\Delta V = V_2 - V_1$ is decreased, the absorption maximum gradually shifts to shorter times. The "window" on the R-axis, defined through the wavelength of the probe laser, is shifted to smaller distances and therefore the wavepacket passes it at ever shorter delay times. The families of absorption signals of the transient molecule can be inverted to yield the R dependence of the dissociative PES (Bersohn and Zewail 1988; Bernstein and Zewail 1990).

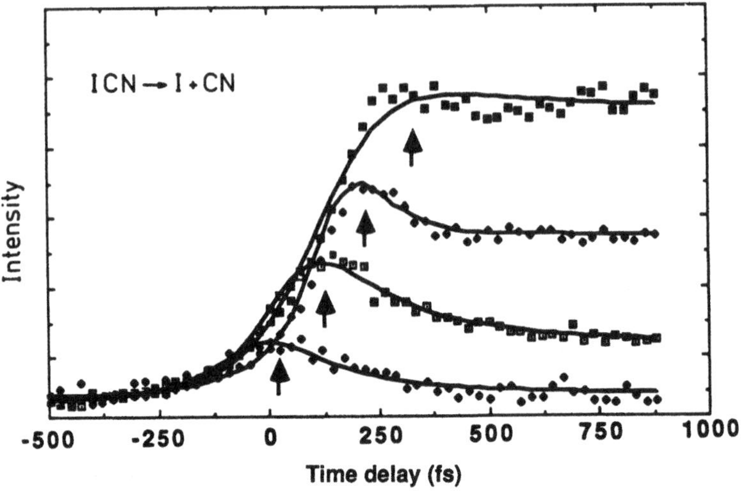

Fig. 16.5. Absorption signal for the dissociating ICN molecule versus delay time. The wavelengths of the probe laser are 389.7 nm, 389.8 nm, 390.4 nm, and 391.4 nm from the top to the bottom. Reproduced from Zewail (1988).

An intriguing example which highlights the idea of *femtosecond chemistry* is the photodissociation of NaI (Rose, Rosker, and Zewail 1988; Rosker, Rose, and Zewail 1988). Figure 16.6 illustrates the potential energy curves involved in the fragmentation. The pump pulse excites NaI to a covalent state which, in the diabatic picture, correlates with Na +

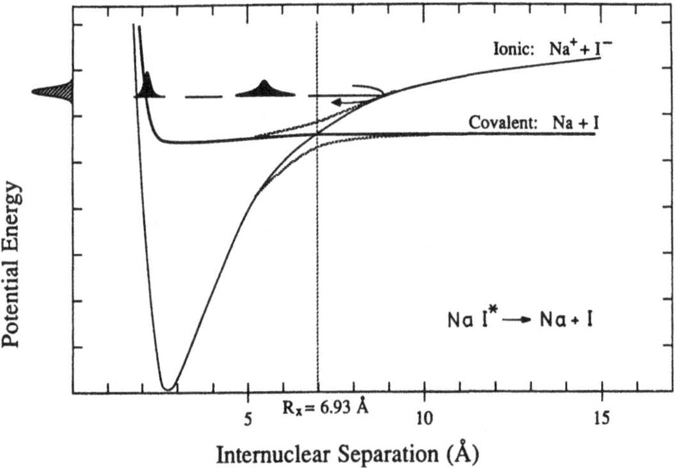

Fig. 16.6. Schematic illustration of the dissociation of NaI. The bell-shaped curve on the vertical axis represents the distribution of stationary states excited by the short pump pulse. Reproduced from Zewail (1988).

I. At an internuclear distance of 6.93 Å, the covalent curve crosses the ionic potential which asymptotically correlates with $Na^+ + I^-$. First, the wavepacket generated by the pump laser at short distances on the repulsive branch of the covalent potential follows the covalent curve until it reaches the region of the crossing. Here, it splits up into two parts, the smaller portion continues to travel outward following the (diabatic) covalent curve and eventually yields free Na and I whereas the larger part follows the (adiabatic) ionic potential. The latter cannot dissociate because the energy is not sufficient to produce Na^+ and I^-. Thus, it turns round and swings back to the Franck-Condon region where a new cycle begins. Because at each passage through the crossing zone some fraction of the wavepacket trapped by the adiabatic curve leaks out, the number of salt molecules gradually diminishes with time.

The oscillatory motion of the wavepacket between the inner and the outer bounds obviously shows up in the absorption signal of the transient salt molecule. Probing with the wavelength of the free Na atom yields the upper trace in Figure 16.7. It roughly measures the produc-

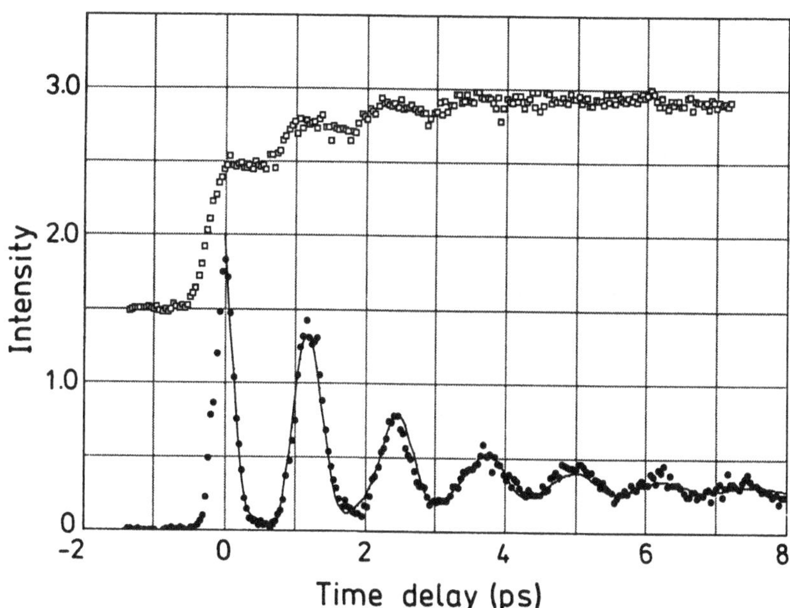

Fig. 16.7. Absorption signals versus delay time for transient NaI. The upper trace is recorded with the probe pulse in resonance with the wavelength of the free Na atom. The lower trace is detected with an off-resonance wavelength. Note the striking similarity with the autocorrelation function whose behavior can be easily inferred from the motion of the wavepacket sketched in Figure 16.6. Reproduced from Zewail (1988).

tion of free Na atoms as a function of the delay time after the photolysis laser is fired. Whenever the wavepacket traverses the region of the avoided crossing a new "pulse of free sodium atoms" is created and the absorption signal makes a distinct step upwards. Probing with a wavelength slightly detuned from the wavelength of the free fragment yields the lower trace in Figure 16.7. It roughly assesses the number of remaining NaI molecules. The oscillations manifest the vibrational motion of the trapped wavepacket. The periods of the lower and the upper signals are virtually the same and reflect the time the wavepacket needs for one round trip from the crossing region to the steep wall at short distances and back to the crossing. The gradual decay of the absorption signal reveals the damping due to dissociation. The experimental curves shown in Figure 16.7 have been successfully reproduced by semiclassical (Marcus 1988) as well as fully quantum mechanical (Engel, Metiu, Almeida, Marcus, and Zewail 1988; Engel and Metiu 1989; Choi and Light 1989) calculations.

It is interesting to imagine how the corresponding absorption spectrum recorded in a conventional experiment with ideal resolution would look. Neglecting rotation, the absorption spectrum would reflect the quasi-bound vibrational states of the upper adiabatic potential curve in Figure 16.7 (Choi and Light 1989). Since the reduced mass of NaI is large and since the potential is rather flat over most of the relevant R range, the average spacing of adjacent absorption lines is very small. The corresponding autocorrelation function $|S(t)|$ depicts an overall shape similar to the lower curve in Figure 16.7 with the maxima manifesting the recurrences of the wavepacket to the Franck-Condon region (see Figure 7.9, for example). The linewidths give the lifetime of each resonance state which is inversely proportional to the square of the nonadiabatic coupling between the ionic and the covalent Born-Oppenheimer states (see Section 7.2).

Pump-probe experiments with ultrashort laser pulses have opened a new dimension in the study of molecular dynamics, especially for dissociative systems, and it is not difficult to surmise that many more beautiful results will emerge in the near future. The general methodology has been applied to a variety of photodissociation systems [ICN, NaI, HgI_2, Bi_2, CH_3I , reviewed by Zewail (1991) and Na_2 investigated by Baumert, Grosser, Thalweiser, and Gerber (1991), Baumert et al. 1991, Engel (1991c), and Baumert et al. (1992)]. They *directly* bring to light how the atoms dance within the polyatomic molecule in the real time-dependent world, whereas the conventional approach inherently requires accompanying calculations in order to make the time dependence visible. Although traditional experiments (long laser pulses and high frequency resolution) and time-resolved experiments (short laser pulses and relatively low frequency resolution) yield — in principle — equivalent information about the system and the multi-dimensional PES, we must underline that time-resolved measurements can solve two major problems. First, the inhomogeneous broadening, which is usually significant for larger systems, can lead to misleading inferences of the dynamics. Second, in reactions there are often different intermediate steps that can be identified only by monitoring the different time scales on which they evolve.

Conventional experiments with high frequency resolution and full specification of the initial and the final states, on one hand, and time-resolved experiments probing the transient states during reaction, on the other hand, should now be in a position to map out the full dynamics of many elementary reactions, as all important steps can be observed and studied.

References

Abramowitz, M. and Stegun, I.A. (1970). *Handbook of Mathematical Functions* (Dover, London).

Alber, G. and Zoller, P. (1991). Laser excitation of electronic wave packets in Rydberg atoms, *Phys. Rep.* **199,** 231–280.

Alberti, F. and Douglas, A.E. (1978). Anomalous populations in the Λ doublets of the $c^1\Pi$ state of NH, *Chem. Phys.* **34,** 399–402.

Alexander, M.H. and Dagdigian, P.J. (1984). Clarification of the electronic asymmetry in Π-state Λ doublets with some implications for molecular collisions, *J. Chem. Phys.* **80,** 4325–4332.

Alexander, M.H., Dagdigian, P.J., and Werner, H.-J. (1991). Potential-energy surface control of the NH product state distribution in the decomposition reaction $HN_3(\tilde{X}^1A') \rightarrow NH(a^1\Delta) + N_2(X^1\Sigma_g^+)$, *Faraday Discuss. Chem. Soc.* **91,** 319–335.

Alexander, M.H., Werner, H.-J., and Dagdigian, P.J. (1988). Energetics and spin- and Λ-doublet selectivity in the infrared multiphoton dissociation $HN_3(\tilde{X}^1A) \rightarrow N_2(X^1\Sigma_g^+) + NH(X^3\Sigma^-, a^1\Delta)$: Theory, *J. Chem. Phys.* **89,** 1388–1400.

Alexander, M.H., Werner, H.-J., Hemmer, T., and Knowles, P.J. (1990). *Ab initio* study of the energetics of the spin-allowed and spin-forbidden decomposition of HN_3, *J. Chem. Phys.* **93,** 3307–3318.

Amatatsu, Y., Morokuma, K., and Yabushita, S. (1991). *Ab initio* potential energy surfaces and trajectory studies of *A*-band photodissociation dynamics: $CH_3I^* \rightarrow CH_3 + I$ and $CH_3 + I^*$, *J. Chem. Phys.* **94,** 4858–4876.

Andresen, P. (1988). Dynamics of the photodissociation of small molecules, in *Frontiers of Laser Spectroscopy of Gases,* ed. A.C.P. Alves et al. (Kluwer, Dordrecht).

Andresen, P., Beushausen, V., Häusler, D., Lülf, H.W., and Rothe, E. (1985). Strong propensity rules in the photodissociation of a single rotational quantum state of vibrationally excited H_2O, *J. Chem. Phys.* **83,** 1429–1430.

Andresen, P., Ondrey, G.S., Titze, B., and Rothe, E.W. (1984). Nuclear and electron dynamics in the photodissociation of water, *J. Chem. Phys.* **80,** 2548–2569.

Andresen, P. and Rothe, E.W. (1985). Analysis of chemical dynamics via Λ doubling: Directed lobes in product molecules and transition states, *J. Chem. Phys.* **82,** 3634–3640.

Andresen, P. and Schinke, R. (1987). Dissociation of water in the first absorption band: A model system for direct photodissociation, in *Molecular Photodissociation Dynamics,* ed. M.N.R. Ashfold and J.E. Baggott (Royal Society of Chemistry, London).

Arnold, V.I. (1978). *Mathematical Methods of Classical Mechanics* (Springer, New York).

Ashfold, M.N.R. and Baggott, J.E. (1987). *Molecular Photodissociation Dynamics* (Royal Society of Chemistry, London).

Ashfold, M.N.R., Bennett, C.L., and Dixon, R.N. (1986). Dissociation dynamics of $NH_3(\tilde{A}^1A_2'')$, *Faraday Discuss. Chem. Soc.* **82,** 163–175.

Ashfold, M.N.R., Lambert, I.R., Mordaunt, D.H., Morley, G.P., and Western, C.M. (1992). Photofragment translational spectroscopy, *J. Phys. Chem.* **96,** 2938–2949.

Ashton, C.J., Child, M.S., and Hutson, J.M. (1983). Rotational predissociation of the Ar·HCl van der Waals complex: Close-coupling scattering calculations, *J. Chem. Phys.* **78,** 4025–4039.

Askar, A. and Rabitz, H. (1984). Action–angle variables for quantum mechanical coplanar scattering, *J. Chem. Phys.* **80,** 3586–3595.

Atabek, O., Beswick, J.A., and Delgado-Barrio, G. (1985). A test of the rotational infinite order sudden approximation in molecular fragmentation, *J. Chem. Phys.* **83,** 2954–58.

Atabek, O., Bourgeois, M.T., and Jacon, M. (1986). Three-dimensional analytical model for the photodissociation of symmetric triatomics. Absorption and fluorescence spectra of ozone, *J. Chem. Phys.* **84,** 6699–6711.

Atabek, O. and Lefebvre, R. (1977). Moment analysis of dynamics of photodissociation of linear triatomics, *Chem. Phys. Lett.* **52,** 29–33.

Atabek, O., Lefebvre, R., and Jacon, M. (1980a). Continuum resonance Raman scattering of light by diatomic molecules. I. The role of radiative crossings between the potentials of the dressed molecule, *J. Chem. Phys.* **72,** 2670–2682.

Atabek, O., Lefebvre, R., and Jacon, M. (1980b). Continuum resonance Raman scattering of light by diatomic molecules. II. Theoretical study of the Q branches of $\Delta n = 1$ profiles of molecular bromine, *J. Chem. Phys.* **72,** 2683–2693.

Atkins, P.W. (1983). *Molecular Quantum Mechanics* (Oxford University Press, Oxford).

Bačić, Z. and Light, J.C. (1986). Highly excited vibrational levels of "floppy" triatomic molecules: A discrete variable representation–distributed Gaussians basis approach, *J. Chem. Phys.* **85,** 4594–4604.

Bačić, Z. and Light, J.C. (1989). Theoretical methods for rovibrational states of floppy molecules, *Ann. Rev. Phys. Chem.* **40,** 469–498.

Bačić, Z., Watt, D., and Light, J.C. (1988). A variational localized representation calculation of the vibrational levels of the water molecule up to 27000 cm^{-1}, *J. Chem. Phys.* **89,** 947–955.

Baer, M. (1983). Quantum mechanical treatment of electronic transitions in atom-molecule collisions, in *Molecular Collision Dynamics,* ed. J.M. Bowman (Springer, Berlin Heidelberg).

Baer, M. (Ed.) (1985a). *Theory of Chemical Reaction Dynamics,* Vol. 1–5 (CRC Press, Boca Raton).

Baer, M. (1985b). The theory of electronic nonadiabatic transitions in chemical reactions, in *Theory of Chemical Reaction Dynamics,* Vol. II, ed. M. Baer (CRC Press, Boca Raton).

Baer, M. and Beswick, J.A. (1979). Electronic transitions in the ion-molecule reaction $(Ar^+ + H_2 \leftrightarrow Ar + H_2^+) \rightarrow ArH^+ + H)$, *Phys. Rev. A* **19,** 1559–1567.

Bai, Y.Y., Ogai, A., Qian, C.X.W., Iwata, L., Segal, G.A., and Reisler, H. (1989). The electronic spectrum of NOCl: Photofragment spectroscopy, vector correlations, and *ab initio* calculations, *J. Chem. Phys.* **90,** 3903–3914.

Balian, R. and Bloch, C. (1972). Distribution of eigenfrequencies for the wave equation in a finite domain: III. Eigenfrequency density oscillations, *Annals of Physics* **69,** 76–160.

Balint-Kurti, G.G. (1974). Potential energy surfaces for chemical reactions, *Adv. Chem. Phys.* **30,** 139–183.

Balint-Kurti, G.G. (1986). Dynamics of OH Λ-doublet production through photodissociation of water in its first absorption band. I. Formal theory, *J. Chem. Phys.* **84,** 4443–4454.

Balint-Kurti, G.G., Dixon, R.N., and Marston, C.C. (1990). Time-dependent quantum dynamics of molecular photofragmentation processes, *J. Chem. Soc., Faraday Trans.* **86,** 1741–1749.

Balint-Kurti, G.G. and Shapiro, M. (1981). Photofragmentation of triatomic molecules. Theory of angular and state distribution of product fragments, *Chem. Phys.* **61,** 137–155.

Balint-Kurti, G.G. and Shapiro, M. (1985). Quantum theory of molecular photodissociation, in *Photodissociation and Photoionization,* ed. K.P. Lawley (Wiley, New York).

Bamford, D.J., Filseth, S.V., Foltz, M.F., Hepburn, J.W., and Moore, C.B. (1985). Photofragmentation dynamics of formaldehyde: $CO(v, J)$ distributions as a function of initial rovibronic state and isotopic substitution, *J. Chem. Phys.* **82,** 3032–3041.

Band, Y.B., Freed, K.F., and Kouri, D.J. (1981). Half-collision description of final state distributions of the photodissociation of polyatomic molecules, *J. Chem. Phys.* **74,** 4380–4394.

Bandrauk, A.P. (Ed.) (1988). *Atomic and Molecular Processes with Short Intense Laser Pulses* (Plenum, New York).

Banichevich, A., Peyerimhoff, S.D., Beswick, J.A., and Atabek, O. (1992). Dynamics of ozone photoabsorption: A theoretical study of the Chappius band, *J. Chem. Phys.* **96,** 6580–6590.

Banichevich, A., Peyerimhoff, S.D., and Grein, F. (1990). *Ab initio* potential surfaces for ozone dissociation in its ground and various electronically excited states, *Chem. Phys. Lett.* **173,** 1–6.

Bar, I., Cohen, Y., David, D., Arusi-Parpar, T., Rosenwaks, S., and Valentini, J.J. (1991). Mode selective bond fission: Comparison between the photodissociation of HOD(0,0,1) and HOD(1,0,0), *J. Chem. Phys.* **95,** 3341–3346.

Bar, I., Cohen, Y., David, R., Rosenwaks, S., and Valentini, J.J. (1990). Direct observation of preferential bond fission by excitation of a vibrational fundamental: Photodissociation of HOD(0,0,1), *J. Chem. Phys.* **93,** 2146.

Barker, J.A. and Auerbach, D.J. (1984). Gas-surface interactions and dynamics: Thermal energy atomic and molecular beam studies, *Surf. Science Reports* **4,** 1–99.

Barrett, R.F., Robson, B.A., and Tobocman, W. (1983). Calculable methods for many-body scattering, *Rev. Mod. Phys.* **55,** 155–243.

Barts, S.A. and Halpern, J.B. (1989). Photodissociation of CLCN between 190 and 213 nm, *J. Phys. Chem.* **93,** 7346–7351.

Baumert, T., Bühler, B., Grosser, M., Thalweiser, R., Weiss, V., Wiedenmann, E., and Gerber, G. (1991). Femtosecond time-resolved wave packet motion in molecular multiphoton ionization and fragmentation, *J. Chem. Phys.* **95,** 8103.

Baumert, T., Engel, V., Röttgermann, C., Strunz, W.T., and Gerber, G. (1992). Femtosecond pump-probe study of the spreading and recurrence of a vibrational wave packet in Na_2, *Chem. Phys. Lett.* **191,** 639–644.

Baumert, T., Grosser, M., Thalweiser, R., and Gerber, G. (1991). Femtosecond time-resolved wave packet motion in molecular multiphoton ionization and fragmentation: The Na_2 system, *Phys. Rev. Lett.* **67,** 3753.

Bell, A.J., Pardon, P.R., and Frey, J.G. (1989). Raman spectra of NOCl and Cl_2 photodissociation dynamics, *Mol. Phys.* **67,** 465–472.

Benoist d'Azy, O., Lahmani, F., Lardeux, C., and Solgadi, D. (1985). State-selective photochemistry: Energy distribution in the NO fragment after photodissociation of the CH_3ONO $n\pi^*$ state, *Chem. Phys.* **94,** 247–256.

Ben-Shaul, A., Haas, Y., Kompa, K.L., and Levine, R.D. (1981). *Lasers and Chemical Change* (Springer, Heidelberg).

Berblinger, M., Pollak, E., and Schlier, Ch. (1988). Bound states embedded in the continuum of H_3^+, *J. Chem. Phys.* **88,** 5643–5656.

Bernstein, E.R. (Ed.) (1990). *Atomic and Molecular Clusters* (Elsevier, Amsterdam).

Bernstein, R.B. and Zewail, A.H. (1990). From femtosecond temporal spectroscopy to the potential by a direct inversion method, *Chem. Phys. Lett.* **170,** 321–328.

Berry, M.V. and Tabor, M. (1976). Closed orbits and the regular bound spectrum, *Proc. Roy. Soc. Lond. A* **349,** 101-123.

Berry, M.V. and Tabor, M. (1977). Calculating the bound spectrum by path summation in action-angle variables, *J. Phys. A* **10,** 371–379.

Bersohn, R. (1984). Final state distributions in the photodissociation of triatomic molecules, *J. Phys. Chem.* **88,** 5145–5149.

Bersohn, R. and Lin, S.H. (1969). Orientation of targets by beam excitation, *Adv. Chem. Phys.* **16,** 67–100.

Bersohn, R. and Zewail, A.H. (1988). Time dependent absorption of fragments during dissociation, *Ber. Bunsenges. Phys. Chem.* **92,** 373–378.

Beswick, J.A. (1991). Photofragmentation dynamics, in *Structure, Interactions, and Reactivity,* ed. S. Fraga (Elsevier, Amsterdam).

Beswick, J.A. and Gelbart, W.M. (1980). Bending contribution to rotational distributions in the photodissociation of polyatomic molecules, *J. Phys. Chem.* **84,** 3148–3151.

Beswick, J.A., Glass-Maujean, M., and Roncero, O. (1993). On the orientation of photofragments produced in highly excited rotational states, *J. Chem. Phys.* **96**, 7514–7527.

Beswick, J.A. and Halberstadt, N. (1993). *Dynamics of weakly bound complexes* (Kluwer, Dordrecht).

Beswick, J.A. and Jortner, J. (1977). Model for vibrational predissociation of van der Waals molecules, *Chem. Phys. Lett.* **49,** 13–18.

Beswick, J.A. and Jortner, J. (1981). Intramolecular dynamics of van der Waals molecules, *Adv. Chem. Phys.* **47,** 363–506.

Biesner, J., Schnieder, L., Ahlers, G., Xie, X., Welge, K.H., Ashfold, M.N.R., and Dixon, R.N. (1989). State selective photodissociation dynamics of $\tilde{A}$ state ammonia. II, *J. Chem. Phys.* **91,** 2901–2911.

Biesner, J., Schnieder, L., Schmeer, J., Ahlers, G., Xie, X., Welge, K.H., Ashfold, M.N.R., and Dixon, R.N. (1988). State selective photodissociation dynamics of $\tilde{A}$ state ammonia.I, *J. Chem. Phys.* **88,** 3607–3616.

Bigio L. and Grant, E.R. (1987a). Two-photon photodissociation dynamics of NO_2, *J. Chem. Phys.* **87,** 360–369.

Bigio, L. and Grant, E.R. (1987b). Polarized absorption spectroscopy of Λ-doublet molecules: Transition moment vs electron density distribution, *J. Chem. Phys.* **87,** 5589–5597.

Billing, G.D. (1984). The semiclassical treatment of molecular roto-vibrational energy transfer, *Com. Phys. Rep.* **1,** 237–296.

Black, J.F., Hasselbrink, E., Waldeck, J.R., and Zare, R.N. (1990). Photofragment orientation as a probe of near-threshold non-adiabatic phenomena in the photodissociation of ICN, *Mol. Phys.* **71,** 1143–1153.

Black, J.F., Waldeck, J.R., and Zare, R.N. (1990). Evidence for three interacting potential energy surfaces in the photodissociation of ICN at 249 nm, *J. Chem. Phys.* **92,** 3519–3538.

Bosanac, S.D. (1988). *Long-Lived States in Collisions* (CRC Press, Boca Raton).

Bower, R.D., Jones, R.W., and Houston, P.L. (1983). State-to-state dissociation dynamics in CF_3NO, *J. Chem. Phys.* **79,** 2799–2807.

Bowman, R.M., Dantus, M., and Zewail, A.H. (1990). Femtosecond transition-state spectroscopy of iodine: From strongly bound to repulsive surface dynamics, *Chem. Phys. Lett.* **161,** 297–302.

Bowman, J.M. and Gazdy B. (1989). A reduced dimensionality L^2 simulation of the photodetachment spectra of $ClHCl^-$ and IHI^-, *J. Phys. Chem.* **93,** 5129–5135.

Bradforth, S.E., Weaver, A., Arnold, D.W., Metz, R.B., and Neumark, D.M. (1990). Examination of the Br+HI, Cl+HI, and F+HI hydrogen abstraction reactions by photoelectron spectroscopy of $BrHI^-$, $ClHI^-$, and FHI^-, *J. Chem. Phys.* **92,** 7205–7222.

Brass, O., Tennyson, J., and Pollak, E. (1990). Spectroscopy and dynamics of the highly excited nonrotating three-dimensional H_3^+ molecular ion, *J. Chem. Phys.* **92,** 3377–3386.

Braunstein, M. and Pack, R.T. (1992). Simple theory of diffuse structure in continuous ultraviolet spectra of polyatomic molecules. III. Application to the Wulff–Chappius band system of ozone, *J. Chem. Phys.* **96**, 6378–6388.

Briggs, R.G., Halpern, J.B., Hancock, G., Shafizadeh, N., Rostas, J., Lemaire, J.L., and Rostas, F. (1989). Photodissociation of D_2O in the second continuum by two-photon absorption at 266 nm, *Chem. Phys. Lett.* **156,** 363–367.

Broeckhove, J., Feyen, B., Lathouwers, L., Arickx, F., and Van Leuven, P. (1990). Quantum time evolution of vibrational states in curve-crossing problems, *Chem. Phys. Lett.* **174,** 504–510.

Brooks, P.R. (1988). Spectroscopy of transition region species, *Chem. Rev.* **88,** 407–428.

Brouard, M., Martinez, M.T., Milne, C.J., Simons, J.P., and Wang, J.-X. (1990). Near threshold stereo-dynamics of molecular photodissociation: The visible and near ultraviolet photodissociation of H_2O_2, *Chem. Phys. Lett.* **165,** 423–428.

Brouard, M., Martinez, M.T., and O'Mahony, J. (1990). Fragment pair correlations in the vibrationally mediated photodissociation of H_2O_2: Rotation-vibration coupling in the third OH stretching overtone, *Mol. Phys.* **71,** 1021–1041.

Brouard, M., Martinez, M.T., O'Mahony, J., and Simons, J.P. (1988). Photofragment vector correlations in the vibrationally mediated photodissociation of H_2O_2, *Chem. Phys. Lett.* **150,** 6–12.

Brouard, M., Martinez, M.T., O'Mahony, J., and Simons, J.P. (1989). Photofragment vector correlations in vibrationally mediated photodissociation, *J. Chem. Soc., Faraday Trans. 2* **85,** 1207–1219.

Brouard, M., Martinez M.T., O'Mahony, J., and Simons, J.P. (1990). Energy and angular momentum disposals in the vibrationally mediated photodissociation of HOOH and HOOD. Intramolecular properties via photofragment mapping, *Mol. Phys.* **69,** 65–84.

Brouwer, L., Cobos, C.J., Troe, J., Dübal, H.-R., and Crim, F.F. (1987). Specific rate constants $k(E, J)$ and product state distributions in simple bond fission reactions. II. Application to HOOH $\rightarrow$ OH+OH, *J. Chem. Phys.* **86,** 6171–6182.

Brown, R.C. and Heller, E.J. (1981). Classical trajectory approach to photodissociation: The Wigner method, *J. Chem. Phys.* **75,** 186–188.

Brudzynski, R.J., Sension, R.J., and Hudson, B. (1990). Resonance raman study of the first absorption band of H_2S, *Chem. Phys. Lett.* **165,** 487–493.

Brühlmann, U., Dubs, M., and Huber, J.R. (1987). Photodissociation of methylnitrite: State distributions, recoil velocity distribution, and alignment effects of the NO($X^2\Pi$) photofragment, *J. Chem. Phys.* **86,** 1249–1257.

Brühlmann, U. and Huber, J.R. (1988). Spin-state ($\Pi_{1/2}, \Pi_{3/2}$) population of NO($X, v'' = 0,1,2,3$) fragments after state-selective photodissociation of CH_3ONO, *Chem. Phys. Lett.* **143,** 199–203.

Bruna, P.J. and Peyerimhoff, S.R. (1987). Excited-state potentials, in *Ab Initio Methods in Quantum Chemistry-I,* ed. K.P. Lawley (Wiley, New York).

Bruno, A.E., Brühlmann, U., and Huber, J.R. (1988). Photofragmentation LIF spectroscopy of NOCL at dissociation wavelengths > 450 nm. Parent electronic spectrum and spin state and Λ-doublet populations of nascent NO and CL fragments, *Chem. Phys.* **120,** 155–167.

Buck, U. (1975). Elastic scattering, *Adv. Chem. Phys.* **30,** 313–388.

Buck, U. (1982). Rotationally inelastic scattering of hydrogen molecules and the nonspherical interaction, *Faraday Discuss. Chem. Soc.* **73,** 187–203.

Buck, U. (1986). State selective rotational energy transfer in molecular collisions, *Comments. At. Mol. Phys.* **17,** 143–162.

Buckingham, A.D. (1967). Permanent and induced molecular moments and long-range intermolecular forces, *Adv. Chem. Phys.* **12,** 107–166.

Buckingham, A.D., Fowler, P.W., and Hutson, J.M. (1988). Theoretical studies of van der Waals molecules and intermolecular forces, *Chem. Rev.* **88,** 963–988.

Buelow, S., Noble, M., Radhakrishnan, G., Reisler, H., Wittig, C., and Hancock, G. (1986). The role of initial conditions in elementary gas-phase processes involving intermediate "complexes", *J. Phys. Chem.* **90,** 1015–1027.

Buenker, R.J., Peric, M., Peyerimhoff, S.D., and Marian, R. (1981). *Ab initio* treatment of the Renner-Teller effect for the X^2B_1 and A^2A_1 electronic states of NH_2, *Mol. Phys.* **43,** 987–1014.

Burak, I., Hepburn, J.W., Sivakumar, N., Hall, G.E., Chawla, G., and Houston, P.L. (1987). State-to-state photodissociation dynamics of *trans*-glyoxal, *J. Chem. Phys.* **86,** 1258–1268.

Burke, P.G. (1989). Electron and photon collisions with molecules, in *Collision Theory for Atoms and Molecules,* ed. F.A. Gianturco (Plenum Press, New York).

Busch, G.E. and Wilson, K.R. (1972a). Triatomic photofragment spectra. I. Energy partitioning in NO_2 photodissociation, *J. Chem. Phys.* **56,** 3626–3638.

Busch, G.E. and Wilson, K.R. (1972b). Triatomic photofragment spectra. II. Angular distributions from NO_2 photodissociation, *J. Chem. Phys.* **56,** 3638–3654.

Butenhoff, T.J., Carleton, K.L., and Moore, C.B. (1990). Photodissociation dynamics of formaldehyde: H_2 rotational distributions and product quantum state correlations, *J. Chem. Phys.* **92,** 377–393.

Butkov, E. (1968). *Mathematical Physics* (Addison-Wesley, Reading, Mass.).

Calvert, J.G. and Pitts, J.N. (1966). *Photochemistry* (Wiley, New York).

Campos-Martínez, J. and Coalson, R.D. (1990). Adding configuration interaction to the time-dependent Hartree grid method, *J. Chem. Phys.* **93,** 4740–2749.

Carrington, T. (1964). Angular momentum distribution and emission spectrum of $OH(^2\Sigma^+)$ in the photodissociation of H_2O, *J. Chem. Phys.* **41,** 2012–2018.

Carrington, A. and Kennedy, R.A. (1984). Infrared predissociation spectrum of the H_3^+ ion, *J. Chem. Phys.* **81,** 91–112.

Čársky, P. and Urban, M. (1980). *Ab Initio Calculations* (Springer, Heidelberg).

Carter, S. and Handy, N.C. (1986). The variational method for the calculation of ro-vibrational energy levels, *Computer Physics Reports* **5,** 115–172.

Cederbaum, L.S. and Domcke, W. (1977). Theoreticals aspects of ionization potentials and photoelectron spectroscopy: a Green's function approach, *Adv. Chem. Phys.* **36,** 205.

Chang, Y.-T., Minichino, C., and Miller, W.H. (1992). Classical trajectory studies of the molecular dissociation dynamics of formaldehyde: $H_2CO \rightarrow H_2 + CO$, *J. Chem. Phys.* **96**, 4341–4355.

Chasman, D., Tannor, D.J., and Imre, D.G. (1988). Photoabsorption and photoemission of ozone in the Hartley band, *J. Chem. Phys.* **89,** 6667–6675.

Child, M.S. (1974). *Molecular Collision Theory* (Academic Press, London).

Child, M.S. (1976). Semiclassical methods in molecular collisions, in *Dynamics of Molecular Collisions, Part B,* ed. W.H. Miller (Plenum Press, New York).

Child, M.S. (1979). Electronic excitation: Nonadiabatic transitions, in *Atom-Molecule Collision Theory,* ed. R.B. Bernstein (Plenum Press, New York).

Child, M.S. (1980). Separable spectroscopic applications, in *Semiclassical Methods in Molecular Scattering and Spectroscopy,* ed. M.S. Child (Reidel, Dordrecht).

Child, M.S. (1985). Local mode overtone spectra, *Acc. Chem. Res.* **18,** 45–50.

Child, M.S. (1991). *Semiclassical Mechanics with Molecular Applications* (Clarendon Press, Oxford).

Child, M.S. and Halonen, L. (1984). Overtone frequencies and intensities in the local mode picture, *Adv. Chem. Phys.* **57,** 1–58.

Child, M.S. and Shapiro, M. (1983). Photodissociation and the Condon reflection principle, *Mol. Phys.* **48,** 111–128.

Choi, S.E. and Light, J.C. (1989). Use of the discrete variable representation in the quantum dynamics by a wave packet propagation: Predissociation of $NaI(^1\Sigma_0^+) \rightarrow NaI(0^+) \rightarrow Na(^2S) + I(^2P)$, *J. Chem. Phys.* **90,** 2593–2604.

Chu, J.-J., Marcus, P., and Dagdigian, P.J. (1990). One-color photolysis-ionization study of HN_3: The N_2 fragment internal energy distribution and μ-v-J correlations, *J. Chem. Phys.* **93,** 257–267.

Clary, D.C. (1986a). A theory for the photodissociation of polyatomic molecules, with application to CF_3I, *J. Chem. Phys.* **84,** 4288–4298.

Clary, D.C. (Ed.) (1986b). *The Theory of Chemical Reaction Dynamics* (Reidel, Dordrecht).

Clary, D.C. (1987). Coupled-channel calculations on energy transfer, photochemistry, and reactions of polyatomic molecules, *J. Phys. Chem.* **91,** 1718–1727.

Clary, D.C., Lovejoy, C.M., ONeil, S.V., and Nesbitt, D.J. (1988). Infrared spectrum of NeHF, *Phys. Rev. Lett.* **61,** 1576–1579.

Cline, J.I., Evard, D.D., Thommen, F., and Janda, K.C. (1986). The laser induced fluorescence spectrum of the $HeCl_2$ van der Waals molecule, *J. Chem. Phys.* **84,** 1165–1170.

Cline, J.I., Reid, B.P., Evard, D.D., Sivakumar, N., Halberstadt, N., and Janda, K.C. (1988). State-to-state vibrational predissociation dynamics and spectroscopy of $HeCl_2$: Experiment and theory, *J. Chem. Phys.* **89,** 3535–3552.

Cline, J.I., Sivakumar, N., Evard, D.D., Bieler, C.R., Reid, B.P., Halberstadt, N., Hair, S.R., and Janda, K.C. (1989). Product state distributions for the vibrational predissociation of $NeCl_2$, *J. Chem. Phys.* **90,** 2605–2616.

Coalson, R.D. (1985). On the computation of two surface properties by coordinate-space propagator techniques, *J. Chem. Phys.* **83,** 688–697.

Coalson, R.D. (1987). Time-domain formulation of photofragmentation involving nonradiatively coupled excited states, and its implementation via wave packet perturbation theory, *J. Chem. Phys.* **86,** 6823–6832.

Coalson, R.D. (1989). Time-dependent wavepacket approach to optical spectroscopy involving nonadiabatically coupled potential energy surfaces, *Adv. Chem. Phys.* **73,** 605–636.

Coalson, R.D. and Karplus, M. (1982). Extended wave packet dynamics; exact solution for collinear atom, diatomic molecule scattering, *Chem. Phys. Lett.* **90,** 301–305.

Coalson, R.D. and Kinsey, J.L. (1986). Time domain formulation of optical spectroscopy involving three potential energy surfaces, *J. Chem. Phys.* **85,** 4322–4340.

Cohen, R.C. and Saykally, R.J. (1991). Multidimensional intermolecular potential surfaces from vibration-rotation tunneling (VRT) spectra of van der Waals complexes, *Ann. Rev. Phys. Chem.* **42,** 369–392.

Cohen-Tannoudji, C., Diu, B., and Laloë, F. (1977). *Quantum Mechanics,* Vols. I and II (Wiley, New York).

Crim, F.F. (1984). Selective excitation studies of unimolecular reaction dynamics, *Ann. Rev. Phys. Chem.* **35,** 657–691.

Crim, F.F. (1987). The dissociation dynamics of highly vibrationally excited molecules, in *Molecular Photodissociation Dynamics,* ed. M.N.R. Ashfold and J.E. Baggott (Royal Society of Chemistry, London).

Dahl, J.P. (1983). Dynamical equations for the Wigner functions, in *Energy Storage and Redistributions in Molecules,* ed. J. Hinze (Plenum Press, New York).

Dantus, M., Rosker, M.J., and Zewail, A.H. (1988). Femtosecond real-time probing of reactions. II. The dissociation reaction of ICN, *J. Chem. Phys.* **89,** 6128–6140.

Das, S., and Tannor, D.J. (1990). Time dependent quantum mechanics using picosecond time steps: Application to predissociation of HeI_2, *J. Chem. Phys.* **92,** 3403–3409.

Dateo, C.E., Engel, V., Almeida, R., and Metiu, H. (1991). Numerical solutions of the time-dependent Schrödinger equation in spherical coordinates by Fourier transform methods, *Computer Physics Communications* **63,** 435–445.

Dateo, C.E. and Metiu, H. (1991). Numerical solution of the time dependent Schrödinger equation in spherical coordinates by Fourier transform methods, *J. Chem. Phys.* **95,** 7392–7400.

Daudel, R., Leroy, G., Peeters, D., and Sana, M. (1983). *Quantum Chemistry* (Wiley, Chichester).

Dayton, D.C., Jucks, K.W. and Miller, R.E. (1989). Photofragmentation angular distributions for HF dimer: Scalar *J-J* correlations in state-to-state photodissociation, *J. Chem. Phys.* **90,** 2631–2638.

Debarre, D., Lefebvre, M., Péalat, M., Taran, J.-P.E., Bamford, D.J., and Moore, C.B. (1985). Photofragmentation dynamics of formaldehyde: $H_2(v, J)$ distributions, *J. Chem. Phys* **83,** 4476–4487.

Delgado-Barrio, G., Mareca, P., Villarreal, P., Cortina, A.M., and Miret-Artés, S. (1986). A close-coupling infinite order sudden approximation (IOSA) to study vibrational predissociation of the HeI_2 van der Waals molecule, *J. Chem. Phys.* **84,** 4268–4271.

Desouter-Lecomte, M., Dehareng, D., Leyh-Nihant, B., Praet, M.Th., Lorquet, A.J., and Lorquet, J.C. (1985). Nonadiabatic unimolecular reactions of polyatomic molecules, *J. Phys. Chem.* **89,** 214–222.

Desouter-Lecomte, M., Dehareng, D., and Lorquet, J.C. (1987). Constructing approximately diabatic states from LCAO-SCF-CI calculations, *J. Chem. Phys.* **86,** 1429–1436.

de Vries, A.E. (1982). Half-collisions, *Comments At. Mol. Phys.* **11,** 157–172.

Dickinson, A.S. (1979). Non-reactive heavy particle collision calculations, *Computer Physics Communications* **17,** 51–80.

Dirac, P.A.M. (1927). The quantum theory of dispersion, *Proc. Roy. Soc. London A* **114,** 710–728.

Dixon, R.N. (1988). The stretching vibrations of ammonia in its $\tilde{A}^1A_2''$ excited state, *Chem. Phys. Lett.* **147,** 377–383.

Dixon, R.N. (1989). The influence of parent rotation on the dissociation dynamics of the $\tilde{A}^1A_2''$ state of ammonia, *Mol. Phys.* **68,** 263–278.

Dixon, R.N., Balint-Kurti, G.G., Child, M.S., Donovan, R., and Simons, J.P. (Eds.) (1986). Dynamics of molecular photofragmentation, *Faraday Discuss. Chem. Soc.* **82,** 1–404.

Dixon, R.N., Marston, C.C., and Balint-Kurti, G.G. (1990). Photodissociation dynamics and emission spectroscopy of H_2S in its first absorption band: A time dependent quantum mechanical study, *J. Chem. Phys.* **93,** 6520–6534.

Dixon, R.N., Nightingale, J., Western, C.M., and Yang, X. (1988). Determination of the pair correlation of the OH rotational states from the 266 nm photolysis of H_2O_2 using velocity-aligned Doppler spectroscopy, *Chem. Phys. Lett.* **151,** 328–334.

Dixon, R.N. and Rieley, H. (1989). State-selected photodissociation dynamics of HONO($\tilde{A}^1A''$): Characterization of the NO fragment, *J. Chem. Phys.* **91,** 2308–2320.

Docker, M.P., Hodgson, A., and Simons, J.P. (1986a). Photodissociation of H_2O_2 at 248 nm: Translational anisotropy and OH product state distributions, *Chem. Phys. Lett.* **128,** 264–269.

Docker, M.P., Hodgson, A., and Simons, J.P. (1986b). Photodissociation Dynamics of H_2O_2 at 248 nm, *Faraday Discuss. Chem. Soc.* **82,** 25–36.

Docker, M.P., Hodgson, A., and Simons, J.P. (1987). High-resolution photochemistry: Quantum-state selection and vector correlations in molecular photodissociation, in *Molecular Photodissociation Dynamics,* ed. M.N.R. Ashfold and J.E. Baggott (Royal Society of Chemistry, London).

Domcke,W. (1991). Theory of resonances in electron molecule collisions: The projection operator approach, *Phys. Rep.* **208,** 97–188.

Drobits, J.C. and Lester, M.I. (1988a). Near threshold photofragmentation dynamics of ICL-Ne *A* state van der Waals complexes, *J. Chem. Phys.* **88,** 120–128.

Drobits, J.C. and Lester, M.I. (1988b). Evidence for final state interactions in the vibrational predissociation of ICL-Ne complexes, *J. Chem. Phys.* **89,** 4716–4725.

Du, M.L. and Delos, J.B. (1988a). Effect of closed classical orbits on quantum spectra: Ionization of atoms in a magnetic field. I. Physical picture and calculations, *Phys. Rev. A* **38,** 1896–1912.

Du, M.L. and Delos, J.B. (1988b). Effect of closed classical orbits on quantum spectra: Ionization of atoms in a magnetic field. II. Derivation of formulas, *Phys. Rev. A* **38,** 1913–1930.

Dubs, M., Brühlmann, U., and Huber, J.R. (1986). Sub-Doppler laser-induced fluorescence measurements of the velocity distribution and rotational alignment of NO photofragments, *J. Chem. Phys.* **84,** 3106–3119.

Dugan, C.H. and Anthony, D. (1987). The origin of fragment rotation in ICN photodissociation, *J. Phys. Chem.* **91,** 3929–3932.

Dunne, L.J., Guo, H., and Murrell, J.N. (1987). The role of the $\tilde{B}-\tilde{X}$ conical intersection in the photodissociation of water, *Mol. Phys.* **62,** 283–294.

Durand, P. and Malrieu, J.-P. (1987). Effective Hamiltonians and pseudo-operator as tools for rigorous modelling, *Adv. Chem. Phys.* **67,** 321.

Dutuit, O., Tabche-Fouhaile, A., Nenner, I., Frohlich, H., and Guyon, P.M. (1985). Photodissociation processes of water vapor below and above the ionization potential, *J. Chem. Phys.* **83,** 584–596.

Dyet, J.A., McCoustra, M.R.S., and Pfab, J. (1988). The spectroscopy, photophysics and photodissociation dynamics of jet-cooled $CF_3NO[\tilde{A}(n,\pi^*)]$, *J. Chem. Soc., Faraday Trans. 2* **84,** 463–482.

Dykstra, C.E. (1988). *Ab Initio Calculations on the Structures and Properties of Molecules* (Elsevier, Amsterdam).

Edmonds, A.R. (1974). *Angular Momenta in Quantum Mechanics* (Princeton University Press, Princeton).

Effenhauser, C.S., Felder, P., and Huber, J.R. (1990). Two-photon dissociation of sulfur dioxide at 248 and 308 nm, *Chem. Phys.* **142,** 311–320.

Eichmann, U., Richter, K., Wintgen, D., and Sandner, W. (1988). Scaled-energy spectroscopy and its relation with periodic orbits, *Phys. Rev. Lett.* **61,** 2438–2441.

Engel, V. (1991a). Excitation of symmetric and asymmetric stretch in a symmetric triatomic molecule: A time-dependent (collinear) analysis of the IHI^- photodetachment spectrum, *J. Chem. Phys.* **94,** 16–22.

Engel, V. (1991b). Excitation of molecules with ultrashort laser pulses: exact time-dependent quantum calculations, *Computer Physics Communications* **63,** 228–242.

Engel, V. (1991c). Femtosecond pump/probe experiments and ionization: The time dependence of the total ion signal, *Chem. Phys. Lett.* **178,** 130.

Engel, V. and Metiu, H. (1989). The study of NaI predissociation with pump-probe femtosecond laser pulses: The use of an ionizing probe pulse to obtain more detailed dynamic information, *Chem. Phys. Lett.* **155,** 77–82.

Engel, V. and Metiu, H. (1990). CH_3ONO predissociation by ultrashort laser pulses: Population transients and product state distribution, *J. Chem. Phys.* **92,** 2317–2327.

Engel, V., Metiu, H., Almeida, R., Marcus, R.A., and Zewail, A.H. (1988). Molecular state evolution after excitation with an ultra-short laser pulse: A quantum analysis of NaI and NaBr dissociation, *Chem. Phys. Lett.* **152,** 1–7.

Engel, V. and Schinke, R. (1988). Isotope effects in the fragmentation of water: The photodissociation of HOD in the first absorption band, *J. Chem. Phys.* **88,** 6831–6837.

Engel, V., Schinke, R., Hennig, S., and Metiu, H. (1990). A time-dependent interpretation of the absorption spectrum of CH_3ONO, *J. Chem. Phys.* **92,** 1–13.

Engel, V., Schinke, R., and Staemmler, V. (1988). Photodissociation dynamics of H_2O and D_2O in the first absorption band: A complete *ab initio* treatment, *J. Chem. Phys.* **88,** 129–148.

Engel, V., Staemmler, V., Vander Wal, R.L., Crim, F.F., Sension, R.J., Hudson, B., Andresen, P., Hennig, S., Weide, K., and Schinke, R. (1992). The photodissociation of water in the first absorption band: A prototype for dissociation on a repulsive potential energy surface, *J. Phys. Chem.* **96,** 3201–3213.

Evard, D.D., Bieler, C.R., Cline, J.I., Sivakumar, N., and Janda, K.C. (1988). The vibrational predissociation dynamics of $ArCl_2$: Intramolecular vibrational relaxation in a triatomic van der Waals molecule?, *J. Chem. Phys.* **89,** 2829–2838.

Ewing, G.E. (1979). A guide to the lifetimes of vibrationally excited van der Waals molecules. The momentum gap, *J. Chem. Phys.* **71,** 3143–3144.

Ewing, G.E. (1981). Vibrational predissociation of van der Waals molecules and intermolecular potential energy surfaces, in *Potential Energy Surfaces and Dynamics Calculations,* ed. D.G. Truhlar (Plenum Press, New York).

Ewing, G.E. (1982). Relaxation channels of vibrationally excited van der Waals molecules, *Faraday Discuss. Chem. Soc.* **73,** 325–338.

Ewing, G.E. (1987). Selection rules for vibrational energy transfer: Vibrational predissociation of van der Waals molecules, *J. Phys. Chem.* **91,** 4662–4671.

Fano, U. (1961). Effects of configuration interaction on intensities and phase shifts, *Phys. Rev.* **124,** 1866–1878.

Fano, U. and Rao, A.R.P. (1986). *Atomic Collisions and Spectra* (Academic Press, Orlando).

Farantos, S.C. and Taylor, H.S. (1991). The photodissociation of O_3: A classical dynamical approach for the interpretation of the recurrences in the autocorrelation function, *J. Chem. Phys.* **94,** 4887–4895.

Faubel, M., Kohl, K.-H., Toennies, J.P., Tang, K.T., and Yung, Y.Y. (1982). The He-N_2 anisotropic van der Waals potential, *Faraday Discuss. Chem. Soc.* **73,** 205–220.

Feit, M.D. and Fleck, J.A., Jr. (1983). Solution of the Schrödinger equation by a spectral method. II. Vibrational energy levels of triatomic molecules, *J. Chem. Phys.* **78,** 301–308.

Feit, M.D. and Fleck, J.A., Jr. (1984). Wave packet dynamics and chaos in the Hénon-Heiles system, *J. Chem. Phys.* **80,** 2578–2584.

Feit, M.D., Fleck, J.A., Jr., and Steiger, A. (1982). Solution of the Schrödinger equation by a spectral method, *J. Comput. Phys.* **47,** 412–433.

Felder, P. (1990). Photodissociation of CF_3I at 248 nm: Internal energy distribution of the CF_3 fragments, *Chem. Phys.* **143,** 141–150.

Feshbach, H. (1958). Unified theory of nuclear reactions, *Annals Phys.* **5,** 357–390.

Feshbach, H. (1962). A unified theory of nuclear reactions. II, *Annals Phys.* **19,** 287–313.

Feshbach, H. (1964). Unified theory of nuclear reactions, *Rev. Mod. Phys.* **36,** 1076–1078.

Feynman, R.B. and Hibbs, A.R. (1965). *Quantum Mechanics and Path Integrals* (McGraw-Hill, New York).

Fisher, W.H., Eng, R., Carrington, T., Dugan, C.H., Filseth, S.V., and Sadowski, C.M. (1984). Photodissociation of BrCN and ICN in the A continuum: Vibrational and rotational distributions of $CN(X^2\Sigma^+)$, *Chem. Phys.* **89,** 457–471.

Ford, K.W. and Wheeler, J.A. (1959). Semiclassical description of scattering, *Annals Phys.* **7,** 259–286.

Forst, W. (1973). *Theory of Unimolecular Reactions* (Academic Press, New York).

Foth, H.-J., Polanyi, J.C., and Telle, H.H. (1982). Emission from molecules and reaction intermediates in the process of falling apart, *J. Phys. Chem.* **86,** 5027–5041.

Founargiotakis, M., Farantos, S.C., Contopoulos, G., and Polymilis, C. (1989). Periodic orbits, bifurcations, and quantum mechanical eigenfunctions and spectra, *J. Chem. Phys.* **91,** 1389–1401.

Foy, B.R., Casassa, M.P., Stephenson, J.C., and King, D.S. (1988). Unimolecular dynamics following vibrational overtone excitation of HN_3 $v_1 = 5$ and $v_1 = 6$: $HN_3(\tilde{X}; v, J, K) \rightarrow HN(X^3\Sigma^-; v, J, \Omega) + N_2(X^1\Sigma_g^+)$, *J. Chem. Phys.* **89,** 608–609.

Freed, K.F. and Band, Y.B. (1977). Product energy distributions in the dissociation of polyatomic molecules, in *Excited States,* Vol. 3, ed. E.C. Lim (Academic Press, New York).

Freeman, D.E., Yoshino, K., Esmond, J.R., and Parkinson, W.H. (1984). High resolution absorption cross-section measurements of ozone at 195 K in the wavelength region 240–350 nm, *Planet. Space. Sci.* **32,** 239–248.

Fried, L.E. and Mukamel, S. (1990). A classical theory of pump-probe photodissociation for arbitrary pulse durations, *J. Chem. Phys.* **93,** 3063–3071.

Friedrich, H. and Wintgen, D. (1989). The hydrogen atom in a uniform magnetic field – An example of chaos, *Physics Reports* **183,** 37–79.

Gazdy, B. and Bowman, J.M. (1989). A three-dimensional L^2 simulation of the photodetachment spectra of $ClHCl^-$ and IHI^-, *J. Chem. Phys.* **91,** 4615–4624.

Gelbart, W.M. (1977). Photodissociation dynamics of polyatomic molecules, *Ann. Rev. Phys. Chem.* **28,** 323–348.

Georgiou, S. and Wight, C.A. (1990). Photodissociation of cobalt tricarbonylnitrosyl and its trialkylphosphine derivatives in the visible region: Evidence for impulsive fragmentation of the nitrosyl ligand, *J. Phys. Chem* **94,** 4935–4940.

Gerber, R.B., Kosloff, R., and Berman, M. (1986). Time-dependent wavepacket calculations of molecular scattering from surfaces, *Computer Physics Reports* **5,** 59–114.

Gerber, R.B. and Ratner, M.A. (1988a). Mean-field models for molecular states and dynamics: New developments, *J. Phys. Chem.* **92,** 3252–3260.

Gerber, R.B. and Ratner, M.A. (1988b). Self-consistent-field methods for vibrational excitations in polyatomic systems, *Adv. Chem. Phys.* **70,** 97–132.

Gericke, K.-H. (1988). Correlations between quantum state populations of coincident product pairs, *Phys. Rev. Lett.* **60,** 561–564.

Gericke, K.-H., Gläser, H.G., Maul, C., and Comes, F.J. (1990). Joint product state distribution of coincidently generated photofragment pairs, *J. Chem. Phys.* **92,** 411–419.

Gericke, K.-H., Gölzenleuchter, H., and Comes, F.J. (1988). Scalar and vector properties of $RO(A^2\Sigma^+)$ formed in the VUV photodissociation of R_2O_2(R=H,D), *Chem. Phys.* **127,** 399–409.

Gericke, K.-H., Grunewald, A.U., Klee, S., and Comes, F.J. (1988). Correlations between angular momenta of coincident product pairs, *J. Chem. Phys.* **88,** 6255–6259.

Gericke, K.-H., Haas, T., Lock, M., Theinl, R., and Comes, F.J. (1991). $HN_3(\tilde{A}^1A'')$ hypersurface at excitation energies of 4.0–5.0 eV, *J. Phys. Chem.* **95,** 6104–6111.

Gericke, K.-H., Klee, S., Comes, F.J., and Dixon, R.N. (1986). Dynamics of H_2O_2 photodissociation: OH product state and momentum distribution characterized by sub-Doppler and polarization spectroscopy, *J. Chem. Phys.* **85,** 4463–4479.

Gericke, K.-H., Theinl, R., and Comes, F.J. (1989). Photofragment energy distribution and rotational anisotropy from excitation of HN_3 at 266 nm, *Chem. Phys. Lett.* **164,** 605–611.

Gericke, K.-H., Theinl, R., and Comes, F.J. (1990). Vector correlations in the photofragmentation of HN_3, *J. Chem. Phys.* **92,** 6548–6555.

Gianturco, F.A. (1979). *The Transfer of Molecular Energies by Collisions* (Springer, Heidelberg).

Gianturco, F.A. (1989). Inelastic molecular collisions at thermal energies, in *Collision Theory for Atoms and Molecules,* ed. F.A. Gianturco (Plenum Press, New York).

Giguèra, P.A. (1959). Revised values of the O–O and the O–H bond dissociation energies, *J. Chem. Phys.* **30,** 322–322.

Gilbert, R.G. and Smith, S.C. (1990). *Theory of Unimolecular and Recombination Reactions* (Blackwell, Oxford).

Glass-Maujean, M. and Beswick, J.A. (1989). Coherence effects in the polarization of photofragments, *J. Chem. Soc., Faraday Trans. 2* **85,** 983–1002.

Goeke, K. and Reinhard, P.-G. (Eds.) (1982). *Time-Dependent Hartree-Fock and Beyond* (Springer, Berlin).

Golden, D.M., Rossi, M.J., Baldwin, A.C., and Barker, J.R. (1981). Infrared multiphoton decomposition: Photochemistry and photophysics, *Acc. Chem. Res.* **14,** 56–62.

Goldfield, E.M., Houston, P.L., and Ezra, G.S. (1986). Nonadiabatic interactions in the photodissociation of ICN, *J. Chem. Phys.* **84,** 3120–3129.

Goldstein, H. (1951). *Classical Mechanics* (Addison-Wesley Press, Reading, Mass.).

Gómez Llorente, J.M. and Pollak, E. (1987). Order out of chaos in the H_3^+ molecule, *Chem. Phys. Lett.* **138,** 125–130.

Gómez Llorente, J.M. and Pollak, E. (1988). Periodic orbit analysis of the photodissociation spectrum of H_3^+, *J. Chem. Phys.* **89,** 1195–1196.

Gómez Llorente, J.M. and Taylor, H.S. (1989). Spectra in the chaotic region: A classical analysis for the sodium trimer, *J. Chem. Phys.* **91,** 953–961.

Gómez Llorente, J.M., Zakrzewski, J., Taylor, H.S., and Kulander, K.C. (1988). Spectra in the chaotic region: A quantum analysis of the photodissociation of H_3^+, *J. Chem. Phys.* **89,** 5959–5960.

Gómez Llorente, J.M., Zakrzewski, J., Taylor, H.S., and Kulander, K.C. (1989). Spectra in the chaotic region: Methods for extracting dynamic information, *J. Chem. Phys.* **90,** 1505–1518.

Gordon, R.G. (1968). Correlation functions for molecular motion, in *Advances in Magnetic Resonance,* Vol. 3, ed. J.S. Waugh (Academic Press, New York).

Gottwald, E., Bergmann, K., and Schinke, R. (1987). Supernumerary rotational rainbows in Na_2-He, Ne, Ar scattering, *J. Chem. Phys.* **86,** 2685–2688.

Goursaud, S., Sizun, M., and Fiquet-Fayard, F. (1976). Energy partitioning and isotope effects in the fragmentation of triatomic negative ions: Monte Carlo Scheme for a classical trajectory study, *J. Chem. Phys.* **65,** 5453–5461.

Gray, S.K. (1991). The nature and decay of metastable vibrations: Classical and quantum studies of van der Waals molecules, in *Advances in Molecular Vibrations and Collision Dynamics 1,* ed. J.M. Bowman (JAI Press, Greenwich).

Gray, S.K. and Child, M.S. (1984). Photodissociation within classical S matrix theory. Hyperbolic umbilic uniform approximation and application to $CH_3I + \hbar\omega \rightarrow CH_3 + I^*$, *Mol. Phys.* **51,** 189–210.

Gray, S.K. and Wozny, C.E. (1989). Wave packet dynamics of van der Waals molecules: Fragmentation of $NeCl_2$ with three degrees of freedom, *J. Chem. Phys.* **91,** 7671–7684.

Gray, S.K. and Wozny, E. (1991). Fragmentation mechanisms from three dimensional wavepacket studies: Vibrational predissociation of $NeCl_2$, $HeCl_2$, and HeICl, *J. Chem. Phys.* **94,** 2817–2832.

Graybeal, J.D. (1988). *Molecular Spectroscopy* (McGraw-Hill, New York).

Greene, C.H. and Zare, R.N. (1982). Photofragment alignment and orientation, *Ann. Rev. Phys. Chem.* **33,** 119–150.

Gregory, T.A. and Lipski, S. (1976). Determination of the Einstein A coefficient for the $\tilde{A} \to \tilde{X}$ transition of ammonia-d_3^*, *J. Chem. Phys.* **65,** 5469–5473.

Grinberg, H., Freed, K.F., and Williams, C.J. (1987). Three-dimensional analytical quantum mechanical theory for triatomic photodissociation: Role of angle dependent dissociative surfaces on rotational and angular distributions in the rotational infinite order sudden limit, *J. Chem. Phys.* **86,** 5456–5478.

Gruebele, M., Roberts, G., Dantus, M., Bowman, R.M., and Zewail, A.H. (1990). Femtosecond temporal spectroscopy and direct inversion to the potential: Application to iodine, *Chem. Phys. Lett.* **166,** 459–469.

Gruebele, M. and Zewail, A.H. (1990). Ultrafast reaction dynamics, *Physics Today,* May issue, 24–33.

Grunewald, A.U., Gericke, K.-H., and Comes, F.J. (1987). Photofragmentation dynamics of hydrogen peroxide: Analysis of two simultaneously excited states, *J. Chem. Phys.* **87,** 5709–5721.

Grunewald, A.U., Gericke, K.-H., and Comes, F.J. (1988). Influence of H_2O_2 internal motion on scalar and vector properties of OH photofragments, *J. Chem. Phys.* **89,** 345–354.

Grunwald, E., Dever, D.F., and Keehn, P.M. (1978). *Megawatt Infrared Laser Chemistry* (Wiley, New York).

Guo, H. (1991). A three-dimensional wavepacket study on photodissociation dynamics of methyl iodide, *Chem. Phys. Lett.* **187,** 360–366.

Guo, H., Lao, K.Q., Schatz, G.C., and Hammerich, A.D. (1991). Quantum nonadiabatic effects in the photodissociation of vibrationally excited CH_3I, *J. Chem. Phys.* **94,** 6562–6568.

Guo, H. and Murrell, J.N. (1988a). A classical trajectory study of the $\tilde{A}$-state photodissociation of the water molecule, *J. Chem. Soc., Faraday Trans. 2* **84,** 949–959.

Guo, H. and Murrell, J.N. (1988b). Dynamics of the $\tilde{A}$-state photodissociation of H_2O at 1983 nm, *Mol. Phys.* **65,** 821–827.

Guo, H. and Schatz, G.C. (1990a). Nonadiabatic effects in photodissociation dynamics: A quantum mechanical study of ICN photodissociation in the A continuum, *J. Chem. Phys.* **92,** 1634–1642.

Guo, H. and Schatz, G.C. (1990b). Time-dependent dynamics of methyl iodide photodissociation in the first continuum, *J. Chem. Phys.* **93,** 393–402.

Gürtler, P., Saile, V., and Koch, E.E. (1977). Rydberg series in the absorption spectra of H_2O and D_2O in the vacuum ultraviolet, *Chem. Phys. Lett.* **51,** 386–391.

Gutzwiller, M.C. (1967). Phase-integral approximation in momentum space and the bound states of an atom, *J. Math. Phys.* **8,** 1979–2000.

Gutzwiller, M.C. (1971). Periodic orbits and classical quantization conditions, *J. Math. Phys.* **12,** 343–358.

Gutzwiller, M.C. (1980). Classical quantization of a Hamiltonian with ergodic behavior, *Phys. Rev. Lett.* **45,** 150–153.

Gutzwiller, M.C. (1990). *Chaos in Classical and Quantum Mechanics* (Springer, New York).

Halberstadt, N., Beswick, J.A., and Schinke, R. (1991). Rotational distributions in the vibrational predissociation of weakly bound complexes: Quasi-classical golden rule treatment, in *Half Collision Resonance Phenomena in Molecules: Experimental and Theoretical Approaches,* ed. M. Garcia-Sucre, G. Raseev, and S.C. Ross (American Institute of Physics, New York).

Halberstadt, N. and Janda, K.C. (1990). *Dynamics of Polyatomic van der Waals Molecules* (Plenum Press, New York).

Halberstadt, N., Roncero, O., and Beswick, J.A. (1989). Decay of vibrationally excited states of the Ne...Cl_2 complex, *Chem. Phys.* **129,** 83–92.

Hale, M.O., Galica, G.E., Glogover, S.G., and Kinsey, J.L. (1986). Emission spectroscopy of photodissociating CH_3I and CD_3I, *J. Chem. Phys.* **90,** 4997–5000.

Hall, G.E. and Houston, P.L. (1989). Vector correlations in photodissociation dynamics, *Ann. Rev. Phys. Chem.* **40,** 375–405.

Hall, G.E., Sears, T.J., and Frye, J.M. (1989). Dissociation of CD_3I at 248 nm studied by diode laser absorption spectroscopy, *J. Chem. Phys.* **90,** 6234–6242.

Halonen, L. (1989). Recent developments in the local mode theory of overtone spectra, *J. Phys. Chem.* **93,** 3386–3392.

Hartke, B. (1991). Continuum resonance Raman scattering in bromine: comparison of time-dependent calculations with time-independent and experimental results, *J. Raman Spec.* **22,** 131–140.

Hase, W.L. (1976). Dynamics of unimolecular reactions, in *Dynamics of Molecular Collisions* Part B, ed. W.H. Miller (Plenum Press, New York).

Hasselbrink, E., Waldeck, J.R., and Zare, R.N. (1988). Orientation of the CN $X^2\Sigma^+$ fragment following photolysis of ICN by circularly polarized light, *Chem. Phys.* **126,** 191–200.

Häusler, D., Andresen, P., and Schinke, R. (1987). State to state photodissociation of H_2O in the first absorption band, *J. Chem. Phys.* **87,** 3949–3965.

Hawkins, W.G. and Houston, P.L. (1982). Rotational distributions in the photodissociation of bent triatomics: H_2S, *J. Chem. Phys.* **76,** 729–731.

Heather, R.W. and Light, J.C. (1983a). Photodissociation of triatomic molecules: Rotational scattering effects, *J. Chem. Phys.* **78,** 5513–5530.

Heather, R.W. and Light, J.C. (1983b). Discrete variable theory of triatomic photodissociation, *J. Chem. Phys.* **79,** 147–159.

Heather, R.W. and Metiu, H. (1986). A numerical study of the multiple Gaussian representation of time dependent wave functions of a Morse oscillator, *J. Chem. Phys.* **84,** 3250–3259.

Heather, R.W. and Metiu, H. (1987). An efficient procedure for calculating the evolution of the wave function by fast Fourier transform methods for systems with spatially extended wave function and localized potential, *J. Chem. Phys.* **86,** 5009–5017.

Heather, R.W. and Metiu, H. (1989). Time-dependent theory of Raman scattering for systems with several excited electronic states: Application to a H_3^+ model system, *J. Chem. Phys.* **90,** 6903–6915.

Heller, E.J. (1975). Time-dependent approach to semiclassical dynamics, *J. Chem. Phys.* **62,** 1544–1555.

Heller, E.J. (1976). Time dependent variational approach to semiclassical dynamics, *J. Chem. Phys.* **64,** 63–73.

Heller, E.J. (1978a). Quantum corrections to classical photodissociation models, *J. Chem. Phys.* **68,** 2066–2075.

Heller, E.J. (1978b). Photofragmentation of symmetric triatomic molecules: Time dependent picture, *J. Chem. Phys.* **68,** 3891–3896.

Heller, E.J. (1981a). The semiclassical way to molecular spectroscopy, *Acc. Chem. Res.* **14,** 368–375.

Heller, E.J. (1981b). Potential surface properties and dynamics from molecular spectra: A time-dependent picture, in *Potential Energy Surfaces and Dynamics Calculations,* ed. D.G. Truhlar (Plenum Press, New York).

Heller, E.J. (1984). Bound-state eigenfunctions of classically chaotic Hamiltonian systems: Scars of periodic orbits, *Phys. Rev. Lett.* **53,** 1515–1518.

Heller, E.J. (1986). Qualitative properties of eigenfunctions of classically chaotic Hamiltonian systems, in *Quantum Chaos and Statistical Nuclear Physics,* ed. T.H. Seligman and H. Nishioka (Springer, Berlin).

Heller, E.J., Sundberg, R.L., and Tannor, D. (1982). Simple aspects of Raman scattering, *J. Phys. Chem.* **86,** 1822–1833.

Hennig, S., Engel, V., and Schinke, R. (1986). Vibrational state distributions following the photodissociation of (collinear) triatomic molecules: The vibrational reflection principle in model calculations for CF_3I, *J. Chem. Phys.* **84,** 5444–5454.

Hennig, S., Engel, V., Schinke, R., Nonella, M., and Huber, J.R. (1987). Photodissociation dynamics of methylnitrite (CH_3O-NO) in the 300–400 nm range: An *ab initio* quantum mechanical study, *J. Chem. Phys.* **87,** 3522–3529.

Hennig, S., Engel, V., Schinke, R., and Staemmler, V. (1988). Emission spectroscopy of photodissociating water molecules: A time-independent *ab initio* study, *Chem. Phys. Lett.* **149,** 455–462.

Hennig, S., Untch, A., Schinke, R., Nonella, M., and Huber, J.R. (1989). Theoretical investigation of the photodissociation dynamics of HONO: Vibrational predissociation in the electronically excited state S_1, *Chem. Phys.* **129,** 93–107.

Henriksen, N.E. (1988). The equivalence of time-independent and time-dependent calculational techniques for photodissociation probabilities, *Comments At. Mol. Phys.* **21,** 153–160.

Henriksen, N.E., Engel, V., and Schinke, R. (1987). Test of the Winger method for the photodissociation of symmetric triatomic molecules, *J. Chem. Phys.* **86,** 6862–6870.

Henriksen, N.E. and Heller, E.J. (1988). Gaussian wave packet dynamics and scattering in the interaction picture, *Chem. Phys. Lett.* **148,** 567–571.

Henriksen, N.E. and Heller, E.J. (1989). Quantum dynamics for vibrational and rotational degrees of freedom using Gaussian wave packets: Application to the three-dimensional photodissociation dynamics of ICN, *J. Chem. Phys.* **91,** 4700–4713.

Henriksen, N.E., Zhang, J., and Imre, D.G. (1988). The first absorption band for H_2O: Interpretation of the absorption spectrum using time dependent pictures, *J. Chem. Phys.* **89,** 5607–5613.

Herzberg, G. (1950). *Molecular Spectra and Molecular Structure I. Spectra of Diatomic Molecules* (Van Nostrand, New York).

Herzberg, G. (1967). *Molecular Spectra and Molecular Structure III. Spectra of Polyatomic Molecules* (Van Nostrand, New York).

Heumann, B., Düren, R., and Schinke, R. (1991). *Ab initio* calculation of the two lowest excited states of H_2S relevant for the photodissociation in the first continuum, *Chem. Phys. Lett.* **180,** 583–588.

Heumann, B., Weide, K., Düren, R., and Schinke, R. (1992). Nonadiabatic effects in the photodissociation of H_2S in the first absorption band: An *ab initio* study, submitted for publication.

Hillery, M., O'Connell, R.F., Scully, M.O., and Wigner, E.P. (1984). Distribution functions in physics: Fundamentals, *Physics Reports* **106,** 121–167.

Hirsch, G., Buenker, R.J., and Petrongolo, C. (1990). *Ab initio* study of NO_2. Part II: Non-adiabatic coupling between the two lowest $^2A'$ states and the construction of a diabatic representation, *Mol. Phys.* **70,** 835–848.

Hirst, D.M. (1985). *Potential Energy Surfaces* (Taylor and Francis, London).

Hirst, D.M. (1990). *A Computational Approach to Chemistry* (Blackwell, Oxford).

Ho, Y.K. (1983). The method of complex coordinate rotation and its application to atomic collision processes, *Phys. Rep.* **99,** 1–68.

Hodgson, A., Simons, J.P., Ashfold, M.N.R., Bayley, J.M., and Dixon, R.N. (1985). Quantum state-selected photodissociation dynamics in H_2O and D_2O, *Mol. Phys.* **54,** 351–368.

Hoinkes, H. (1980). The physical interaction potential of gas atoms with single-crystal surfaces, determined from gas-surface diffraction experiments, *Rev. Mod. Phys.* **52,** 933–970.

Holdy, K.E., Klotz, L.C. and Wilson, K.R. (1970). Molecular dynamics of photodissociation: Quasidiatomic model for ICN, *J. Chem. Phys.* **52,** 4588–4599.

Holle, A., Wiebusch, G., Main, J., Hager, B., Rottke, H., and Welge, K.H. (1986). Diamagnetism of the hydrogen atom in the quasi-Landau regime, *Phys. Rev. Lett.* **56,** 2594–2597.

Holle, A., Wiebusch, G., Main, J., Welge, K.H., Zeller, G., Wunner, G., Ertl, T., and Ruder, H. (1987). Hydrogenic Rydberg atoms in strong magnetic fields: Theoretical and experimental spectra in the transition region from regularity to irregularity, *Z. Phys. D* **5,** 279–285.

Houston, P.L. (1987). Vector correlations in photodissociation dynamics, *J. Phys. Chem.* **91,** 5388–5397.

Houston, P.L. (1989). Correlated photochemistry, the legacy of Johann Christian Doppler, *Acc. Chem. Res.* **22,** 309–314.

Howard, B.J. and Pine, A.S. (1985). Rotational predissociation and libration in the infrared spectrum of Ar-HCl, *Chem. Phys. Lett.* **122,** 1–8.

Huber, J.R. (1988). Photodissociation of molecules: The microscopic path of a molecular decay, *Pure and Appl. Chem.* **60,** 947–952.

Huber, D. and Heller, E.J. (1988). Hybrid mechanics: A combination of classical and quantum mechanics, *J. Chem. Phys.* **89,** 4752–4760.

Huber, D., Ling, S., Imre, D.G., and Heller, E.J. (1989). Hybrid mechanics. II, *J. Chem. Phys.* **90,** 7317–7329.

Hudson, R.D. (1971). Critical review of ultraviolet photoabsorption cross sections for molecules of astrophysical and aeronomic interest, *Rev. Geophys. Space Phys.* **9,** 305.

Hunt, R.H., Leacock, R.A., Peters, C.W., and Hecht, K.T. (1965). Internal-rotation in hydrogen peroxide: The far-infrared spectrum and the determination of the hindering potential, *J. Chem. Phys.* **42,** 1931–1946.

Hutson, J.M. (1990). Intermolecular forces from the spectroscopy of van der Walls molecules, *Ann. Rev. Phys. Chem.* **41,** 123–154.

Hutson, J.M. (1991). An introduction to the dynamics of van der Waals molecules, in *Advances in Molecular Vibrations and Collision Dynamics 1,* ed. J.M. Bowman (JAI Press, Greenwich).

Imre, D.G., Kinsey, J.L., Field, R.W., and Katayama, D.H. (1982). Spectroscopic characterization of repulsive potential energy surfaces: Fluorescence spectrum of ozone, *J. Phys. Chem.* **86,** 2564–2566.

Imre, D.G., Kinsey, J.L., Sinha, A., and Krenos, J. (1984). Chemical dynamics studied by emission spectroscopy of dissociating molecules, *J. Phys. Chem.* **88,** 3956–3964.

Imre, D.G. and Zhang, J. (1989). Dynamics and selective bond breaking in photodissociation, *Chem. Phys.* **139,** 89–121.

Jackson, W.H. and Okabe, H. (1986). Photodissociation dynamics of small molecules, in *Advances in Photochemistry,* Vol. 13, ed. D.H. Volman, G.S. Hammond, and K. Gollnick (Wiley, New York).

Jacob, M. and Wick, G.C. (1959). On the general theory of collisions for particles with spin, *Ann. Phys. (N.Y.)* **7,** 404–428.

Jacobs, A., Kleinermanns, K., Kuge, H., and Wolfrum, J. (1983). OH($X^2\Pi$) state distribution from HNO_3 and H_2O_2 photodissociation at 193 nm, *J. Chem. Phys.* **79,** 3162–3163.

Jacobs, A., Wahl, M., Weller, R., and Wolfrum, J. (1987). Rotational distribution of nascent OH radicals after H_2O_2 photolysis at 193 nm, *Appl. Phys. B* **42,** 173–179.

Janda, K.C. (1985). Predissociation of polyatomic van der Waals molecules, in *Photodissociation and Photoionization,* ed. K.P.Lawley (Wiley, New York).

Janda, K.C. and Bieler, C.R. (1990). Rotational rainbows, quantum interference, intramolecular vibrational relaxation and chemical reactions: All in rare gas-halogen molecules, in *Atomic and Molecular Clusters,* ed. E.R. Bernstein (Elsevier, Amsterdam).

Janssen, M.H.M., Bowman, R.M., and Zewail, A.H. (1990). Femtosecond temporal spectroscopy of ICl: Inversion to the $A^3\Pi_1$ state potential, *Chem. Phys. Lett.* **172,** 99–108.

Jaquet, R. (1987). Investigations with the finite element method. II. The collinear F + H_2 reaction, *Chem. Phys.* **118,** 17–23.

Jiang, X-.P., Heather, R., and Metiu H. (1989). Time dependent calculation of the absorption spectrum of a photodissociating system with two interacting excited electroni states, *J. Chem. Phys.* **90,** 2555–2569.

Johnson, B.R. and Kinsey, J.L. (1987). Time-dependent analysis of the Hartley absorption band and resonance Raman spectra in ozone, *J. Chem. Phys.* **87,** 1525–1537.

Johnson, B.R. and Kinsey, J.L. (1989a). Dynamical interpretation of the Hartley-absorption oscillations in O_3, *Phys. Rev. Lett.* **62,** 1607–1610.
Johnson, B.R. and Kinsey, J.L. (1989b). Recurrences in the autocorrelation function governing the ultraviolet absorption spectra of O_3, *J. Chem. Phys* **91,** 7638–7653.
Johnson, B.R., Kinsey, J.L., and Shapiro, M. (1988). A three-mode large amplitude model for the ground electronic state of CH_3I, *J. Chem. Phys.* **88,** 3147–3158.
Joswig, H., O'Halloran, M.A., Zare, R.N., and Child, M.S. (1986). Photodissociation dynamics of ICN, *Faraday Discuss. Chem. Soc.* **82,** 79–88.
Junker, B.R. (1982). Recent computational developments in the use of complex scaling in resonance phenomena, *Adv. At. Mol. Phys.* **18,** 207–263.
Kasper, J.V.V. and Pimentel, G.C. (1964). Atomic iodine photodissociation laser, *Appl. Phys. Lett.* **5,** 231.
Kerman, A.K. and Koonin, S.E. (1976). Hamiltonian formulation of time-dependent variational principles for the many-body system, *Annals Phys.* **100,** 332–358.
Khundkar, L.R. and Zewail, A.H. (1990). Ultrafast molecular reaction dynamics in real-time: Progress over a decade, *Ann. Rev. Phys. Chem.* **41,** 15–60.
Kidd, I.F. and Balint-Kurti, G.G. (1985). Theoretical calculation of photodissociation cross sections for the Ar-H_2 van der Waals complex, *J. Chem. Phys.* **82,** 93–105.
Kidd, I.F. and Balint-Kurti, G.G. (1986). Infrared predissociation of the Ar-HD van der Waals complex, *Faraday Discuss. Chem. Soc.* **82,** 241–250.
King, G.W. and Moule, D. (1962). The ultraviolet absorption spectrum of nitrous acid in the vapor state, *Can. J. Chem.* **40,** 2057–2065.
Kirby, K.P. and van Dishoeck, E.F. (1988). Photodissociation processes in diatomic molecules of astrophysical interest, *Adv. At. Mol. Phys.* **25,** 437–476.
Klee, S., Gericke, K.-H., and Comes, F.J. (1988). Photodissociation of H_2O_2/D_2O_2 from the lowest excited state: The origin of fragment rotation, *Ber. Bunsenges. Phys. Chem.* **92,** 429–434.
Kleinermanns, K., Linnebach, E., and Suntz, R. (1987). Emission spectrum of dissociating H_2S, *J. Phys. Chem.* **91,** 5543–5545.
Klessinger, M. and Michl, J. (1989). *Lichtabsorption und Photochemie organischer Moleküle* (VCH Verlagsgesellschaft, Weinheim).
Kleyn, A.W. and Horn, T.C.M. (1991). Rainbow scattering from solid surfaces, *Physics Reports* **199,** 191–230.
Köppel, H., Domcke, W., and Cederbaum, L.S. (1984). Multimode molecular dynamics beyond the Born-Oppenheimer approximation, *Adv. Chem. Phys.* **57,** 59–246.
Korsch, H.J. and Schinke, R. (1980). A uniform semiclassical sudden approximation for rotationally inelastic scattering, *J. Chem. Phys.* **73,** 1222-1232.
Korsch, H.J. and Wolf, F. (1984). Rainbow catastrophes in inelastic molecular collisions, *Comments At. Mol. Phys.* **15,** 139–154.
Kosloff, R. (1988). Time-dependent quantum-mechanical methods for molecular dynamics, *J. Phys. Chem.* **92,** 2087–2100.
Kosloff, R. and Kosloff, D. (1983). A Fourier method solution for the time dependent Schrödinger equation: A study of the reaction $H^+ + H_2$, $D^+ + HD$, and $D^+ + H_2$, *J. Chem. Phys.* **79,** 1823–1833.
Kotler, Z., Nitzan, A., and Kosloff, R. (1988). Multiconfiguration time-dependent self-consistent field approximation for curve crossing in presence of a bath. A fast Fourier transform study, *Chem. Phys. Lett.* **153,** 483–489.
Kouri, D.J. (1979). Rotational excitation II: Approximation methods, in *Atom-Molecule Collision Theory,* ed. R.B. Bernstein (Plenum Press, New York).
Kouri, D.J. and Mowrey, R.C. (1987). Close coupling-wave packet formalism for gas phase nonreactive atom-diatom collisions, *J. Chem. Phys.* **86,** 2087–2094.
Kramers, H.A. and Heisenberg, W. (1925). Über die Streuung von Strahlung durch Atome, *Z. Phys.* **31,** 681–708.
Krause, J.L., Shapiro, M., and Bersohn, R. (1991). Derivation of femtosecond pump-probe dissociation transients from frequency resolved data, *J. Chem. Phys.* **94,** 5499–5507.
Krautwald, H.J., Schnieder, L., Welge, K.H., and Ashfold, M.N. (1986). Hydrogen-atom photofragment spectroscopy, *Faraday Discuss. Chem. Soc.* **82,** 99–110.

Kresin, V.Z. and Lester, W.A., Jr. (1986). Quantum theory of polyatomic photodissociation, in *Advances in Photochemistry,* Vol. 13, ed. D.H. Volman, G.S. Hammond, and K. Gollnick (Wiley, New York).

Kučar, J. and Meyer, H.-D. (1989). Exact wave packet propagation using time-dependent basis sets, *J. Chem. Phys.* **90,** 5566–5577.

Kučar, J., Meyer, H.-D., and Cederbaum, L.S. (1987). Time-dependent rotated Hartree approach, *Chem. Phys. Lett.* **140,** 525–530.

Kühl, K. and Schinke, R. (1989). Time-dependent rotational state distributions in direct photodissociation, *Chem. Phys. Lett.* **158,** 81–86.

Kukulin, V.I., Krasnopolsky, K.M., and Horáček,J. (1989). *Theory of Resonances, Principles and Applications* (Kluwer, Dordrecht).

Kulander, K.C. (Ed.) (1991). Time-dependent methods for quantum dynamics, *Computer Phys. Comm.* **63,** 1–577.

Kulander, K.C., Cerjan, C., and Orel, A.E. (1991). Time-dependent calculations of molecular photodissociation resonances, *J. Chem. Phys.* **94,** 2571–2577.

Kulander, K.C. and Heller, E.J. (1978). Time dependent formulation of polyatomic photofragmentation: Application to H_3^+, *J. Chem. Phys.* **69,** 2439–2449.

Kulander, K.C. and Light, J.C. (1980). Photodissociation of triatomic molecules: Application of the R-matrix propagation methods to the calculation of bound-free Franck-Condon factors, *J. Chem. Phys.* **73,** 4337–4346.

Kulander, K.C. and Light, J.C. (1986). Theory of polyatomic photodissociation in the reactive infinite order sudden approximation: Application to the Rydberg states of H_3, *J. Chem. Phys.* **85,** 1938–1949.

Kuntz, P.J. (1976). Features of potential energy surfaces and their effect on collisions, in *Dynamics of Molecular Collisions,* Part B, ed. W.H. Miller (Plenum Press, New York).

Kuntz, P.J. (1979). Interaction potentials II: Semiempirical atom-molecule potentials for collision theory, in *Atom-Molecule Collision Theory,* ed. R.B. Bernstein (Plenum Press, New York).

Kuppermann, A. (1981). Reactive scattering resonances and their physical interpretation: The vibrational structure of the transition state, in *Potential Energy Surfaces and Dynamics Calculations,* ed. D.G. Truhlar (Plenum Press, New York).

Lahmani, F., Lardeux, C., and Solgadi, D. (1986). Rotational and electronic anisotropy in NO $X^2\Pi$ from the photodissociation of CH_3ONO, *Chem. Phys. Lett.* **129,** 24–30.

Lao, K.Q., Person, M.D., Xayariboun, P., and Butler, L.J. (1989). Evolution of molecular dissociation through an electronic curve crossing: Polarized emission spectroscopy of CH_3I at 266 nm, *J. Chem. Phys.* **92,** 823–841.

Launay, J.-M (1976). Body-fixed formulation of rotational excitation: Exact and centrifugal decoupling results for CO-He, *J. Phys. B* **9,** 1823–1838.

Lavi, R. and Rosenwaks, S. (1988). Dynamics of photofragmentation of dimethylnitrosamine from its first two excited singlet states, *J. Chem. Phys.* **89,** 1416–1426.

Lawley, K.P. (Ed.) (1987). *Ab Initio Methods in Quantum Chemistry,* Parts I and II (Wiley, Chichester).

Lax, M. (1952). The Franck-Condon principle and its application to crystals, *J. Chem. Phys.* **20,** 1752–1760.

Lee, S.-Y. and Heller, E.J. (1979). Time-dependent theory of Raman scattering, *J. Chem. Phys.* **71,** 4777–4788.

Lee, S.-Y. and Heller, E.J. (1982). Exact time-dependent wave packet propagation: Application to the photodissociation of methyl iodide, *J. Chem. Phys.* **76,** 3035–3044.

Lee, L.C. and Judge, D.L. (1973). Population distribution of triplet vibrational levels of CO produced by photodissociation of CO_2, *Can. J. Phys.* **51,** 378–381.

Lee, L.C. and Suto, M. (1986). Quantitative photoabsorption and fluorescence study of H_2O and D_2O at 50–190 nm, *Chem. Phys.* **110,** 161–169.

Lee, L.C., Wang, X. and Suto M. (1987). Quantitative photoabsorption and fluorescence spectroscopy of H_2S and D_2S at 49–240 nm, *J. Chem. Phys.* **86,** 4353–4361.

Lefebvre, R. (1985). Box quantization and resonance determination. The multidimensional case, *J. Phys. Chem.* **89,** 4201–4206.

Lefebvre-Brion, H. and Field, R.W. (1986). *Perturbations in the Spectra of Diatomic Molecules* (Academic Press, Orlando).

Leforestier, C. (1991). Grid representation of rotating triatomics, *J. Chem. Phys.* **94,** 6388-6397.

Leforestier, C., Bisseling, R., Cerjan, C., Feit, M.D., Friesner, R., Guldberg, A., Hammerich, A., Jolicard, G., Karrlein, W., Meyer, H.-D., Lipkin, N., Roncero, O., and Kosloff, R. (1991). A comparison of different propagation schemes for the time dependent Schrödinger equation, *J. Comput. Phys.* **94,** 59–80.

Legon, A.C. and Millen; D.J. (1986). Gas-phase spectroscopy and the properties of hydrogen-bonded dimers: HCN···HF as the spectroscopic prototype, *Chem. Rev.* **86,** 635–657.

Lengsfield III, B.H., Saxe, P., and Yarkony, D.R. (1984). On the evaluation of nonadiabatic coupling matrix elements using SA-MCSCF/CI wavefunctions and analytic gradient methods. I, *J. Chem. Phys.* **81,** 4549–4553.

Lengsfield III, B.H. and Yarkony, D.R. (1986). On the evaluation of nonadiabatic coupling matrix elements using SA-MCSCF/CI wavefunctions and analytic gradient methods. III: Second derivative terms, *J. Chem. Phys.* **84,** 348–353.

Leone, S.R. (1982). Photofragment dynamics, *Adv. Chem. Phys.* **50,** 255–324.

Le Quéré, F. and Leforestier, C. (1990). Quantum exact three-dimensional study of the photodissociation of the ozone molecule, *J. Chem. Phys.* **92,** 247–253.

Le Quéré, F. and Leforestier, C. (1991). Hyperspherical formulation of the photodissociation of ozone, *J. Chem. Phys.* **94,** 1118–1126.

Le Roy, R.J. (1984). Vibrational predissociation of small van der Waals molecules, in *Resonances in Electron-Molecule Scattering, van der Waals Molecules, and Reactive Chemical Dynamics,* ed. D.G. Truhlar (American Chemical Society, Washington, D.C.).

Le Roy, R.J. and Carley, J.S. (1980). Spectroscopy and potential energy surfaces of van der Waals molecules, *Adv. Chem. Phys.* **42,** 353–420.

Lester, W.A., Jr. (1976). The N coupled-channel problem, in *Dynamics of Molecular Collisions,* Part A, ed. W.H. Miller (Plenum Press, New York).

Letokhov, V.S. (1983). *Nonlinear Laser Chemistry* (Springer, Berlin).

Levene, H.B., Nieh, J.-C., and Valentini, J.J. (1987). Ozone visible photodissociation dynamics, *J. Chem. Phys.* **87,** 2583–2593.

Levene, H.B. and Valentini, J.J. (1987). The effect of parent motion on photofragment rotational distributions: Vector correlation of angular momenta and C_{2v} symmetry breaking in dissociation of AB_2 molecules, *J. Chem. Phys.* **87,** 2594–2610.

Levine, I.N. (1975). *Molecular Spectroscopy* (Wiley, New York).

Levine, R.D. and Bernstein, R.B. (1987). *Molecular Reaction Dynamics and Chemical Reactivity* (Oxford University Press, Oxford).

Levy, D.H. (1980). Laser spectroscopy of cold gas-phase molecules, *Ann. Rev. Phys. Chem.* **31,** 197–225.

Levy, D.H. (1981). Van der Waals molecules, *Adv. Chem. Phys.* **47,** 323–362.

Light, J.C. (1979). Inelastic scattering cross sections I: Theory, in *Atom-Molecule Collision Theory,* ed. R.B. Bernstein (Plenum Press, New York).

Likar, M.D., Sinha, A., Ticich, T.M., Vander Wal, R.L., and Crim, F.F. (1988). Spectroscopy and photodissociation dynamics of highly vibrationally excited molecules, *Ber. Bunsenges. Phys. Chem.* **92,** 289–295.

Loisell, W.H. (1973). *Quantum Statistical Properties of Radiation* (Wiley, New York).

Loison, J.-C., Kable, S.H., Houston, P.L., and Burak, I. (1991). Photofragment excitation spectroscopy of the formyl (HCO/DCO) radical: Linewidths and predissociation rates of the $\tilde{A}(A'')$ state, *J. Chem. Phys.* **94,** 1796–1802.

Loudon, R. (1983). *The Quantum Theory of Light* (Oxford University Press, Oxford).

Lovejoy, C.M. and Nesbitt, D.J. (1990). Mode specific internal and direct rotational predissociation in HeHF, HeDF, and HeHCl: van der Waals complexes in the weak binding limit, *J. Chem. Phys.* **93,** 5387–5407.

Lovejoy, C.M. and Nesbitt, D.J. (1991). Rotational predissociation, vibrational mixing, and van der Waals intermolecular potentials of NeDF, *J. Chem. Phys.* **94,** 208–223.

Lowe, J.P. (1978). *Quantum Chemistry* (Academic Press, New York).

Luo, X. and Rizzo, T.R. (1990). Rotationally resolved vibrational overtone spectroscopy of hydrogen peroxide at chemically significant energies, *J. Chem. Phys.* **93,** 8620–8633.

Luo, X. and Rizzo, T.R. (1991). Unimolecular dissociation of hydrogen peroxide from single rovibrational states near threshold, *J. Chem. Phys.* **94,** 889–898.

Lupo, D.W. and Quack, M. (1987). IR-laser photochemistry, *Chem. Rev.* **87,** 181–216.

Macpherson, M.T. and Simons, J.P. (1978). Spectroscopic study of the predissociation $H(D)CN(\tilde{C}^1A') \rightarrow H(D) + CN(B^2\Sigma^+)$, *J. Chem. Soc., Faraday Trans. 2* **74,** 1965–1977.

Main, J., Holle, A., Wiebusch, G., and Welge, K.H. (1987). Semiclassical quantization of three-dimensional quasi-Landau resonances under strong-field mixing, *Z. Phys. D* **6,** 295–302.

Main, J., Wiebusch, G., Holle, A., and Welge, K.H. (1986). New quasi-Landau structure of highly excited atoms: The hydrogen atom, *Phys. Rev. Lett.* **57,** 2789–2792.

Makri, N. and Miller, W.H. (1987). Time-dependent self-consistent field (TDSCF) approximation for a reaction coordinate coupled to a harmonic bath: Single and multiple configuration treatments, *J. Chem. Phys.* **87,** 5781–5787.

Manthe U. and Köppel, H. (1990a). New method for calculating wave packet dynamics: Strongly coupled surfaces and the adiabatic basis, *J. Chem. Phys.* **93,** 345–356.

Manthe, U. and Köppel, H. (1990b). Dynamics on potential energy surfaces with a conial intersection: Adiabatic, intermediate, and diabatic behavior, *J. Chem. Phys.* **93,** 1658–69.

Manthe, U. and Köppel, H. (1991). Three-dimensional wave-packet dynamics on vibronically coupled dissociative potential energy surfaces, *Chem. Phys. Lett.* **178,** 36–42.

Manthe, U., Köppel, H., and Cederbaum, L.S. (1991). Dissociation and predissociation on coupled electronic potential energy surfaces: A three-dimensional wave packet dynamical study, *J. Chem. Phys.* **95,** 1709–1720.

Manz, J. (1985). Molecular dynamics along hyperspherical coordinates, *Comments At. Mol. Phys.* **17,** 91–113.

Manz, J. (1989). Mode selective bimolecular reactions, in *Molecules in Physics, Chemistry, and Biology,* Vol. III, ed. J. Maruani (Kluwer, Dordrecht).

Manz, J., Meyer, R., Pollak, E., Römelt, J., and Schor, H.H.R. (1984). On spectroscopic properties and isotope effects of vibrationally stabilized molecules, *Chem. Phys.* **83,** 333–343.

Manz, J. and Parmenter, C.S. (Eds.) (1989). Mode selectivity in unimolecular reactions, *Chem. Phys.* **139,** 1–506.

Manz, J. and Römelt, J. (1981). On the collinear I + HI and I + MuI reactions, *Chem. Phys. Lett.* **81,** 179–184.

Marcías, A. and Riera, A. (1978). Calculation of diabatic states from molecular properties, *J. Phys. B* **11,** L489–L492.

Marcus, R.A. (1988). Semiclassical wavepackets in the angle representation and their role in molecular dynamics, *Chem. Phys. Lett.* **152,** 8–13.

Margenau, H. and Kestner, N.R. (1969). *Theory of Intermolecular Forces* (Pergamon Press, Oxford).

Marinelli, W.J., Sivakumar, N., and Houston, P.L. (1984). Photodissociation dynamics of nozzle-cooled ICN, *J. Phys. Chem.* **88,** 6685–6692.

Marston, C.C. and Wyatt, R.E. (1984*a*). Resonant quasi-periodic and periodic orbits for the three-dimensional reaction of fluorine atoms with hydrogen molecules, in *Resonances in Electron-Molecule Scattering, van der Waals Molecules, and Reactive Chemical Dynamics,* ed. D.G. Truhlar (American Chemical Society, Washington, D.C.).

Marston, C.C., and Wyatt, R.E. (1984*b*). Semiclassical theory of resonances in 3D chemical reations I:Resonant periodic orbits for $F+H_2$, *J. Chem. Phys.* **81,** 1819.

Marston, C.C. and Wyatt, R.E. (1985). Semiclassical theory of resonances in 3D chemical reactions II:Resonant quasiperiodic orbits for F and H_2, *J. Chem. Phys.* **83,** 3390.

Maul, C., Gläser, H., and Gericke, K.-H. (1989) *J. Chem. Soc., Faraday Trans. 2* **85,** 1297–1300.

McCurdy, C.W. and Miller, W.H. (1977). Interference effects in rotational state distributions: Propensity and inverse propensity, *J. Chem. Phys.* **67,** 463–468.

McGuire, P. and Kouri, D.J. (1974). Quantum mechanical close coupling approach to molecular collisions. j_z-conserving coupled states approximation, *J. Chem. Phys.* **60,** 2488–2499.

Mead, C.A. and Truhlar, D.G. (1982). Conditions for the definition of a strictly diabatic electronic basis for molecular systems, *J. Chem. Phys.* **77,** 6090–6098.

Meier, C.H., Cederbaum, L.S., and Domcke, W. (1980). A spherical-box approach to resonances, *J. Phys. B.* **13,** L119–L124.

Merzbacher, E. (1970). *Quantum Mechanics* (Wiley, New York).

Messiah, A. (1972). *Quantum Mechanics,* Vols. I and II (North-Holland, Amsterdam).

Metiu, H. (1991). Coherence and transients in photodissociation with short pulses, *Faraday Discuss. Chem. Soc* **91,** 249–258.

Metiu, H.M. and Engel, V. (1990). Coherence, transients, and interference in photodissociation with ultrashort pulses, *J. Opt. Soc. Am. B.* **7,** 1709–1726.

Metz, R.B., Kitsopoulos, T., Weaver, A., and Neumark, D.M. (1988). Study of the transition state region in the Cl + HCl reaction by photoelectron spectroscopy of $ClHCl^-$, *J. Chem. Phys.* **88,** 1463–1465.

Metz, R.B., Weaver, A., Bradforth, S.E., Kitsopoulos, T.N., and Neumark, D.M. (1990). Probing the transition state with negative ion photodetachment: The Cl + HCl and Br + HBr reactions, *J. Phys. Chem.* **94,** 1377–1388.

Meyer, H.-D., Kučar, J., and Cederbaum, L.S. (1988). Time-dependent rotated Hartree: Formal development, *J. Math. Phys.* **29,** 1417–1430.

Meyer, H.-D., Manthe, U., and Cederbaum, L.S. (1990). The multi-configurational time-dependent Hartree approach, *Chem. Phys. Lett.* **165,** 73–78.

Miller, R.E. (1988). The vibrational spectroscopy and dynamics of weakly bound neutral complexes, *Science* **240,** 447–453.

Miller, R.L., Kable, S.H., Houston, P.L., and Burak, I. (1992). Product distributions in the 157-nm photodissociation of CO_2, *J. Chem. Phys.* **96,** 32.

Miller, W.H. (1974). Classical limit quantum mechanics and the theory of molecular collisions, *Adv. Chem. Phys* **25,** 69–177.

Miller, W.H. (1975). The classical S-matrix in molecular collisions, *Adv. Chem. Phys.* **30,** 77–136.

Miller, W.H. (1985). Semiclassical methods in chemical dynamics, in *Semiclassical Descriptions of Atomic and Nuclear Collisions,* ed. J. Bang and J. de Boer (Elsevier, Amsterdam).

Miller, W.H. (1990). Recent advances in quantum mechanical reactive scattering theory, including comparison of recent experiments with rigorous calculations of state-to-state cross sections for the H/D + $H_2 \rightarrow H_2$/HD + H reactions, *Ann. Rev. Phys. Chem.* **41,** 245–281.

Mitchell, R.C. and Simons, J.P. (1967). Energy distribution among the primary products of photo-dissociation, *Faraday Discuss. Chem. Soc.* **44,** 208–217.

Mladenović, M. and Bačić, Z. (1990). Highly excited vibration-rotation states of floppy triatomic molecules by a localized representation method: The HCN/HNC molecule, *J. Chem. Phys.* **93,** 3039–3053.

Mohan, V. and Sathyamurthy, N. (1988). Quantal wavepacket calculations of reactive scattering, *Computer Physics Reports* **7,** 213–258.

Moiseyev, N. (1984). The Hermitian representation of the complex coordinate method: Theory and application, in *Lecture Notes in Physics,* Vol. 211, ed. S. Albeverino, L.S. Ferreira, and L. Streit (Springer, Heidelberg).

Moore, C.B. and Weisshaar, J.C. (1983). Formaldehyde photochemistry, *Ann. Rev. Phys. Chem.* **34,** 525–555.

Moore, D.S., Bomse, D.S., and Valentini, J.J. (1983). Photofragment spectroscopy and dynamics of the visible photodissociation of ozone, *J. Chem. Phys.* **79,** 1745–1757.

Morse, M.D. and Freed, K.F. (1980). Rotational distributions in photodissociation: The bent triatomic molecule, *Chem. Phys. Lett.* **74,** 49–55.

Morse, M.D. and Freed, K.F. (1981). Rotational and angular distributions from photodissociations. III. Effects of dynamic axis switching in linear triatomic molecules, *J. Chem. Phys.* **74,** 4395–4417.

Morse, M.D. and Freed, K.F. (1983). Rotational distributions from photodissociations. IV. The bent triatomic molecule, *J. Chem. Phys.* **78,** 6045–6065.

Morse, M.D., Freed, K.F., and Band, Y.B. (1979). Rotational distributions from photodissociations. I. Linear triatomic molecules, *J. Chem. Phys.* **70,** 3604–3619.

Mowrey, R.C. and Kouri, D.J. (1985). On a hybrid close-coupling wave packet approach to molecular scattering, *Chem. Phys. Lett.* **119,** 285–289.

Mowrey, R.C., Sun, Y., and Kouri, D.J. (1989). A numerically exact full wave packet approach to molecule-surface scattering, *J. Chem. Phys.* **91,** 6519–6524.

Murrell, J.N., Carter, S., Farantos, S.C., Huxley, P., and Varandas, A.J.C. (1984). *Molecular Potential Energy Functions* (Wiley, Chichester).

Myers, A.B., Mathies, R.A., Tannor, D.J., and Heller, E.J. (1982). Excited state geometry changes from preresonance Raman intensities: Isoprene and hexatriene, *J. Chem. Phys.* **77,** 3857–3866.

Nadler, I., Mahgerefteh, D., Reisler, H., and Wittig, C. (1985). The 266 nm photolysis of ICN: Recoil velocity anisotropies and nascent E,V,R,T excitations for the CN + I($^2P_{3/2}$) and CN + I($^2P_{1/2}$) channels, *J. Chem. Phys.* **82,** 3885–3893.

Nadler, I., Noble, M., Reisler, H., and Wittig, C. (1985). The monoenergetic vibrational predissociation of expansion cooled NCNO: Nascent CN(V,R) distributions at excess energies 0–5000 cm^{-1}, *J. Chem. Phys.* **82,** 2608–2619.

Negele, J.W. (1982). The mean-field theory of nuclear structure and dynamics, *Rev. Mod. Phys.* **54,** 913–1015.

Nesbitt, D.J. (1988a). High-resolution infrared spectroscopy of weakly bound molecular complexes, *Chem. Rev.* **88,** 843–870.

Nesbitt, D.J. (1988b). Spectroscopic signature of floppiness in molecular complexes, in *The Structure of Small Molecules and Ions,* ed. Naaman, R. and Vager, Z. (Plenum, New York).

Nesbitt, D.J., Child, M.S., and Clary, D.C. (1989). Rydberg-Klein-Rees inversion of high resolution van der Waals infrared spectra: An intermolecular potential energy surface for Ar+HF($v = 1$), *J. Chem. Phys.* **90,** 4855–4864.

Nesbitt, D.J., Lovejoy, C.M., Lindemann, T.G., ONeil, S.V., and Clary, D.C. (1989). Slit jet infrared spectroscopy of NeHF complexes: Internal rotor and J-dependent predissociation dynamics, *J. Chem. Phys.* **91,** 722–731.

Neumark, D.M. (1990). Transition state spectroscopy of hydrogen transfer reactions, in *Electronic and Atomic Collisions: Invited Papers of the XVI ICPEAC,* ed. A. Dalgarno, R.S. Freund, P. Koch, M.S. Lubell, and T.B. Lucatorto (American Institute of Physics, New York).

Noid, D.W., Koszykowski, M.L., and Marcus, R.A. (1981). Quasiperiodic and stochastic behavior in molecules, *Ann. Rev. Phys. Chem.* **32,** 267–309.

Nonella, M. and Huber, J.R. (1986). Photodissociation of methylnitrite: An MCSCF calculation of the S_1 potential surface, *Chem. Phys. Lett.* **131,** 376–379.

Nonella, M., Huber, J.R., Untch, A., and Schinke, R. (1989). Photodissociation of CH_3ONO in the first absorption band: A three-dimensional classical trajectory study, *J. Chem. Phys.* **91,** 194–204.

Ogai A., Brandon, J., Reisler, H., Suter, H.U., Huber, J.R., von Dirke, M., and Schinke, R. (1992). Mapping of parent transition-state wavefunctions into product rotations: An experimental and theoretical investigation of the photodissociation of FNO, *J. Chem. Phys.* **96,** 6643–6653.

Ogai, A., Qian, C.X.W., and Reisler, H. (1990). Photodissociation dynamics of jet-cooled ClNO on $S_1(1^1A'')$: An experimental study, *J. Chem. Phys.* **93,** 1107–1115.

Okabe, H. (1978). *Photochemistry of Small Molecules* (Wiley, New York).

ONeil, S.V., Nesbitt, D.J., Rosmus, P., Werner, H.-J., and Clary, D.C. (1989). Weakly bound NeHF, *J. Chem. Phys.* **91,** 711–721.

Ondrey, G., van Veen, N., and Bersohn, R. (1983). The state distribution of OH radicals photodissociated from H_2O_2 at 193 and 248 nm, *J. Chem. Phys.* **78,** 3732–3737.

Pacher, T., Cederbaum, L.S., and Köppel, H. (1988). Approximately diabatic states from block diagonalization of the electronic Hamiltonian, *J. Chem. Phys.* **89,** 7367–7381.

Pack, R.T. (1974). Space-fixed vs body-fixed axes in atom-diatomic molecule scattering. Sudden approximations, *J. Chem. Phys.* **60,** 633–639.

Pack, R.T. (1976). Simple theory of diffuse vibrational structure in continuous uv spectra of polyatomic molecules. I. Collinear photodissociation of symmetric triatomics, *J. Chem. Phys.* **65,** 4765–4770.

Parker, G.A. and Pack, R.T. (1978). Rotationally and vibrationally inelastic scattering in the rotational IOS approximation. Ultrasimple calculation of total (differential, integral, and transport) cross sections for nonspherical molecules, *J. Chem. Phys.* **68,** 1585–1601.

Pattengill, M.D. (1979). Rotational excitation III: Classical trajectory methods, in *Atom-Molecule Collision Theory,* ed. R.B. Bernstein (Plenum Press, New York).

Pauly, H. (1979). Elastic scattering cross sections I: Spherical potentials, in *Atom-Molecule Collision Theory,* ed. R.B. Bernstein (Plenum Press, New York).

Pechukas, P. (1976). Statistical approximations in collision theory, in *Dynamics of Molecular Collisions, Part B,* ed. W.H. Miller (Plenum Press, New York).

Pechukas, P. and Light, J.C. (1965). On detailed balancing and statistical theories of chemical kinetics, *J. Chem. Phys.* **42,** 3281–3291.

Pechukas, P., Light, J.C., and Rankin, C. (1966). Statistical theory of chemical kinetics: Application to neutral-atom-molecule reactions, *J. Chem. Phys.* **44,** 794–805.

Pernot, P., Atabek, O., Beswick, J.A., and Millié, Ph. (1987). A quantum mechanical model study of CH_3NO_2 fragmentation dynamics, *J. Phys. Chem.* **91,** 1397–1399.

Person, M.D., Lao, K.Q., Eckholm, B.J., and Butler, L.J. (1989). Molecular dissociation dynamics of H_2S at 193.3 nm studied via emission spectroscopy, *J. Chem. Phys.* **91,** 812–820.

Petrongolo, C., Buenker, R.J., and Peyerimhoff, S.D. (1982). Nonadiabatic treatment of the intensity distribution in the V-N bands of ethylene, *J. Chem. Phys.* **76,** 3655–3667.

Petrongolo, C., Hirsch, G., and Buenker, R.J. (1990). Diabatic representation of the $\tilde{A}^2A_1/\tilde{B}^2B_2$ conical intersection in NH_2, *Mol. Phys.* **70,** 825–834.

Philippoz, J.-M., Monot, R., and van den Bergh, H. (1990). Multiple oscillations observed in the rotational state population of $I_2(B)$ formed in the photodissociation of $(I_2)_2$, *J. Chem. Phys.* **92,** 288–291.

Pilling, M.J. and Smith, I.W.M. (1987). *Modern Gas Kinetics* (Blackwell Scientific Publications, Oxford).

Pollak, E. (1981). Periodic orbits, adiabaticity and stability, *Chem. Phys.* **61,** 305–316.

Pollak, E. (1985). Periodic orbits and the theory of reactive scattering in *Theory of Chemical Reaction Dynamics*, Vol. III, ed. M. Baer (CRC Press, Boca Raton).

Pollak, E. and Child, M.S. (1981). A simple classical prediction of quantal resonances in collinear reactive scattering, *Chem. Phys.* **60,** 23–32.

Pollak, E., Child, M.S., and Pechukas, P. (1980). Classical transition state theory: A lower bound to the reaction probability, *J. Chem. Phys.* **72,** 1669–1678.

Pollard, W.T., Lee, S.Y., and Mathies, R.A. (1990). Wave packet theory of dynamic absorption spectra in femtosecond pump-probe experiments, *J. Chem. Phys.* **92,** 4012–4028.

Porter, R.N. and Raff, L.M. (1976). Classical trajectory methods in molecular collisions, in *Dynamics of Molecular Collisions,* Part B, ed. W.H. Miller (Plenum Press, New York).

Pritchard, H.O. (1985). State-to-state theory of unimolecular reactions, *J. Phys. Chem.* **89,** 3970–3976.

Qian, C.X.W., Noble, M., Nadler, I., Reisler, H., and Wittig, C. (1985). NCNO $\rightarrow$ CN + NO: Complete NO(E,V,R) and CN(V,R) nascent population distributions from well-characterized monoenergetic unimolecular reactions, *J. Chem. Phys.* **83,** 5573–5580.

Qian, C.X.W., Ogai, A., Brandon, J., Bai, Y.Y., and Reisler, H. (1991). The spectroscopy and dynamics of fast evolving states, *J. Phys. Chem.* **95,** 6763.

Qian, C.X.W., Ogai, A., Iwata, L., and Reisler, H. (1988). State-selective photodissociation dynamics of NOCl: The influence of excited state bending and stretching vibrations, *J. Chem. Phys.* **89,** 6547–6548.

Qian, C.X.W., Ogai, A., Iwata, L., and Reisler, H. (1990). The influence of excited-state vibrations on fragment state distributions: The photodissociation of NOCl on $T_1(1^3A'')$, *J. Chem. Phys.* **92,** 4296–4307.

Qian, C.X.W. and Reisler, H. (1991). Photodissociation dynamics of the nitrosyl halides: The influence of parent vibrations, in *Advances in Molecular Vibrations and Collision Dynamics 1,* ed. J.M. Bowman (JAI Press, Greenwich).

Quack, M. and Troe, J. (1974). Specific rate constants of unimolecular processes II. Adiabatic channel model, *Ber. Bunsenges. Phys. Chem.* **78,** 240–252.

Quack, M. and Troe, J. (1975). Complex formation in reactive and inelastic scattering: Statistical adiabatic channel model of unimolecular processes III, *Ber. Bunsenges. Phys. Chem.* **79,** 170–183.

Quack, M. and Troe, J. (1977). Unimolecular reactions and energy transfer of highly excited molecules, in *Gas Kinetics and Energy Transfer,* ed. P.G. Ashmore and R.J. Donovan (The Chemical Society, London).

Quack, M. and Troe, J. (1981). Statistical methods in scattering, in *Theoretical Chemistry,* Vol. 6b: *Theory of Scattering,* ed. D. Henderson (Academic Press, New York).

Quinton, A.M. and Simons, J.P. (1981). Λ-doublet population inversions in $NH(c^1\Pi)$ excited through vacuum ultraviolet dissociation of NH_3, *Chem. Phys. Lett.* **81,** 214–217.

Rabalais, J.W., McDonald, J.M., Scherr, V., and McGlynn, S.P. (1971). Electronic spectroscopy of isoelectronic molecules. II Linear triatomic groupings containing sixteen valence electrons, *Chem. Rev.* **71,** 73–108.

Rabitz, H. (1976). Effective Hamiltonians in molecular collisions, in *Dynamics of Molecular Collisions,* Part A, ed. W.H. Miller (Plenum Press, New York).

Raff, L.M. and Thompson, D.L. (1985). The classical trajectory approach to reactive scattering, in Theory of Chemical Reaction Dynamics, Vol. III, ed. M. Baer (CRC Press, Bow Ruton).

Rama Krishna, M.V. and Coalson, R.D. (1988). Dynamic aspects of electronic excitation, *Chem. Phys.* **120,** 327–333.

Rebentrost, F. (1981). Nonadiabatic molecular collisions, in *Theoretical Chemistry,* Vol. 6b: *Theory of Scattering,* ed. Henderson (Academic Press, New York).

Reid, B.P., Janda, K.C., and Halberstadt, N. (1988). Vibrational and rotational wave functions for the triatomic van der Waals molecules $HeCl_2$, $NeCl_2$, and $ArCl_2$, *J. Phys. Chem.* **92,** 587–593.

Reinhardt, W.P. (1982). Complex coordinates in the theory of atomic and molecular structure and dynamics, *Ann. Rev. Phys. Chem.* **33,** 223–255.

Reinhardt, W.P. (1983). A time-dependent approach to the magnetic-field-induced redistribution of oscillator strength in atomic photoabsorption, *J. Phys. B* **16,** L635–L641.

Reisler, H., Noble, M., and Wittig, C. (1987). Photodissociation processes in NO-containing molecules, in *Molecular Photodissociation Dynamics,* ed. M.N.R. Ashfold and J.E. Baggott (Royal Society of Chemistry, London).

Reisler, H. and Wittig, C. (1985). Multiphoton ionization of gaseous molecules, *Adv. Chem. Phys.* **60,** 1.

Reisler, H. and Wittig, C. (1986). Photo-initiated unimolecular reactions, *Ann. Rev. Phys. Chem.* **37,** 307–349.

Reisler, H. and Wittig, C. (1992). State-resolved simple bond-fission reactions: Experiment and theory, in *Advances in Chemical Kinetics and Dynamics,* ed. J.A. Barker (JAI Press, Greenwich).

Rizzo, T.R., Hayden, C.C., and Crim, F.F. (1984). State-resolved product detection in the overtone vibration initiated unimolecular decomposition of $HOOH(6\nu_{OH})$, *J. Chem. Phys.* **81,** 4501–4509.

Robin, M.B. (1974, 1975, 1985). *Higher Excited States in Polyatomic Molecules,* Vols. I, II, and III (Academic Press, New York).

Robra, U., Zacharias, H., and Welge, K.H. (1990). Near threshold channel selective photodissociation of NO_2, *Z. Phys. D* **16,** 175–188.

Rodwell, W.R., Sin Fain Lam, L.T., and Watts, R.O. (1981). The helium-hydrogen fluoride potential surface, *Mol. Phys.* **44,** 225–240.

Römelt, J. (1983). Prediction and interpretation of collinear reactive scattering resonances by the diagonal corrected vibrational adiabatic hyperspherical model, *Chem. Phys.* **79,** 197–209.

Roncero, O., Beswick, J.A., Halberstadt, N., Villarreal, P., and Delgado-Barrio, G. (1990). Photofragmentation of the Ne$\cdots$ ICL complex: A three-dimensional quantum mechanical study, *J. Chem. Phys.* **92,** 3348–3358.

Rose, T.S., Rosker, M.J., and Zewail, A.H. (1988). Femtosecond real-time observation of wavepacket oscillations (resonances) in dissociation reactions, *J. Chem. Phys.* **88,** 6672–6673.

Rosker, M.J., Rose, T.S., and Zewail, A.H. (1988). Femtosecond real-time dynamics of photofragment-trapping resonances on dissociative potential energy surfaces, *Chem. Phys. Lett.* **146,** 175–179.

Rosmus, P., Botschwina, P., Werner, H.-J., Vaida, V., Engelking, P.C., and McCarthy, M.I. (1987). Theoretical $A^1A_2'' - X^1A_1$ absorption and emission spectrum of ammonia, *J. Chem. Phys.* **86,** 6677–6692.

Rousseau, D.L. and Williams, P.F. (1976). Resonance Raman scattering of light from a diatomic molecule, *J. Chem. Phys.* **64,** 3519–3537.

Russell, J.A., McLaren, I.A., Jackson, W.M., and Halpern, J.B. (1987). Photolysis of BrCN between 193 and 266 nm, *J.Phys. Chem.* **91,** 3248–3253.

Salem, L. (1982). *Electrons in Chemical Reactions: First Principles* (Wiley, New York).

Satchler, G.R. (1990). *Nuclear Reactions*, second edition (Oxford University Press, New York).

Sawada, S.-I., Heather, R., Jackson, B., and Metiu, H. (1985). A strategy for time dependent quantum mechanical calculations using a Gaussian wave packet representation of the wave function, *J. Chem. Phys.* **83,** 3009–3027.

Saxe, P. and Yarkony, D.R. (1987). On the evaluation of nonadiabatic coupling matrix elements for MCSCF/CI wave functions, IV. Second derivative terms using analytic gradient methods, *J. Chem. Phys.* **86,** 321–328.

Schaefer III, H.F. (1979). Interaction potentials I: Atom-molecule potentials, in *Atom-Molecule Collision Theory,* ed. R.B. Bernstein (Plenum Press, New York).

Schatz, G.C. (1983). Quasiclassical trajectory studies of state to state collisional energy transfer in polyatomic molecules, in *Molecular Collision Dynamics,* ed. J.M. Bowman (Springer, Berlin).

Schatz, G.C. (1989a). A three dimensional reactive scattering study of the photodetachment spectrum of $ClHCl^-$, *J. Chem. Phys.* **90,** 3582–3589.

Schatz, G.C. (1989b). A three dimensional quantum reactive scattering study of the I + HI reaction and of the IHI^- photodetachment spectrum, *J. Chem. Phys.* **90,** 4847–4854.

Schatz, G.C. (1990). Quantum theory of photodetachment spectra of transition states, *J. Phys. Chem.* **94,** 6157–6164.

Schinke, R. (1985). Rainbows in CO rotational distributions following photofragmentation of formaldehyde, *Chem. Phys. Lett.* **120,** 129–134.

Schinke, R. (1986a). Semiclassical analysis of rotational distributions in scattering and photodissociation, *J. Phys. Chem.* **90,** 1742–1751.

Schinke, R. (1986b). Rotational state distributions of H_2 and CO following the photofragmentation of formaldehyde, *J. Chem. Phys.* **84,** 1487–1491.

Schinke, R. (1986c). The rotational reflection principle in the direct photodissociation of triatomic molecules. Close-coupling and classical calculations, *J. Chem. Phys.* **85,** 5049–5060.

Schinke, R. (1988a). Rotational distributions in direct molecular photodissociation, *Ann. Rev. Phys. Chem.* **39,** 39–68.

Schinke, R. (1988b). Rotational state distributions following direct photodissociation of triatomic molecules: Test of classical models, *J. Phys. Chem.* **92,** 3195–3201.

Schinke, R. (1988c). Angular momentum correlation in the photodissociation of H_2O_2 at 193 nm, *J. Phys. Chem.* **92,** 4015–4019.

Schinke, R. (1989a). Rotational excitation in direct photodissociation and its relation to the anisotropy of the excited state potential energy surface: How realistic is the impulsive model?, *Comments At. Mol. Phys.* **23,** 15–44.

Schinke, R. (1989b). Dynamics of molecular photodissociation, in *Collision Theory for Atoms and Molecules,* ed. F.A. Gianturco (Plenum Press, New York).

Schinke, R. (1990). Rotational state distributions following the photodissociation of Cl-CN: Comparison of classical and quantum mechanical calculations, *J. Chem. Phys.* **92,** 2397–2400.

Schinke, R. and Bowman, J.M. (1983). Rotational rainbows in atom-diatom scattering, in *Molecular Collision Dynamics,* ed. J.M. Bowman (Springer, Berlin).

Schinke, R. and Engel, V. (1986). The rotational reflection principle in photodissociation dynamics, *Faraday Discuss. Chem. Soc.* **82,** 111–124.

Schinke, R. and Engel, V. (1990). Periodic orbits and diffuse structures in the photodissociation of symmetric triatomic molecules, *J. Chem. Phys.* **93,** 3252–3257.

Schinke, R., Engel, V., Andresen, P., Häusler, D., and Balint–Kurti, G.G. (1985). Photodissociation of single H_2O quantum states in the first absorption band: Complete characterization of OH rotational and Λ-doublet state distributions, *Phys. Rev. Lett.* **55,** 1180–1183.

Schinke, R., Engel, V., and Staemmler, V. (1985). Rotational state distributions in the photolysis of water: Influence of the potential anisotropy, *J. Chem. Phys.* **83,** 4522–4533.

Schinke, R., Hennig, S., Untch, A., Nonella, M., and Huber, J.R. (1989). Diffuse vibrational structures in photoabsorption spectra: A comparison of CH_3ONO and CH_3SNO using two-dimensional *ab initio* potential energy surfaces, *J. Chem. Phys.* **91,** 2016–2029.

Schinke, R., Nonella, M., Suter, H.U., and Huber, J.R. (1990). Photodissociation of ClNO in the S_1 state: A quantum-mechanical *ab initio* study, *J. Chem. Phys.* **93,** 1098–1106.

Schinke, R. and Staemmler, V. (1988). Photodissociation dynamics of H_2O_2 at 193 nm: An example of the rotational reflection principle, *Chem. Phys. Lett.* **145,** 486–492.

Schinke, R., Untch, A., Suter, H.U., and Huber, J.R. (1991). Mapping of transition-state wave functions: I. Rotational state distributions following the decay of long-lived resonances in the photodissociation of HONO(S_1), *J. Chem. Phys.* **94,** 7929–7936.

Schinke, R., Vander Wal, R.L., Scott, J.L., and Crim, F.F. (1991). The effect of bending vibrations on product rotations in the fully state-resolved photodissociation of the $\tilde{A}$ state of water, *J. Chem. Phys.* **94,** 283–288.

Schinke, R., Weide, K., Heumann, B., and Engel, V. (1991). Diffuse structures and periodic orbits in the photodissociation of small polyatomic molecules, *Faraday Discuss. Chem. Soc.* **91,** 31-46.

Schmidtke, H.-H. (1987). *Quantenchemie* (Verlag Chemie, Weinheim).

Schneider, R., Domcke, W., and Köppel, H. (1990). Aspects of dissipative electronic and vibrational dynamics of strongly vibronically coupled systems, *J. Chem. Phys.* **92,** 1045–1061.

Schreider, Y.A. (1966). *The Monte Carlo Method* (Pergamon Press, Elmsford, NY).

Schulz, G.J. (1973). Resonances in electron impact on diatomic molecules, *Rev. Mod. Phys.* **45,** 423–486.

Schulz, P.A., Sudbø, Aa.S., Krajnovich, D.J., Kwok, H.S., Shen, Y.R., and Lee, Y.T. (1979). Multiphoton dissociation of polyatomic molecules, *Ann. Rev. Phys.* **30,** 379–409.

Schwartz-Lavi, D., Bar, I., and Rosenwaks, S. (1986). Rotational alignment and nonstatistical Λ doublet population in NO following $(CH_3)_3CONO$ photodissociation, *Chem. Phys. Lett.* **128,** 123–126.

Secrest, D. (1975). Theory of angular momentum decoupling approximations for rotational transitions in scattering, *J. Chem. Phys.* **62,** 710–719.

Secrest, D. (1979a). Rotational excitation I: The quantal treatment, in *Atom-Molecule Collision Theory,* ed. R.B. Bernstein (Plenum Press, New York).

Secrest, D. (1979b). Vibrational Excitation I: The quantal treatment, in *Atom-Molecule Collision Theory,* ed. R.B. Bernstein (Plenum Press, New York).

Seel, M. and Domcke, W. (1991). Femtosecond time-resolved ionization spectroscopy of ultrafast internal conversion dynamics in polyatomic molecules: Theory and computational studies, *J. Chem. Phys.* **95,** 7806.

Segev, E. and Shapiro, M. (1982). Three-dimensional quantum dynamics of H_2O and HOD photodissociation, *J. Chem. Phys.* **77,** 5604–5623.

Segev, E. and Shapiro, M. (1983). Energy levels and photopredissociation of the He-I_2 van der Waals complex in the IOS approximation, *J. Chem. Phys.* **78,** 4969–4984.

Sension, R.J., Brudzynski, R.J., and Hudson, B.S. (1988). Resonance Raman studies of the low-lying dissociative Rydberg-valence states of H_2O, D_2O and HOD, *Phys. Rev. Lett.* **61,** 694–697.

Sension, R.J., Brudzynski, R.J., Hudson, B.S., Zhang, J., and Imre, D.G. (1990). Resonance emission studies of the photodissociating water molecule, *Chem. Phys.* **141,** 393–400.

Shafer, N., Satyapal, S., and Bersohn, R. (1989). Isotope effect in the photodissociation of HDO at 157.5 nm, *J. Chem. Phys.* **90,** 6807–6808.

Shan, J.H., Vorsa, V., Wategaonkar, S.J., and Vasudev, R. (1989). Influence of intramolecular vibrational dynamics on state-to-state photodissociation: *Trans* DONO ($\tilde{A}$) versus HONO ($\tilde{A}$), *J. Chem. Phys.* **90,** 5493–5500.

Shan, J.H., Wategaonkar, S.J., and Vasudev, R. (1989). Vibrational state dependence of the $\tilde{A}$ state lifetime of HONO, *Chem. Phys. Lett.* **158**, 317–320.

Shapiro, M. (1972). Dynamics of dissociation. I. Computational investigation of unimolecular breakdown processes, *J. Chem. Phys.* **56,** 2582–2591.

Shapiro, M. (1977). Exact collinear calculations of the $N_2O(X\,^1\Sigma^+) \rightarrow N_2(^1\Sigma_g^+) + O(^1S)$ photodissociation process, *Chem. Phys. Lett.* **46,** 442–449.

Shapiro, M. (1981). Photofragmentation and mapping of nuclear wavefunctions, *Chem. Phys. Lett.* **81,** 521–527.

Shapiro, M. (1986). Photophysics of dissociating CH_3I: Resonance-Raman and vibronic photofragmentation maps, *J. Phys. Chem.* **90,** 3644–3653.

Shapiro, M. and Balint-Kurti, G.G. (1979). A new method for the exact calculation of vibrational-rotational energy levels of triatomic molecules, *J. Chem. Phys.* **71,** 1461–1469.

Shapiro, M. and Bersohn, R. (1980). Vibrational energy distribution of the CH_3 radical photodissociated from CH_3I, *J. Chem. Phys.* **73,** 3810–3817.

Shapiro, M. and Bersohn, R. (1982). Theories of the dynamics of photodissociation, *Ann. Rev. Phys. Chem.* **33,** 409–442.

Shepard, R. (1987). The multiconfiguration self-consistent field method, in *Ab Initio Methods in Quantum Chemistry – II,* ed. K.P.Lawley (Wiley, New York).

Sheppard, M.G. and Walker, R.B. (1983). Wigner method studies of ozone photodissociation, *J. Chem. Phys.* **78,** 7191–7199.

Sidis,V (1989*a*). Non-adiabatic molecular collisions, in *Collision Theory for Atoms and Molecules,* ed. F.A. Gianturco (Plenum Press, New York).

Sidis, V. (1989*b*). Vibronic phenomena in collisions of atomic and molecular species, *Adv. At., Mol., Opt. Phys.* **26,** 161–208.

Siebrand, W. (1976). Nonradiative processes in molecular systems, in *Dynamics of Molecular Systems,* Part A, ed. W.H. Miller (Plenum Press, New York).

Simons, J.P. (1977). The dynamics of photodissociation, in *Gas Kinetics and Energy Transfer,* ed. P.G. Ashmore and R.J. Donovan (The Chemical Society, London).

Simons, J.P. (1984). Photodissociation: A critical survey, *J. Phys. Chem.* **88,** 1287–1293.

Simons, J.P. (1987). Dynamical stereochemistry and the polarization of reaction products, *J. Phys. Chem.* **91,** 5378–5387.

Simons, J.P., Smith, A.J., and Dixon, R.N. (1984). Rotationally resolved photofragment alignment and dissociation dynamics in H_2O and D_2O, *J. Chem. Soc., Faraday Trans. 2* **80,** 1489–1501.

Simons, J.P. and Tasker, P.W. (1973). Energy partitioning in photodissociation and photosensitization Part 2. Quenching of Hg(6^3P_1) by CO and NO and the near ultra-violet photodissociation of ICN, *Mol. Phys.* **26,** 1267–1280.

Simons, J.P. and Tasker, P.W. (1974). Further comments on energy partitioning in photodissociation and photosensitization: Vacuum u.v. photolysis of cyanogen halides, *Mol. Phys.* **27,** 1691–1695.

Simons, J.P. and Yarwood, A.J. (1963). Decomposition of hot radicals. Part 2.– Mechanisms of excitation and decomposition, *Trans. Faraday. Soc.* **59,** 90–100.

Sivakumar, N., Burak, I., Cheung, W.-Y., Houston, P.L., and Hepburn, J.W. (1985). State-resolved photofragmentation of OCS monomers and clusters, *J. Phys. Chem.* **89,** 3609–3611.

Sivakumar, N., Hall, G.E., Houston, P.L., Hepburn, J.W., and Burak, I. (1988). State-resolved photodissociation of OCS monomers and clusters, *J. Chem. Phys.* **88,** 3692–3708.

Skene, J.M., Drobits, J.C., and Lester, M.I. (1986). Dynamical effects in the vibrational predissociation of ICl-rare gas complexes, *J. Chem. Phys.* **85,** 2329–2330.

Skodje, R.T. (1984). Gaussian wave packet dynamics expressed in the classical interaction picture, *Chem. Phys. Lett.* **109,** 227–232.

Smith, E.B., Buckingham, A.D., Fish, Y.A., Howard, B.J., Maitland, G.C., and Young, D.A. (Eds.) (1982). Van der Waals molecules, *Faraday Discuss. Chem. Soc.* **73,** 7–423.

Smith, I.W.M. (1980). *Kinetics and Dynamics of Elementary Gas Reactions* (Butterworths, London).

Smith, I.W.M. (1990). Exposing molecular motions, *Nature* **343,** 691–692.

Smith, N. (1986). On the use of action-angle variables for direct solution of classical nonreactive 3D (Di) atom-diatom scattering problems, *J. Chem. Phys.* **85,** 1987–1995.

Sölter, D., Werner, H.-J., von Dirke, M., Untch, A., Vegiri, A., and Schinke, R. (1992). The photodissociation of ClNO through excitation in the T_1 state: An *ab initio* study, *J. Chem. Phys.* **97,** 3357–3374.

Sorbie, K.S. and Murrell, J.N. (1975). Analytical potentials for triatomic molecules from spectroscopic data, *Mol. Phys.* **29,** 1387–1407.

Sparks, R.K., Shobatake, K., Carlson, L.R., and Lee, Y.T. (1981). Photofragmentation of CH_3I: Vibrational distribution of the CH_3 fragment, *J. Chem. Phys.* **75,** 3838–3846.

Spiglanin, T.A. and Chandler, D.W. (1987). Rotational state distributions of NH($a^1\Delta$) from HNCO photodissociation, *J. Chem. Phys.* **87,** 1577–1581.

Spiglanin, T.A., Perry, R.A., and Chandler, D.W. (1987). Internal state distributions of CO from HNCO photodissociation, *J. Chem. Phys.* **87,** 1568–1576.

Staemmler, V. and Palma, A. (1985). CEPA calculations of potential energy surfaces for open-shell systems. IV. Photodissociation of H_2O in the $\tilde{A}^1B_1$ state, *Chem. Phys.* **93,** 63–69.

Stechel, E.B. and Heller, E.J. (1984). Quantum ergodicity and spectral chaos, *Ann. Rev. Phys. Chem.* **35,** 563–589.

Stephenson, J.C., Casassa, M.P., and King, D.S. (1988). Energetics and spin- and Λ-doublet selectivity in the infrared multiphoton dissociation $DN_3 \rightarrow DN(X^3\Sigma^-, a^1\Delta) + N_2(X^1\Sigma_g^+)$: Experiment, *J. Chem. Phys.* **89,** 1378–1387.

Stock, G. and Domcke, W. (1990). Theory of femtosecond pump-probe spectroscopy of ultrafast internal conversion processes in polyatomic molecules, *J. Opt. Soc. Am. B* **7,** 1971.

Sun, Y., Mowrey, R.C., and Kouri, D.J. (1987). Spherical wave close coupling wave packet formalism for gas phase nonreactive atom-diatom collisions, *J. Chem. Phys.* **87,** 339–349.

Sundberg, R.L., Imre, D., Hale, M.O., Kinsey, J.L., and Coalson, R.D. (1986). Emission spectroscopy of photodissociating molecules: A collinear model for CH_3I and CD_3I, *J. Phys. Chem.* **90,** 5001–5009.

Suter, H.U., Brühlmann, U., and Huber, J.R. (1990). Photodissociation of CH_3ONO by a direct and indirect mechanism, *Chem. Phys. Lett.* **171,** 63–67.

Suter, H.U. and Huber, J.R. (1989). S_1 potential energy surface of HONO: Absorption spectrum and photodissociation, *Chem. Phys. Lett.* **155,** 203–209.

Suter, H.U., Huber, J.R., Untch, A., and Schinke, R. (1992). The photodissociation of HONO (S_1): An *ab initio* study, unpublished.

Suter, H.U., Huber, J.R., von Dirke, M. Untch, A., and Schinke, R. (1992). A quantum mechanical, time-dependent wavepacket interpretation of the diffuse structures in the

$S_0 \to S_1$ absorption spectrum of FNO: Coexistence of direct and indirect dissociation, *J. Chem. Phys.* **96,** 6727–6734.

Suzuki, T., Kanamori, H., and Hirota, E. (1991). Infrared diode laser study of the 248 nm photodissociation of CH_3I, *J. Chem. Phys.* **94,** 6607–6619.

Swaminathan, P.K., Stodden, C.D., and Micha, D.A. (1989). The eikonal approximation to molecular photodissociation: Application to CH_3I, *J. Chem. Phys.* **90,** 5501–5509.

Szabo, A. and Ostlund, N.S. (1982). *Modern Quantum Chemistry* (Macmillan, New York).

Tabor, M. (1981). The onset of chaotic motion in dynamical systems, *Adv. Chem. Phys.* **46,** 73–151.

Tabor, M. (1989). *Chaos and Integrability in Molecular Dynamics* (Wiley, New York).

Tadjeddine, M., Flament, J.P., and Teichteil, C. (1987). Non-empirical spin-orbit calculation of the CH_3I ground state, *Chem. Phys.* **118,** 45–55.

Tal-Ezer, H. and Kosloff, R. (1984). An accurate and efficient scheme for propagating the time dependent Schrödinger equation, *J. Chem. Phys.* **81,** 3967–.....

Tannor, D.J. and Heller, E.J. (1982). Polyatomic Raman scattering for general harmonic potentials, *J. Chem. Phys.* **77,** 202–218.

Tannor, D.J. and Rice, S.A. (1988). Coherent pulse sequence control of product formation in chemical reactions, *Adv. Chem. Phys.* **70,** part I, 441–523.

Taylor, H.S. (1970). Models, interpretations, and calculations concerning resonant electron scattering processes in atoms and molecules, *Adv. Chem. Phys.* **18,** 91–147.

Taylor, H.S. and Zakrzewski, J. (1988). Dynamic interpretation of atomic and molecular spectra in the chaotic regime, *Phys. Rev. A* **38,** 3732–3748.

Taylor, J.R. (1972). *Scattering Theory: The Quantum Theory on Nonrelativistic Collisions* (Wiley, New York).

Tellinghuisen, J. (1985). The Franck-Condon principle in bound-free transitions, in *Photodissociation and Photoionization,* ed. K.P. Lawley (Wiley, New York).

Tellinghuisen, J. (1987). The Franck-Condon principle, in *Photons and Continuum States,* ed. N.K. Rahman, C. Guidotti, and M. Allegrini (Springer, Berlin).

Tennyson, J. (1986). The calculation of the vibration-rotation energies of triatomic molecules using scattering coordinates, *Computer Physics Reports* **4,** 1–36.

Tennyson, J. and Henderson, J.R. (1989). Highly excited rovibrational states using a discrete variable representation: The H_3^+ molecular ion, *J. Chem. Phys.* **91,** 3815–3825.

Theodorakopoulos, G. and Petsalakis, I.D. (1991). Asymmetric dissociation and bending potentials of H_2S in the ground and excited electronic states, *Chem. Phys. Lett.* **178,** 475–482.

Theodorakopoulos, G., Petsalakis, I.D., and Buenker, R.J. (1985). MRD CI calculations on the asymmetric stretch potentials of H_2O in the ground and the first seven singlet excited states, *Chem. Phys.* **96,** 217–225.

Thomas, L.D., Alexander, M.H., Johnson, B.R., Lester, W.A. Jr., Light, J.C., McLenithan, K.D., Parker, G.A., Redmon, M.J., Schmalz, T.G., Secrest, D., and Walker, R.B. (1981). Comparison of numerical methods for solving the second-order differential equations of molecular scattering theory, *J. Comp. Phys.* **41,** 407–426.

Thompson, S.D., Carroll, D.G., Watson, F., O'Donnell, M., and McGlynn, S.P. (1966). Electronic spectra and structure of sulfur compounds, *J. Chem. Phys.* **45,** 1367.

Ticich, T.M., Likar, M.D., Dübal, H.-R., Butler, L.J., and Crim, F.F. (1987). Vibrationally mediated photodissociation of hydrogen peroxide, *J. Chem. Phys.* **87,** 5820–5829.

Ticktin, A., Bruno, A.E., Brühlmann, U., and Huber, J.R. (1988). $NO(X^2\Pi)$ rotational distributions from the photodissociation of NOCl and NOBr at 450 and 470 nm, *Chem. Phys.* **125,** 403–413.

Top, Z.H. and Baer, M. (1977). Incorporation of electronically nonadiabatic effects into bimolecular reaction dynamics. II. The collinear $(H_2 + H^+, H_2^+ + H)$ systems, *Chem. Phys.* **25,** 1.

Troe, J. (1988). Unimolecular reaction dynamics on *ab initio* potential energy surfaces, *Ber. Bunsenges. Phys. Chem.* **92,** 242–252.

Troe, J. (1992). Statistical aspects of ion-molecule reactions, in *State-Selected and State-to-State Ion-Molecule Reaction Dynamics,* Part 2: *Theory,* ed. M. Baer and C.Y. Ng (Wiley, New York).

Truhlar, D.G. (1981). *Potential Energy Surfaces and Dynamics Calculations* (Plenum Press, New York).

Truhlar, D.G. (Ed.) (1984). *Resonances in Electron-Molecule Scattering, van der Waals Complexes, and Reactive Chemical Dynamics* (American Chemical Society, Washington, D.C.).

Truhlar, D.G. and Muckerman, J.T. (1979). Reactive scattering cross sections III: Quasi-classical and semiclassical methods, in *Atom-Molecule Collision Theory,* ed. R.B. Bernstein (Plenum Press, New York).

Tuck, A.F. (1977). Molecular beam studies of ethyl nitrite photodissociation, *J. Chem. Soc., Faraday Trans. 2* **5,** 689–708.

Tully, J.C. (1976). Nonadiabatic processes in molecules, in *Dynamics of Molecular Collisions,* Part B, ed. W.H. Miller (Plenum Press, New York).

Turro, N.Y. (1965). *Molecular Photochemistry* (Benjamin, Reading).

Turro, N.Y. (1978). *Modern Molecular Photochemistry* (Benjamin, Menlo Park).

Untch, A. (1992). *Theoretische Untersuchung zur Photodissoziation Kleiner polyatomarer Moleküle der Form R-NO mittels Zeitabhängiger quantenmechanischer Methoden* (Max-Planck-Institute für Strömungsforschung, Bericht 8/1992, Göttingen).

Untch, A., Hennig, S., and Schinke, R. (1988). The vibrational reflection principle in direct photodissociation of triatomic molecules: Test of classical models, *Chem. Phys.* **126,** 181–190.

Untch, A. and Schinke, R. (1992). Time-dependent three-dimensional study of the photodissociation of $CH_3ONO(S_1)$: Comparison with classical and experimental results. To be published.

Untch, A., Weide, K., and Schinke, R. (1991a). 3D wavepacket study of the photodissociation of $CH_3ONO(S_1)$, *Chem. Phys. Lett.* **180,** 265–270.

Untch, A., Weide, K., and Schinke, R. (1991b). The direct photodissociation of $ClNO(S_1)$: An exact 3D wavepacket analysis, *J. Chem. Phys.* **95,** 6496–6507.

Vaida, V., McCarthy, M.I., Engelking, P.C., Rosmus, P., Werner, H.-J., and Botschwina, P. (1987). The ultraviolet absorption spectrum of the $\tilde{A}^1A_2'' \leftarrow \tilde{X}^1A_1'$ transition of jet-cooled ammonia, *J. Chem. Phys.* **86,** 6669–6676.

van der Avoird, A., Wormer, P.E.S., Mulder, F., and Berns, R.M. (1980). *Ab Initio Studies of the Interaction in van der Waals Molecules.* Topics in Current Chemistry 93 (Springer, Berlin).

Vander Wal, R.L. and Crim, F.F. (1989). Controlling the pathways in molecular decomposition: The vibrationally mediated photodissociation of water, *J. Phys. Chem.* **93,** 5331–5333.

Vander Wal, R.L., Scott, J.L., and Crim, F.F. (1990). Selectively breaking the O-H bond in HOD, *J. Chem. Phys.* **92,** 803–805.

Vander Wal, R.L., Scott, J.L., and Crim, F.F. (1991). State resolved photodissociation of vibrationally excited water: Rotations, stretching vibrations, and relative cross sections, *J. Chem. Phys.* **94,** 1859–1867.

Vander Wal, R.L., Scott, J.L., Crim, F.F., Weide, K., and Schinke, R. (1991). An experimental and theoretical study of the bond selected photodissociation of HOD, *J. Chem. Phys.* **94,** 3548–3555.

van Dishoeck, E.F., van Hemert, M.C., Allison, A.C., and Dalgarno, A. (1984). Resonances in the photodissociation of OH by absorption into coupled $^2\Pi$ states: Adiabatic and diabatic formulations, *J. Chem. Phys.* **81,** 5709–5724.

van Veen, G.N.A., Baller, T., De Vries, A.E., and Shapiro, M. (1985). Photofragmentation of CF_3I in the A band, *Chem. Phys.* **93,** 277–291.

van Veen, G.N.A., Baller, T., De Vries, A.E., and van Veen, N.J.A. (1984). The excitation of the umbrella mode of CH_3 and CD_3 formed from photodissociation of CH_3I and CD_3I at 248 nm, *Chem. Phys.* **87,** 405–417.

van Veen, G.N.A., Mohamed, K.A., Baller, T., and De Vries, A.E. (1983). Photofragmentation of H_2S in the first continuum, *Chem. Phys.* **74,** 261–271.

Vasudev, R., Zare, R.N., and Dixon, R.N. (1983). Dynamics of photodissociation of HONO at 369 nm: Motional anisotropy and internal state distribution of the OH fragment, *Chem. Phys. Lett.* **96,** 399–402.

Vasudev, R., Zare, R.N., and Dixon, R.N. (1984). State-selected photodissociation dynamics: Complete characterization of the OH fragment ejected by the HONO $\tilde{A}$ state, *J. Chem. Phys.* **80,** 4863–4878.

Vegiri, A., Untch, A., and Schinke, R. (1992). Mapping of transition-state wavefunctions. II. A model for the photodissociation of ClNO(T_1), *J. Chem. Phys.* **96,** 3688–3695.

Vien, G.N., Richard-Viard, M., and Kubach, C. (1991). Origin of the resonances in IH+I dynamics from a Born-Oppenheimer type separation, *J. Phys. Chem.* **95,** 6067–6070.

Vinogradov, I.P. and Vilesov, F.I. (1976). Luminescence of the OH($A^2\Sigma^+$) radical during photolysis of water vapor by vacuum uv radiation, *Opt. Spectrosc.* **40,** 32–34.

von Bünau, G. and Wolff, T. (1987). *Photochemie* (VCH Verlagsgesellschaft, Weinheim).

Waite, B.A. and Dunlap, B.I. (1986). The photodissociation of ClCN: A theoretical determination of the rotational state distribution of the CN product, *J. Chem. Phys.* **84,** 1391–1396.

Waldeck, J.R., Campos-Martínez, J., and Coalson, R.D. (1991). Application of a coupled-surface time-dependent Hartree grid method to excited state optical spectroscopy, *J. Chem. Phys.* **94,** 2773–2780.

Waller, I.M., Kitsopoulos, T.N., and Neumark, D.M. (1990). Threshold photodetachment spectroscopy of the I + HI transition-state region, *J. Phys. Chem.* **94,** 2240–2242.

Wang, H.-t., Felps, W.S., and McGlynn, S.P. (1977). Molecular Rydberg states. VII. Water, *J. Chem. Phys.* **67,** 2614–2627.

Wannenmacher, E.A.J., Lin, H., and Jackson, W.M. (1990). Photodissociation dynamics of C_2N_2 in the threshold region for dissociation, *J. Phys. Chem.* **94,** 6608–6615.

Wardlaw, D.M. and Marcus, R.A. (1988). On the statistical theory of unimolecular processes, *Adv. Chem. Phys.* **70,** 231–263.

Waterland, R.L., Lester, M.I., and Halberstadt, N. (1990). Quantum dynamical calculations for the vibrational predissociation of the He-ICl complex: Product rotational distribution, *J. Chem. Phys.* **92,** 4261–4271.

Waterland, R.L., Skene, J.M., and Lester, M.I (1988). Rotational rainbows in the vibrational predissociation of ICl-He complexes, *J. Chem. Phys.* **89,** 7277–7286.

Wayne, R.P. (1988). *Principles and Applications of Photochemistry* (Oxford University Press, Oxford).

Weaver, A., Metz, R.B., Bradforth, S.E., and Neumark, D.M. (1988). Spectroscopy of the I + HI transition-state region by photodetachment of IHI^-, *J. Phys. Chem.* **92,** 5558–5560.

Weaver, A. and Neumark, D.M. (1991). Negative ion photodetachment as a probe of bimolecular transition states: the F + H_2 reaction, *Faraday Discuss. Chem. Soc.* **91,** 5–16.

Weber, A. (Ed.) (1987). *Structure and Dynamics of Weakly Bound Molecular Complexes* (Reidel, Dordrecht).

Weide, K., Hennig, S., and Schinke, R. (1989). Photodissociation of vibrationally excited water in the first absorption band, *J. Chem. Phys.* **91,** 7630–7637.

Weide, K., Kühl, K., and Schinke, R. (1989). Unstable periodic orbits, recurrences, and diffuse vibrational structures in the photodissociation of water near 128 nm, *J. Chem. Phys.* **91,** 3999–4008.

Weide, K. and Schinke, R. (1987). Photodissociation dynamics of water in the second absorption band. I. Rotational state distributions of OH($^2\Sigma$) and OH($^2\Pi$), *J. Chem. Phys.* **87,** 4627–4633.

Weide, K. and Schinke, R. (1989). Photodissociation dynamics of water in the second absorption band. II. *Ab initio* calculation of the absorption spectra for H_2O and D_2O and dynamical interpretation of "diffuse vibrational" structures, *J. Chem. Phys.* **90,** 7150–7163.

Weide, K., Staemmler, V., and Schinke, R. (1990). Nonadiabatic effects in the photodissociation of H_2S, *J. Chem. Phys.* **93,** 861–862.

Weiner, B.R., Levene, H.B., Valentini, J.J., and Baronavski, A.P. (1989). Ultraviolet photodissociation dynamics of H_2S and D_2S, *J. Chem. Phys.* **90,** 1403–1414.

Weissbluth, M. (1978). *Atoms and Molecules* (Academic Press, New York).

Weissbluth, M. (1989). *Photon-Atom Interactions* (Academic Press, Boston).

Werner, H.-J. (1987). Matrix-formulated direct multiconfiguration self-consistent field and multiconfiguration reference configuration-interaction methods, in *Ab Initio Methods in Quantum Chemistry-II,* ed. K.P. Lawley (Wiley, New York).

Werner, H.-J., Follmeg, B., and Alexander, M.H. (1988). Adiabatic and diabatic potential energy surfaces for collisions of CN($X^2\Sigma^+, A^2\Pi$) with He, *J. Chem. Phys.* **89,** 3139–3151.

Werner, H.-J., Follmeg, B., Alexander, M.H., and Lemoine, P. (1989). Quantum scattering studies of electronically inelastic collisions of CN($X^2\Sigma^+$, $A^2\Pi$) with He, *J. Chem. Phys.* **91,** 5425–5439.

Werner, H.-J. and Meyer, W. (1981). MCSCF study of the avoided curve crossing of the two lowest $^1\Sigma^+$ states of LiF, *J. Chem. Phys.* **74,** 5802–5807.

Whetten, R.L., Ezra, G.S., and Grant, E.R. (1985). Molecular dynamics beyond the adiabatic approximation: New experiments and theory, *Ann. Rev. Phys. Chem.* **36,** 277–320.

Wigner, E. (1932). On the quantum correction for thermodynamic equilibrium, *Phys. Rev.* **40,** 749–759.

Williams, S.O. and Imre, D.G. (1988a). Raman spectroscopy: time-dependent pictures, *J. Phys. Chem.* **92,** 3363–3374.

Williams, S.O. and Imre, D.G. (1988b). Time evolution of single- and two-photon processes for a pulse-mode laser, *J. Phys. Chem.* **92,** 6636–6647.

Williams, S.O. and Imre, D.G. (1988c). Determination of real time dynamics in molecules by femtosecond laser excitation, *J. Phys. Chem.* **92,** 6648–6654.

Williams, C.J., Qian, J., and Tannor, D.J. (1991). Dynamics of triatomic photodissociation in the interaction representation. I. Methodology, *J. Chem. Phys.* **35,** 1721–1737.

Wilson, E.B., Decius, J.C., and Cross, P.C. (1955). *Molecular vibrations. The Theory of Infrared and Raman Vibrational Spectra* (McGraw-Hill, New York).

Wintgen, D. and Friedrich, H. (1987). Correspondence of unstable periodic orbits and quasi-Landau modulations, *Phys. Rev. A* **36,** 131–142.

Wintgen, D., Holle, A, Wiebusch, G., Main, J., Friedrich, H., and Welge, K.-H. (1986). Precision measurements and exact quantum mechanical calculations for diamagnetic Rydberg states in hydrogen, *J. Phys. B* **19,** L557–L561.

Wittig, C., Nadler, I., Reisler, H., Noble, M., Catanzarite, J., and Radhakrishnan, G. (1985). Nascent product excitations in unimolecular reactions: The separate statistical ensembles method, *J. Chem. Phys.* **83,** 5581–5588.

Woodbridge, E.L., Ashfold, M.N.R., and Leone, S.R. (1991). Photodissociation of ammonia at 193 nm: Rovibrational state distribution of the $NH_2(\tilde{A}^2A_1)$ fragment, *J. Chem. Phys.* **94,** 4195–4204.

Wu, T.Y. and Ohmura, T. (1962). *Quantum Theory of Scattering* (Englewood Cliffs, Prentice-Hall).

Xie, X., Schnieder, L., Wallmeier, H., Boettner, R., Welge, K.H., and Ashfold, M.N.R. (1990). Photodissociation dynamics of $H_2S(D_2S)$ following excitation within its first absorption continuum, *J. Chem. Phys.* **92,** 1608–1616.

Xu, Z., Koplitz, B., and Wittig, C. (1987). Kinetic and internal energy distributions via velocity-aligned Doppler spectroscopy: The 193 nm photodissociation of H_2S and HBr, *J. Chem. Phys.* **87,** 1062–1069.

Yabushita, S. and Morokuma, K. (1990). *Ab initio* potential energy surfaces for rotational excitation of CN product in the *A*-band photodissociation of ICN, *Chem. Phys. Lett.* **175,** 518.

Yamashita, K. and Morokuma, K. (1990). *Ab initio* study of transition state spectroscopy: $ClHCl^-$ photodetachment spectrum, *J. Chem. Phys.* **93,** 3716–3717.

Yamashita, K. and Morokuma, K. (1991). *Ab initio* molecular orbital and dynamics study of transition-state spectroscopy, *Faraday Discuss. Chem. Soc.* **91,** 47–61.

Yang, S.-C. and Bersohn, R. (1974). Theory of the angular distribution of molecular photofragments, *J. Chem. Phys.* **61,** 4400–4407.
Yang, S.C., Freedman, A., Kawasaki, M., and Bersohn, R. (1980). Energy distribution of the fragments produced by photodissociation of CS_2 at 193 nm, *J. Chem. Phys.* **72,** 4058–4062.
Zare, R.N. (1972). Photoejection dynamics [1], *Mol. Photochem.* **4,** 1–37.
Zare, R.N. (1988). *Angular Momentum* (Wiley, New York).
Zare, R.N. and Hershbach, D.R. (1963). Doppler line shape of atomic fluorescence excited by molecular photodissociation, *Proc. IEEE* **51,** 173–182.
Zewail, A.H. (1988). Laser femtochemistry, *Science* **242,** 1645–1653.
Zewail, A.H. (1991). Femtosecond transition-state dynamics, *Faraday Discuss. Chem. Soc.* **91,** 207–237.
Zewail, A.H. and Bernstein, R.B. (1988). Real-time laser femtochemistry, *Chemical and Engineering News* **66,** 24–43.
Zhang, J., Abramson, E.H., and Imre, D.G. (1991). $\tilde{C} \to \tilde{A}$ emission in H_2O following two-photon excitation: Dissociation dynamics in the $\tilde{A}$ state for different initial states, *J. Chem. Phys.* **95,** 6536–6543.
Zhang, J., Heller, E.J., Huber, D., and Imre, D.G. (1988). CH_2I_2 photodissociation: Dynamical modeling, *J. Chem. Phys.* **89,** 3602–3611.
Zhang, J. and Imre, D.G. (1988a). CH_2I_2 photodissociation: Emission spectrum at 355 nm, *J. Chem. Phys.* **89,** 309–313.
Zhang, J. and Imre, D.G. (1988b). OH/OD bond breaking selectivity in HOD photodissociation, *Chem. Phys. Lett.* **149,** 233–238.
Zhang, J. and Imre, D.G. (1989). Spectroscopy and photodissociation dynamics of H_2O: Time-dependent view, *J. Chem. Phys.* **90,** 1666–1676.
Zhang, J., Imre, D.G., and Frederick, J.H. (1989). HOD spectroscopy and photodissociation dynamics: Selectivity in OH/OD bond breaking, *J. Phys. Chem.* **93,** 1840–1851.
Zhang, J.Z.H. (1990). New method in time-dependent quantum scattering theory: Integrating the wave function in the interaction picture, *J. Chem. Phys.* **92,** 324–331.
Zhang, J.Z.H. and Miller, W.H. (1990). Photodissociation and continuum resonance Raman cross sections and general Franck-Condon intensities from *S*-matrix Kohn scattering calculations with application to the photoelectron spectrum of $H_2F^- + h\nu \to H_2 + F, HF + H + e^-$, *J. Chem. Phys.* **92,** 1811–1818.
Zhang, J.Z.H., Miller, W.H., Weaver, A., and Neumark, D. (1991). Quantum reactive scattering calculations of Franck-Condon factors for the photodetachment of H_2F^- and D_2F^- and comparison with experiment, *Chem. Phys. Lett.* **182,** 283–289.
Ziegler, L.D. and Hudson, B. (1984). Resonance rovibronic Raman scattering of ammonia, *J. Phys. Chem.* **88,** 1110–1116.

Index

ab initio methods, 19, 35
absorption cross section, 27, 32–33, 144ff
  as Fourier transformation of the autocorrelation function, 74
  classical, 102–105
  partial, 33
absorption spectrum, 10ff
action-angle variables, 94
adiabatic
  basis, 62ff
  decay, 217
  potential, 135, 156
  representation, 34, 349–352
  separation, 36, 117ff, 155–159, 171, 188ff, 193ff
adiabatic approximation
  vibrational excitation, 61–67
Airy function, 115
alignment, 285
angular distribution, 15, 284ff
angular momentum
  orbital, 8, 56
  rotational, 8, 56
  total, 8, 56, 256, 264
anisotropy
  of potential energy surface, 60, 125ff, 222ff
  parameter, 15, 159, 270, 284
ArH_2, 296, 305
ArHCl, 2
ArHF, 302
artificial channel method, 71
asymmetric top, 267, 302
autocorrelation function, 73ff, 181ff, 339
  damping of, 160ff
  short-time behavior, 113, 116
  superposition of sinusoidal functions, 143ff
avoided crossing, 36, 134, 351ff

basis size
  of coupled equations, 54
Beer's law, 27
Bi_2, 378
Boltzmann distribution, 9, 233, 257, 282
bond coordinates, 223, 254
Born-Oppenheimer approximation, 19, 33–37, 61, 347, 369
bound-state energies
  calculation of, 41–42
boundary conditions
  for radial functions, 55, 70
  general form, 45
  in full collisions, 48
  in photodissociation, 45ff, 268
boxing method, 104–105
branching ratio
  chemical, 12
  electronic, 13, 347, 357

BrCN, 127
Breit-Wigner expression, 159

CF_3I, 132ff, 210–213
Chebychev polynomial, 82
chemical channels, 12ff
chemical lasers, 6
CH_2I_2, 337
CH_3I, 6, 210–213, 233, 331, 333, 337, 357–359, 378
$(CH_3)_3CONO$, 276
$(CH_3)_2NNO$, 276, 287
CHOCHO, 287
CH_3ONO, 20–23, 62, 127, 147–152, 156, 207, 217ff, 249, 259, 276, 286, 308
CH_3SNO, 119
classical path approximation, 89–90
classical trajectory, 95–97
ClCN, 60, 96, 124, 127, 254ff
ClNO, 2, 107, 119, 127ff, 163–166, 217ff, 223ff, 249, 254, 337
close-coupling method, 69ff
  for electronic excitation, 350
  for rotational excitation, 56–60
  for rotational excitation, $\mathbf{J} \neq 0$, 265
  for vibrational excitation, 53–56
  time-dependent, 84ff, 355
closed channel, 44, 70
closure relation, 28, 44
CO_2, 216
$Co(CO)_3NO$, 127
coherent excitation, 368–370
collisions
  full, 8ff, 159ff
  half, 8ff
complex scaling, 174
conical intersection, 360
continuum basis, 43–48
continuum wavefunction
  calculation of, 69–71
coordinate system
  body-fixed, 262
  space-fixed, 262
Coriolis coupling, 266, 304, 306
coupled equations
  for coherent excitation, 369
  for electronic transitions, 34
  for excitation by a photon, 28ff
  for rotational excitation, 58ff
  for the decay of excited states, 140–141
  for vibrational excitation, 53–55
  in the adiabatic representation, 63
  time-dependent, 85
cross-correlation function, 336, 339
CS_2, 216

decay of excited states, 138–143
decay rate
  partial, 142, 307, 311
  total, 142, 144, 307
diabatic
  basis, 62
  potential, 135, 348
  representation, 296, 352–356
diffuse vibrational structures, 118ff, 177–201
  due to excitation of large-amplitude bending, 193–200
  due to excitation of symmetric and anti-symmetric stretch motion, 179–193
dipole selection rule, 9, 160, 170
Dirac delta-function, 31, 44, 145
discrete variable representation, 85
dissociation channels, 42ff
dissociation wavefunction
  partial, 46, 58, 64, 71
  relation to the evolving wavepacket, 77ff, 154
  total, 50–51, 60, 71, 152, 243
driven equations method, 71
dynamical mapping, 210

Ehrenfest's theorem, 61, 87, 97
elastic limit
  for vibrational excitation, 131
electric dipole
  approximation, 29
  operator, 27ff

electronic quenching, 347
energy
  excess, 3
  internal, 3, 96
  mean, 14
  photon, 3
  relative, 14
  translational, 3, 43, 95
  vibrational, 43
energy gap law, 300
exchange reaction, 168
excitation function
  rotational, 97, 121, 126ff, 245ff
  vibrational, 98, 129
exit channel dynamics, 222

Fermi's Golden Rule, 49
Feynman's path integral, 188
finite elements method, 71
FNO, 242–248
force, 95, 97
  intermolecular, 19
  intramolecular, 19
Fourier transformation, 183, 227
  between coordinate and momentum spaces, 83ff
  between time and energy domains, 74, 78ff
Franck-Condon distribution, 203–206
Franck-Condon factor
  one-dimensional, 65, 117, 157
  rotational, 226, 228
  vibrational, 204
Franck-Condon mapping, 206
free wave, 44

glyoxal, 13

H_3^+, 200
H atom
  in homogenous magnetic field, 79, 189, 199ff
Hamilton equations, 87, 94–97
Hamilton function, 94
Hamiltonian
  classical path, 89
  mean field, 88
  molecular, 28, 34, 349
HCN, 168
HCO, 168
H_2CO, 127, 207, 249, 288
$HeCl_2$, 300, 309ff, 313
HeHF, 2, 301ff
HeICl, 310
helicity quantum number, 264, 303
$(HF)_2$, 2
HgI_2, 378
HN_3, 127, 276, 288, 292
HNCO, 127, 288
H_2O, 2, 10–16, 23–26, 178, 223, 228ff, 285ff, 337
  first absorption band, 189–191, 213–215, 230–233, 258ff, 272–275, 278–282, 319–324, 338–344
  rotational levels, 267
  second absorption band, 106, 193–200, 238–241
  vibrational states, 319–320
H_2O_2, 2, 127, 207, 235–238, 256–257, 286ff, 289ff
HOD, 12ff, 106, 324–329
HONO, 127, 207, 220ff, 233, 249, 276, 286
H_2S, 215ff, 233, 337, 344–345, 359–365
hyperspherical coordinates, 171

I_2, 373
ICN, 127, 207, 276, 286, 375, 378
IHI, 168–172
importance sampling, 105
impulsive model, 251–255
interaction picture, 83
interaction potential, 40, 95, 202ff, 304, 311
internal vibrational energy redistribution, 5, 149

Jacobi coordinates, 38–41, 57, 224ff, 254, 263, 293ff
  mass-scaled, 40, 171ff

kinetic coupling, 40, 171, 224, 353
Kohn variational principle, 71
Kramers-Heisenberg-Dirac
expression, 334

Laguerre polynomial, 100
Legendre polynomial, 59, 284, 294
lifetime, 142, 145, 157ff
local coordinates, 38
local modes, 319ff
Lorentzian, 145
line shape, 143–147

mixing angle, 353
momentum gap law, 300
Monte Carlo sampling, 104–105
Mulliken difference potential, 316
multiphoton dissociation, 3
multiple collision, 240

Na_2, 378
NaI, 376ff
NCNO, 288
$NeCl_2$, 310ff
NeDF, 302
NeHF, 302
NeICl, 310
NH_3, 166–168, 337
nonadiabatic
coupling, 140, 219
decay, 219
normal coordinates, 38

O_3, 127, 331, 337
Chappuis band, 216
Hartley band, 192ff
OCS, 127
OH
Λ-doublet states, 271ff
rotational levels, 271ff
open channel, 44, 70
orientation, 285

pair-pair correlation, 287–292
parity, 264, 304
partial width, 146
periodic orbit
anti-symmetric stretch, 186
bending, 196ff
hyperspherical, 186
symmetric stretch, 185
unstable, 79, 184–189
perturbation theory
first-order, 30, 371
for rotational excitation, 126
time-dependent, 28–31, 140ff
phase-space, 95
phase-space distribution function,
98–102
for normal coordinates, 101
Wigner, 99
phase-space theory, 250
photochemistry, 6
photodetachment spectroscopy, 168ff
photodissociation
direct, 4, 109ff
indirect, 4, 109, 134ff
of single quantum states, 15ff,
277–282
photodissociation amplitude
partial, 49, 60, 64, 68, 335
photodissociation cross section,
16–18, 48–50
differential, 15, 270
integral, 15, 17, 270
partial, 18, 49, 65, 81ff, 117, 121,
146, 164, 323ff
state-to-state, 267–270
total, 18, 49, 68
photofragment yield spectra, 163, 217
potential energy surface, 19–26
potential matrix
for rotational excitation, 59, 265
for vibrational excitation, 54ff, 202,
296–297
predissociation
electronic, 4, 363
rotational, 301–307
vibrational, 5, 296–301, 363
principal value, 141
prior distribution, 250

probability current, 45

quantum beats, 149

radiationless transition, 347
radiative recombination, 331
Raman scattering, 331
  time-dependent view, 335–337
  time-independent view, 334–335
Rayleigh scattering, 333
recoil velocity, 283
recurrence, 79, 134ff, 149, 161ff, 181ff
reflection principle, 110, 316–319
  multi-dimensional, 115–118
  one-dimensional, 67, 110–115
  rotational, 120–125, 236
  vibrational, 128–132, 209
resonance, 134, 137ff, 152ff
  assignment, 152ff
  condition, 33
  due to excitation of bending motion, 163–168
  due to excitation of symmetric and anti-symmetric stretch motion, 168–173
  Feshbach, 298, 303, 305
  in full collisions, 159–160
  reactive, 168
  shape, 303, 305
  stationary wavefunction, 154ff
resonant transition, 31
revival, 373
rotating wave approximation, 30
rotation matrix, 264
rotational constant, 58
rotational excitation
  elastic case, 225–233
  inelastic case, 234–241
  sources of, 222ff
rotational rainbow, 123, 125
  supernumerary, 123, 125
rotational sudden approximation, 67–69
RRKM theory, 250

scalar properties, 222
scattering
  theory, 43, 137, 152
scattering-coordinates, 38–41
scattering matrix **S**, 45, 48, 70, 159
Schrödinger equation
  electronic, 19, 34
  nuclear, 35
selection rules
  in photoabsorption, 269ff
shift
  partial, 142
  total, 142, 144
skewing angle, 171
SO_2, 216
spectral resolution
  relation to resolution in the time domain, 160ff
spherical harmonic, 58, 68, 226, 264
  semiclassical approximation, 227
spin-orbit states, 271ff
split operator method, 83
stabilization procedure, 174
state distributions
  rotational, 13ff
  vibrational, 13ff
stationary phase approximation, 69
statistical adiabatic channel model, 250
sum rule, 51

thermal averaging, 17, 282
time-dependent SCF approximation, 88–89
time-evolution operator, 75
torque, 25, 95ff, 196, 222ff, 234
transition
  bent-bent, 128
  bent-linear, 193ff
  linear-bent, 60, 127
  parallel, 269, 284, 357
  perpendicular, 269, 284, 357
transition dipole moment function, 36, 269, 283
transition frequency, 27, 29, 371
transition rate, 31

for van der Waals molecules, 299
transition state, 168, 241
spectroscopy, 173
transition-state wavefunction, 242ff
triatomic molecule
general, 262ff
linear, 37–48, 53–56, 73ff, 115ff, 128ff, 179ff
rigid rotor model, 56–60, 121ff, 193ff
symmetric, 168ff
turning point, 111, 114, 121

uncertainty relation
coordinate-momentum, 99
time-energy, 78, 92, 114, 371
unimolecular reaction, 5

van der Waals molecule, 2, 174, 293–313
vector correlations, 15, 283–287
vector properties, 222
vibrational excitation
elastic case, 203–207
inelastic case, 208–213
symmetric molecules, 213–216

wavenumber, 44, 59
wavepacket, 73–78, 143ff
as coherent superposition of stationary states, 73ff, 367
evolution of, 76–78
Gaussian, 86–87
generated by a δ-pulse in time, 80ff
initial condition, 74
short-time behavior, 113
spatial propagation, 83ff
temporal propagation, 82ff
wavetrain, 369
weighting function, 121
Wigner $3j$-symbol, 59
WKB approximation, 114
WKB phase, 316

Printed in the United States
By Bookmasters